The Evolutionary Ecology
of Plants

The Evolutionary Ecology of Plants

EDITED BY

Jane H. Bock
and Yan B. Linhart

Routledge
Taylor & Francis Group

LONDON AND NEW YORK

First published 1989 by Westview Press

Published 2019 by Routledge
52 Vanderbilt Avenue, New York, NY 10017
2 Park Square, Milton Park, Abingdon, Oxon OX14 4RN

Routledge is an imprint of the Taylor & Francis Group, an informa business

Library of Congress Cataloging in Publication Data
The Evolutionary ecology of plants / edited by Jane H. Bock and Van B.
    Linhart: with an introduction by G. Ledyard Stebbins.
        p. cm.
    Papers presented as a tribute to Herbert G. Baker by his students.
    Includes bibliographies.
    Includes index.
    1. Botany—Ecology. 2. Ecology. 3. Evolution. I. Bock, Jane H.
II. Linhart, Van B. III. Baker, Herbert G.
QK901.E89   1989
581.5'24—dc19   87-35199
                                                                    CIP

ISBN 13: 978-0-367-29197-6 (hbk)
ISBN 13: 978-0-367-30743-1 (pbk)

CONTENTS

PREFACE

This book presents a broad view of contemporary research in evolutionary plant ecology. Two major themes provide links between the chapters. One is that an evolutionary perspective is necessary and useful at all levels of ecology. The other is that animals are of utmost importance in the ecology and evolution of plants. Ecology is the study of interactions between organisms and their environments, and ecological studies can focus on individuals, populations of organisms of the same species, or biological communities made up of individuals of several to many species. The study of ecological interactions is becoming increasingly refined. Many early analyses concentrated on immediate or short-term responses of individuals or populations to environmental change. These short-term responses were thought to be relevant in an "ecological" or "proximate" time frame. Long-term responses were thought to occur in an "evolutionary" or "ultimate" time frame and were, therefore, the domain of evolutionists and population geneticists. In addition, many ecologists tended to treat both individuals and populations as entities that could be defined by simply assigning them their scientific names. Inter-individual variation was an annoyance whose impact needed to be ignored or minimized. Gradually, the recognition that ecological versus evolutionary time scales and proximate versus ultimate factors were false dichotomies, led biologists such as L.C. Birch and E.B. Ford to seek an integration of ecology with genetics and evolution. This integration is now being carried out by many biologists who are in the process of helping to develop a vigorous discipline of evolutionary ecology.

Most of the contributors to this volume are intellectual descendents of the 19th century naturalists; but unlike our predecessors, we have had the advantage of being trained additionally in genetics and biometry. There is a tendency in some circles to argue that ecology and evolution are such complex fields that only the reductionist approach can be productive. We intend these studies to demonstrate that no one approach will provide all the answers. We need reductionist and holistic studies, in the field and in the laboratory, using short-lived "model" species with simple life styles, and long-lived wild species that are complex and refractory to simplistic experiments.

In EVOLUTIONARY DYNAMICS, examples are provided of ecological settings within which evolutionary changes can occur. The settings range from the multi-celled bodies of higher plants, where plastids have evolved, to broad geographic regions that have helped shape patterns of variation. The characteristics being studied include chromosomes, plastids, flowers, and genetic architecture of populations. In BREEDING SYSTEMS, we describe some of the elegant solutions elaborated by plants when faced with the problem of being sessile, yet needing to move their gametes about in order to reproduce. These solutions become particularly elaborate when animal pollen vectors are not reliably available. In REPRODUCTIVE BIOLOGY, we illustrate the broad spectrum of life history stages which affect plant reproductive success in some fashion. These stages include the complexities of pollen coverings, developmental pathways leading to seed production, and the impact of yearly variation in insect parasites and rainfall conditions upon total seed output.

PLANT-ANIMAL INTERACTIONS is one of this book's recurrent themes. In addition, we address two specific facets of this phenomenon. One has to do with herbivory and its consequences, and the other is a potential solution to the vexing problem of being eaten: attraction by plants of mercenary arthropod bodyguards. Traditionally, an evolutionary perspective in ecology has been especially sought in studies of single-species populations. In the section on EVOLUTION AND ORGANIZATION IN COMMUNITIES, we illustrate how networks of interactions among species, played out against a background of variable physical conditions, can generate predictable evolutionary consequences.

A favorite activity for students of evolution since Darwin's time has been the study of domesticated species to illustrate the power of selection and its complex outcomes. In EVOLUTION IN CULTIVATED PLANTS, we not only pursue this theme by illustrating the interplay between human and natural selection, but we go beyond it to show that the evolution of cultivated plants may involve an interplay among biological, geographic, historical, and political factors. Finally, in the BIOGEOGRAPHY AND CONSERVATION section, we illustrate how knowledge obtained from studies in evolutionary ecology can be used to suggest appropriate strategies for the conservation of biological resources.

All of the senior authors except for Dr. Stebbins have shared the experience of studying with Professor Herbert G. Baker. This has been of considerable importance to all of us. The breadth of subject matter in this volume reflects his catholic tastes and interests as well as his encouragement to follow our particular interests while working under his guidance. To each of us, he has supplied standards to emulate, if not to achieve. These include intellectual curiosity, tolerance for differing viewpoints, a recognition of the need for breadth of training and field experience, and, at a personal level, patience, wisdom, and kindness.

Most of these papers were presented at a symposium in honor of Herbert G. Baker that was sponsored by the Botanical Society of America and the Ecological Society of America at their joint meeting in August 1987 at Columbus, Ohio. We are grateful to the program committee of the Botanical Society and to Ohio State University for their help in arranging the program. We give special thanks to the many outside reviewers who helped prepare these papers for final publication. Carl Bock, Robert Cruden, Michael Grant, and Meredith Lane deserve particular thanks for their help. Phyllis O'Connell and Carl Bock, along with our Westview editor, Kellie Masterson, helped generously with manuscript preparation. Finally, we thank the contributors for their enthusiasm for this project and for their fine articles.

Jane H. Bock
Yan B. Linhart

CONTRIBUTORS

Barrett, S.C.H. Department of Botany, University of Toronto, Toronto, Ont. M5S 1Al, Canada

Bird, R. McK. P.O. Box 5142, Raleigh, NC 27650, U.S.A.

Bock, C.E. Department of Environmental Population and Organismic Biology, University of Colorado, Boulder, CO 80309, U.S.A.

Bock, J.H. Department of Environmental Population and Organismic Biology, University of Colorado, Boulder, CO 80309, U.S.A.

Chiang, Y.C. Department of Plant Science, University of New Hampshire, Durham, NH 03824, U.S.A.

Cook, S.A. Department of Biology, University of Oregon, Eugene, OR 97403, U.S.A.

Copsey, A.D. Department of Biology, Central College, Pella, IA 50219, U.S.A.

Cox, P.A. Department of Botany and Range Science, Brigham Young University, Provo, UT 84602, U.S.A.

Cruden, R.W. Department of Botany, University of Iowa, Iowa City, IA 52242, U.S.A.

Dickman, A.W. Department of Biology, University of Oregon, Eugene, OR 97403, U.S.A.

Dobson, H.E.M. Department of Botany, University of California, Berkeley, CA 94720, U.S.A.

Fone, A.L. Department of Botany, University of Georgia, Athens, GA 30602, U.S.A.

Frankie, G.W. Department of Entomology, University of California, Berkeley, CA 94720, U.S.A.

Fukuda, I. Division of Biology, Tokyo Woman's Christian University, Zempukuji, Suginami, Tokyo, Japan 167

Guerrant, E.O. Department of Botany and Plant Pathology, Oregon State University, Corvallis, OR 97331, U.S.A.

Hamrick, J.L. Departments of Botany and Genetics, University of Georgia, Athens, GA 30602, U.S.A.

Hendrix, S.D. Department of Biology, University of Iowa, Iowa City, IA 52242, U.S.A.

Janzen, D.H. Department of Biology, University of Pennsylvania, Philadelphia, PA 19104, U.S.A.

Keeler, K.H. School of Biological Sciences, University of Nebraska, Lincoln, NB 68588, U.S.A.

Kiang, Y.-T. Department of Plant Science, University of New Hampshire, Durham, NH 03824, U.S.A.

Koptur, S. Department of Biological Sciences, Florida International University, Miami, FL 33199, U.S.A.

Kwankin, B. School of Biological Sciences, University of Nebraska, Lincoln, NB 68585, U.S.A.

Linhart, Y.B. Department of Environmental Population and Organismic Biology, University of Colorado, Boulder, CO 80309, U.S.A.

Loveless, M.D. Department of Biology, College of Wooster, Wooster, OH 44691, U.S.A.

Lyon, D.L. Department of Biology, Cornell College, Mt. Vernon, IA 52314, U.S.A.

Newstrom, L.E. Entomology Department, University of California, Berkeley, CA 94720, U.S.A.

Opler, P. Office of Information Transfer, U.S. Fish and Wildlife Service, 1025 Penock Place, Ft. Collins, CO 80524, U.S.A.

Pemberton, R.W. Rangeland Insect Laboratory, USDA-ARS, Montana State University, Bozeman, MT 58717, U.S.A.

Schlising, R.A. Department of Biology, California State University, Chico, CA 95929, U.S.A.

Spira, T.P. Department of Biology, Georgia Southern College, Statesboro, GA 30460, U.S.A.

Stebbins, G.L. Genetics Department, University of California, Davis, CA 95616, U.S.A.

Turner, C.E. USDA Regional Research Center, 800 Buchanan Street, Albany, CA 99710, U.S.A.

Uno, G.E. Department of Botany and Microbiology, University of Oklahoma, 770 Van Vleet Oval, Norman, OK 73019, U.S.A.

Vincent, S.B. Department of Entomology, Texas A & M University, College Station, TX 27843, U.S.A.

Wagner, L.K. Department of Biology, Georgia Southern College, Statesboro, GA 30460, U.S.A.

Whatley, J.M. Department of Plant Sciences, University of Oxford, South Parks Road, Oxford, 0X1 3RA, England

Williams, H. Department of Entomology, Texas A & M University, College Station, TX 27843, U.S.A.

INTRODUCTION

HERBERT BAKER: A RENAISSANCE BOTANIST

G. Ledyard Stebbins

In our culture, we often refer to someone with broad
cultural background and interests as a "Renaissance Person."
With this designation, we recognize a broad cultural interest, that
may not have led to preeminent fame and glory in any narrow field
of endeavor, but has made the bearer a welcome addition to any
cultural group because of his breadth of interest, his enthusiasm
for his listeners, both neophytes and old pros alike, to develop a
similar attitude. A renaissance person is one who opens the doors
to the temple of wisdom and understanding, both for his own
pleasure and satisfaction and, more importantly, for that of
others. Without a sprinkling of renaissance minds and attitudes,
the world of knowledge and silence would be dull indeed.

Can we speak of a renaissance botanist? In a literal sense,
no, since botany as a science was unknown during the historical
renaissance centuries. Nevertheless, some botanists have genuine
characteristics that we associate with the "renaissance" epithet.
They have the breadth, knowledge of several fields, and enthusiasm
that make them recognized and revered by all who know them well,
and they inspire others with their genuine thirst for knowledge
and understanding.

In my opinion, Herbert Baker is the very model of a
renaissance botanist. He has achieved this distinction in three
different ways.

First, a renaissance botanist is not satisfied with
achievements in pure science for its own sake. He seeks to spread
the good news to those outside of his field. This can be done both by
brilliant lecturing and teaching, and by promoting institutions
dedicated to building bridges between scientists and the public.
Herbert Baker has been outstanding in both of these
accomplishments. Even while working on his Ph.D. thesis, he
aided his country while it was passing through the crisis of World
War II, serving as a volunteer warden in a civil defense post. His
first scientific paper was published during this period (1942),
under the title "A red coloring matter from the green leaves of
spinach beet." His interest in bringing science to public attention

grow during four years (1953-57) as Professor and Lecturer at the University of Ghana. He already realized that tropical botany is a wave of the future, largely because tropical plants, now in grave danger of decimation and extinction, make many major contributions to the well being of all men. After becoming Professor at the University of California, he became a prime mover in the Organization for Tropical Studies. Nevertheless, he was not blinded by the glamor of the rich biota in tropical rain- and cloud-forests. He realized that the study of tropical weeds could at the same time provide a major source of information for understanding mechanisms of evolution, and also increase the efficiency of crop production in tropical climates. One of the most delightful associations that many of us, and I in particular, had with Herbert was at the Symposium on the Genetics of Colonizing Species, held at Asilomar, California, in 1964. When the then President of the International Union of Biological Sciences, C. H. Waddington, decided that a major function of the Union should be to organize symposia dealing with topics of major scientific and economic interest, such as the effect of human activity in spreading many kinds of animals and plants throughout the world, he gave me the task, as General Secretary of the Union, of organizing the symposium. I immediately realized that I would need expert help, and that Herbert Baker was the key person to provide it. The result was a twofold reward: first, one of the liveliest and most stimulating conferences that I have ever attended, and, second, the origin of a volume that has become a minor classic in the field of organismic evolution and human affairs. Since then, Professor Baker has continued to develop the field of colonizing plant species, both tropical and temperate, via public addresses, talks at symposia, journal papers, and a book "Plants and Civilization."

Through all of this activity, he has never forgotten the importance of direct contact with amateur plant lovers. He expressed this interest during his service as Director of the Botanical Garden at the University of California, Berkeley, where he did much to make it both a beauty spot and a center of scientific information about plants. He continues this activity by teaching a course on "Plants and Civilization," and by showing a real interest in a purely amateur organization, the California Native Plant Society.

Another feature of a renaissance botanist is that he does not let the glamor of public service and approval overshadow his attention to the nuts and bolts of his profession: research that is directed toward other botanists and provides them with principles on which to build sounder edifices of basic knowledge. Throughout his career, Dr. Baker has remained faithful to this mission. During his earliest period of scientific investigation, he made a major contribution by investigating natural populations of wild pinks (*Melandrium* or *Lychnis*) where he demonstrated the importance of hybridization in plant speciation and evolution, and at the same time showed that a particular character, hairiness on leaves, has the function of protecting the plant from attacks of aphids that can transmit lethal diseases. Later during this period, he followed in Darwin's footsteps by investigating floral dimorphism in the genera *Armeria* and *Limonium,* and uncovered that facts that underlie a generalization often termed "Baker's Law." Populations found near the original center of distribution of a genus usually are sexual and predominantly cross fertilized, while those that have migrated the farthest from this center are usually self fertilized or have reverted to asexual reproduction.

His sojourn in Africa inspired Herbert and his co-workers to gather hard data on a topic that was then controversial: the pollination of flowers by bats. Largely because of their carefully recorded observations, bat pollination became a generally recognized phenomenon in many tropical species. During his tenure at the University of California, his growing interest in weeds enabled him to recognize and characterize plants that have "general purpose" genotypes: the same genotype can be successful in a variety of environments. Finally, his research in recent years, in collaboration with his wife Irene, who is an excellent botanist in her own right, has uncovered many highly important facts about the nectar that in many flowers cross pollinated by animals, serves as the major attractant and reward for the vector. These investigations of pollination biology will for years to come stand as a landmark in unravelling the complexities of plant evolution.

Still another admirable trait of a renaissance man, in any field, is a sense of history. Herbert Baker has this appreciation of his forerunners in large measure, as is evident from reading several of his papers.

Finally, the crowning characteristic of a renaissance botanist is his ability to reach for the stars. He has successfully exploited the aphorism of Theodosius Dobzhansky, "Nothing in biology makes sense without evolution." My first acquaintance with Herbert Baker was when he was, at the invitation of Jens Clausen, a postdoctoral fellow of the Division of Plant Biology of the Carnegie Institution of Washington, at Stanford University. There he exploited and sharpened his ability to work with other scientists in a synthetic approach to the basic problems of evolution. His subsequent career has included participation in various synthetic and cooperative endeavors. Such cooperation is possible only if a scientist recognized that his won competence, however great it may be, is matched by the equal or greater competence of others. Critical respect for the achievements of scientists working in similar or related fields is essential for making real progress in understanding evolution or any other major scientific problem. These are necesary qualities of a renaissance botanist. Herbert Baker has them to a high degree. All of us who are participating in this symposium held in his honor salute him both for his outstanding stature as a scientist, and his ability to remain a productive, sympathetic and cooperative member of the world's scientific community.

EVOLUTIONARY DYNAMICS

The dynamics of the evolutionary process are studied at many levels. No single method, or species, or line of evidence is likely to provide the critical demonstration of the existence of a complex series of events required for evolutionary change to take place. In this section the authors describe evolutionary pathways favoring new structures and functions in plastids, and new flower shapes. They also describe factors affecting genomic architecture. This architecture can be affected at the level of multiplication of whole chromosome sets, i.e. polyploidy, or at the level of ecological variables that determine genetic recombination and genome fluidity.

The origin of plastids is one of plant evolution's intriguing mysteries. The most likely scenario suggests that they are derived from unicellular algae originally "acquired" as endosymbionts by protists. As multicellular plants evolved, their internal environments provided new opportunities for plastids. Whereas most plastids in the more primitive plants retained their photosynthetic abilities, in more recently evolved multi-celled plants, especially angiosperms, plastids have evolved new functions. Jean Whatley has studied plastid structure and function extensively. Her essay describes the role of plastids as accumulation sites of secondary metabolites that are probably involved in activities far removed from photosynthesis, including developmental control, attraction of pollinators, and deterrence of herbivores.

Evolution is most often detectable because of changes in the phenotypic features of organisms. These features have a morphological and developmental history of their own that often is shaped by changes in ecological conditions. Ecological forces that can affect plant growth and structure include interactions with other species such as pollinators, and the uncertainties of temporally variable resources. Factors affecting the shape of individual plant organs have been of interest to many botanists. It is a common observation that, for most plants, relative amounts and predictability of resources will affect both overall plant size and shape and the size and shape of individual vegetative organs. The shape of flowers is commonly perceived as predictable, hence the dependence of systematics upon floral characters. Yet flowers can vary in shape within species, and this variation often has a genetic basis. This provides the ingredients for evolutionary change.

Because competition affects competitors negatively, it is not surprising that the flowers of closely related, sympatric species often differ from one another in shape and/or color. This results in the attraction of different pollinators and reduces the likelihood of deposition of allospecific pollen. The details of the evolutionary changes associated with the process of adapting to a new set of pollinators are poorly understood. One reason for this may be that many pollinators are insects. They have characteristic sensory systems, physiologies, and behaviors that are very different from ours. In addition, there are important differences among insect groups in these features. Also, many insect-flower interactions have a long geologic history. In contrast, hummingbirds consist of a relatively small number of species with similar physiologies and behaviors. Their search behaviors depend on a vision system more similar to ours than to that of insects. Finally, adaptation to hummingbird pollination is a relatively recent phenomenon. For these reasons, hummingbirds as pollinators have attracted much attention. The flowers they visit tend to have a certain "standardized" tubular shape and are typically red and odorless. In this volume, Stebbins addresses the evolution of flower shape in hummingbird flowers, treating these shapes as striking innovations. He suggests that such innovations, which lead species in new directions, are important, but very rare and difficult to observe. He surveys the flora of California and demonstrates how a suite of characteristics including flower shape, color, and the ability to grow where hummingbirds are abundant reflect selection by hummingbirds as pollinators. He concludes that various components of flower shape play the most important role in the adaptive process. Adaptation to hummingbirds as pollinators has occurred very often in some families, and Stebbins suggests that these families have in their genomes a latent potential for responding in specific directions in response to these pollinators.

Guerrant has undertaken careful studies of floral development to illustrate how evolution acts very specifically at the morphological and developmental levels. In the current essay, he argues for caution in interpreting evolutionary changes in floral shape. He analyzes reduction in flower size in an autogamous species of *Limnanthes*. Such reduction is often seen in autogamous species and is usually interpreted in adaptive terms. Autogamous species, that do not depend on pollinators for fertilization, can produce small, inconspicuous flowers, and shunt the energy resources they save to another function. Guerrant argues that, in *Limnanthes*, both flower size and autogamy are incidental consequences of selection for faster development and earlier maturation.

The study of polyploids has a long and distinguished history among students of plant evolution. These analyses have demonstrated the central role played by polyploidization in species formation. Polyploids can arise as a result of "spontaneous" multiplication of chromosome sets, in which case there can be variation in number of sets within species. Keeler and Kwankin's contribution highlights this fascinating aspect of polyploidy: the presence of intra-specific variation in chromosome numbers. Such variation is quite common in higher plants, but much more common in some taxa than in others. In the grasses of the Great Plains of central North America, it is especially common, and Keeler and Kwankin attempt to explain why. Their answer is complex and suggests that we may be observing a transitional period in the evolutionary history of certain genera in which one chromosomal complement is being replaced by another.

In a detailed comparative study of the genus *Trillium*, Fukuda demonstrates how different ecological and genetic circumstances in North America and Japan have generated very different patterns of variation and species formation. In North America, speciation appears to have been primarily by allopatric differentiation. Patterns of variation in chromosome morphology are also strongly associated with the extent of environmental heterogeneity. In Japan, polyploid formation in the context of hybridization has been much more important than in North America. Differences in geographic distribution, paleogeologic history, and breeding systems have interacted to generate very different patterns of speciation in the two regions.

The genetic organization of plant populations is the product of interactions between genome organization, spatial and temporal distribution of individuals, mating patterns among these individuals, seed dispersal, germination, and a combination of selective and random events, that determine which plants survive to maturity. These interactions have been studied in temperate plants in this context for centuries. The existence of geographic patterns of variability in forest trees was described in pre-Darwinian times, and was cited by Darwin as evidence for evolution. With increasing concern about the loss of tropical forests, serious efforts are being made to document genetic organization in tropical plants. Hamrick and Loveless make a significant contribution in their essay. They demonstrate that tropical trees are at least as genetically variable as temperate trees, and that trees in general tend to be more genetically variable than other life forms. This tendency for woody plants to be highly variable has been noted on a number of occasions. Linhart's essay on community interactions provides one possible explanation for this observed pattern. Long-lived woody plants are exposed to extremely variable diversifying selection because of their interactions with large numbers of species of herbivores, parasites and diseases, each of which may generate its own patterns of selection on a plant host. A second finding of Hamrick and Loveless is that the organization of genetic variability can be predicted to some extent by patterns of pollen and seed dispersal: amount of gene flow, usually deduced from pollinators' and seed dispersers' flight capabilities, is correlated with the extent of inter-population genetic differentiation. This finding has been anticipated at the theoretical level but documenting it in a heterogeneous tropical forest is very important.

PLASTIDS IN A CHANGING ENVIRONMENT

Jean M. Whatley

INTRODUCTION

My current research on ultrastructural aspects of plastid evolution and development is far removed from the ecological research that I carried out under the guidance of Dr. Herbert Baker. Nevertheless, I have found over many years that ideas acquired during my ecology days can often provide an unexpected insight or slant to the questions that arise with respect to ultrastructure. The supposed endosymbiotic origin of plastids makes it particularly appropriate to consider the cell as a habitat. Successive stages of plastid development can be regarded as a parallel to successive stages of colonization; productivity (rate of photosynthesis) increases to a peak and then declines during the life of a chloroplast, just as it does during the life of an ecosystem (Kirt and Tilney-Bassett, 1978). The various symbiotic and commensal interrelationships among organisms in an ecological setting can give insight into subjects as different as the control of plastid populations on the one hand, or plastid structure and function in relation to pollination on the other. The forms or functions of the plastids in plant cells at a particular site or at a particular time can have important ecological consequences by contributing, for example to the synthesis of plant protectants as well as attractants or, indirectly, by influencing photoperiodic controls. It therefore seems appropriate here to review some of the ways in which plastids may have evolved in response to a changing land environment.

ENDOSYMBIOSIS

Towards the end of the last century, Schimper (1883) suggested that chloroplasts might have evolved from photosynthetic organisms that were once free-living, but were later taken up by unicellular hosts whose endosymbionts they became. Following the discovery by Ris and Plaut (1962) that chloroplasts had their own (prokaryotic-type) DNA, interest in the Schimper hypothesis revived and stimulated many ultrastructural and biochemical research projects and symposia. As a result of these, the original hypothesis has been somewhat modified but it has now become widely accepted that chloroplasts had an endosymbiotic origin, though many more detailed aspects of the general evolutionary scheme still have to be resolved. Briefly, it is now thought that the chloroplasts of red (rhodophyte) and green (chlorophyte) algae and land plants may have evolved from prokaryotic algal symbionts. By contrast, the more complex chloroplasts of the remaining algal phyla (Chromophyta sensu Christensen [1962] and Euglenophyta) may have evolved from a range of eukaryotic algal symbionts in which prokaryotic-type chloroplasts were already established. The similarity in pigmentation, ultrastructure, and sequencing data clearly point to a cyanobacterium (blue-green alga) as the ancestor of the red algal chloroplast. There is some difference of opinion as to whether the ancestral "green" chloroplast was a cyanobacterium with chlorophyll a and phycobilins as its photosynthetic pigments or another prokaryotic algal that already had both chlorophylls a and b, but lacked phycobilins 1/M the ancestor of the modern *Prochloron* (Dodson, 1979; Gibbs, 1981; Whatley and Whatley, 1981; Stackenbrandt, 1983; Cavalier-Smith, 1982).

THE CELL AS A HABITAT

The establishment of an alga as an endosymbiont would immediately provide the host with a new and continuing source of food through algal photosynthesis, and the alga, having avoided being digested in a phagocytic vacuole, would have found a new, relatively stable habitat, free of other predators, where nutrients could be intercepted from some of the host "waste" products. Some metabolites produced in excess of immediate requirements could be stored temporarily in the plastid or in the cytoplasm. Others of no potential use would be concentrated in vacuoles or eliminated directly from the cell, and, as Church (1919) suggested, some of these would have contributed to newly evolving external structures such as scales and cell walls. In modern corals, for example, up to 60% of the total carbon fixed by algal symbionts is exported as mucolipid and mucopolysaccharide (Crossland et al., 1980).

Taylor (1983) has suggested that, for permanent endosymbionts, continued isolation from their former gene pool would have been likely to lead to divergent genetic change similar to that found among small island communities. The evolution of species-specific strains among modern symbionts seem to reflect such a change. Through the transformation of an algal endosymbiont into a chloroplast must also have been accompanied by considerable physiological and structural modification, the chloroplast, once established, seems to have been a highly conservative organelle. The number of new structural forms that have evolved is small. An important part of the integration of the chloroplast into the cell life cycle has been the loss by the plastid of much of its own DNA; this has surely been a major factor in the apparent structural stability of the chloroplast. Although plastids retain the capacity to code for some of their own polypeptides, it is the cell nucleus which exerts overall control. As the majority of plastid modifications are limited to the angiosperms, it seems that the nucleus has been very slow to influence plastid forms.

Most green algae and some lower land plants have a single, large, most commonly cup-shaped chloroplast in each cell. However, there may be a major and ancient distinction between those green algal species whose chloroplasts have irregularly stacked thylakoids (the prokaryotic *Prochloron* and members of the Ulvophyceae and Chlorophyceae, as classified by Steward and Mattox in 1984) and those which have grana (members of the Charophyceae including the Zygnematales and Coleochaetales); it is from the latter that land plants are believed to have evolved (Figs. 1, 2, and 3). Modifications in the structure of plastids seem to have been mainly influenced by two major trends; (1) the evolution of multicellular algae with the appearance of meristems and with the potential for cell specialization; (2) colonization of the land and the evolution of a wide range of adaptive structures which assisted survival under terrestrial conditions.

MULTICELLULAR ALGAE AND MERISTEMS

In endosymbiotic partnerships, the symbionts are liable to be eliminated by the host under conditions of stress (Cook, 1983). By contrast, plastids very seldom suffer "predation", though those from one parent are often eliminated during sexual reproduction (Whatley, 1982). Indeed it is very difficult to eliminate plastids even from organisms like euglenoids (Kivic and Vesk, 1974), which can feed themselves heterotrophically. Instead when nutrients, particularly nitrogen, are in short supply or when growing conditions are otherwise unfavorable, the plastids tend to dispense with their thylakoids, a response which may save energy. In unicellular algae which had come to depend on their chloroplasts as their sole source of nutrition, any loss of photosynthetic activity must of necessity have been of only short duration or would have required a major reduction in other metabolic activity as, for example, during dormancy. Plastid dedifferentiation may well have arisen in the first place as a response by the cell to stress conditions irregularly imposed by the unpredictable environment.

Plastid dedifferentiation as a regular phenomenon within the algal life cycle perhaps first appeared in the male gamete (plastid degeneration) or in zygotes (becoming dormant or possibly

12

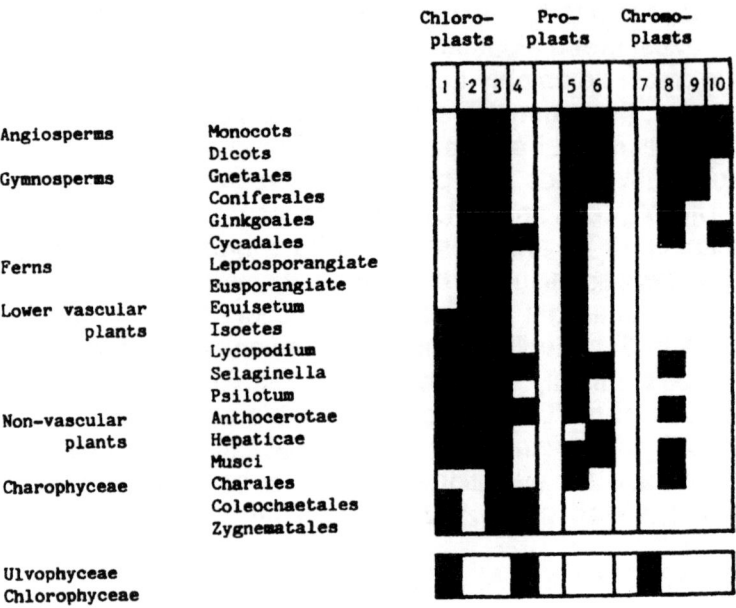

		Chloro-plasts				Pro-plasts		Chromo-plasts			
		1	2	3	4	5	6	7	8	9	10

Angiosperms — Monocots
Dicots
Gymnosperms — Gnetales
Coniferales
Ginkgoales
Cycadales
Ferns — Leptosporangiate
Eusporangiate
Lower vascular plants — Equisetum
Isoetes
Lycopodium
Selaginella
Psilotum
Non-vascular plants — Anthocerotae
Hepaticae
Musci
Charophyceae — Charales
Coleochaetales
Zygnematales

Ulvophyceae
Chlorophyceae

1. Cup-shaped chloroplast (or plate)
2. Discoid chloroplast
3. Grana present
4. Pyrenoid present
5. Non-green plastids in reproductive cells and meristems
6. Secretory non-green plastids
7. Eyespots
8. Chloro-chromoplasts
9. Globulous chromoplasts
10. Other chromoplasts

Figure 1. Some modifications in plastid structure.

undergoing meiosis) and must have required the evolution of new internal control mechanisms. The introduction of proplastids to the egg cells, young embryos and meristems, first, of the sporophyte and, later, to some specialized cells of the gametophyte generation may represent several successive evolutionary steps. The fragmentary information presently available suggests that today chloroplasts rather than proplastids are found in the gametophytic meristems of hornworts, liverworts, mosses, and ferns though proplastids are characteristic of some sporophytic meristems (Duckett and Renzaglia, 1988). Proplastids are also characteristic of meristems in some red and brown algae. Thus the evolution of proplastids is not a modification unique to the ancestors of land plants. Nevertheless, the proplastid may well have been the original key to the development of new and functionally more flexible forms of plastid in specialized cells which represent adaptations to the land environment (Fig. 1). Certainly as land plants have evolved, the proportion of the plant body given over to photosynthesis has conspicuously declined and most of the new plastid forms found today in land plants lack chlorophyll.

PLASTID DEVELOPMENT

Before considering the evolutionary responses by plastids to colonization of the land, it is helpful to describe the basic pattern of plastid development as it is found today in the photosynthetic tissues of angiosperms. With this as background it is easier to describe the various "new" forms of plastic that seem to represent adaptations to the changing environment.

The chloroplast is only one of several different types of plastid found in angiosperms. Some of the less familiar forms are found only as stages during development; others are mature forms with a function which is non-photosynthetic. Assuming that plastids evolved from photosynthetic endosymbionts, then the photosynthetically active chloroplast is undoubtedly the most ancient plastid form. From an historical point of view, the non-photosynthetic forms of plastid, including the proplastids, should therefore be considered as derived. However, for practical purposes, it is usually convenient to consider plastid development, particularly in angiosperms, as beginning with the proplastid stages present in the ripe seed and proceeding to the chloroplast stages of mature leaves or other green organs. The pattern of plastid development from the simplest proplastid stages to the various mature (climax) and, finally, senescent types can be followed both in time, during synchronous development, and in space, from a basal meristem upwards along a file of cells, in a

Figure 2. Part of a filament of the green alga, *Enteromorpha* sp. (Ulvophyceae) showing the large cup-shaped chloroplast with its irregularly stacked thylakoids and pyrenoid (P).

Figure 3. A typical chloroplast from *Fittonia verschaffeltii* showing the grana in both longitudinal and transverse section.

Figure 4. Part of the endosperm mother cell of *Phaseolus vulgaris* showing the extensive development of endoplasmic reticulum around the amyloplasts.

Figure 5. Globulous chromoplasts in the upper epidermis of a petal of *Caltha palustris*.

Figure 6. Sieve element plastid from the fern, *Matteucia struthiopteris*, showing osmiophilic material within a fragment of the thylakoid sac.

Figure 7. Sieve element plastid from the fern, *Salvina* sp., showing proteinaceous crystals in the depleted stroma.

Figure 8. Sieve element plastid from the angiosperm, *Phaseolus vulgaris*, showing the specialized starch in the depleted stroma.

Figure 9. Electron dense material within the thylakoid sac and within a membrane-bound-body of a developing chloroplast of *Impatiens* sp.

Fig. 7, scale bar = 0.5μm; Figs. 2-6, 8 & 9 = 1. 0 μm

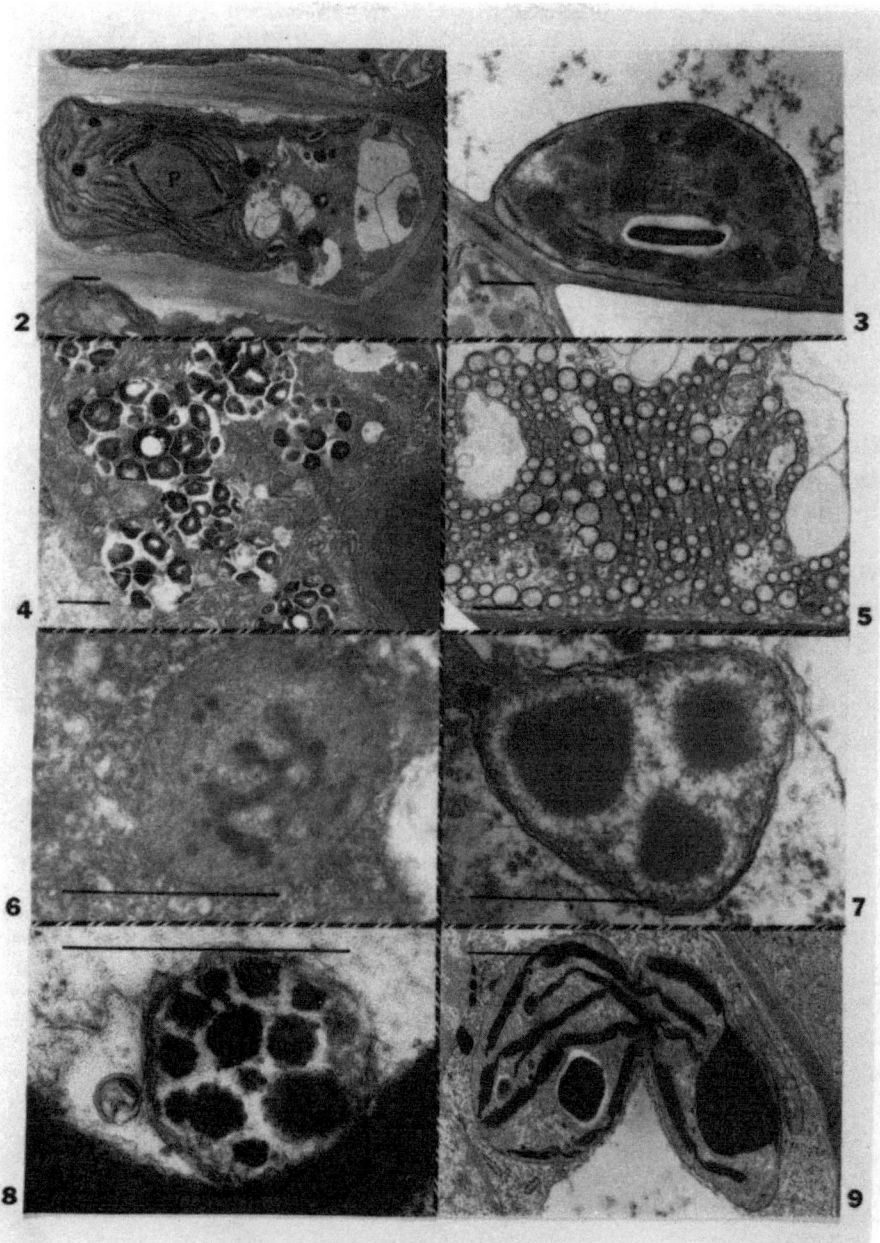

Figure 10. A crystalline membrane-bound body in a young chloroplast of *Spinacia oleracea.*

Figure 11. A chloroplast of *Equisetum telmateia* showing osmiophilic material in the plastid envelope and in the associated endoplasmic reticulum.

Figure 12. A chloroplast from the tern, *Pilularia globulifera.* There are no obvious modifications in plastid structure, but note the osmiophilic deposits in the developing vacuoles (→).

Figure 13. Chromoplast segments in yellow flowers of *Narcissus pseudonarcissus,* showing the concentric membranes (→) of one and the accumulation of strongly electron dense material in the envelope sac of another.

Figure 14. Part of a plastid in a white petal of *Phaseolus vulgaris.* Note the osmiophilic globules and the tubular complex in the stroma.

Figure 15. Degraded tapetal cells and pollen grains in a pollen chamber of *Phaseolus vulgaris.*

Figure 16. Secretory and non-secretory cells in a leaf of *Theobroma cacoa.* In the secretory cells, the osmiophilic material lies in the cytoplasm, not in the vacuoles.

Figure 17. Pleomorphic plastids (P) with electron dense stroma, in the mucilage-secreting cells at the periphery of the root cap of *Phaseolus vulgaris.*

Figs. 15-16, scale bar = 10.0 μm; Fig. 10 = 0.5μm; Figs. 11-14 & 17 = 1.0μm.

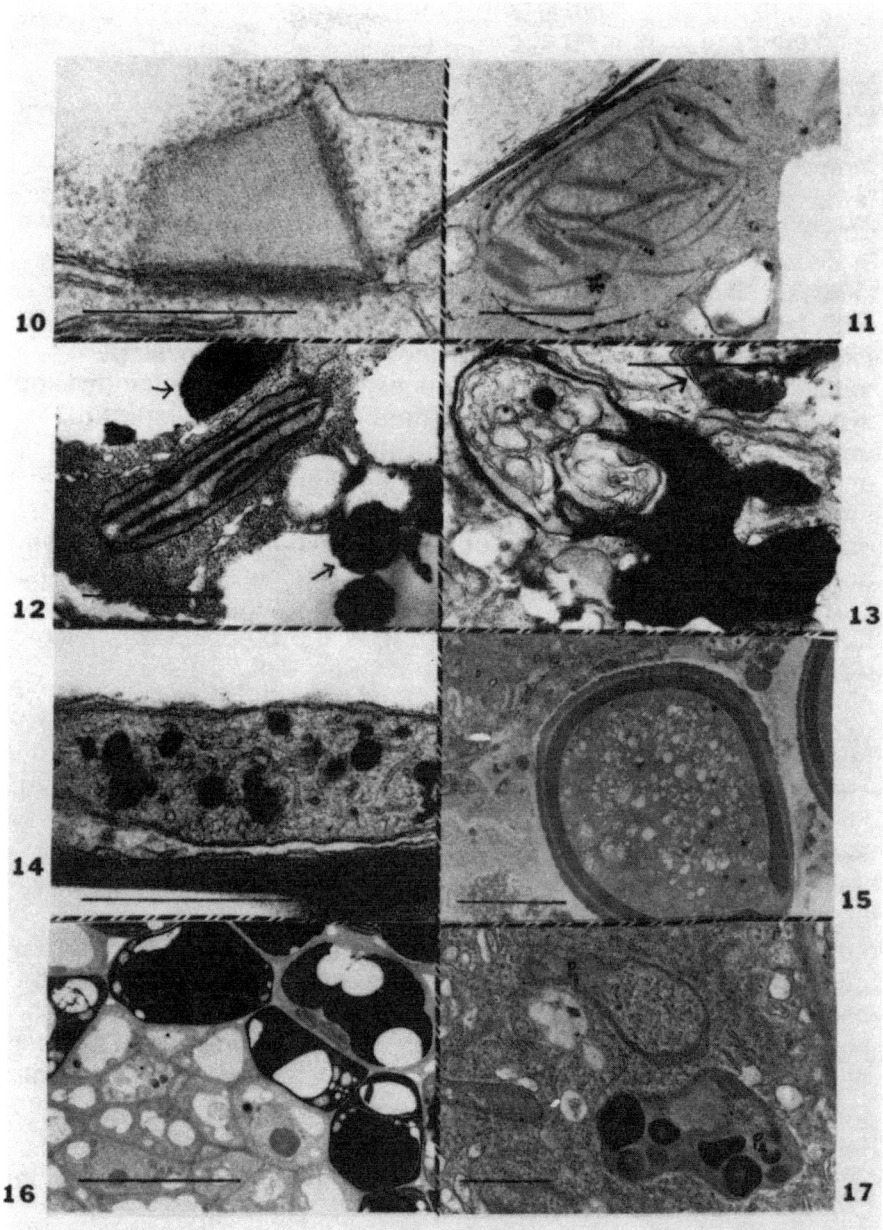

way that is analogous to following the successional stages of plant communities, either at various time intervals from the onset of colonization to the climax community, or along a transect of, say, sand dunes of increasing age.

Five basic stages have been identified as leading to the formation of chloroplasts in angiosperms (Whatley, 1977). The first stage is the simplest form of proplastid, essentially a double membrane-bound sac containing stroma and a fragment (a template?) of thylakoid ribbon (Whatley, 1983a). As development begins, the plastid starts to accumulate starch which is derived from reserves mobilized from older photosynthetic tissues. This amyloplast phase is usually brief and the plastid soon begins to change in shape from spheroidal to plemorphic. At this third stage, invaginations of the inner plastid envelope become common and there is some slight extension of the thylakoid system. During the period when starch is accumulating, and probably also when it is being lost, the plastid is temporarily ensheathed by rough endoplasmic reticulum (Fig. 4). Following the pleomorphic third stage, the plastid assumes its mature discoid shape. There is marked extension of the thylakoid system and the number of incipient grana or bithylakoids increases. The formation of true grana leads to the final stage of development, that of the mature chloroplast (Fig. 3). The progress of plastid development can be interrupted or diverted before the mature chloroplast stage is reached and so lead to the development of variant forms. It is by such diversion that distinctive forms of plastid (e.g., etioplasts or chromoplasts) develop. The inhibition and reestablishment of the normal pattern of development may be caused by factors that are either external, reflecting local environmental conditions (etioplasts formed in darkness) or internal, reflecting the overall nuclear control (most chromoplasts).

In annual plants or in deciduous species, the chloroplasts and their thylakoid systems continue to increase in size for some time prior to the onset of senescense, at which time the photosynthetic apparatus becomes degraded and increased amounts of carotenoids are synthesized. In species in which the leaves or other photosynthetic organs persist for several years, the chloroplasts sometimes undergo dedifferentiation during a regular cold or dry season, followed by redifferentiation when active growth is again possible, i.e. the chloroplasts revert temporarily to a proplastid state before once more developing into chloroplasts (Gaff, Zee and O'Brien, 1976; Senser, Schötz and Beck, 1975; Whatley, 1978). Thus chloroplast development is not an unidirectional system which invariably ends with senescence but rather a cyclical phenomenon which is, perhaps, best represented as a series of developmental waves. The onset of senescence and divergence to variant forms are not restricted to the chloroplast stage but can take place at any point within the cycle. Although it was suggested above that ripe seed was a convenient starting point for developmental studies, it should be recognized that this "starting point" is merely the trough of a wave in which the plastids may have already undergone various degrees of differentiation during seed growth and of dedifferentiation during seed dehydration and ripening while still on the parent plant. From the point of view of the evolution of new plastid forms and functions, the most important aspect of the plastid developmental cycle may well be the capacity to dedifferentiate and redifferentiate.

Colonization of the Land. As land plants have evolved, the proportion of tissues devoted to photosynthesis has declined. Thus, within each plant, an increasing proportion of the cells and their plastids have become available for other metabolic activities. For what functions have the plastids become specialized and when did these modifications take place?

It has long been recognized that the evolution of land plants has involved the development of new structures which could provide, inter alia, a means of anchorage and of taking up nutrients and water from the soil, a vascular system and mechanisms for waterproofing, structural strengthening, and exchanging gases with the atmosphere. Essentially, land plants had to fulfill these requirements by using and modifying the components and biochemical pathways already available to their green algal ancestors. And the process of modification has had to continue to meet the changing conditions of the land habitat to which the evolving plants and accompanying animals have themselves contributed. It might be thought that a survey of the plant species extant today might reveal a sharp break between the plastids in green algae and those in land plants. This is certainly not so. Characteristically algal features like cup-shaped plastids (Figs. 1 and 2) are retained by some land plants; "new" forms of plastid, like chromoplats, can be identified in the alga *Chara* (Pickett-Heaps, 1975; Whatley, 1983b). Among modern land plants, the great majority of the known modified plastid forms are restricted to the angiosperms, but we do not know if similar modified plastids were present in lower plants which are now extinct or if their evolution has gone hand in hand with evolution of the angiosperms and perhaps even, indirectly, of the insects and other animals. The overall control by the nucleus of most chloroplast biosynthesis would presumably allow chloroplast metabolism to be diverted to activities in keeping with those required by the plant as a whole.

In addition to their main phosotynthetic activity, green algal chloroplasts have two important abilities; (1) to accumulate in their stroma excess photosynthetic reserves like starch, as well as pyrenoid protein and globules of lipids and carotenoids and (2) to exchange metabolites with the cell cytoplasm. The evolution of new forms of non-green plastid has apparently been accompanied by the exploitation of these two abilities. As land plants have become increasingly complex, some algal features have been lost (*e.g.* the capacity to form pyrenoids), but in general, there seem to have been (1) changes in the use to which established metabolites are put; (2) a noticeable increase in the range of metabolites, particularly of secondary plant products, to which the plastids contribute; (3) an increase in the number of sites within the plastid where reserves of these metabolites can accumulate; (4) an increase in area of the surface across which the plastid may exchange metabolites with other cell compartments.

The Changing Use of Metabolites

Plant geotropism is believed to be mediated by sedimenting particles or statoliths. In rhizoids of the green alga, *Chara*, the gravity-perceiving particles are crystals of barium sulfate; in land plants the statoliths are amyloplasts. The initial function of the amyplast was presumably solely that of stage of photosynthetic reserves (Fig. 4). Sedimenting amyloplasts that are associated with a geotropic response are found in localized groups of cells in a range of different organs, and in taxa as diverse as bryophytes and angiosperms - in the stems of mosses, in the setae of moss and liverwort sporophytes, and perhaps also in the underground axes of some liverwort gametrophytes (Osborne, pers. comm.; Duckett and Renzaglia, 1988), in the root cortex of *Selaginella* and in the root caps of *Equisetum, Isoetes*, ferns and gymnosperms. In different species of angiosperms, geoperceptive amyloplasts can be found in the root cap, the root cortex, the stem endodermis and in leaf nodes (Osborne and Wright, 1977; Whatley, 1983a). Nor is gravity perception the only specialized function of amyloplasts. In petals of some members of the Ranunculaceae, starch is only accumulated in the plastids of the subepidermal layer. As these cells mature most of their contents disappear, but the numerous large starch grains remain and apparently serve as attractants for pollinating insects by causing light to be reflected back from the petal surface (Kay, Daoud, and Stirton, 1981; Whatley, 1984; Brett and Sommerard, 1986).

Although phycobilins and the chlorophylls are the main photosynthetic pigments, carotenoids may also participate in the process, both directly in light absorption and indirectly, by preventing photooxidation. These carotenoids are localized in the chloroplast membranes. In many algae which are motile, or have motile stages in their life cycle (gametes or zoospores), additional carotenoids are assembled in localized arrays of globules, the so-called eyespot, at a well-defined site often close to a flagellum, either within the cell's single plastid or within one only of its several plastids. It has frequently been assumed that the eyespot plays a role in flagellar movement, possibly as a photoreceptor or, more possibly, by shading the flagellar base, but its function, if any, remains unclear. Although eyespots are characteristic of many green algae like *Chlamydomonas*, they are never found in those taxa which are thought to be ancestral to land plants. However, during sexual reproduction in *Chara,* carotenoid globules accumulate in all the chloroplasts in each cell within a single layer of cells of the antheridial jacket. Thus, *Chara* can produce plastids which have characteristics of both chloroplasts and chromoplasts. This form of chloro-chromoplast can also be found in land plant taxa ranging from bryophytes to angiosperms (Fig. 1), usually in ancillary reproductive tissues (Duckett, 1975; Pickett-Heaps, 1975; Whatley, 1985). Though carotenoids in the chromoplasts of *Chara* may retain a protective role those in chromoplasts of most land plants (Figs. 5 and 13) have assumed the very different function of attracting spore, seed, and pollen dispersing animals.

The Introduction of New Metabolites

Plastids synthesize or contribute to the synthesis of many metabolites, including carbohydrates, amino acids and some proteins, photosynthetic pigments, lipids, and isoprenoids. Many of these metabolites undergo further modification within other cytoplasmic organelles and some are later excreted from the cell. In algae the excreted products can be used in several ways. Polysaccharides can be incorporated into organic scales, cellulose fibrils, and mucilages, for example, and deposited as a protective covering immediately outside the plasma membrane; mucilage can be used to aid movement as well as to provide a physical barrier against predation. An extruded glycoprotein helps to promote sexual reproduction in *Chamydomonas*; phenolic and various compounds can act as protectants against predators, as growth promoters or as growth inhibitors (Hellebust, 1974).

Some of the metabolites excreted by the cells of land plants are similar to, but others are very different from, and more varied than, those released by the cells of green algae; the latter clearly reflect the changing requirements imposed by dynamic terrestrial ecosystems. Nevertheless, the general functions of land plant plastids remain essentially similar to those of algal plastids. Mucilages are produced by many land plants, including bryophytes; they may provide a useful antidesiccant. Some of those released by vascular plants can still assist movement, but by lubricating roots. Polysaccharides continue to be incorporated into wall celluloses, but many of the walls are strengthened by the inclusion of lignin and exterior walls of aerial cells are coated by waxes and cutins which reduce water loss. The evolution of insects and other herbivores has been accompanied by the evolution of new and increasingly different plant protectants; the biochemical diversity of inhibitory, stimulatory, attractant, and recognition compounds has also increased.

The evolution of land plants has been accompanied by a conspicuous increase in the range and complexity of secondary plant metabolites, and the plastids have been implicated in the synthesis of many of these or of their associated carbohydrates (Jensen, 1985). Indeed, Kirk and Tilney-Bassett (1978) raise the question of whether the plastids in higher plants might not be the sites of synthesis of all terpenoid secondary products and, perhaps, even of virtually all phenolic secondary products. As these authors point out, such a capacity would imply for the plastids an "extraodinary range of biosynthetic capabilities."

Among the secondary products to which plastids may contribute are the flavonoids (Jensen, 1985). These are perhaps best known in the forms of anthocyanins and flavonols, pigments which normally accumulate in the vacuoles, and of lignin. The flavonoids function as attractants or insect guides in many flower petals, usually being concentrated in cells of the upper epidermis (Kay et al., 1981). Swain (1975), for example, has reviewed the main aspects of flavonoid evolution in relation to the evolution of green plants, starting with the appearance of glycoflavones in the algae, *Chara* and *Nitella*, continuing with the later diverging biochemical steps which may have led to the introduction on the one hand of the lignins and on the other of the pro-anthocyanidins (colorless to man) in lower land plants and leading to the complex flavonoids in advanced angiosperms.

New Sites of Forms of Metabolite Accumulation

The Stroma. The metabolites accumulated by the plastids of green algae all appear to be deposited in the stroma. Though the capacity to form pyrenoids has been lost by most land plants, starch and globules of lipid and carotenoid continue to be laid down, but sometimes in modified form. In the Mullerian bodies (specialized storage organs) of *Cecropia peltata*, for example, glycogen accumulates in the plastid stroma instead of starch; this glycogen provides food for "guardian" ants. In cells elsewhere in the plant, starch is formed in the usual way (Rickson, 1976). In epidermal cells of some members of the Liliaceae, Amaryllidaceae and Orchidaceae and in the cacti *Echinocereus* and *Echinocactus* photosynthetic reserves are stored in large oil droplets (summarized in Kirk and Tilney-Bassett, 1978).

Chloro-chromoplasts belong to the simplest class of chromoplast; they are found in some algae and in many lower land plants. During evolution on the land, the range of carotenoids synthesized has increased enormously (Goodwin, 1980) and the structure of chromoplasts has become more varied. Sitte (1974) has identified five additional classes of chromoplast. Of these the globulous chromoplasts are the most common (Figs. 1 and 5), being found in the gymnosperms, *Taxus baccata* and *Ephedra aphylla* (personal observation; specimens of *Ephedra*, courtesy of Prof. A. Fahn of the Hebrew University, Jerusalem) and in many angiosperms. They resemble chloro-chromoplasts in that the carotenoids are concentrated in globules in the stroma, but they lack thylakoids. In other classes of chromoplast, the carotenoids are assembled as fibrils or tubules in the stroma, within the sace of other tubules, in proliferating envelope membranes or as crystals within thylakoid sace. The ripe red seeds of the cycads, *Cycas revoluta, Macrozamia lucida* and *Zamia purpurea* are the only sites in non-angiosperm species where chromoplasts other than the two globulous classes have so far been found (Whatley, 1985).

In young chloroplats, the stroma also appears to be the site of accumulation of some thylakoid and other precursors materials either produced in excess of immediate requirements or held temporarily in a suitably accessible form when the normal developmental processes become imbalanced. Stromacenters, possibly comprising large subunits of ribulose-bis-phosphate carboxylase, may form in the plastid stroma under conditions of stress (Kirk and Tilney-Bassett, 1978; Wellburn, 1982). Usually large clusters of phytoferritin particles have been observed in the stroma in plants suffering from iron deficiency (Platt-Aloia, Thomson and Terry, 1983) and in the hypocotyls of bean seedlings grown in darkness (Whatley, 1977).

In many cells engaged in the secretion and/or transport of specialized and often secondary products, the unusually high electron density of the stroma points to the accumulation there of other as yet undetermined metabolites. Many ultrastructural investigations of secretory tissues (reviewed in Fahn, 1979 and in Werker and Zamski, 1985) concenrate on the endoplasmic reticulum and the Golgi apparatus. Few of the published micrographs show plastids at a sufficiently high magnification to allow a detailed analysis of their structure. Thus, the range of secretory tissues for which adequate information about the plastids is available is unfortunately restricted.

Intra-membrane Sacs. There are two different types of intra-membrane space which have been exploited by land plants as sites of metabolite accumulation, (1) between the two membranes of the plastid envelope and (2) the intrathylakoid sace. The extent to which these two sites are structurally contiguous remains uncertain. It is, however, probable that continuity is more frequent between the envelope space and the apparently envelope-derived, non-photosynthetic, concentric lamellae characteristic of some angiosperm chromoplasts, than it is between the envelope space and "true" photosynthetic thylakoids (Thomson and Whatley, 1980; Liedvogel and Falk, 1980; Spurr and Harris, 1968). Even in the absence of structural continuity, material may be moved intermittently between the envelope and the thylakoid spaces by way of vesicles and tubules (Figs. 13 and 14). Thus, channels may be made available for the movement of metabolites. The intramembrane compartments of the plastid also have the potential to provide a large space for storage, and an extensive surface area for contact with the plastid stroma.

The sacs of both the plastid envelope and the thylakoids have been stated to be the site of accumulation of essential oils, the volatile attractants produced, inter alia, in the "sprouting" glands of the angiosperm, *Dictammus albus* (Amelunxen and Arbeiter, 1967). The fruits of *Phellodendron amurense* also have plastids that are believed to be the site of biosynthesis of the essential oil (Heinrich and Schultze, 1985). In young glands, the site of accumulation is thought to be in tubuli within the stroma, but in older glands the osmiophilic deposits lie in the space between the membranes of the plastid envelope. In plastids in the digestive glands of the carnivorous plant, *Dionaea muscipular* Ellis (Robins, 1978), the intramembrane spaces accumulate an electron-lucent material of unknown composition.

The intramembrane space of the plastid envelope seems to be even more effectively exploited in the petals of some white and also pigmented cultivars of *Narcissus* (Whatley and Whatley, 1987). The orange and yellow flowers of *Narcissus* species contain chromoplasts in which the carotenoids are concentrated in envelope membranes which have proliferated to form several concentric rings. In spite of the absence of carotenoids, the structure of the plastids in white floral tissues of some cultivars is indistinguishable from the of the membranous chromoplasts of the pigmented flowers. In both the pigmented and the nonpigmented cultivars the envelope sac becomes greatly dilated and filled with a strongly electron-dense deposit as the petals age (Fig. 13). The product stored may well be an essential oil and the volume must be considerable. It has been suggested that such osmiophilic deposits in secretory cells may be artifacts of fixation (Galatis, Katsoras and Apostalakos, 1978). However in leaves of *Equisetum telmateia* (Fig. 11), dilations of the plastid envelope which are associated with osmiophilic deposits following chemical fixation are also present when the tissue is subjected to rapid-freezing and freeze-fracturing (McLean, Whatley and Juniper, 1988).

Though the intramembrane spaces of the thylakoids and the plastid envelope may both be used for the accumulation of some metabolites, the accumulation of others appears to be restricted to sacs provided by the thylakoids, sensu stricto. Today, the simplest plants which use intra-thylakoid spaces in this way are apparently the hornworts and mosses, but similar deposits are also found in some lower vascular plants (Fig. 6). The stored material is of moderate electron density, sometimes amorphous, sometimes granular, sometimes crystalline, but of unknown composition. In lower land plants, the cells containing these modified plastids are all concerned with conduction - transfer cells in gametophytic tissues of some hornworts (Duckett and Renzaglia, 1988), the leptoids of the moss, *Polytrichum*, and the sieve elements in some lower vascular plants (e.g., Burr and Evert, 1973; Hebant, 1974; Dute and Evert, 1977; Warmbrodt and Evert, 1979a and 1979b; personal observation). In other ferns, in gymnosperms and in angiosperms, sieve element starch and/or protein are stored as granules, crystals or fibrils, but these are in the stroma not in thylakoid sacs (Fig. 7 and 8). The modifications in plastid structure in conducting tissues of some bryophytes and of lower vascular plants may have been among the first, and hence the longest-established to be specifically associated with particular types of cell in land plants. It is therefore interesting that the sieve elements of gymnosperms and angiosperms have the only plastids in which structural modification has been successfully correlated with taxonomy (Behnke, 1972). The presence of modified plastids in so many different types of cell engaged in the secretion or transport of specialized metabolites may be a further reflection of the early evolution of new forms of plastid into conductive tissues.

The thylakoid sac can also be used for the accumulation of other metabolites (Figs. 9 and 10). In algae, the ribulose-bis-phosphate carboxylase, which seems to be the main constituent of pyrenoids (Holdsworth, 1971), lies in the plastid stroma (Fig. 2); the deposit is almost always granular. As described above, this protein may also be deposited in the stroma of plastids of sieve elements in some species of land plant, or as stromacenters, under conditions of stress. However, in mesophyll chloroplasts of spinach (Fig. 10), ribulose-bis-phosphate carboxylase has been identified as the main constituent of initially granular but subsequently crystalline deposits, lying not in the stroma but within the thylakoid sac, i.e. in the so-called membrane-bound bodies (Sprey and Lambert, 1977; Platt-Aloia and Thomson, 1977). Proteinaceous membrane-bound bodies have been observed in the plastids of many angiosperms and in some lower plant groups. However, cytochemical analyses of their contents in plastids of a few of the angiosperm species have suggested the probable presence of peroxidase and polyphenol oxidase rather than of ribulose-bis-phosphate carboxylase. Thus, the intrathylakoidal contents of the membrane-bound bodies may vary (Henry, 1975a, b; Hurkman and Kennedy, 1977; Lazorovits and Singh, 1986). It has already been mentioned that in some angiosperm chromoplasts, the carotenoids form crystals which are initiated within the thylakoid sac (Harris and Spurr, 1969). May this intrathylakoid site provide, for these carotenoids and for some proteins, a protected surface which is particularly suitable for the crystallization process?

Increase in Transport of Exchange Capacity

Some substances accumulated in plastids are only released when the cells and their plastids break down. These include essential oils, carotenoids and the precursors of pollen sporopollenin (Fig. 15), which originate in the tapetum (Stanley and Linskens, 1974). Other substances including sugars and components of phenolics, are regularly released through the plastid envelope.

Association with Endoplasmic Reticulum. The endoplasmic reticulum and Golgi apparatus are important in the further metabolism and transport across the cell of sugars, proteins and other metabolites originating in the plastids. In many bryophytes and lower vascular plants, the plastids are surrounded by a sheath of endoplasmic reticulum; occasionally the plastid envelope can be seen to be contiguous not only with the endoplasmic reticulum, but also with the envelopes of other organelles (Crotty and Ledbetter, 1973; Whatley, 1977; Duckett and Renzaglia, 1988). However, recent work on rapidly-frozen, freeze-fractured tissues suggest that the endoplasmic reticulum and the plastid envelope may become contiguous much more frequently than had been indicated by chemical fixation techniques (McLean, Whatley and Juniper, 1988). The establishment of such intra-membrane channels would certainly promote the exchange of metabolites between the two organelles. In some ferns and in *Equisetum* (Fig. 11), strongly electron dense deposits (possibly tannins of phenolic compounds) accumulate within the plastid envelope, in the endoplasmic reticulum and often, also, in the vacuoles. In the other ferns there is no obvious association between plastids and endoplasmic reticulum, although phenolic compounds are accumulated in the vacuoles of many species (Fig. 12). An association between plastids and endoplasmic reticulum has been reported in a few green algae (Gibbs, 1981) and in many lower land plants but in gymnosperms only in *Ginkgo biloba, Juniperus communis* L. and the resin canal cells of *Pinus pinea* (Stewart and Gifford, 1967; Whatley, 1977; Wooding and Northcote, 1965). In angiosperms the association has been observed in several different organs (including leaves, pollen and secretory tissues), but in almost every case it seems to be limited to a brief period of starch accumulation and loss during early stages of plastid development (Fig. 4).

Pleomorphic Plastids. As proplastids develop into chloroplasts, they regularly go through a phase when they become pleomorphic. This phase can be identified in chloroplasts of species ranging from the alga, *Chara*, to angiosperms and so seems to have been long established. Newcomb (1967) has suggested that the pleomorphic form may represent a "feeding" stage. the semipermeable plastid envelope allows the entry of some selected metabolites and the release of others to the cytoplasm where they may undergo further processing; the pleomorphic form conspicuously increases the plastid surface area over which such exchange of metabolites can take place. Also, in many types of pleomorphic plastid, the normal system of invaginations of the inner envelope membrane and its associated vesicles proliferates considerably. This results in a further increase in the potential exchange surface, but in this case, between the plastid stoma and the intra-membrane envelope channel (Whatley and Whatley, 1987). In C_4 plants, the somewhat similar invagination system called the peripheral reticulum is believed to be important in CO_2 exchange (Gracen et al., 1972). In some secretory cells and in some petals, much of the interior of each plastid is occupied by a complex tubular network (Whatley, 1977, Cheniclet and Carde, 1985; Brett and Sommerard, 1986) which may, in part, be assembled from vesicles derived from the plastid envelope. The precise forms of these tubular networks vary considerably, but in general they resemble the "thylakoid plexes" (assemblages of thylakoid or tubular elements) described by Spurr and Harris (1988). In plastids with such plexes the capacity for storage of metabolites within the tubules, and perhaps also for their transfer to the envelope membrane space, must be considerable.

Pleomorphic plastids with many inner membrane invaginations and vesicles are particularly well-developed in cells engaged in the secretion of an unexpectedly wide range of unrelated substances, from salt, sugars, and digestive enzymes on the one hand to essential oils, lignin, gums, and resins on the other (Whatley and Whatley, 1987). Pleomorphic plastids are found in many different types of cell (Figs. 5, 13, 14, 15, 16 and 17); in transfer cells of some moss sporophytes, in cotyledon bundle sheath cells in *Welwitschia*, in the mucilage-secreting cells of the *Isoetes* Ligule and of root caps, in epidermal cells of petals and in a wide variety of secretory cells (Paolillo, 1962; Fahn, 1979; Whatley, 1983a; Werker and Zamski, 1985; Whatley and Whatley, 1987; Duckett and Renzaglia, 1988). The apparently numerous metabolites of pleomorphic plastids have not, for the most part, been analyzed, nor have their specific roles in secretion been determined.

The only appropriate review of this subject of which I am aware, is a recent one in which plastid structure and volatile compounds were analyzed in specialized secretory structures of 45 species (Cheniclet and Carde, 1985). When the essential oil contained significant amounts of monoterpene hydrocarbons or oxygenated compounds, the plastid stroma was homogeneous and lacked ribosomes and thylakoids. In the absence or near absence of monoterpenes, ribosomes were present in an electron dense stroma which all contained some thylakoids and a well-developed tubular complex. The published micrographs suggest that most of the plastids are pleomorphic. The authors report a quantitative relationship between plastid volume and the percentage of monterpenes in the oil, but no such relationship for that of the sequiterpenes, phenylpropanoid or aliphatic compounds. Nevertheless, they suggest that, in some species, the involvement of plastids in sesquiterpene biosynthesis cannot be ruled out. No comparable information is available about the possible role of plastids in the biosynthesis of other secreted metabolites.

Chloroplasts. In contrast to the non-green plastids, the chloroplasts of land plants appear to have undergone little structural or other modification (Fig. 3). Nevertheless they do show consistent quantitative variations from one type of cell to another, in their size and numbers, in the extent of their thylakoid systems and in the number and size of their starch grains. Although all chloroplasts follow the same basic pathway during their development, those in each cell file in a root or a leaf, say, will show characteristic quantitative differences. In leaves, for example, spongy mesophyll cells tend to have larger plastids with bigger franal stacks and much more starch than palisade cells and the plastids in the epidermal layers tend to be fewer and smaller and to have a less extensive thylakoid system.

The only major modification in chloroplast structure is in the dimorphic plastids in some C_4 plants, in which the bundle sheath chloroplasts accumulate large amounts of starch and thylakoid system is reorganized (Kirt and Tilney-Bassett, 1978). As indicated above, the development of a peripheral reticulum has been equated with an increase in capacity for CO_2 exchange. The well-known C_4 syndrome appears to be restricted to angiosperms where it has been found in some 500 species from phylogenetically diverse angiosperm families (Downton, 1975). CAM plants (Crassulacean Acid Metabolism) resemble C_3 plants in that they, too, store CO_2 temporarily as organic acid; in both types of plant, modification in the mode of photosynthesis is thought to be an adaptation to an environment where water is a limiting factor. However chloroplasts of CAM plants show no consistent structural modification. Most CAM plants are angiosperms, but this form of metabolism has also been found in *Isoetes* and *Welwitschia* (summarized in Whatley, 1983b).

Responses to Light. Light of different wavelengths is required not only for chlorophyll synthesis, chloroplast development and photosynthesis, but also for many other plant responses. The wavelengths of light received by a particular cell are obviously influenced by both the quantity and the wavelengths of the light absorbed by overlying tissues. The evolution of new pigments, synthesized partly or entirely by the plastids, has accompanied the evolution of new responses to light controlled both directly, by the plants themselves and indirectly, by way of animal vision. It has been suggested that the evolution of glycoflavones was important initially in providing a screen against ultra-violet light. In ferns have the specific growth responses to blue light developed in response to the evolution of their flavonoids? True anthocyanins are common in angiosperms but elsewhere have been found only in the receptacles of *Podocarpus* and the cones of some other conifers (Swain, 1975). Do anthocyanins in pine cones attract insects as they do in angiosperms?

The prevailing pigmentation of different cell layers within an organ at, say, different seasons can have important developmental or ecological consequences. In citrus the developing seeds are deeply embedded in the fruits, but it is not unusual for the young cotyledons to turn green. In lemons, a large proportion of the light received at the surface is absorbed by the carotenoids and flavonoids of the peel and internal tissues. The wavelengths of light which are not absorbed by the parent fruit, and which in consequence reach the seeds and their embryos, are those very wavelengths which are necessary for the conversion of protochlorohyll to chlorophyll; the quantity of light received is low but nevertheless adequate for the essential photochemical step (Whatley and Price, 1983). In organs where the overlying tissues contain other pigments, different responses will be induced. Of particular ecological importance is the absorption of light by the photosynthetic pigments and its effect on the resulting transmission of the red and the far-red wavelengths to the cells below. Thus, although covered over by many layers of green scales or young leaves, cells in meristems, in dormant buds and seeds, and in pre-flowering shoot apices can all respond to photoperiodic and other phytochrome-based controls (Vince-Prue, 1975).

CONCLUSIONS

From unicellular green algae to angiosperms, phosotynthesis has remained the principal function of the majority of plastids, the chloroplasts. Changes in plastid structure and function have almost all been confined to non-green plastids; most have involved modifications of structural components and biosynthetic pathways already present in algal plastids. Though few such changes are found in the algae themselves or even in lower land plants, their introduction may have been linked to the evolution (a) or proplastids in meristems, (b) of multicellular plants requiring conductive tissues and (c) of adaptation to the often dry and unpredictable land environment and to the limitation of predation.

As land plants have evolved, the tendency towards the elimination of comparatively few "waste products" from the nonspecialized cells seems to have become modified. Instead an increasing range of tissue-specific, specialized secretory products are accumulated in localized groups of cells (Fig. 16) as well as in a range of precisely formed cavities and ducts; other such products are ejected from distinctively organized trichomes, glands and other such "new" structures. In lower land plants these cells or structures are uncommon (and the form of their plastids is generally not known); the few recorded examples include mucilage-secreting cells or ducts in some bryophytes and in *Lycopodium, Selaginella* and *Isoetes,* hydathodes in *Equisetum,* extrafloral nectaries (some "non-structural") in a few fern genera, salt-, lemon-scented- and flavonoid-secreting-glands in some other ferns and latex-producing cells in the fern, *Regnellidum* (Darwin, 1877, Johnson, 1937; Chapham, Tutin and Warburg, 1962; Smith, Craig and Santarosa, 1971; Horner *et al.* 1975; Bruce, 1976; Fahn 1979; Duckett and Renzaglia, 1988. By contrast, glands and other secretory structures with modified plastids are common in the angiosperms and the evolution of flowers has led to even greater changes in plastid structure. It is particularly unfortunate that almost no information is available about plastid structure in the secretory cells of ferns, as it is there that important clues to the progress of plastid evolution in land plants may well be waiting to be found.

Secretory substances to which plastids are believed to contribute (Kirk and Tilney-Bassett, 1978) are phenolics (lignin, tannin, anthocyanins and flavonoids), isoprenoids (carotenoids, essential oils, terpenoids, resins, lates and the hormones, abscissic acid and giberellin), sugars (especially mucilages and nectar) and lipids (cutin, suberin and waxes). The sugars released as the Stage 2 amyloplasts develop into the Stage 3 pleomorphic plastids may additionally provide an energy source for further synthesis within the endomembrane system. During the evolution of land plants, the major shift in plastid structure and function seems to have been the increased exploitation of the various sub-forms of pleomorphic plastid. These plastids have become characteristic of many different types of non-photosynthetic cell in which they have apparently acquired two important roles; 1) by exploiting existing biochemical pathways, they seem to have extended their biosynthetic capacity and to have become a major source of precursors for an increased range of secondary metabolites and 2) by greatly increasing their surface area and intra-membrane spaces in several different ways, they have increased their potential for the exchange and storage of these metabolites. It therefore seems reasonable to consider pleomorphic plastids not only as major sites for secondary metabolism, but also as the organellar equivalent of transfer cells (Whatley and Whatley, 1987)

Plastids have many different functions. In addition to their photosynthetic activities they can contribute to metabolites which protect the plant against desiccation and other environmental hazards, and against predators; they contribute to some growth promoting, growth inhibiting, and recognition substances; they provide a range of attractants for insects and other animals which are essential for successful plant reproduction and long-term survival. The pigments to which they contribute modify light in such a way as to allow the operation of photoperiodic and other control mechanisms which permit a form of fine-tuning to the local climatic conditions. The chloroplast, through photosynthesis, has always had a major ecological role as a primary food source. In modern land plants and particularly in angiosperms, many of the varied forms of non-green plastids have also acquired features of significant ecological importance.

SUMMARY

Plastids are believed to have evolved from algae that were once free-living but later were taken up as endosymbionts by unicellular protists. In their new habitat, the photosynthetic symbionts provided their (phagocytic) hosts with new and continuing source of food, but lost many of their other synthetic activities, and much of their control was taken over by the cell nucleus. Once the symbionts had become established and transformed into chloroplasts, there seems to have been a lengthy phase of evolution of green algae and land plants during which there was little change in chloroplast structure and function; the basic cup-shape is retained by many green algae and some lower land plants. However, as multicellular organisms evolved, there appears to have been an increase in the proportion of the plant devoted to a nonphotosynthetic function. The nonphotosynthetic cells virtually all retained their plastids but in a dedifferentiated state. Most subsequent changes in plastid structure and function seem to be associated with these nongreen plastids. Some of them are protoplastids in meristems. Others, specializing in the synthesis or accumulation of a range of storage products or secondary metabolites, represent diversions from the basic pathway of chloroplast development; it is these that show the most conspicuous structural modifications. Lower land plants have few such modified plastids; those that do occur may reflect adaptation of the plant to the land environment. Most of the "new" forms of plastid are to be found in angiosperms and many of these seem to be associated with specialized functions linked to interactions between plants and animals.

LITERATURE CITED

Amelunxen, F., and H. Arbeiter. 1967. Untzersuchungen an den Spritzdrüsen von *Dictamus albus* L. Z. Plfanzenphysiol. 58: 49-69.

Behnke, H. D. 1972. Sieve-tube plastids in relation to angiosperm systematics. Bot. Rev. 38: 155-197.

Brett, D. W., and A. P. Sommerard. 1986. Ultrastructural development of plastids in the epidermis and starch layer of glossy *Ranunculus* petals. Ann. Bot. 58: 903-910.

Bruce, J. G. 1976. Development and distribution of mucilage canals in *Lycopodium*. Amer. J. Bot. 63: 481-491.

Burr, F. A., and R. F. Evert. 1973. Some aspects of sieve element structure and development in *Selaginella kraussiana*. Protoplasma 78: 81-97.

Cavalier-Smith, T. 1982. The origins of plastids. J. Linn. Soc. Biol. 17: 289-306.

Cheniclet, C., and J.-P. Carde. 1985. Presence of leucoplasts in secretory cells and of monoterpenes in the essential oil; a correlated study. Isr. J. Bot. 54: 219-238.

Christensen, T. 1962. Algae. *In* T. W. Böcher, M. Lange, and T. Sorensen [eds.], Systematic Botanik, Vol. II, no. 2, pp. 128-146. Munksgaard, Copenhagen.

Church, A. H. 1919. The building of an autotrophic flagellate. Oxf. Bot. Mem. 1: 1-27.

Clapham, A.R., T.G. Tutin, and E. F. Warburg. 1962. Flora of the British Isles. The University Press, Cambridge.

Cook, C. B. 1983. Metabolic interchange in algae - invertebrate symbiosis. *In* K. W. Jeon [ed.], Intercellular symbiosis, pp. 177-210. Academic Press, New York.

Crossland, C. J., D. J. Barnes, T. Cox, and M. Devereux. 1980. Compartmentation and turnover of organic carbon in the staghorn coral *Acropora formosa*. Mar. Biol. 59: 181-187.

Crotty, W. J., and M. C. Ledbetter. 1973. Membrance continuities involving chloroplasts and other organelles in plant cells. Science 182: 839-841.

Darwin, F. 1877. On the nectar glands of the common brakefern. J. Linn. Soc. Bot. 15: 407-409.

Dodson, E.D. 1979. Crossing the procaryote-eucaryote border: endosymbiosis or continuous development? Can. J. Microbiol. 25: 652-674.

34

Downton, W. J. S. 1975. The occurrence of C$_4$ photosynthesis among plants. Photosynthetica 9: 96-105.

Duckett, J. G. 1975. An ultrastructural study of the differentiation of antheridial plastids in *Anthoceros laevis* L. Cytobiologie 10: 432-448.

Duckett, J. G., and K. S. Renzaglia. 1988. Adv. in Bryol. 3: 33-93.

Dute, R.R., and R. F. Evert. 1977. Sieve element ontogeny in the root of *Equisetum hyemale*. Amer. J. Bot. 64: 421-438.

Fahn, A. 1979. Secretory Tissues in Plants. Academic Press, London.

Gaff, D.F., Z. Y. Zee, and T. P. O'Brien. 1976. The fine structure of dehydrated and reviving leaves of *Borya nitida* Labill. - a disiccation tolerant plant. Austr. J. Bot. 24: 225-236.

Galatis, B., C. Katsaros, and P. Apostolakos. 1978. Ultrastructural studies on the oil bodies of *Marchantia paleacea* Bert. II. Advanced stages of oil-body cell differentation: synthesis of lipophilic material. Can.J. Bot. 56: 2268-2285.

Gibbs, S.P. 1981. The chloroplast endoplasmic reticulum: structure, function and evolutionary significance. Int. Rev. Cytol. 72: 49-99.

Goodwin, T. W. 1980. The Biochemistry of the Carotenoids: 1: Plants. Chapman and Hall, London.

Gracen, V. E. Jr., J. H. Hilliard, R. H. Brown, and S. H. West. 1972. Peripheral reticulum in chloroplasts of plants differing in CO$_2$ fixation. Planta 107: 189-204.

Harris, W. M. and A. R. Spurr. 1969. Chromoplasts in tomato fruits. I.Ultrastructure of low-pigment and high-beta mutants. Carotene analyses. Amer. J. Bot. 56: 369-379.

Hebant, C. 1974. Polarized accumulation of endoplasmic reticulum and other ultrastructural features of leptoids in *Polytrichadelphus magellanicus* gametophytes. Protoplasma 81: 373-382.

Heinrich, G., and W. Schultze. 1985. Composition and site of biosynthesis of the essential oil in fruits of *Phellodendron amurense* Rupr. (Rutaceae). Isr. J. Bot. 34: 205-217.

Hellebust, J. A. 1974. Extracellular products. *In* W. D. .P. Stewart [ed.], Algal physiology and biochemistry. Bot. Mono. 10. Blackwell Scientific Pub., Oxford.

Henry, E. W. 1975a. Polyphenol oxidase activity in thylakoids and membrane-bound granular components of *Nicotiana tabacum* chloroplasts. J. Microsc. 22: 109-116.

Henry, E.. W. 1975b. Peroxidases in tobacco abscission zone tissue. III. Ultrastructural localization in thylakoids and membrance-bound bodies of chloroplasts. J. Ultrastruct. Res. 52: 289-299.

Holdsworth, R. H. 1971. The isolation and partial characterization of the pyrenoid protein of *Eremosphaera viridis.* J. Cell Biol. 51: 499-513.

Horner, H. T. Jr., C. K. Beltz, R. Jagels, and R. E. Boudreau. 1975. Ligule development and fine structure in two heterophyllous species of *Selaginella.* Can. J. Bot. 53: 127-143.

Hurkman, W. J., and G. S. Kennedy. 1977. Development and cytochemistry of the thalakoid body in tobacco chloroplasts. Amer. J. Bot. 64: 86-95.

Jensen, R. A. 1985. Tyrosine and phenylalanine biosynthesis: relationships between alternative pathways, regulation and subcellular location. In E. E. Conn [ed.] The shikimic acid pathway. Recent Advances in Phytochemistry 20. Plenum Press, New York.

Johnson, M. A. 1937. Hydathodes in the genus *Equisetum.* Bot. Gaz. 98: 598-608.

Kay, Q. O. N., H. S. Daoud, and C. H. Stirton. 1981. Pigment distribution, light reflection and cell structure in petals. J. Linn. Soc. Bot. 83: 57-84.

Kirk, J. T. O., and R. A. E. Tilney-Bassett. 1978. The Plastids. Elsevier, North Holland.

Kivic, P. A. and M. Vesk. 1974. An electron microscope search for plastids in bleached *Euglena gracilis* and in *Astasia longa.* Can. J. Bot. 52: 695-699.

Lazarovits, G. and B. Singh. 1986. Localization of polyphenol oxidase activity in the lamellae and membrance-bound inclusions of etiolated soybeans. Can. J. Bot. 64: 1675-1681.

Liedvogel, B. and H. Falk. 1980. Leucoplasts mimicking membranous chloroplasts. Z. Pflanzenphysiol 98: 371-375.

McLean, B., J. M. Whatley, and B. E. Juniper. 1988. Continuity of chloroplast and endoplasmic reticulum membranes in *Chara* and *Equisetum.* New Phytol. 109: 59-65.

Newcomb, E. H. 1967. Fine structure of protein-storing plastids in bean root tips. J. Cell. Biol. 33: 143-163.

Osborne, D. J., and M. Wright. 1977. Gravity-induced cell elongation. Proc. R. Soc. Lond. B. 199: 551-564.

Paolillo, D. J., Jr. 1962. The plastids of *Isoetes howellii.* Amer. J. Bot. 49: 590-598.

Pickett-Heaps, J. D. 1975. Green Algae. Sinnauer Assoc., Sunderland, Mass.

Platt-Aloia, K. A., and W. W. Thomson. 1977. Chloroplast development in young sesame plants. New Phytol. 78: 599-605.

36

Platt-Aloia, K. A., W. W. Thomson, and N. Terry. 1983. Changes in plastid ultrastructure during iron nutrition-mediated chloroplast development. Protoplasma 114: 85-92.

Rickson, F. R. 1976. Ultrastructural differentiation of the Müllerian body glycogen plastid of *Cecropia peltata* L. Amer. J. Bot. 63: 1272-1279.

Ris, H., and W. Plaut. 1962. Ultrastructure of DNA-containing areas in the chloroplast of *Chlamydomonas*. J. Cell Biol. 13: 383-391.

Robins, R. J. 1978. Studies in Secretion and Absorption in *Dionaea muscipula Ellis*. D. Phil. Thesis. Univ. Oxford.

Schimper, A. F. W. 1883. Über die Entwickelung der Chlorophyllkörner und Farbkörper. Bot. Z. 41: 105-114.

Senser, M., F. Schøotz, and E. Beck. 1975. Seasonal changes in structure and function of spruce chloroplasts. Planta 126:1-10.

Sitte, P. 1974. Plastiden-Metamorphose und Chromooplasten bei *Chrysoplenium*. Z. Pflanzenphysiol. 73:243-265.

Smith, D. M., S. P. Craig, and J. Santarosa. 1971. Cytological and chemical variation in *Pityogramma triangularis*. Amer. J. Bot. 58: 292-299.

Sprey, B., and C. Lambert. 1977. Lamellae-bound inclusions in isolated spinach chloroplasts. II. Identification and composition. Z. Pflanzenphysiol. 83: 227-247.

Spurr, A. R., and W. M. Harris. 1968. Ultrastructure of chloroplasts and chromoplasts in *Capsicum annuum*. I. Thylakoid membrane changes during fruit ripening. Amer. J. Bot. 55: 1210-1224.

Stackebrandt, E. 1983. A phylogenetic analysis of *Prochlorom*. *In* H. E. A. Schenk and W. Schwemmler [eds.] Endocytobiology II. Intracellular Space as an Oligogenetic Ecosystem. de Gruyter, Berlin.

Stanley, R. G., and H. F. Linskens. 1974. Pollen: Biology, Biochemistry, Management. Springer-Verlag, Berlin.

Stewart, K. D., and E. M. Gifford. 1967. The ultrastructure of the developing megasproe mother cell of *Ginkgo biloba*. Amer. J. Bot. 54: 375-383.

Stewart, K. D., and K. R. Mattox. 1984. Classification of the green algae: a concept based on comparative cytology. *In* D. E. G. Irvine, and D. M. John [eds], Systematics Assoc. Special Vol. 27: 29-72. Academic Press, London.

Swain, T. 1975. Evolution of flavinoid compounds. *In* J. B. Harborne, T. J. Mabry, and H. Mabry [eds.] The Flavinoids. Chapman and Hall, London.

Taylor, F. J. R. 1983. Some eco-evolutionary aspects of intracellular symbiosis. *In* K. W. Jeon [ed.] Intracellular Symbiosis. Academic Press, New York.

Thomson, W. W., and J. M. Whatley. 1980. Development of nongreen plastids. *Ann. Rev. Plant Physiol.* 31: 373-394.

Vince-Prue, D. 1975. Photoperiodism in Plants. McGraw-Hill, London.

Warmbrodt, R. D., and R. F. Evert. 1979a. Comparative leaf structure of several species of homosporous leptosporangiate ferns. Amer. J. Bot. 66a: 412-440.

Warmbrodt, R. D., and R. F. Evert. 1979b. Comparative leaf structure of six species of eusporangiate and protoleptosporangiate ferns. Bot. Gaz. 140-167.

Wellburn, A. R. 1982. Bioenergetic and ultrastructural changes associated with chloroplast development. Int. Rev. Cytol. 80: 133-191.

Werker, W., and E. Zamski [eds.]. 1985. Secretion and secretory structures in plants. Isr. J. Bot. 34: 67-395.

Whatley, J. M. 1977. Variations in the basic pathway of chloroplast development. New Phytol. 78: 407-420.

Whatley, J. M. 1978. A suggested cycle of plastid developmental inter-relationships. New Phytol. 80: 489-502.

Whatley, J. M. 1982. Ultrastructure of plastid inheritance: green algae to angiosperms. Biol. Rev. 57: 527-569.

Whatley, J. M. 1983a. The ultrastructure of plastids in roots. Int. Rev. Cytol 85: 175-220.

Whatley, J. M. 1983b. Plastids - past, present and future. *In* K. W. Jeon [ed.], Intracellular Symbiosis. Academic Press, New York.

Whatley, J. M. 1984. The ultrastructure of plastids in the petals of *Caltha palustris* L. New Phytol. 94: 19-27.

Whatley, J. M. 1985. Chromoplasts in some cycads. New Phytol. 101: 595-604.

Whatley, J. M., and D. N. Price. 1983. Do lemon coytledons green in the dark? New Phytol. 94: 19-27.

Whatley, J. M., and F. R. Whatley. 1981. Chloroplast evolution. New Phytol. 87: 233-247.

Whatley, J. M., and F. R. Whatley. 1987. When is a chromoplast? New Phytol. 106: 667-678.

Wooding, F. B. P., and D. H. Northcote. 1965. Association of the indoplasmic reticulum and the plastids in *Acer* and *Pinus*. Amer. J. Bot. 52: 526-531.

ADAPTIVE SHIFTS TOWARD HUMMINGBIRD POLLINATION

G. Ledyard Stebbins

INTRODUCTION

I take great pleasure in dedicating this contribution to one of my oldest and most revered scientific friends, Herbert G. Baker. For a generation, he has set the example of how to exploit facts extracted directly from nature to illustrate and explain general principles of evolution.

One of the least understood of the major features of evolution is the nature and explanation of events that lead to striking innovations or anagenetic events; those that initiate entirely new directions. One reason for this relative lack of understanding is that such events are rare, and many of them are unique turning points in the three billion year history of organismic evolution. Hard data on these unique events cannot be obtained, since they all occurred in the distant past, and the conditions that brought them about will always remain to a certain extent speculative. An indirect way of approaching the problem is to analyze as carefully as possible intermediate situations; those that are unusual with respect to the origin of the great majority of species belonging to a particular group, but are still common enough, and have occurred recently enough so that the events that brought them about can be reconstructed with reasonable confidence. Particularly favorable are situations involving differently adapted species that are sympatric or nearly so and can be crossed so that segregating progeny from hybrids can be analyzed. When such situations exist, analysis is possible at all levels, from the ecological problem of organism-environment interactions through the problems of conventional cytogenetics, such as the number of gene differences involved, down to the molecular developmental problem of how the genes produce their effects.

Another circumstance favorable for this intermediate, indirect approach would be the presence of several similar examples in the same geographic region. This permits us to test on other examples hypotheses that arise from the first situation to be examined. In no case can we expect to prove conclusively that an hypothesis concerning past events is valid. Nevertheless, the larger is the number of similar but unrelated examples that fit a particular hypothesis, the higher is the probability that this hypothesis is correct.

To botanists living in the western United States, a problem of immediate importance that meets this criteria is the shift in several unrelated genera from adaptation to pollination by Hymenoptera (bees or wasps) or occasionally from Lepidoptera (butterfly) pollination, to pollination by hummingbirds, Trochilidae. This topic has been explored by several botanists. An excellent review by Karen and Verne Grant (1968) is the source of much of the data presented below.

PARALLEL SHIFTS TO HUMMINGBIRD POLLINATION

The Grants list as adapted to hummingbird pollination 129 species belonging to 39 genera and 18 families. Of these species, 21 do not have in California related species that are pollinated by Hymenoptera, so that modern counterparts of the species involved in the adaptive shift are not available, and their evolution must be investigated elsewhere, if at all. This leaves 108 species belonging to 30 genera as favorable material for at least partial analysis. Although in most of these genera only one shift per genus is clearly indicated, the shift probably took place twice in *Delphinium, Lilium*, and *Mimulus* (incl. *Diplacus)*, and three times in *Penstemon* (incl. *Keckiella*). In the western American flora, therefore, reasonably good evidence exists for 33 separate shifts from hymenopteran or lepidopteran to hummingbird pollination. If other areas, particularly tropical America, were considered, the number would be much larger. For eleven examples: *Aquilegia, Delphinium* (2), *Mimulus* (2), *Penstemon* (2), *Gilia splendens, Lonicera involucrata* and *Lilium* (2), exchange of genes via hybridization between hummingbird-pollinated species or varieties and those pollinated by insects is clearly possible, and for several other examples it is highly likely. Evolutionists, therefore, have available a rich source of material for exploring this important problem.

THE SYNDROME AND ITS VARIANTS

As a descriptive model of the syndrome, the brilliant analysis by E.O. Guerrant (1982) in Dr. Herbert Baker's laboratory of the red larkspur, *Delphinium nudicaule*, serves ideally. This species grows in moist canyons and stream banks in northern California. It is closely related to a series of partly sympatric species centering about the purple-flowered, bee-pollinated *D. decorum*, between many of which, including *D. nudicaule*, natural hybrids have been recorded. Chromosomally, all are diploid, having the somatic number of $2n = 16$.

The differences between *D. nudicaule* and its relatives are of three kinds, each of which is probably governed by a different gene system. First, the leaves of *D. nudicaule* are less deeply lobed than are those of its relatives, and are completely glabrous, whereas those of its most nearly sympatric congener, *D. decorum*, are somewhat pubescent. Where these species are sympatric in the northern Coast Ranges, *D. nudicaule* favors moister habitats, grows most actively in the moister weeks of early spring, and blooms earlier than *D. decorum*. Second, the red flower color of *D. nudicaule* contrasts sharply with the dark blue-purple of *D. decorum*. Third, the non-spurred sepals of *D. nudicaule* are smaller than those of its relatives, and are directed forward so that they form a continuation of the nectar bearing floral tube formed by the sepal spur. Guerrant (1982) showed that these sepals grow more slowly compared to the rest of the flower than do those of *D. decorum* and characterizes their growth are neotenic.

Nearly all hummingbird-pollinated species as perennials. A few are shrubby, but the majority are herbaceous. Annuals are exceptional as hummingbird flowers, the only example of the shift from insects to hummingbirds in the western flora being *Gilia splendens*. Hence, in later sections of this review that estimate the probability of an insect-pollinated group giving rise to a hummingbird flower, only perennial species are considered. Hummingbird flowers usually produce large amounts of nectar, in which sucrose predominates, while amino acids, if present, exist in low concentrations (Baker and Baker, 1982). This shift in nectar production is apparently accomplished so easily by genetic change that it is only a minor barrier to the completion of the adaptive shift.

The three major components of the syndrome [shift in vegetative characters that control adaption to habitat, in flower pigmentation, and in floral structure,] have all been involved, via parallel changes, in most of the examples to be considered. This shows that the shift cannot be brought about by one or two mutations, since the genetic basis of these three components is certainly different, as is borne out by the results of hybridization experiments to be reviewed below. In each of them, three different kinds of selection pressures must have been involved: (1) vegetative adaptation to habitats in which hummingbird pollinators are more abundant, at least seasonally, than are large bees or other pollinators to which the ancestors are adapted; (2) a shift in color that makes the flowers more conspicuous and easily approached by hummingbirds than are the ancestors; (3) a floral structure that fits the hummingbird beak better than the insect proboscis.

Nevertheless, some of the examples reviewed by Grant and Grant include only one or two of these three elements. Most often, the habitat differences does not appear to be pronounced, at least without carefully collected data on the species involved. This is true for *Monardella macrantha* vs. *M. nana, Trichostema lanatum* vs. *T. parishii, Silene californica* vs. other western American species of the genus, *Astragalus coccineus* vs. other related species, *Fritillaria recurva* vs. *F. lanceolata*, and both *Lilium maritimum* and *L. parvum* vs. their nearest relatives. Intermediate or less pronounced differences with respect to flower color and structure are less frequent. Color differences in *Castilleja* are usually pronounced, but in at least two examples they are less so. These are the hummingbird flower of *C. Payneas* vs. the insect pollinated *C. pilosa* in northern California, and the hummingbird (probably) *C. coccinea* vs. insect pollinated *C. pallida* and *C. septentrionalis* in northern United States and Canada. In *Lilium parvum*, an entire spectrum of color differences exists from orange-yellow to pinkish red. This may reflect a difference within the species of the most common pollinators, from Lepidoptera to hummingbirds. In their review, the Grants include a photograph of a hummingbird pollinating a normally Lepidopteran flower, *Lilium humboldtii*. With respect to flower structure, there is little if any difference between the corolla of typical *Lonicera involucrata*, that is insect (probably Lepidopteran) pollinated, and its var. *ledebourrii*, that is pollinated by hummingbirds.

Apparently, many flowers are visited by vectors other than the predominant one, but their occasional visits have little or no effect on selection for the adaptive syndromes.

These partial and intermediate situations illustrate two important points. First the shift from insect to hummingbird pollination is not necessarily saltational, although, given strong enough selection pressures, it could be rapid in terms of the geological time scale. Intermediate situations, in which occasional pollination by either of the two vectors is possible, may be a normal transitional stage. Second, the completion of the three parallel trends of selection, giving rise to fully developed differential syndromes, is unlikely to occur in sympatry. The striking differences observed in *Delphinium, Aquilegia, Mimulus, Penstemon* and other genera were probably at least completed, if not initiated, when the hummingbird pollinated species were completely isolated spatially from pollinators carrying pollen from the insect pollinated ancestor.

THE IMPORTANCE OF ECOLOGICAL DIFFERENCES

The Grants have presented a thorough account of the geographical distribution of breeding hummingbirds and of hummingbird-pollinated flowers. The diversity of habitats is great, but some habitats in the western USA are devoid of both breeding hummingbirds and of hummingbird flowers. These are principally seashore, marsh and aquatic habitats, sagebrush and savanna. Areas that hummingbirds visit but in which they do not breed are lacking or poor in hummingbird flowers. This suggests that hummingbirds have been able to favor the occurrence and frequency of those plant species that bear flowers which they visit, but only if the selective pressure that they exert is continuous, being associated with breeding and raising young. Moreover, the examples that they present show that species which probably evolved adaptations to hummingbird pollination a long time ago, such as *Fouqueria splendens* (ocotillo) and various species of paintbrush (*Castilleja*), are less strictly restricted to breeding sites than are species which are closely related to insect-pollinated relatives. This topic is developed further in a later section.

These ecological relationships show first that natural selection for hummingbird pollination involves adaptation to hummingbird habitats, and increases the probability that, as already stated, hummingbird flowers evolved in geographic and ecological isolation from insect pollinated ancestors.

THE SIGNIFICANCE OF FLOWER COLOR DIFFERENCES

Red coloration of flowers and nearby reproductive organs is characteristic of hummingbird pollinated flowers occurring in western North America. Exceptions occur in other parts of the range of Trochilidae.

The extent of red coloration varies greatly from one species to another. The majority of genera are those having either only the corolla or both corolla and calyx pigmented. In some, (*Monardella, Salvia, Ribes, Epilobium* sect. *Zauschneria*) the pigment extends to the floral bracts or calyx tube. In two species the reddish or pinkish color is more restricted. In *Lonicera involucrata* var. *ledebourii* it is confined to the floral bracts and in *Trichostema lanatum* it occurs only on the hairs that densely cover the inflorescence. The genus *Castilleja* is remarkably variable in this respect. In the majority of its species, red color exists on the tips of floral bracts, calyx lobes, and the margins of the corolla. In some, however, (*C. hololeuca*), the corolla lacks red color, in others, (*C. Lemmonii, C. Culbertsonii*), the floral parts are pinkish or purplish rather than scarlet. In still others (*C. Payneae, C. pilosa, C. nana*) both the extent and hue of color vary among floral organs and parts of them, in still other species (*C. longispica, C. arachnoidea, C. mollis, C. plagiotoma*) reddish or purplish pigmentation is completely lacking from the floral parts. These color deviants include species in which floral structures closely resemble those of the red flowered species plus others that, with respect to both calyx and corolla, approach the insect-pollinated genus *Orthocarpus*, and were included in this genus by some botanists. The possible evolutionary implications of this situation are discussed below.

The significance of red coloration in bird-pollinated species has been the subject of much controversy, well reviewed by the Grants. Several experiments have shown that the somewhat naive and anthropomorphic hypothesis, that hummingbirds "prefer" red flowers is incorrect. Like many other animals, these birds establish associations between visual cues and sources of food. The Grants, in my opinion correctly, assume that such associations are most easily made if all flowers, regardless of taxonomic affinity, that have structures and nectar supplies suitable to hummingbirds are similarly colored. But why red, rather than yellow, white, pink, or any other color? In this connection, one must note the fact that red is the predominant color not only in New World species pollinated by hummingbirds, but also in Old World plant species visited by birds having very different affinities (Meeuse 1961). Three reasons can be deduced. First, red or scarlet, though found in insect-pollinated genera, such as *Papaver* and *Lychnis*, is uncommon among them. Hence the association of red with structural suitability is a distinctive one for birds and avoids the confusion that would exist if a more common color were associated. Wright (1979), reviewing many experiments on visual sensitivity in the pigeon, says that it resembles humans in the long wave length end of the spectrum. To birds, therefore, red is a conspicuous color as it is among humans. Once the association, red with a source of food, was firmly established, the bird could detect this source more easily from a distance. I and many other naturalists have been amazed by the speed with which a hummingbird will "dive-bomb" a flower of *Castilleja* from ten or scores of meters away. Finally, many birds, including hummingbirds, possess red signals in their plumage that either enable males to advertise and defend their territory more easily, or females to locate more easily a potential mate. Once a bird is conditioned to associated red with various adaptively advantageous behaviors, the association between red and a source of food could, presumably, be more easily made.

CHEMICAL BASIS OF FLOWER COLOR DIFFERENCES

The chemical basis of flower pigments is somewhat complex, and has been extensively reviewed (Harborne, 1964, Goodwin,

1976). The principal compounds involved in those species that have given rise to hummingbird pollination are carotenoids and anthocyanins. Carotenoids are usually found in the chromoplasts embedded in the cytoplasm of epidermal cells or hairs of the corolla, while anthocyanins are dissolved in the vacuoles. The carotenoids are responsible for yellow colors, while the three principal anthocyanidins, delphinidin, cyanidin and pelargonidin, are responsible for colors ranging from blue to purple, lavender, pink and red plus various intermediate shades. Pelargonidin has been identified as the basis of red color in *Mimulus* (Pollock et al., 1967) as well as *Cantua, Collomia* and *Ipomopsis* in the Polemoniaceae and *Penstemon* in the Scrophulariaceae (Harborne and Smith 1978). Delphinidin and cyanidin occur also in these families and many more; they are among the most ubiquitous phenolic compounds in angiosperms. Hence, although precise identifications of these compounds have not been made in species believed to be related to ancestors of hummingbird flowers, the fact that all of these postulated ancestral groups have flower colors normally produced by delphinidin or cyanidin suggests that one or both of these compounds were present and a prerequisite for the evolution of hummingbird flower colors. The three compounds have very similar formulae, the ring structure characteristic of flavone (Harborne, 1964, p. 89) unmethylated, and with hydroxyl (OH) groups at various positions. The three differ in that delphinidin is the most hydroxylated (OH at 6 different positions), cyanidin next (5 hydroxyl groups) and pelargonidin the least (4 OH groups). This difference is of evolutionary importance, since genetical experiments on several genera have shown that alleles coding for the more hydroxylated anthocyanidins are dominant over those that code for less hydroxylated compounds (Alston, 1964). The above statement explains only part of a complex genetic picture, that is discussed more fully in a later section.

THE NATURE AND SIGNIFICANCE OF STRUCTURAL DIFFERENCES

With respect to the structure of the flowers and inflorescence, hummingbird-pollinated flowers have been said to possess four different modifications: (1) calyx and/or corolla forming a tube that is long enough to accommodate the bird's beak, so that when probing for nectar at the bottom of the tube, the bird brushes its head or the base of its beak against the stamens and stigma of the flower; (2) a thickening of tissues in those parts of the flower most likely to be damaged by the bird's beak; (3) a plentiful supply of nectar, of which the nutritive components include a low content of amino acids (Baker and Baker, 1982), and (4) flowers solitary or at least separated from each other in the inflorescence, and pendant, the latter position being in agreement with the hovering position of the bird in seeking nectar.

The third prerequisite is far from precise. In fact, the array of inflorescence types represented in the list of hummingbird flowers presented by the Grants includes almost every type except the flat topped umbel as found in the Apiaceae. The largest of genera have either simple or compound racemes. Solitary or nearly solitary flowers occur in *Aquilegia* and *Ribes speciosum*; a cymose inflorescence in *Brodiaea* (*Dichelostemma*) *ida-maia*, corymbs in *Silene*, dense spikes in *Pedicularis densiflora* and *Castilleja* spp., and a capitate inflorescence in *Monardella macrantha*. Apparently, if other factors favor the shift, no type of inflorescence is a serious barrier to the evolution of hummingbird pollination.

The evolution of a suitable floral tube is, however, much more of a problem. None of the relatives or presumed ancestors of hummingbird flowers have flat or bowl shaped flowers. Small, bell-shaped flowers are also lacking among them, as are corollas having very narrow throats, such as those of many Boraginaceae. Among tubular flowers there appears to be a lower limit to tube length, below which modification to a size large enough for the beak is highly improbable. The floral shape that is most easily modified into a suitable tube via gene-controlled repatterning of growth is either broadly tubular or narrowly bell-or vase-shaped. Such flowers are usually pollinated by medium-sized or large bees or wasps, and require the pollinator to enter the flowers when seeking nectar, rather than merely inserting its proboscis.

The conversion to the hummingbird syndrome usually involves lengthening the tube, and shortening the lobes at its apex, particularly those that in the ancestral flower may have served as a landing platform for the insect. Ancestrally bilabiate corollas are sometimes modified by turning back the lobes (Hiesey, Nobs and Björkman, 1971).

The above description applies to the majority of hummingbird flowers, in which the ancestral flowers have united petals and the calyx is not involved in the formation of the tube. The situation in flowers having separate petals or (in Liliales, tepals) is quite different. In them, either the calyx, or a combination of calyx-corolla (tepals) is involved. In *Silene*, *Astragalus* and *Epilobium-Zauschneria* the sepals are united into a tubular calyx in the ancestral flower. In them, the tube is formed by elongation of the calyx tube, but the petals remain, evolve scarlet color, and serve as attractants. In *Delphinium* and *Aquilegia*, the chief source of the tube is a spur-bearing petal or sepal. In *Fritillaria recurva* and *Lilium* the tepals are relatively narrow and directed forward, so that they form a broad pseudo-tube, and in both ancestral and hummingbird species the stamens and stigmas protrude far enough from the nectar source so that they brush against bird's head. In the last two genera, the modifications for hummingbird pollination are minimal. They are also distinctive in that the ancestral flowers are pollinated not by Hymenoptera but by large butterflies.

The examples in the last paragraph illustrate an important point. Flowers that become adapted to hummingbird pollinator do not converge toward a similar construction. Given a large enough difference in the structure of the ancestral flower, natural selection for adaptation to hummingbird pollination, occurring simultaneously in different unrelated lines, may cause these lines to diverge rather than converge with respect to structural organization.

TAXONOMIC AND GEOGRAPHIC DISTRIBUTION OF SEPARATE HUMMINGBIRD FLOWERS

Although the number of species in the Western American flora that are regularly cross pollinated by insects has not been precisely estimated, it is surely well over 1000, and may approach 2000 species. The 33 separate conversions to hummingbird pollination, therefore, have arisen from not more than 2 to 3 percent of the species that make up the flora. To what extent has this conversion depended upon the chance presence of hummingbirds, and to what extent have been involved either gene-controlled characteristics of the ancestral populations, or certain attributes, many of which probably also have a genetic basis, that have adapted them to habitats in which hummingbirds are relatively numerous compared to other pollinators? The factors that could have caused deviations from chance are here reviewed.

An important point brought out by the Grants is the scarcity of hummingbird-pollinated species among native annuals. They list only one annual, *Gilia splendens*. They attribute the scarcity to the fact that hummingbirds are attracted to flowers that produce an abundant supply of nectar over a long period of time, while annuals usually bloom only for short periods, since they must ripen their seeds before the summer drought. Another drawback to the annual habit is that it is often associated with self-pollination, and self-pollinated flowers tend to be small and produce little nectar.

One might ask whether *Gilia splendens* is no different from other annual species of *Gilia* with respect to characters that would favor the transition, or whether it occupies a favorable position in this respect. The latter is probably true. All species of *Gilia* have narrowly bowl-shaped, vase-shaped or tubular corollas, that by becoming larger would form the requisite tube. In contrast to several other species, *Gilia splendens* and its relatives bear flowers in an open inflorescence, held laterally, as do many hummingbird species. Flower colors in the genus include blue, purple and yellow, probably based upon the common favonoids (see above). Finally, the distribution of *G. splendens* is in southern California, where hummingbirds are particularly abundant. This species suggests that the transition can occur even if a major potential (perennial habit) is lacking, provided other factors are consistently favorable. The widespread distribution of *Ipomopsis aggregata*, a hummingbird flower that has in the past been united with *Gilia*, but is perennial or biennial, shows that given this growth habit, the complex of characters found in *Gilia* is very likely to form the basis of the transition.

Turning to the perennials, we can ask whether those that lack united petals and either a corolla tube or a shape that can be easily converted into one, nevertheless possess, along with their insect pollinated relatives, compensating factors that would facilitate the transition.

Looking at the two genera involved in the family Ranunculaceae, *Delphinium* and *Aquilegia*, they are distantly related to each other, and possess in common only one feature, floral parts that bear nectar-containing spurs. In *Aquilegia* the five spurs are elongations of petals, in *Delphinium* the single spur is formed from a sepal and two petals. Even though both of these genera have close relatives in their family (*Isopyrum, Aconitum*, etc.) none of these other genera have been ancestral to hummingbird flowers. The red-flowered, hummingbird-pollinated species of *Aquilegia* (*A. canadensis, A. elegantula, A. formosa*, etc.) are all closely related to each other, and, so far as it is known, completely interfertile, so that they might be regarded as subspecies of a single widespread and highly polymorphic species (Miller, 1978). The ancestor of this complex was most probably a member of the bee-pollinated *A. vulgaris* complex, having purple flowers and relatively short, curved spurs. While this complex is principally Eurasian, a few species occur in North America, of which one, *A. brevistyla*, is widespread in western Canada and the northwestern United States. It may be the ancestor of the *A. canadensis-formosa* complex, but this origin occurred so far in the remote past that little can be deduced about it.

The situation is different with respect to the two hummingbird-pollinated species of *Delphinium*. Since all other species of this genus resemble each other in floral structure and color, one can ask only whether the hummingbird-pollinated species, *D. nudicaule* and *D. cardinale*, plus their probable ancestors, live in habitats where hummingbirds are relatively common compared to bumblebees, that are the common pollinators of the remaining 29 species. *Delphinium nudicaule*, in northern California, lives in a region where bumblebees are probably more abundant than hummingbirds, but inhabits mostly deep, shady canyons, that are frequented by the birds. Although hard data are needed, one may suspect that in the *D. nudicaule* habitat, hummingbirds are more available than bees. The species favors cool, shady canyons and flowers in early spring, when days are relatively short, and the skies are often cloudy. The observations of Cruden (1972) in Mexico showed that in cool, shady habitats, such as montane forests, hummingbirds are much more consistent pollinators than bees. *D. cardinale* shares with the purplish-blue flowered *D. parryi*, its possible ancestor, a distribution in southern California and northern Lower California, the most southerly of the genus, and a region where hummingbirds are particularly abundant. I am not aware of the relative ecological distribution of these species, but believe it to be reasonably probable that *D. cardinale* acquired its hummingbird syndrome in some deep canyon area where both *D. parryi* and bees were less abundant than elsewhere.

Among the other choripetalous dicotyledons and the monocotyledons, *Astragalus coccineus* inhabits an area in which both bees and hummingbirds are abundant. In the absence of ecological data, no hypothesis can be suggested to explain why it is the only one of more than 100 species of its genus that has evolved the hummingbird syndrome. *Ribes speciosum* extends the farthest southward of any species belonging to the subg. *grossularia* that have a well developed floral tube, but overlaps with the polymorphic *R. menziesii* complex, from which it may have been derived. Both species favor deep shady canyons, typical hummingbird flower habitats.

The species complex, *Epilobium* sect. *Zauschneria* (Raven, 1976), is the only hummingbird-pollinated group of *Onagraceae* north of Mexico, although several others exist farther south. It is the only perennial section of *Epilobium* that is abundant in regions where hummingbirds commonly breed, and that regularly inhabits the shady canyons that are hummingbird habitats. Its nearest from an insect-pollinated group that approaches the hummingbird syndrome in more ways than does any other species group of Onagraceae living in the area.

Among the Caryophyllaceae, *Silene* is the only genus native to the western area in which the sepals are united to form a calyx tube, that could be modified to fit the hummingbird syndrome. Of the 20 native species, only *Silene californica* and *S. laciniata* have evolved the hummingbird syndrome. The latter is the only one abundant on shady slopes in southern California, where hummingbirds are most abundant. *Silene californica* was probably derived from *S. laciniata* after acquisition of the syndrome.

Among the monocotyledons, *Brodiaea ida-maia* and the rare *B. venusta*, belonging to the subgenus (or genus) *Dichelostemma*, are the only species of monocotyledons in the western area that have the fully evolved hummingbird syndrome. Compared to other sections or genera of the *Brodiaea* complex, only *Dichelostemma* has a floral tube that might be easily modified for hummingbird pollination, and among its six species, *ida-maia* and *venusta* are relatively mesic and the most likely to occur in association with hummingbirds. Except for some chromosomal races of *B. pulchella*, all species of sect. *Dichelostemma* are polyploids that have no recognizable diploid ancestors. Presumably, therefore, many species of this group have become extinct, including the insect pollinated ancestors of *B. ida-maia* and *B. venusta*.

The remaining choripetalous monocotyledons, in *Lilium* and *Fritillaria*, have become little modified for hummingbird pollination from their lepidopteran pollinated relatives. All of them may still be pollinated by both kinds of vectors. They need no further discussion.

Origin of Hummingbird Pollination among Sympetalous Dicotyledons

By far the largest number of hummingbird flowers in western North America are sympetalous, having acquired the nectar bearing floral tube as a modification of a corolla adapted to insects. I shall consider only those found in California, about which I have some knowledge.

Fouquieria splendens belongs to a small endemic family, of which the relationships are obscure. Its insect-pollinated ancestors, if any, are almost surely extinct. Likewise, *Beloperone*, *Ipomaea*, *Lobelia* and *Lonicera ciliosa* are not considered, because of lack of knowledge about their relationships.

Among the other sympetalous families, the temperate North American species of Ericaceae, Hydrophyllaceae, Boraginaceae,

Beloperone, *Ipomaea*, *Lobelia* and *Lonicera ciliosa* are not considered, because of lack of knowledge about their relationships.

Among the other sympetalous families, the temperate North American species of Ericaceae, Hydrophyllaceae, Boraginaceae, Rubiaceae and Asteraceae are conspicuously lacking in hummingbird adaptations. Their absence in Ericaceae is somewhat surprising, as noted by the Grants, since flowers of *Arctostaphylos* are visited, though rarely if ever pollinated by hummingbirds, and both Old World members of the family (particularly in the large genus *Erica* and its relatives) and tropical New World ones, have evolved bird pollination. The most likely explanation is that the only Ericaceae found in the breeding area of hummingbirds in western North America are the genera *Arbutus*, *Arctostaphylos* and their relatives, plus a few species of *Vaccinium*, all of which have relatively small floral tubes, and white or nearly white flowers, that lack flavonoid pigments from which pelargonidin and other red producing pigments are derived.

The remaining four families, with respect to species that occur in our area, have the following characteristics in common: (1) dense inflorescences, either scorpioid cymes (Hydrophyllaceae, Boraginaceae), umbels (*Galium*) or dense heads (Asteraceae), (2) small corolla tubes, often with narrow throats, and (3) few seeds per flower or floret. In these species transition to the hummingbird syndrome requires developmental modification of both inflorescence and flower. The value of occasional mistake pollination is restricted, since only a few, if any, seeds would be produced by each mistake. Also, compared to the total number of perennial species in these families, only a small percentage occupy sites where breeding hummingbirds might be abundant. In them, therefore, the barriers to the shift apparently outweigh any possible advantage.

This series of restrictions serve to emphasize the degree to which various phyletic constraints plus habitat restrictions reduce in most families the probability of evolving a successful hummingbird pollination syndrome.

The families that remain to be considered are Lamiaceae, Polemoniaceae, and Scrophulariaceae. Of these, the fewest examples in our area are in Lamiaceae. Among them, *Trichostema lanatum*, *Stachys chamissonis* and *S. ciliata* have poorly-developed hummingbird syndromes, and may well be often pollinated by insects. *Salvia spathacea* belongs to a genus in which most species are insect-pollinated, but many Mexican species are strongly modified for hummingbird pollination. The affinities of *S. spathacea* are obscure and no obvious insect pollinated relative exists in California. *Monardella macrantha* is distinctive in having the longest floral tube in the genus, and inhabiting the hummingbird-rich southern California region.

In the Polemoniaceae, the only genus that contains many perennial species is *Phlox*, that is pollinated chiefly by Lepidoptera rather than Hymenoptera, and has rotate corollas, of which the tube is very narrow, and apparently not easily modified into the hummingbird adapted structure. Neither *Phlox* nor the related *Leptadactylon* have evolved hummingbird pollination (Harborne and Smith, 1978). The perennial hummingbird flowers of California are *Collomia rawsoniana* and three species of *Ipomopsis*, of which insect pollinated relatives are absent or difficult to imagine as similar to their ancestors. This family, therefore, contains little helpful information regarding the evolution of the syndrome.

The final family to consider is the Scrophulariaceae, which has given rise to the largest number of hummingbird flowers in our area. Except for *Veronica*, in which floral tubes are lacking, every genus that contains several perennial species in our area is represented. The largest number belong to *Castilleja*, the most distinctive hummingbird genus in the flora. The Grants list 48 hummingbird-adapted species, found throughout western North America. Nevertheless, three species of California, *C. arachnoidea, C. longispica* and *C. pilosa* are not included. These all have relatively short corollas, and were placed by some botanists, such as W.L. Jepson, in the insect-pollinated genus *Orthocarpus*. Are these to be regarded as relictual survivors of the insect pollinators of ancestral *Castilleja*? The affinities of the genus, except for its close connection to annual *Orthocarpus*, are obscure. At any rate, *Castilleja* is probably the oldest group of hummingbird-pollinated species in temperate North America, except for isolated relics such as *Fouquieria*. The conditions under which it evolved will never become clear.

Galvezia speciosa is probably another species of relatively ancient derivation, distantly related to perennial Old World species of *Antirrhinum* and *Linaria*, but only in a general way. The genus *Pedicularis*, one of the largest in the Scrophulariaceae, contains several species in western North America, most of them in mountain areas above the breeding range of hummingbirds. An interesting fact is, therefore, the presence of a species having the hummingbird syndrome, *P. densiflora*, even though, at least in northern California, it is pollinated by bees as often as by hummingbirds (L. Macior, pers. comm.). Its range is entirely within the hummingbird breeding range, including southern California, where it frequents the favored moist canyon habitat.

Its affinities within this large and complex genus cannot be determined without careful research.

Closer connections can be traced between hymenopteran and hummingbird-pollinated species in the large genera *Penstemon* (including *Keckiella*) and *Mimulus*. The three species groups in *Penstemon* that have evolved separately the hummingbird syndrome all support the hypothesis that favorable habitats and flower structures promote its evolution. *P. newberryi* and *P. rupicola* belong to a relatively mesic, high montane group, of which *P. newberryi* descends to the lowest altitudes. Its floral tube differs little from that of its bee pollinated relatives, and its color, also, is a pinkish red rather than tomato red or scarlet. It may be a relatively recent adaptation, associated with favorable conditions in lowland northern California during the Pleistocene pluvial period. *P. rupicola* is hardly more than a subspecies of *P. newberryi,* and probably has a similar origin. The origin of the largest group of hummingbird species, centering about *P. centranthifolius,* is more of a problem. Nevertheless, they are abundant in southern California plus adjacent southwestern Utah and northwestern Arizona, where breeding hummingbirds are abundant in many deep canyons. Since *Penstemon* species hybridize freely to form partly or fully fertile F_1 hybrids, the first successful hummingbird species may have spread its genes in this manner to many others inhabiting the same or adjacent regions. This course of events is partly supported by the investigations of Straw (1955). Finally, the seven species recognized by Straw (1967) as *Keckiella* serve as a model for the evolution of hummingbird pollination from bee pollinated species that have structures and color favorable for the transition. The four bee pollinated species, *P. rothrockii, P. breviflorus, P. lemmonii* and *P. antirrhinoides* all have strongly bilabiate, short tubed corollas in which such colors as pink, purplish, brownish and brownish red occur to some extent, and *P. antirrhioides* occurs on the edges of the southwestern area that harbors the richest hummingbird fauna. *P. cordifolius* and *P. ternatus* both centered in this area, inhabit shady habitats favored by hummingbirds and have long-tubed scarlet corollas. The final species, *P. corymbosus,* occurs further north along the coast, still in hummingbird country, favors shady, rocky slopes and has similar corollas. It probably originated after the syndrome had evolved.

The genus *Mimulus* is of particular interest, because of the detailed investigations of its cell structure and genetics (Hiesey, Nobs, and Björkman, 1971). Among the 77 species found in

California, 14 are perennials. They belong in four highly distinctive sections, between which gene exchange is impossible. Only two of these sections have given rise to hummingbird flowers. Those that have not are (1) rhizomatous species including *M. mochatus* and *M. primuloides*. These are low in stature, a condition unfavorable for hummingbird pollination, and have consistently yellow flowers, apparently without delphinidin or cyanidin. The other perennial species (2) center about the ubiquitous *M. guttatus*, also with flowers consistently yellow, and with a bilabiate corolla that would require great modification to form a hummingbird-adapted tube.

The two sections that include hummingbird flowers are *Diplacus*, often recognized as a separate genus, and *Erythranthe*. The seven species of *Diplacus* all have the same chromosome number, many of them intergrade in nature, and garden hybrids are common. Three, *M. aurantiacus*, *M. flemingii* and *M. puniceus*, are listed by the Grants as hummingbird pollinated. They are most abundant in the relatively moist coastal areas, where they occur together with other hummingbird flowers of the genera *Delphinium*, *Silene*, *Salvia*, *Pedicularis* and *Castilleja*. *Mimulus aridus*, *M. bifidus* and *M. clevleandii*, the insect pollinated species, occur chiefly in the drier interior, where both hummingbirds and flowers adapted to them are less abundant. Finally *M. longiflorus*, which is widespread and both overlaps and hybridizes with other species, has populations that contain a range of flower colors, from white or orange-yellow to red. It is definitely pollinated by cyrtid flies, but its visitors may also include hummingbirds. The section *Diplacus* of *Mimulus* appears to be a species group in which all stages of evolution of the hummingbird syndrome exist in the same general area of southern California. Various kinds of quantitative data on this group would be particularly valuable.

The section *Erythranthe* of *Mimulus* has been studied from the systematic, cytogenetic, ecological and physiological point of view. The detailed monograph by Heisey et al., (1971) is one of the most important scientific investigations included in this review. It contains one species, *M. lewisii*, that is bee pollinated and is widespread in high montane and subalpine regions of western America, plus four hummingbird pollinated species. Of these, *M. cardinalis* occurs in California, where it is adjacent to *M. lewisii* but at lower elevations, and in southeastern Arizona. *M. verbenacea* is distributed across Arizona and northern Mexico, while the other three are highly localized in Utah, Arizona and Mexico.

Only *M. cardinalis* and *M. lewisii* have been carefully studied. These two species differ with respect to three separate character complexes, each of which is controlled by several gene loci. Physiological differences exist with respect to photosynthetic rate under different temperatures and light intensities, growth rates of both seedlings and adult plants under different temperatures, and various other parameters. Few specific data are available on the genetic basis of these differences, but it appears to be complex. Differences with respect to parameters of floral structure, reflexed vs. spreading corolla lobes, width of lower corolla lobe, corolla aperture and style length also are inherited according to complex patterns, that are different for each parameter. Transgressive segregation was recorded for both lower lobe width and style length. Differences in flower color were inherited in a more simple fashion, including one simple mendelian 3:1 segregation, for presence vs. absence of yellow chromoplasts on the upper (but not the lower) corolla lobe. Nevertheless, the overall color of the corolla exhibited dominance for the *M. lewisii* condition, both in F_1 and to a lesser degree in later generations. The results are compatible with the hypothesis of two "switch" loci, one of them affecting overall color via the soluble anthocyanidins, and the other other presence vs. absence of a yellow upper lip. For the latter, *M. lewisii* apparently carried a single dominant suppressor, while the data of Pollock et al. (1967) suggest that the apparent recessiveness of red color is due to the usual condition of a recessive allele for pelargonidin and dominant for cyanidin.

A final remark must be made about the latter point. In his extensive review of the extended literature on genetics of phenolic compounds in plants, Alston (1964) notes that in every published investigation involving the anthocyanidins delphinidin, cyanidin and pelargonidin, the three compounds are determined by alternative alleles at a single locus, and dominance is correlated with higher numbers of OH groups on the molecules; i.e. delphinidin>cyanidin>pelargonidin. Most probably, therefore, the red colors found in the various hummingbird flowers reviewed are homozygous for a single recessive allele that determines the presence of pelargonidin.

DISCUSSION AND SUMMARY

The data reviewed in previous sections show that in western North America, 33 separate origins can be recognized of a series of morphological and physiological traits that adapt plant species to hummingbird pollination. These include three kinds of alterations: (1) growth characteristics that adapt them to regions and habitats where hummingbirds are most abundant, (2) modifications of floral structure that result in delivery and acquisition of pollen while the bird is seeking nectar, and (3) bright colors, particularly red, that enable the bird to see the flower from a distance, and to make the association with it and obtaining nectar as easy as possible. The degree of alteration varies greatly from one example to another of the adaptive shift. In some instances pollination by both hummingbirds and other vectors is known or probable. At the same time, variation within many populations or species with respect to the critical characters is great enough so as to make possible natural selection for character combinations that more nearly resemble the fully developed adaptive syndrome than do the majority of individuals in the population. Although in the majority of examples, clear phyletic connections between hummingbird flowers and related forms pollinated by insects cannot be recognized, (in most instances because the insect pollinated ancestors have become extinct) twelve examples in eight genera include both insect pollinated and closely related hummingbird flowers. Among nine of them, genetic compatibility is such that either in nature or in the garden, partly or fully fertile hybrids can be obtained. These examples open the door to complete analysis of the pathway over which the hummingbird pollination syndrome can evolve.

Another important point that emerges from the comparisons made is that the insect pollinated relatives of plants cross pollinated by hummingbirds are by no means a random sample of insect pollinated flowers in the flora. They are distinctive with respect to several characteristics that by further divergence would fit in with the hummingbird syndrome. Even large families such as Asteraceae, having character complexes that do not include any, or at most only a few approaches to the syndrome, have not given rise in North America to hummingbird pollinated species. The few South American Asteraceae that are pollinated by hummingbirds belong chiefly or entirely to the tribe Mutisieae, that contains a number of insect pollinated species having long tubed florets.

These results indicate that among insect pollinated species, a small number contain combinations of characters that do not in themselves adapt the species to predominant pollination by hummingbirds, but facilitate the transition by reducing the number of selectively based changes needed to acquire the syndrome. They may be said to possess a latent potential for responding to selection pressures favoring the hummingbird syndrome.

The term latent potential was first used by Hartl and Dykuizen (1981, 1985) to characterize a situation in which certain enzymes, coded to known genes, facilitated the adaptive shift of a population of *Escherichia coli* from one substrate to another. Although this situation appears at first sight to be radically different from the adaptive shifts recorded in this review, the principle is similar. This principle can be stated in general terms as follows.

Either minor of major adaptive shifts can be facilitated by the presence in the ancestral genotypes of alleles or combinations of them that aid certain individuals or populations to evolve in the direction of the shift. Populations that possess such genes or alleles can be said to have a latent selection potential for completing the shift. The greater is the number of genetic changes required for the shift, the more important is the presence of latent selection potential (LSP). The potential is a cumulative and quantitative condition. For complex adaptive shifts, the potential must exist with respect to several characters.

The condition here called latent selection potential is similar to conditions called by some zoologists, such as Cuénot, preadaptation. Mayr (1960) and others have rightly pointed out that this word is misleading, since the organisms possessing the distinctive characteristics are not actually adapted to the new environment: they only have genotypes that are more susceptible to modification via the usual evolutionary processes to the final adaptation. Gould and Vrba (1982) have suggested a way out of this dilemma by coining two new terms: preadaptation for individual traits that could contribute to a new adaptation, but do not because they exist in organisms that do not have other traits necessary for it; and exaptation for characters that could potentially be selected for the new adaptation. These terms are awkward, and their definition is somewhat recondite. Moreover, they tend to focus attention upon particular characters, while the more flexible and clearly understood term selection potential or potential for selection, can be used either for individual characters of the total adaptive complex to which they contribute.

Returning to the adaptive syndrome for hummingbird pollination, the following characters found in insect pollinated aniosperms have the greatest potential to evolve, via further mutation and/or recombination and selection into characters that can contribute directly to the syndrome.

Vegetative: various shifts in growth rate, winter dormancy (usually a reduction of this property), shade tolerance and others that adapt the plants to the somewhat mesic, equable conditions favored by breeding hummingbirds.

Inflorescence: elongation of the axia and its branches, nodding flowers, lengthening of the total flowering season.

Floral structure: development of nectar bearing spurs on sepals, petals of both; reduction and straightening of sepals, petals or corolla lobes; development and elongation, as well as either narrowing or widening a cayx or corolla tube; evolution of higher nectar production; alteration of corolla structure that in many insect pollinated flowers serve as landing platforms or guide lines.

Flower color: Existence of genes coding for the synthesis of delphinidin or cyanidin in the cells of the flower; mutations to the recessive allele coding for pelargonidin; mutations of various regulator genes that affect the density and distribution of these and other sources of color.

Not all of these characters are essential for the syndrome; some are complementary alternatives to each other. The genetic analysis of *Mimulus* sect. *Erythanthe* revealed segregation patterns that favor the hypothesis that selection potential for structural characteristics of the flower is of prime importance, that for adaptation to hummingbird habitats somewhat less so, and flower color is the least significant, at least within the wide spectrum of species having colors that are based upon the presence of delphinidin, cyanidin or both.

The principal objective of this paper has been to point out the value of investigating the adaptive syndrome for hummingbird pollination as a means of understanding better major adaptive shifts that give rise to evolutionary novelty. If it has shown only that investigations of this sort can be carried out in a meaningful way, and might lead to similar approaches to adaptive syndromes in both plants and animals, it will have served its purpose.

ACKNOWLEDGMENTS

The author is pleased to acknowledge several comments on the manuscript by Y. Linhart and R. W. Cruden, that have greatly improved its accuracy.

LITERATURE CITED

Alston, R. E. 1964. The genetics of phenolic compounds. In J. B. Harborne [ed.]. Biochemistry of Phenolic Compounds. Academic Press, London and New York.

Baker, I., and H. G. Baker. 1982. Some chemical constituents of floral nectars of *Erythrina* in relation to pollinators and systematics. Allertonia 3: 25-37.

Cruden, R. W. 1982. Pollinators in high elevation ecosystems: relative effectiveness of birds and bees. Science 176: 1439-1440.

Cruden, R. W. 1982. Pollinators in high elevation ecosystems: relative effectiveness of birds and bees. Science 176: 1439-1440.

Goodwin, T. W. [ed.]. 1976. Chemistry and Biochemistry of Plant Pigments. Academic, London, New York.

Gould, S. J. and E. S. Vrba. 1982. Exaptation -- a missing term in the science of form. Paleobiol. 8: 4-15.

Grant, K. A. and V. Grant. 1968. Hummingbirds and their Flowers. Columbia University Press, New York.

Guerrant, E. D. Jr. 1982. Neotenic evolution of *Delphinium nudicaule* (Ranunculaceae): A hummingbird pollinated larkspur. Evol. 36 (4): 699-712.

Harborne, J. B. [ed.]. 1964. Biochemistry of Phenolic Compounds. Academic Press, London.

Harborne, J. B., and D. M. Smith. 1978. Correlations between anthocyanin chemistry and pollination ecology in the Polemoniaceae. Biochem. Syst. Ecol. 6: 127-130.

Hartl, D. L. and D. E. Dykuizen. 1981. Potential for selection among nearly neutral allozymes of 2-phosphogluconate dehydrogenase in *Escherichia coli*. Proc. Natl. Acad. Sci. U.S.A. 78: 6344-6348.

Hartl, D. L. and D. E. Dykuizen. 1985. The neutral theory and the molecular basis of preadaptation. Population Genetics and Molecular Evolution, T. Ohta and K. Aoki eds. Japan Sci. Soc. Press, Tokyo; Springer, Berlin pp. 107-124.

Hiesey, W. M., M. A. Nobs, and O. Björkman. 1971. Experimental Studies on the Nature of Species V. Biosystematics, Genetics and Physiological Ecology of the *Erythranthe* section of the *Erythranthe* section of *Mimulus*. Carnegie Inst. Wash. Publ. 628, Washington, D.C.

Mayr, E. 1960. The emergence of evolutionary novelties. In S. Tax [ed.]. Evolution after Darwin 1: The Evolution of Life.

Meeuse, B. 1961. The Story of Pollination. Ronald Press, New York.

Miller, R.B. 1978. The pollination ecology of *Aquilegia elegantula* and *A. caerulea* (Ranunculaceae) in Colorado. Amer. J. Bot. 65: 406-414.

Pollock, G., R. K. Vickery, and K. G. Wilson. 1967. Flavonoid pigments in *Mimulus cardinalis* and its related species. 1. Anthocyanins. Amer. J. Bot. 54: 695-701.

Raven, P. H. 1976. Generic and sectional delimination in Onagraceae, tribe Epilobieae. Ann. Missouri Bot. Gard. 63: 326-340.

Straw, R. M. 1955. Hybridization, homogamy, and sympatric speciation. Evol. 9: 441-444.

Straw, R. M. 1967. *Keckiella*: new name for *Keckia* Straw. Brittonia 19: 203-204.

Vickery, R.K. and R.L. Olson. 1956. Flower color inheritance in the *Minulus cardinalis* complex. J. Hered. 47:195-199.

Wright, A.A. 1979. Color-vision psychophysics: a comparison of pigeon and human. In A. Granda and J.H. [eds.]. Neural Mechanisms of Behavior in the Pigeon. Plenum, New York.

EARLY MATURITY, SMALL FLOWERS AND AUTOGAMY: A DEVELOPMENTAL CONNECTION?

Edward O. Guerrant, Jr.

INTRODUCTION

Darwin's legacy looms even more brightly due to the work of Herbert and Irene Baker. While the power of natural selection is celebrated in the work of all three, they are also keenly aware of phylogenetic, structural and other constraints within which natural selection operates in nature. Theirs is a rich view of life. The multifaceted place of breeding systems, with its myriad causes and consequences, is a theme that appears throughout the Bakers' work (cf Baker, 1959; Baker and Baker, 1979). The present paper follows in and is inspired by the Bakers' eclectic path. Specifically, it explores the potential for a developmental connection among rapid flowering, small flowers and autogamy in some herbaceous wildflowers. More generally, it also considers the usefulness of developmental information in attempts to distinguish adaptive features from incidental consequences in an organism's phenotype.

The flowers of many autogamous wildflower species are conspicuously smaller than those of their presumed xenogamous ancestors (Ornduff, 1969; Jain, 1976; Wyatt, 1983). From the standpoint of adaptation, it seems that small flower size is commonly thought (at least implicitly) to be a consequence of the evolution of autogamy. For example, Solbrig and Rollins (1977) stated that... "once a plant has acquired the ability to self habitually, changes that decrease the cost of pollinator attraction may be favored...". Note the hypothetical sequence of events: the change in breeding system is assumed to occur first, precipitating selection for reduction in flower size. This is an energy optimization argument about the adaptive significance of small flower size inbreeders. It postulates that energy 'saved' may be reallocated to features enhancing competitive ability, reproductive output, or both. Implicit also is the notion that small flower size and early flowering are developmentally independent.

The present paper explores an alternative though not mutually exclusive hypothesis. Reduced flower size and associated shape changes in some inbreeders are hypothesized to be but one manifestation of a broader pattern of evolutionary juvenilization,

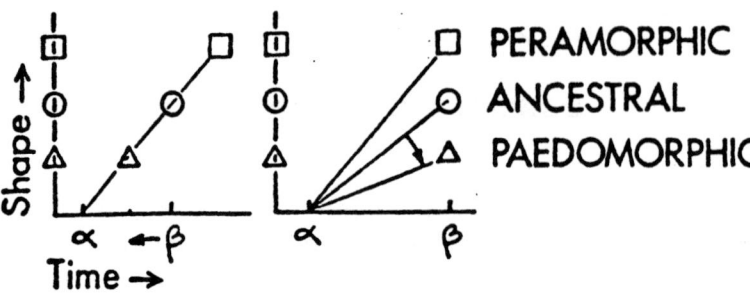

Figure 1. Diagramatic representation of the effect of changes in developmental rates and timing on morphological evolution. The abscissa is time, with ∂ and ß designating the onset and offset times of arbitrarily chosen developmental events; the ordinate is a non-dimensional shaped axis traversed by both the ancestral type (adult morphology indicated by circles) and evolutionarily juvenilized, or paedomorphic descendant (adult indicated by triangles). Paedomorphosis by the process of progenesis is indicated in the figure on the left and by neoteny on the right. Figure adapted from several in Alberch *et al.*, 1979.

or paedomorphosis, by the process of progenesis (Fig. 1). According to this view, if there were a selective premium on the ability to make flowers in a restricted amount of time, then, as a developmental correlate, small flowers with a juvenilized morphology would be expected to result. The progenesis hypothesis is based on theory concerning the relationship of ontogeny and phylogeny (Gould, 1977; Alberch et al., 1979). It establishes a common currency - time - that allows ontogenetic information to be incorporated into established ecological and evolutionary theory. For example, the age at maturity is of considerable ecological and evolutionary interest. Ontogeny-phylogeny theory provides a means to assess morphological and other consequences of natural selection for changes in the age at maturity.

The ability of self-pollinating species to flower and mature seed earlier in the spring than their outbreeding ancestors is considered to have been of primary selective importance for the evolution of inbreeding in at least three genera of herbaceous wildflowers. Evolutionarily derived, inbreeding species of *Leavenworthia* (Lloyd, 1965; Solbrig and Rollins, 1977), *Clarkia* (Moore and Lewis, 1965) and *Limnanthes* (Arroyo, 1973; Ritland and Jain, 1984) all live in habitats in which the summer drought arrives earlier and more unpredictably in the spring than it does in those of their outbreeding ancestors.

Arroyo (1973) concluded that the main selective force leading to the evolution of autogamy in *Limnanthes floccosa* was for earlier flowering and seed set. Furthermore, she argued that *L. floccosa* is descended from the extant species *L. alba.* The ability of *L. floccosa* to flower earlier in the year than does *L. alba*, where the two grow in relative proximity, is genetically determined (Ritland and Jain, 1984). This property appears to be an important factor in allowing this species to persist and flourish in the habitats it now occupies, which are edaphically and climatically poorer than those of *L. alba* (Arroyo, 1973; Ritland and Jain, 1984).

Narrowly framed, the progenesis hypothesis rests on the premise that *L. alba* is the immediate ancestor of *L. floccosa.* It also presumes *L. alba* retains the floral form from which the latter evolved. Unfortunately, the phylogenetic relationship underpinning this strict view of the progenesis hypothesis is not supported by recent electrophoretic and infertility data (McNeill and Jain, 1983). Rather, it appears that *L. floccosa* and *L. alba* each have diverged independently from their most recent common ancestor. Nevertheless, at a more general level, the progenesis hypothesis predicts that if the age at maturity were reduced by natural selection during evolution, then the size and the morphological form of *L. floccosa* flowers will have been affected in ways described by the Gould (1977) and Alberch et al. (1979) models.

MATERIALS AND METHODS

A single population of each taxon was used. The Agate population of *Limnanthes floccosa* ssp. *floccosa* (ca 42°31'4"N., 122°49'9"W), and the Oroville population of *L. alba* var. *alba* (ca39°28'8"N, 121°41'3"W), were chosen from among those surveyed electrophoretically by McNeill and Jain (1983). The *L. alba* population is located in the Richvale vernal pool area of the Central Valley of California, and the *L. floccosa* population is located in a vernally moist roadside ditch adjacent to vernal pools in the Agate desert region of south-central Oregon, just north of Medford.

All floral measurements, except those pertaining to growth rates and timing of developmental events, were obtained from buds of flowers fixed in FAA (Radford et al., 1974). These samples were collected in the field at various times throughout the growing seasons of 1981 and 1982. Estimates of growth rates and timing of events were obtained from plants grown in a glasshouse from field-collected seed. The glasshouse was maintained at a temperature between 18°C to 28°C, and receives ca 1500 μmol $m^{-2}s^{-1}$ (maximal) of sunlight on a clear day in May (Robichaux, pers. com.).

Twenty-one floral characters were measured on ca 30 flowers and up to 25 characters on about 80 buds per species. To determine the size of buds when meiosis occurred (and later by inference the 'age' at meiosis), semi-permanent slides stained with acetocarmine were made of anthers from both whorls of stamens (Radford, et al., 1974).

Because allometric (log-log) plots of length to width for buds and petals appeared to be bent-lines (as opposed to straight lines), a two-step non-linear estimation technique was developed to determine the best bent-line through the data (A.P. Frick, pers. comm.). The first step was to run a series of regressions on adjoining subsets across a wide range of the data, to determine the best bent-line through the data using the model:

$$y = a_0 d + a_1(a-d) + b_0 x d + b_1 x(1-d)$$

which was subject to the constraint that

$$a_1 + b_1 x_0 = a_0 + b_0 x_0$$

where the slope changes (and thus is the point of intersection of the straight line segments); a is the y-intercept and b the slope; subscripts 1 and 0 refer to the pre- and post-bend segments of the line respectively; d is a dummy variable that takes on the value 0 when x is less than x_0, and 1 when x is equal to or greater than 1. For the first step, the bending point (x_0) was arbitrarily designated as being mid-way between the highest data value of the lower set, and lowest value of the upper. The values from the regression with the highest r^2 were used as the starting values for the second step, a non-linear estimate using SAS Proc NLIN (SAS, 1982). The results of this analysis were used only to locate the best bending point for the data set. In subsequent interspecific comparisons the respective bending points were used only to divide the data into appropriate pre- and post-bend segments. For purposes of interspecific comparisons, linear regressions of the data on either side of the bend are used separately.

Time Scales

The best description of the relationship between bud size and time that could be obtained for each species independently was used to calculate the scales, from which growth rates and 'ages' of specific events were inferred. The Richards function was used to describe the increase in length through time for 39 buds of *L. alba* (Richards, 1959, 1969; see Causton et al., 1978; Causton and Venus, 1981; Hunt, 1982; Guerrant, 1984). It is a generalized form of the logistic curve in which there is an additional term that allows the point of inflection (where the curve changes from 'concave up' to 'concave down') to vary. Bud lengths and widths were measured approximately every other day from when first observed until anthesis. Due to the structure of the inflorescence, bud width could be measured one to several days before bud length. Lengths were estimated for these early widths from allometric relationships of bud length and width in buds covering the appropriate range of sizes.

Attempts to describe the growth of *Limnanthes floccosa* buds with the Richards function were not successful. As an alternative, ln-transformed and untransformed data were subjected to ANCOVA (SAS, 1982) to obtain the single best slope for length versus time for 39 buds of *L. floccosa* measured daily from their first appearance until anthesis. A linear regression of the non-transformed data on time provided a better fit ($r^2 = 0.97$) than did the regression for ln-transformed data ($r^2 = 0.92$).

Data were subjected to a discriminant function analysis with eight designated groups (flowers of both species, and three size classes of buds: less than 3 mm, between 3 and 6 mm, and greater than 6 mm) using the BMDP7M program available on the University of California IBM computer. Linear regression and descriptive statistics were done with SAS package on the same computer (SAS, 1982). Slopes and intercepts were compared using the appropriate t-tests.

RESULTS

At maturity, the flowers of *L. alba* are substantially larger than those of *L. floccosa* for all characters measured except individual mericarps (Figs. 2 and 3). Neither the length nor width of the single-seeded mericarps differs significantly between

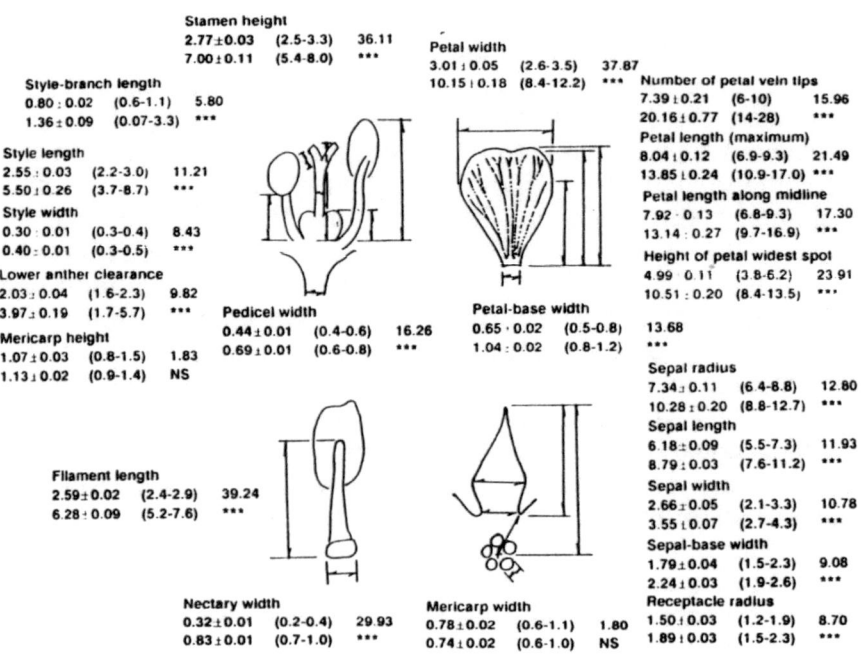

Figure 2. Schematic illustration of characters measured. All measurements are in mm, except for the number of petal vein tips. Mean values and standard errors are given along with range. Values for Lff are above those for Laa. Results of *t*-tests are given to the right, with *t*-statistic above symbol for p-value.

*** = p < 0,001, N.S. = difference not statisically significant.

Figure 3. Photos of flowers Laa (larger) and Lff, with some perianth parts removed to reveal staments and pistils. Scale bar is 1 cm long.

species. The floral primordia of *L. floccosa* appear smaller at 'inception' than those of *L. alba* (Fig. 4).

The results of a discriminant function analysis of 17 characters (Fig. 5) indicate that the flowers of *L. floccosa* do not particularly resemble any bud stage or the flowers of *L. alba*. Rather, the species are most similar to one another in the youngest of three sizes classes of buds, and become increasingly more dissimilar with age.

Allometric analyses, which examine the relationship between size and shape (Huxley, 1932; see Gould, 1966, for review) indicate that in external dimensions the buds of both taxa grow through similar size-shape trajectories (Table 1, Fig. 6); that is, the slopes and intercepts are indistinguishable. The relationships of bud size to the diameter of its supporting petiole are equivalent as well. However, not all homologous floral parts grow through the same size-shape series.

In the buds, allometric relationships (ratio of relative rates of growth in length to width) change during development, apparently fairly abruptly, when they are ca 1 mm wide (Table 1, Fig. 6). Early in growth the buds are wider than they are long(Figs. 4, 6), and initially, the relative growth rates in length are greater than in width, yielding a slope greater than one. This produces a bud whose length exceeds its width at the time of allometric transition. From then on, relative rates of growth in length and width are subequal (i.e., the slope approaches 1.0), and

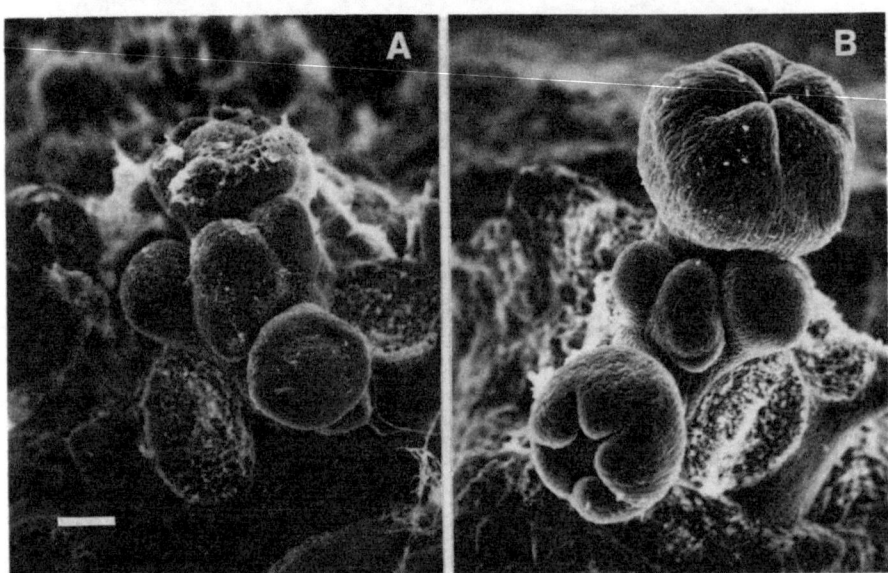

Figure 4. Scanning electron micrographs of shoot apical region of Laa (A.) and Lff (B.). In both, the shoot apical meristem and five floral primordia in various stages of development can be seen. Note that the phyllotactic spiral is in the opposite direction in the two photographs. Scale bar is ca. 0.1 mm long.

Table 1. Summary statistics of allometric relationships: mean and standard error of slope and intercepts of regressions of log transformed data are provided along with r^2, t-statistics and p values.

Character relationships	r^2			Slope				Intercept			
	Laa	Lff	df	Laa	Lff	t	p	Laa	Lff	t	p
Bud len x wid											
Pre-bend	.97	.99	50	1.66±.06	1.71±.04	-0.634	NS	-1.14±.10	-1.18±.06	0.334	NS
Post-bend	.97	.96	108	0.93±.03	0.93±.03	0.055	NS	0.03±.06	0.37±.06	-0.252	NS
Bud len x Petiole wid	.84	.67	56	1.43±.12	1.42±.19	0.065	NS	0.36±.20	0.39±.30	-0.084	NS
Sepal wid x len	.97	.95	55	0.90±.03	0.86±.04	0.975	NS	-0.10±.08	-0.01±.09	0.752	NS
Mericarp wid x len	.86	.84	57	0.67±.05	0.62±.05	0.627	NS	-0.17±.02	-0.18±.02	0.296	NS
Petal wid x len											
Pre-bend	.99	.97	62	0.92±.02	0.87±.03	1.506	NS	0.23±.03	0.28±.04	-1.203	NS
Post-bend	.96	.97	63	0.75±.03	0.60±.02	4.614	***	-0.62±.08	0.76±.04	-1.535	NS
Small anther wid x len	.99	.96	66	0.74±.02	0.90±.03	4.500	***	0.29±.03	0.17±.05	-2.116	*

* = p<0.05, *** = p<0.001

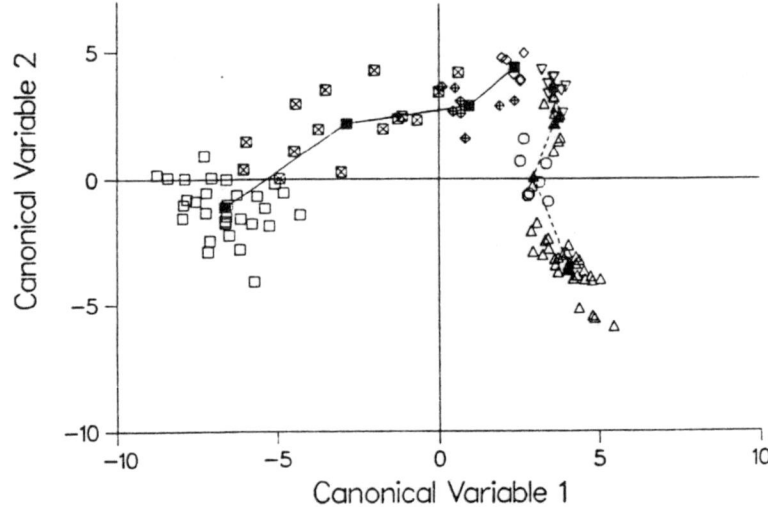

Figure 5. Results of a discriminant function analysis of flowers and three size classes of buds. Solid squares connected by solid lines indicate centroids of Laa flowers and buds, likewise solid trianges connected by dashed lines for Lff. The flowers of Laa are denoted by hollow squares on the left, and Lff by hollow upward-pointing trianges on the lower right of the figure. Buds of each are connected in chronological order beginning in the upper right, and indicated by a variety of symbols to distinguish the size classes. They are most similar to one another during the youngest stage in each, and most dissimilar as flowers.

growth contributes more to change in size than shape (Table 1, Fig. 6).

The sepals of both species have a common size-shape trajectory, and no shift in the allometric relationship was detected in the sepals during development (Table 1). Petals of the two taxa begin growth along a common size-shape trajectory, yet later they diverge along different paths (Table 1, Figs. 7,8). The allometric shift during development occurs at different sizes and ages in the two taxa, but not necessarily at different stages of maturity. In both species the change in relative growth rates in length and width occurs at approximately the same amount of time after meiosis and tetrad formation in the anthers.Unlike the buds, sepals and petals, the anthers of the two species do not grow through a similar size-shape series (Table 1). Their size-shape trajectories are different throughout growth. This holds true for anthers in both the inner and outer whorls of stamens.

Finally, the individual mericarps also grow through comparable size shape trajectories (Fig. 3, Table 1). These are

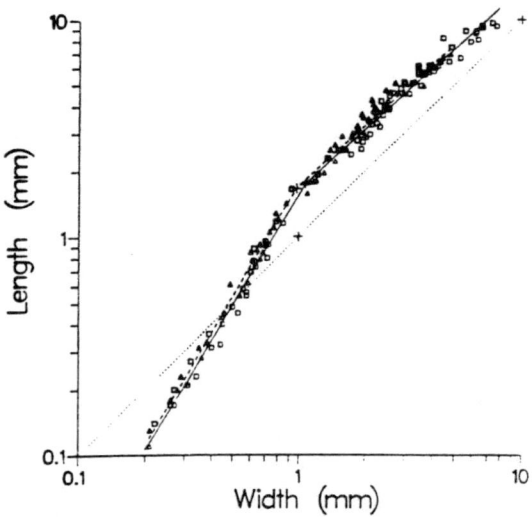

Figure 6. Allometric (log-log) plot of bud length vs. width. Squares represent individuals of Laa, and triangles Lff. Solid and dashed lines are regressions for Laa and Lff respectively: the two are statistically indistinguishable. Dotted line along the diagonal represents the line of isometry, where length and width are the same.

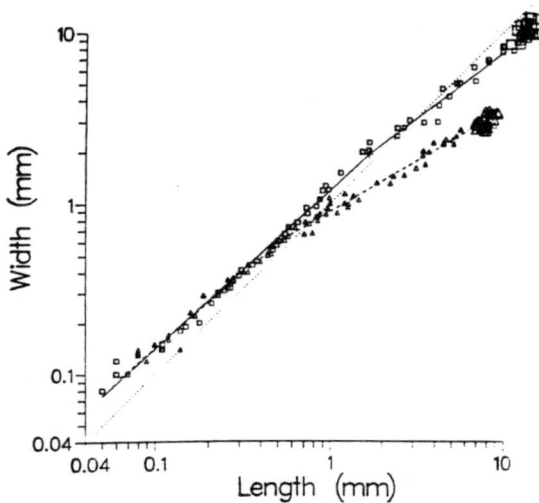

Figure 7. Allometric (log-log) plot of petal width vs. length. Symbols as in Fig. 6, except that the larger symbols designate flowers, and the smaller buds. The two are initially indistinguishable, but diverge at point where the slope of Lff changes.

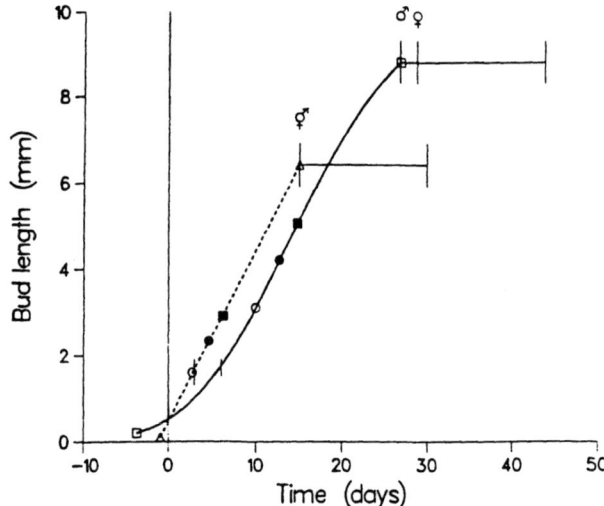

Figure 8. Plot of relationship between bud length and time. Squares and solid line represents Laa, and triangles and dashed line Lff. Along each of the growth curves the vertical line represents the time of bud 'bend', hollow circle meiosis in anthers, solid circle tetrad stage of pollen development, and solid square petal 'bend'. Horizontal lines at upper end of curves indicate the time of fruit growth and maturity. Sexual stages of flower are indicated above.

the only floral parts that did not differ in size between species at anthesis (Fig. 2). Allometry provides no information about actual, relative, or absolute·rates of growth (Tables 1, 2). To establish this, a relationship between size and time must be determined for each species, and from this the rates may be inferred.

The best description of the relationship between bud size to time that could be obtained for each species was used to calculate the time scales (Fig. 8). In turn, the relationships of size and time allowed the absolute and relative rates of growth and the timing of events, which could not be observed directly, to be calculated (Table 2). The functional form of the relationship of bud length to time differed between species. In *L. alba*, the data from each of 39 buds for which sizes had been measured repeatedly during development were fitted to a Richards curve. Asingle estimate of the relationship between size and time was obtained for the population from the six of these that were best described (as judged by r^2). However growth in length through time of *L. floccosa* could not be described by the Richards function; the best description that could be obtained was provided by a linear regression of untransformed bud length on time.

Table 2. Relative and absolute growth rates in length of various structures as estimated by the slopes of regressions through time of log (n) transformed data and untransformed data respectively.

Character	Relative Growth						Absolute Growth					
	r^2		df	(mean ± s.e.)				r^2		(mean ± s.e.)		
	Laa	Lff		Laa	Lff	t	p	Laa	Lff	Laa	Lff	t	p
Petal													
Pre-bend	.94	.93	62	.23±.01	.34±.02	-5.319	***	.70	.89	.08±.01	.07±.01	0.996	NS
Post-bend	.62	.87	50	.11±.02	.25±.02	-4.919	***	.70	.84	.54±1.0	.52±.04	0.245	NS
Large Anther													
Pre-meiosis	.97	.78	34	.26±.01	.41±.07	-2.289	*	.96	.76	.13±.01	.08±.01	3.690	***
Post-meiosis	.63	.91	66	.04±.01	.05±.003	-2.493	*	.64	.90	.07±.001	.03±.001	-3.733	***
Bud													
Pre-bend	.99	.90	51	.24±.003	.68±.05	-9.743	***	.93	1.00	13±.007	.40±.00	-38.258	***
Post-bend	.94	.97	96	.08±.003	.11±.003	-5.673	***	.99	1.00	.37±.004	.40±.00	-7.243	***
Mericarp	.83	.90	47	.10±.01	.13±.01	-2.649	*	.80	.92	.03±.004	.06±.003	-5.117	***
Sepal	.94	.94	47	.09±.01	.13±.01	-4.771	***	.98	.97	.36±.001	.42±.01	-3.747	***

* = p <0.05, *** = p <0.001

Despite this difference in how the time scales were generated, an internal test suggests that both estimations are adequate. When the ratios of relative growth between pairs of parameters (e.g. sepal length and width) were calculated, they were generally close to the slopes of the allometric plots (Tables 1,2). These represent two independent estimations of the same relationship, one (allometry) that did not incorporate the time dimension, and one that did (direct calculation from estimated values of relative growth rate of comparable components).

From the estimated relationship between size and time, not only can growth rates be calculated, but also the relationship in time of comparable events can be compared between species. Time zero was arbitrarily chosen to be when the buds were as long as they were wide. This early shape common to both is within the range of bud sizes sampled, but before any organs other than sepals were measurable with the techniques used in this study. The size-shape relationship of the buds before they reach 'time-zero' are similar between species, and the rate at which each increased in size can be compared. Both before and after time zero, *L. alba* buds grow more slowly than do those of *L. floccosa* in both relative and absolute terms (Table 2).

In bud, *Limnanthes floccosa* reached each stage sooner than did *L. alba* (Fig. 8). In *L. floccosa* the bud 'bend' occurred in half the time it took *L. alba* (ca 3 vs. 6 days). Meiosis in the pollen mother cells occurs shortly before bud 'bend' in *L. floccosa*, while in *L. alba* it was nearly 4 days after. Although more than 7 days separate these two species at the age of petal 'bend', in both the allometric reorientation of the petal size-shape trajectory takes place within 2 days after the tetrads are formed following meiosis. The flowers open about 15 and 27 days after time zero in *L. floccosa* and *L. alba* respectively.

Even after the flowers open, *L. floccosa* continues to be more precocious. Usually all of the anthers open the first day of anthesis in *L. floccosa*, whereas in *L. alba* they dehisce over a span of 2-3 days. The onset of stigmatic receptivity as inferred from the presence of peroxidase activity (Zeisler, 1938; and I. Baker pers. comm.) overlaps in time with anther dehiscence in *L. floccosa*. By contrast, in *L. alba* there is a 1-2 day delay after all of the anthers have dehisced before the stigma becomes receptive. Subsequent to stigma receptivity, the maturation of ripe fruit takes about 15 days in both taxa (Fig. 8). Therefore, embryogenesis appears to take the same amount of time in both species, and mature 'seeds' (including mericarp wall) are of approximately equal mass at maturity (Ritland and Jain, 1983).

The relative rate of growth is greater in *Limnanthes floccosa* than in *L. alba* for all floral structures measured (Table 2). In terms of absolute growth rate, however, floral components of *L. floccosa* may grow either more rapidly or slowly than *L. alba*, or as with the petals, the absolute growth rate does not distinguish the taxa.

DISCUSSION

This study had two goals. The more narrowly framed one concerns the processes by which the morphological and phenological phenotype of *Limnanthes floccosa* flowers originated. Empirical data on floral development presented here, however, have little meaning in isolation. Interpretation is necessarily dependent on the phylogenetic and ecological context from which the flowers arose. Moreover, even though flowers are clearly homologous between taxa, they are borne on shoots, which themselves have properties that may also affect floral form. Hence, the initial part of this discussion will be devoted to these topics. A broader goal of this study concerns the value of ontogenetic data with respect to questions concerning the origin of phenotypes and how ontogenetic information may bear on the study of adaptation.

Context of Floral Development

Phylogenetically, it appears that the most exclusive group of species to which both *L. alba* and *L. floccosa* belong also includes *L. montana*, and the apparently paraphyletic *L. gracilis* (Mason, 1952; Ornduff and Crovello, 1968; McNeill and Jain, 1983). These four species comprise the presumably monophyletic section Inflexae. There are two modern studies of the group, and they differ only slightly in their estimates of overall similarity of taxa within Inflexae. One is based on vegetative and reproductive morphological characters (Ornduff and Crovello, 1968), and the other on allozymes coded for by 18 loci (McNeill and Jain, 1983). Within Inflexae, both morphological and electrophoretic data place all five *L. floccosa* subspecies together as the apparent sister group of the *L. alba-gracilis-montana* complex (after first mention hereafter, taxa may be referred to by their initials). The discrepancy between the studies is within this latter group. *Limnanthes alba* var. *alba* (Laa) is morphologically distant from *L. gracilis* var. *gracilis* (Lgg), *L. gracilis* var. *parishii* (Lgp) and *L. montana* (Lm), which closely resemble one another. In contrast, the geographically peripheral Lgg and Lgp are very close to Laa in terms of Nei's genetic distance, and the three are sharply separated from *Lm.* Indeed, both *Lgg* and *Lgp* are genetically closer to *Laa* than they are to each other.

Relative interfility of intertaxon hybrids involving *Laa, Lgg, lgp* and *Lm* (estimated by the proportion of viable pollen in hybrid flowers) decreases precipitously and almost linearly with increasing genetic distance (McNeill and Jain, 1983). The congruence of electrophoretic and interfertility data supports the hypothesis that *Laa, Lgg,* and *Lgp* are phylogenetically more closely related to one another than any are to *Lm.* On the strength of this information, and of the high degree of intersterility and amount of genetic distance between *L. floccosa* and the rest of the Infelexae, *L. floccosa* and *L. alba* appear not to be the closest of relatives. However, the lack of a rigorous cladistic analysis limits phylogenetic conclusions and therefore interpretation of the development data.

The one discrepancy between branching diagrams based on morphological and apparent phylogenetic relationships is resolved if the large flowers of *Laa*, which in Inflexae are unique to it, are considered derived within the section. From this perspective, the greater morphological resemblance among the flowers of *Lm, Lgg* and *Lgp* reflects their pleisiomorphic, or shared ancestral condition. It has always been assumed and there is currently no reason to doubt that small flowers associated with autogamy in *Lff* are also derived within Inflexae.

If so, then the flowers of *Lff* and *Laa* have each evolved independently and divergently from the presumably intermediate floral form of their most recent common ancestor. Flower size increased during the evolution of *L. alba*, and decreased in *L. floccosa.* The rest of Inflexae have flowers of an intermediate size.

Although Arroyo's (1973) phylogenetic assessment of *Lff* and *Laa* has not withstood the test of time, her ecological characterization has proved remarkably perceptive. In an extensive theoretical and empirical study comparing life histories of these two taxa, Ritland and Jain (1984) corroborated and extended the notion that in comparison with *Laa*, earlier flowering and smaller plant size are adaptive in *Lff.* The more montane habitats of *Lff* are of generally lower quality in terms of soil nutrients and moisture availability (Ritland and Jain, 1984). They are also less predictable from year to year than the vernal pools of the Central Valley grasslands inhabited by *Laa* (Arroyo, 1973; Ritland and Jain, 1984).

The ability of *Lff* to flower earlier and initially increase in biomass at a slower rate are genetically determined traits (Ritland and Jain, 1984). In localities where both taxa occur the naturally, *Lff* plants are approximately one-fifth to one-fourth the size of *Laa* plants. The average relative growth rate in biomass of young *Lff* plants is less than that of those of *Laa*, but in later life they appear to have the same intrinsic ability to increase in biomass despite their initial differences (Ritland and Jain, 1984). There appears to be a surprisingly simple morphogenetic basis for differences in absolute growth rates and final sizes that distinguish plants of *Lff* from those of *Laa* (Guerrant, 1984, in press). Individual leaves of both taxa have the same mean relative growth rate (a measure equivalent to compound interest rate), and grow for the same length of time. However, in absolute terms *Lff* leaves grow more slowly and mature at a smaller size than do those of *Laa*. By analogy to the growth of funds in a bank accounts, it is hypothesized that a difference only in 'initial capital' (shoot apical meristem and leaf primodrium sizes) is sufficient to account for absolute growth rate and mature size differences. Subsequent analyses have shown that the shoot apical meristem in seeds of *Lff* is indeed smaller than that in *Laa* (Guerrant, unpub. data).

Floral Development in Space and Time

   Developmental data presented in this study may now be interpreted in the context of several factors, each of which impinges on floral ontogeny in its own way. Both *Lff* and *Laa* appear to have evolved independently and divergently from their most recent common ancestor. Consequently, any phenetic resemblance they share, either as adults or during development, can reasonably be assumed to have been inherited unchanged from their most recent common ancestor. It appears adaptive for *Lff* to be smaller and to flower earlier than *Laa* does. These genetically determined differences are correlated with, and may have been mediated in part by the smaller apical meristem size of *Lff*. In turn, such a structural difference of the vegetative plant body should affect floral form.

   Unlike the neotenic flowers of *Delphinium nudicaule*, which resemble the buds of *D. decorum* more closely than they do their flowers, *Lff* flowers are not obviously juvenilized versions of *Laa* flowers (compare Fig. 5 with Fig. 5 in Guerrant, 1982). Rather, the relationship is more like that envisioned by von Baer's law (see Gould, 1977): related taxa resemble each other most closely when youngest and become increasingly more dissimilar with age. This progressive morphological divergence may be due in part to the different initial sizes, and dramatic differences in relative timing of events. These differences appear to have had a cascading effect, compounding their morphological effect as development proceeds.

In external appearance, the buds of *Lff* and *Laa* do grow through a similar and complex size-shape series. Buds of both taxa are initially wider than long, and grow in length considerably faster than width. When the buds of both taxa are about 1 mm wide, there is a dramatic and fairly abrupt shift in the ratio of relative growth rates in length and width. Thereafter, bud growth is largely in size with shape remaining roughly the same. This allometric 'shift' occurs at dramatically different states of floral maturity in the two taxa. It is just after meiosis in *Lff*, whereas in *Laa* it occurs nearly a week before.

Two preliminary conclusions may now be drawn. One is that to a considerable extent whatever controls overall bud size-shape growth in both taxa apparently has been inherited unchanged from their most recent common ancestor. The second, also an inference, is that whatever controls bud size-shape growth is independent of 'reproductive' or gametophytic development.

Like the buds, petals of both species initially grow through a common size-shape trajectory, and undergo an allometric shift during development. However, unlike the buds, petal shape diverges from its initial path to different degrees, and at different sizes and ages in the two taxa. It may be significant that this event occurs approximately the same amount of time after the tetrad stage of pollen development in both taxa. If appears, therefore, that either the state of anther maturity affects the timing of the shift, or both it and the petal shift have a common temporal cause.

Of the floral components considered, the taxa differ most dramatically in anther growth. During development, the anthers grow through entirely separate size-shape trajectorites. The two differ dramatically in amounts of pollen produced, a difference clearly related to their current breeding systems. The highly inbred and genetically depauperate *Lff* produces nearly two orders of magnitude fewer pollen grains per flower (ca 9,000 vs 900,000; Arroyo, 1973; Guerrant unpub. data) than the genetically more heterogeneous outbreeding *Laa* (McNeill and Jain, 1983)

In the preceding section, growth of homologous structures between taxa have been compared: petals with petals, anthers with anthers, and so on. However, the currency of comparison between structures are characteristics of development having to do with time; examples are relative growth rates, allometric relationships, and relative timing of comparable events. Phenomena that are even partially described by variables such as these appear to be important in generating form. Comparing taxa for these and other parameters provides a means to explore pathways of evolutionary transformation of form.

The greater relative growth rate of *Lff* floral tissue is the one developmental parameter that distinguishes it from *Laa*. This difference also distinguishes floral from foliar tissues, in which the taxa differed not in relative growth rate but seemingly only in initial size.

At the level of the vegetative plant body, such 'characters' as relative growth of the leaves, and shoot apical meristem size are seen as homologous 'elements of construction' (Guerrant, 1984, in press). The ecologically adaptive traits of slower absolute growth rate and smaller ultimate size of *Lff* plants appear to be incidental developmental consequences of their differences in shoot apex size. Initial size is an important variable in theory relating ontogehy and phylogeny (Gould, 1977; Alberch et al. 1979).

By distilling plant structure into such presumably genetically controlled developmental determinates of form as these, it may be possible, ultimately, to separate the adaptive effects of selection from its incidental consequences.

Inference of Adaptation and Incidental Consequences

Assume that during the evolution of *Limnanthes floccosa*, natural selection resulted in smaller plants that reproduced at a younger age. Upon what might natural selection have acted? What would be the expected morphological consequences?

There are at least two theoretical possibilities, one of which is more likely. *Lff* could achieve sexual maturity at a younger age merely by initiating otherwise identical flowers at nodes produced earlier in shoot development. Direct control of the first node to flower, however, appears not to be a major cause of the observed difference. Both initiate flowers in response to increased duration of photoperiod in growth chambers (Joliff et al., 1984; McNeill, pers. comm; Guerrant pers. obs.). Furthermore, the time of seedling establishment also is environmentally determined, but by autumn rains, which vary in timing and amount from year to year (Ritland and Jain, 1984). Consequently, a possible intrinsic difference in first node to flower appears to be an unlikely source of consistent differences in flowering times observed between taxa.

A more likely feature upon which selection may have acted to reduce the age at maturity is the time interval between floral initiation and fruit maturity (or some portion of it). There is a dramatic difference between the taxa in this interval: *Lff* is able to mature fruits in only about two-thirds the time it takes *Laa*.

The difference in overall timing between species is not, however, proportionately divided among all comparable developmental intervals from floral initiation to fruit maturation.

Flowers of *Lff* reach anthesis in just over half the time it takes those of *Laa*. The most dramatic difference between taxa occurs during the interval of time leading up to meiosis in the anthers -- a developmental state reached by *Lff* in just slightly over one-quarter the time taken by *Laa*. It is possible that selection acted only to move up the onset of reproductive maturity itself, which would be a cause of pure progenesis (see Fig. 1). Alternatively, earlier maturity might be a developmental consequence of selection for increased relative growth rate of floral tissues. Both these and other factors may be involved. In any case, meiosis occurs at a younger age in *Lff*, and its floral tissues grow at a greater relative rate than do those of *Laa.*

In contrast, the interval of time from stigmatic receptivity (corresponding to the onset of ovule or 'female' reproductive maturity) to fruit maturation is similar in the two species. Embryo maturity in both is reached in about 15 days, which is nearly as long as it takes *Lff* buds to achieve anthesis. The mature mericarp, or so-called 'seed' mass is about the same in both taxa. The difference between them is that *Lff* produces only one mericarp per flower, while *Laa* produces more (Ritland and Jain, 1984). The size of the individual mericarps is the only morphological feature that did not distinguish flowers of the two taxa. Embryo size may be subject to stabilizing selection for ecological reasons relating to seedling establishment (Baker, 1972; Harper, 1977), and so may not be as labile evolutionarily as other traits.

CONCLUSIONS

The hypothesis of progenesis can be addressed in a variety of ways, even though its interpretations are dictated by phylogenetic circumstances. Arroyo's (1973) conclusion, that natural selection for early flowering operated during (and may have precipitated) the origin of *L. floccosa* is still a reasonable hypothesis. However, *Laa* does not appear to be the immediate ancestor of *Lff*. Rather, each appears to have evolved independently and in divergent directions from their most recent common ancestor. Despite these differences, considerable similarity remains between the two in patterns of floral development. This is probably because both inherited these developmental features unchanged from their most common ancestor. If so, these are developmentally homologous and show the ancestral, or plesiomorphic condition.

Assuming selection acted to reduce the age at maturity of individual flowers, then it appears both to have modified the relative rate at which floral tissues grow, and to have lowered the age at which various developmental events occur. The nature of the interaction, if any, among these characters cannot be assessed at this time. Also having an effect is the difference in shoot apical meristem size, which appears to mediate the slower growth of the vegetative plant bodies on which the flowers are borne.

The following is a plausible adaptive and evolutionary scenario that accounts for the phenotypic differences in flowers between *Lff* and *Laa*. In the presumably larger flowered, later maturing, and intrinsically slower growing most recent outbreeding ancestor of *Lff*, there were selective advantages to early floral maturity and small plant size at maturity (or slow absolute growth). Slower growth of the whole plant and smaller mature size appear to be the consequences of a smaller shoot apical meristem. Small flowers appear to show the effects of at least two adaptations: a floral modification for earlier maturation, within the context of an overriding vegetative reduction in size.

Increased levels of autogamy also may have evolved as an incidental consequence of selection for more rapid floral maturation. Given that genetic self-compatibility appears to be ubiquitous in *Limnathes* (Mason, 1952; and I am aware of no reports of genetic self-incompatibility), the rate of self-pollination should reflect the relative timing of anther dehiscence and stigma receptivity. As natural selection reduced the time allotted to make a flower, these two events would be expected to converge in time, and the rate of autogamy to increased proportionately. This does not seem to be included in any of the pathways leading to the evolution of autogamy outlined by Jain (1976), but is complementary with several of them.

While many of the evolutionary scenarios and ideas discussed here are necessarily speculative, the proposed mechanisms are experimentally testable. Insofar as artificial selection to reduce the interval from floral initiation to anthesis in *L. alba* flower is successful, the theory of Gould (1977) and Alberch et al. (1979) leads to the expectation that the mature size and shape of the resultant plants should lie along their parental growth trajectory. Preliminary evidence indicates that there is heritable variation in this interval, and that the first plants to flower have smaller flowers than the population as a whole (Guerrant, unpub. data). If successful, and continued over a sufficient number of generations, the rate of auto-fertility is expected to increase as well.

SUMMARY

The small flowers of the autogamous *Limnanthes floccosa* are hypothesized to show the effects of paedomorphosis, or evolutionary juvenilization by the progress of progenesis, relative to the large flowered, xenogamous *L. alba*. The hypothesis, based in theory concerning the relationship of ontogeny and phylogeny, postulates a developmental connection between the amount of time between floral initiation and anthesis (which may be affected by natural selection), and floral size and shape. Ontogenetic data on these two species are interpreted within the context of phylogenetic history, ecological relationships, and aspects of vegetative morphology. It is suggested that increased levels of autogamy in *L. floccosa* may have evolved as an incidental developmental consequence of selection for earlier or more rapid floral maturation.

LITERATURE CITED

Alberch, P., S. J. Gould, G. Oster, and D. B. Wake 1979. Size and shope in ontogeny and phylogeny. Paleobiology 5: 296-317.

Arroyo, M. T. Kalin 1973. Chiasma frequency evidence on the evolution of autogamy in *Limnanthes floccosa* (Limnanthaceae). Evolution 27: 679-688.

Baker, H. G. 1959. Reproductive methods as factors in speciation in flowering plants. Cold Spring Harbor Symp. Quant. Biol. 24: 177-191.

Baker, H. G. 1972. Seed weight in relation to environmental conditions in California. Ecology 53(6): 997-1010.

Baker, H. G. and I. Baker 1979. Starch in angiosperm pollen grains and its evolutionary signnificance. Amer.J. Bot. 66(5): 591-600.

Causton, D. R., C. O. Elias, and P. Hadley. 1978. Biometrical studies of plant growth. 1. The Richards function, and its application in analysing the effects of temperature on leaf growth. Plant, Cell and Environment. I: 163-184.

Causton, D. R. and J. C. Venus 1981. The Biometry of Plant Growth. Edward Arnold, London.

Gould, S. J. 1966. Allometry and size in ontogeny and phylogeny. Biol. Rev. 41: 587-640.

Gould, S. J. 1977. Ontogeny and Phylogeny. Harvard Univ. Press, Cambridge.

Guerrant, E. O., Jr. 1982. Neotenic evolution of *Delphinium nudicale* (Ranunculaceae): a hummingbird pollinated larkspur. Evolution 36(4): 699-712.

Guerrant, E. O., Jr. 1984. The role of ontogeny in the evolution and ecology of selected species of *Delphinium* and*Limnanthes*. PhD dissertation. University of California, Berkeley.

Guerrant, E. O., Jr. In press. Heterochrony in plants: the intersection of evolution, ecology, and ontogeny. In Heterochory in Evolution: a Multidisciplinary Approach. [ed.] M. McKinney. Plenum Press, New York.

Harper, J. L. 1977. Population Biology of Plants. Academic Press, London.

Hunt, R. 1982. Plant Growth Curves - the Functional Approach to Plant Growth Analysis. University Park Press, Baltimore.

Huxley, J. S. 1932. Problems of Relative Growth. MacVeagh, London.

Jain, S. K. 1976. The evolution of inbreeding in plants. Ann. Rev. Ecol. Syst. 7: 468-495.

Jolliff, G. D., W. Calhoun, and J. M. Crane 1984. Development of a self-pollinated meadowfoam from interspecific hybridization. Crop Sci. 24: 369-370.

Lloyd, D. G. 1965. Evolution of self-compatibility and racial differentiation in Leavenworthia (Cruciferae). Contr. Gray Herb. 195: 3-195.

McNeill, C. I., and S. K. Jain 1983. Genetic differentiation studies and phylogenetic inference in the plant genus Limnanthes (Section Inflexae). Theor. Appl. Genet. 66: 257-269.

Mason, C. T. 1952. A systematic study of the genus Limnanthes. Univ. of Calif. Publ. Bot. 25: 455-512.

Moore, D. M., and H. Lewis. 1965. The evolution of self-pollination in Clarkia xantiana. Evolution 19: 104-114.

Ornduff, R. 1969. Reproductive biology in relation to systematics. Taxon 18: 121-133.

Ornduff, R., and T. J. Crovello. 1968. Numerical taxonomy of Limnanthaceae. Amer. J. Bot. 55: 178-182.

Radford, A. E., W. C. Dickenson, J. R. Massey, and C. R. Bell. 1974. Vascular Plant Systematics. Harper and Row, New York.

Richards, F. J. 1959. A flexible growth function for impirical use. J. of Exp. Bot. 10: 290-300.

Richards, F. J. 1969. The quantitative analysis of growth. In Plant Physiology - A Treatise. VA. Analysis of Growth: Behavior of Plants and their Organs. [ed.] F. C. Steward, pp. 1-76. Academic Press, London.

Ritland, K. M., and S. K. Jain. 1984. The comparative life histories of two annual Limnanthes species in a temporally variable environment. Am. Nat. 124(5):656-679.

SAS Institute Inc. 1982. SAS Users Guide: Statistics. SAS Institute Inc. Cary, NC, USA.

Solbrig, O. T., and R. C. Rollins. 1977. The evolution of autogamy in species of the mustard genus Leavenworthia. Evolution 31: 265-281.

Wyatt, R. 1983. Pollinator-plant interactions and the evolution of breeding systems. In Pollination Biology. [ed.] L. Real. Academic Press, London.

Zeisler, M. 1938. Uber die Abgrenzung der eigentlichen Narbenflache mit Hilfe von Reaktionen. Beih. Bot. Zentralb. A. 58: 308-318.

CHROMOSOME VARIATION AND EVOLUTION IN AMERICAN AND ASIAN *TRILLIUM* SPECIES

Ichiro Fukuda

INTRODUCTION

The genus *Trillium* is common in eastern and western North America and eastern Asia. The plants have ten large chromosomes (2n=10) which exhibit cold-induced banding patterns as shown in Figure 1. Population studies of chromosome banding for *Trillium* species with disjunct distributions have been done for several representative species in the past three decades (Fukuda and Kozuka, 1958; Kurabayashi, 1960; Fukuda, 1962, 1967, 1969, 1988; Fukuda and Channel, 1975; Fukuda and Grant,1980).

In this article, I will compare modes of speciation of American and Asian *Trillium* species, adding more recent chromosomal data and ecological observations of the reproductive system. I then will discuss how the different mechanisms of speciation may have occurred in North America and Asia from both genetic and ecological standpoints.

MATERIALS AND METHODS

Samples were collected from natural populations for each species of *Trillium* throughout the distribution range as shown in Table 1. Collected materials were grown in pots. After sufficient root development, plants were placed in a refrigerator at 0°C for 96 hours for pre-treatment of the Feulgen reaction. Fixation was done in LaCour 2BE for 15 minutes. After hydrolysis at 60°C for 15 minutes, a Feulgen solution was used for staining the chromosomes. Chromosomal analyses were performed on the meristematic cells of root tips; the preparations were made by a squash method and mounted in 45% acetic acid. The cold-induced banding pattern was determined for each individual in the population sample (Table 1). There is variation of chromosome banding patterns in five chromosomes (Chromosome A-E) in each species; only patterns found in Chromosome A were used to compare the different species (Fig. 2-4), because they were the most consistent.

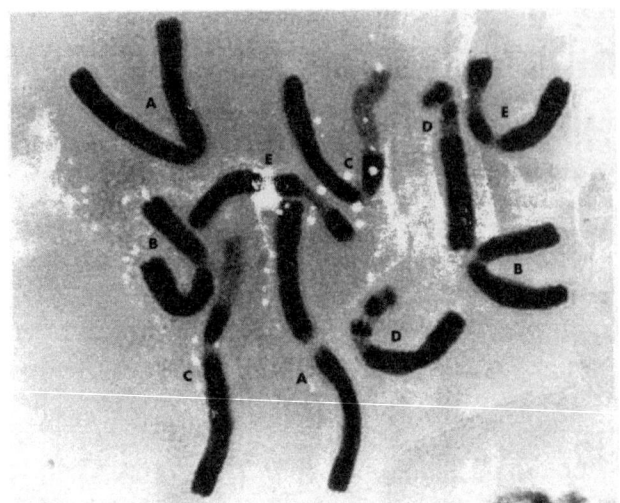

Figure 1. Cold induced chromosome banding in *Trillium erectum*. Dark parts show euchromatic segments (after Darlington and LaCour, 1940).

Table 1. Collected Material for the *Trillium* survey.

Species	Actual population sizes (N=No. of plants/population)	Number of analyzed populations	Number of individuals analyzed
North America			
T. grandiflorum	large (N=100-10,000)	22 (N=13-33)	630
T. erectum	small (N=10-50)	14 (N=5-32)	196
T. ovatum	medium (N=50-100)	13 (N=20-54)	514
Asia			
T. kamtschaticum	large (N=100-10,000)	61 (N=10-100)	1,931
		Total plants	3,271

RESULTS
Variation of Banding Patterns in *Trillium* Species in North America

In eastern North America, *Trillium grandiflorum* (Michx.) Salisb. grows throughout the Great Lakes region, extending southward through the Appalachian Mountain region. Populations are comparatively large (Table 1) in these deciduous forests composed mainly of maple (*Acer*) and oak (*Quercus*) species. Analyses of populations from the whole distribution range show that banding patterns are fairly homogeneous both within and among populations, although the northern populations in the Great Lakes region are more homogeneous than southern populations located in the Appalachian Mountains (Fig. 2). Another eastern species, *Trillium erectum* L., is distributed through the Appalachian Mountains from Quebec in the north to Alabama in the south. The plants are widely scattered and have comparatively small populations within deciduous forests. Banding patterns are highly heterogeneous among populations but homogeneous within each population (Fig. 3; Fukuda, 1984, 1988). These results suggest that topographic relief has helped shape variability in these *Trillium species.*

In western North America, *T. ovatum* Pursh. occurs along the Pacific Coast and in the Rocky Mountains. The plants grow in *Pseudotsuga-Sequoia-Pinus ponderosa-Acer-Alnus* forests. Populations of this species are apparently sub-divided into two regions with different chromosomal banding patterns characteristic to each region. Populations have homogeneous banding patterns in the Pacific coast region. In contrast, in the Rocky Mountains, where climatic and topographic conditions are much more variable, the patterns are extremely heterogeneous. (Fig. 2; Fukuda and Channel, 1975).

These three examples, two from eastern and one from western North America indicate that, within a given species of *Trillium*, there is a strong association between chromosomal variation and environmental heterogeneity.

Chromosomal Variation in *Trillium* Species in Asia

Trillium kamchaticum Pall. grows in the *Acer, Ulmus* , and *Cryptomeria* mixed forests of northern Japan. This species is the only diploid *Trillium* in Asia. Populations are chromosomally heterogeneous, though on the small islands some are homogeneous (Fig. 4). The species is divided into three geographical races - south, north and east (Kurabayashi 1960; Fukuda, 1962, 1967). The most important difference between North American and Asian *Trillium* species is the development of polyploid species (3x, 4x,

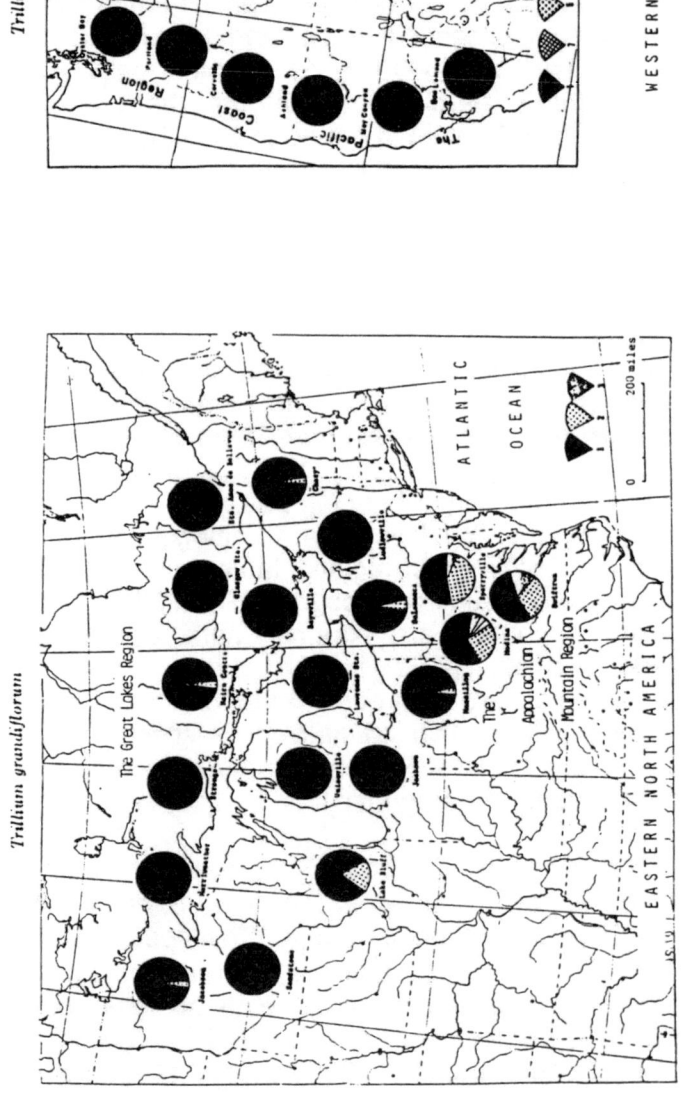

Fig. 2 Left map; circle charts showing poulation structures by chromosome banding in *T. grandiflorum*. [Chromosome A. Standard pattern (A1) and variants (A2, A3).]

Right map; circle charts showing population structures by chromosome banding in *T. ovatum*. [Chromosome A. Standard pattern (A1) and variants (A2-A6).]

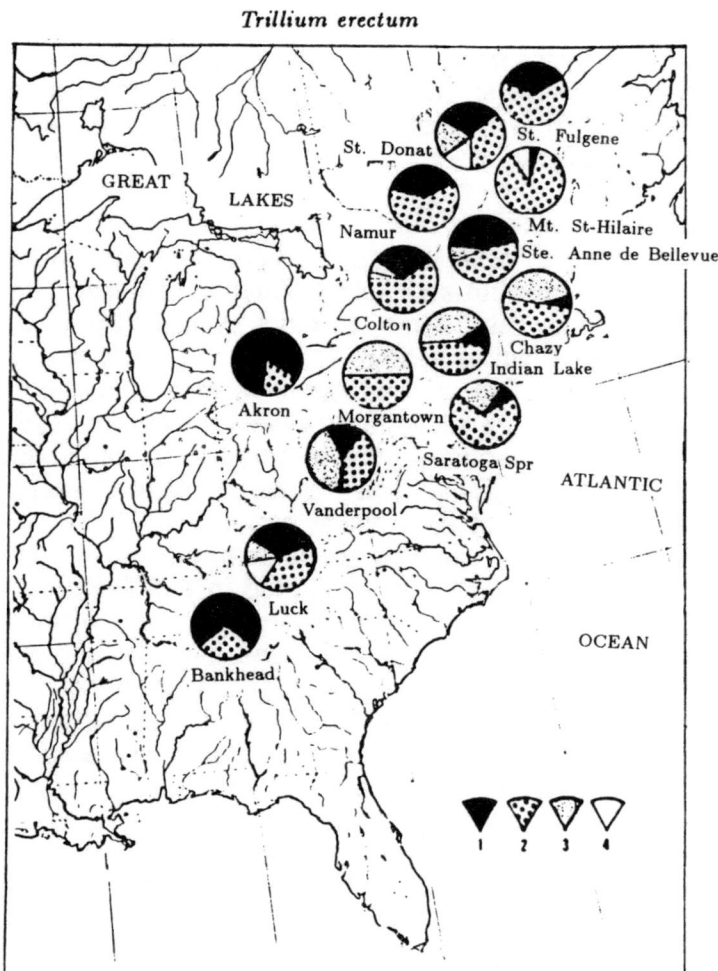

Figure 3. Circle charts showing population structures by chromosome banding in *T. erectum*. [Chromosome A. Standard pattern (A1, A2) and variants (A3, A4).]

6x and 8x) in Asia. For example, *T. hagae* (3x) has resulted from hybridization between *T. kamtschaticum* (2x) and *T. tschonoskii* (4x). This triploid species is found frequently in the same habitats in which both parents grow. The highest polyploid found to date is the octaploid (8x = 40), *Kinugasa japonica* Tatewaki et Suto, which is distributed in a mountainous region in central Japan. The chromosome banding patterns show characteristics of both *Trillium* and *Paris*. It is believed that this species might have originated from hybridizations between *Trillium tschonoskii* and *Paris polyphylla*.

DISCUSSION

Chromosomal variability shows very different patterns between North American and Asiatic *Trillium* species. There are several reasons for this. First, we must consider the overall geographic distribution of habitats in these two regions. In North America, individual *Trillium* species have extensive geographic distributions and are all allopatric. However, the Japanese and other Asian *Trillium* species grow in heterogeneous habitats in restricted montane regions. They frequently have overlapping distributions and can easily produce hybrids.

Second, we must consider contrasts in paleo-geographical conditions. After *Trillium* had originated in the Tertiary period, the North American continent experienced several Pleistocene glaciations. On the other hand, the Japanese islands and other Asian *Trillium* habitat regions escaped glaciation, although the region experienced much volcanic activity. It may be that glaciers generated gradual changes for plants over extensive areas while volcanic activities generated abrupt, locally heterogeneous changes over relatively small areas.

Third, the most important factor is that different reproductive systems of *Trillium* have developed in North America and Asia perhaps as a result of contrasting environmental conditions. Here, I would like to discuss in more detail these reproductive systems. Using the data from chromosome banding patterns, the inbreeding coefficient (f) was calculated according to the following formula:

$$f = \frac{\text{observed homozogotes (\%)} - \Sigma iPi}{1 - \Sigma iPi}$$

where Pi is the frequency of chromosome pattern Ai in a population and the zygotic proportions are $2PiPj(1-f)$ in the heterozygotes $AiAj(i=j)$ and $fPl + (1-f) Pi^2$ in the homozygotes

AiAi. (Hiraizumi *et al.*,1961; Fukuda, 1967; Fukuda and Channell, 1975). Table 2 shows the Inbreeding Coefficient (f) for American and Japanese *Trillium* species. The American species, *T. grandiflorum*, *T. erectum* and *T. ovatum* have high inbreeding (0.23 to 0.39), although mountain populations in *T. ovatum* have a slightly lower value of 0.19. In contrast, the value for Japanese *T. kamtschaticum* is 0.04, which indicates a very high out-crossing rate.

I next compared these species in terms of field characteristics (Table 3). Although *Trillium grandiflorum* and *T. erectum* usually grow allopatrically in eastern North America, their distributions overlap in some parts of the Appalachian Mountain region (Figs. 2 and 3). Their flowering seasons also overlap, but they do not form hybrids. For example, Figure 5 indicates a 6 x 4 m² quadrat contained 476 *T. grandiflorum* plants and 135 *T. erectum* plants in the forest at St. Anne de Bellevue, Quebec. No hybrids were found there, presumably because both species are primarily self-pollinating. Observations of the flowers at anthesis showed that the stigmas had been self-fertilized, presumably because stigma and stamens develop synchronously, allowing self-pollination in the bud. The flowers were visited by *Bombus bimacula, B. vagans, Apis mellifera etc.*, but their visits were infrequent. At a later flowering stage, the stigmas expanded rapidly and, when shaken by wind, the flower could undergo further self-pollination.

Seventeen species of sessile-flowered *Trillium* grow in the southeastern United States (Freeman, 1975). In these species, variation in chromosome banding is low within populations. For example, samples of *T. recurvatum* Bock from the Meeman Shelby Forest in Tennessee, have few chromosome variants. The peak of flowering in this species is late March, and their flowering season spans about two weeks. At flowering time the stamens easily touch the stigma. Flowers are cleistogamous. Pollen fertility and seed reproduction are reduced as a result of inbreeding, and they reproduce asexually by slender rhizomes. There appeared to be no hybridization among these species even though they occur sympatrically, and grow in mixed populations with pedicellate-flowered ones (Freeman, 1984). On the other hand, the Japanese species, *T. katschaticum* , blooms for one week in May. The flower opens when upright, in contrast to many American species in which the flowers are nodding. At anthesis, the stigma is mature but stamens are immature, and dehisce after three days. If outcross pollen reaches the stigma during these three days, fertilization occurs. Visiting insects are more active during the

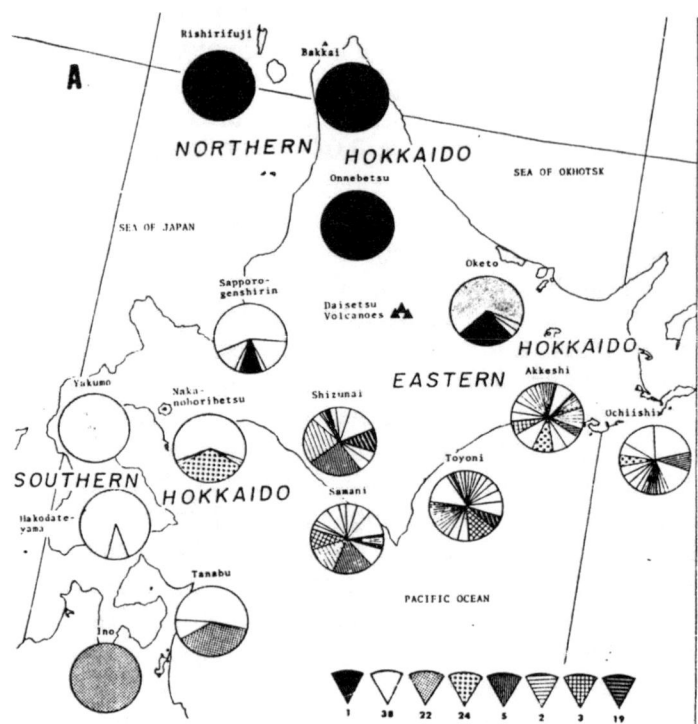

Figure 4. Circle charts showing population structures by chromosome banding in *T. kamtschaticum*. [Chromosome A, no standard]

Table 2. Inbreeding coefficients (f) in *Trillium* species in North America and Asia.

	Species	f
North America	*Trillium grandiflorum*	0.3866
	Trillium erectum	0.2295
	Trillium ovatum coast	0.3301
	mountain	0.1853
Asia	*Trillium kamtschaticum*	0.0410

flowering season in Japan than in North America where the blooming season is longer.

Insects observed included flies and bees, *Scopexma stercoratium, Melanostomo scalare, Tucowyia frigidae, Teuthredo* spp. *etc.* Moreover, evidence of wind pollination was obtained from field observations, and it appeared to lead to hybridization between *T. kamtschaticum* (2x) and *T. tschonoskii* (4x). When these two species coexist, *T. kamtschaticum* grows on slopes while *T. tschonoskii* grows in the flatlands. A triploid hybrid species (*T. hagae*) was found in intermediate locations, but closer to *T. kamtschaticum* (Fig. 6).From these examples, we can say in general that Japanese *Trillium* species have developed a reproductive system with extensive outcrossing and that polyploidy is common. American species are predominantly homogamous, and have enlarged stigmas (Fig. 7). In contrast, Japanese *Trillium* are predominantly heterogamous, and have short stigmas. The large stigmas of American *Trillium* species are notable (e.g., *T. grandiflorum* and *T. erectum* in Fig. 7). These stigmas are small at anthesis, but enlarge quickly later. A large stigma can accept much pollen. This feature of American *Trillium* may be due to the paucity of visiting insects. In contrast, Japanese *Trillium* are abundantly visited by insects, except in the case of *T. apetalon* , an apetalous species (Fukuda, 1961), with its very small stigmas. These characteristics of the reproductive system may have contributed to speciation by genic change in North America (especially in sessiled-flowered *Trillium*); but in Asia, there has been hybridization and associated polyploid formation.

Trillium may have originated in North America in the later Tertiary. Before the Pleistocene glaciations, several species of *Trillium* arrived in the Japanese islands from North America across the Bering Sea (Flint, 1957). Glaciations likely brought on severe climatic conditions in North America, leading to a paucity of visiting insects. The plants may have adapted by switching to inbreeding from a probable ancestral outbreeding system. Meanwhile, the Japanese island region in Asia was not glaciated. Japanese *Trillium* have produced several amphiploid taxa; two species (4x) migrated to the Himalayan mountain region. It appears that the reproductive system had an important effect upon plant speciation in this genus.

SUMMARY AND CONCLUSIONS

Trillium species are widely distributed in North America and Asia. In eastern North America, *T. grandiflorum* populations throughout the Great Lakes region show a fairly uniform

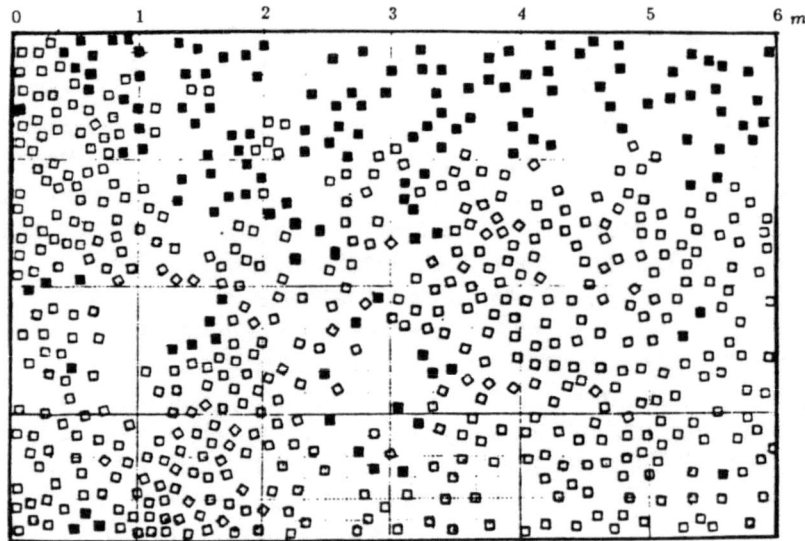

Figure 5. Map showing locations of *T. grandiflorum* (white square) and *T. erectum* (dark square) in a 6 x 4m² quadrate at St. Anne de Bellevue, Quebec, Canada.

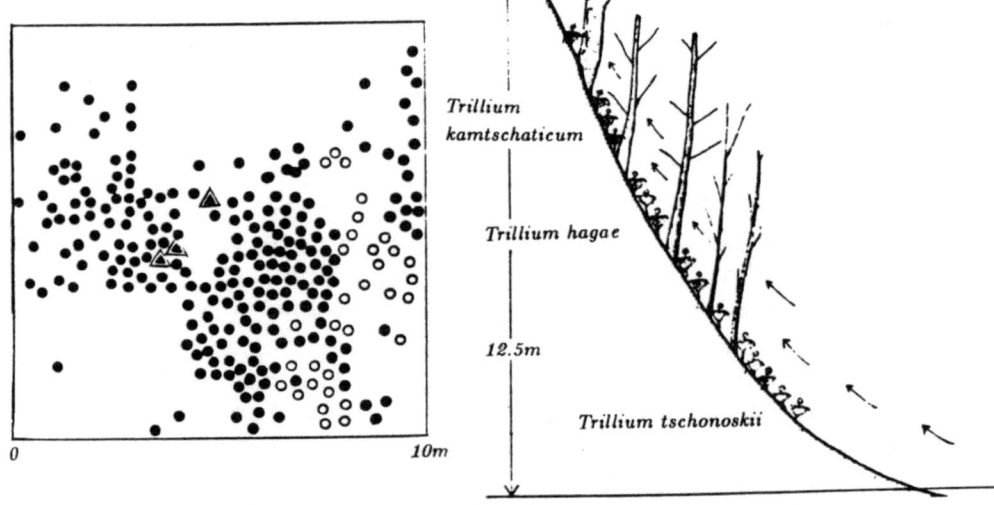

Figure 6. Map showing locations of *T. kamtschaticum* (dark circle), *T. tschonoskii* (white circle) and *T. hagae* (triangle) in a 10 m² quadrat. Drawing shows topographical conditions of *Trillium* habitat. Arrows indicate a direction of the wind at Nanae, Aomori, Japan.

T. kamtschaticum   *T. tschonoskii*   *T. apetalon*   *T. grandiflorum*   *T. erectum*

Figure 7. The shape of the ovary in several *Trillium* species. Short stigmas in Japanese *T. kamtschaticum, T. tschonoskii* and *T. apetalon*; largs stigmas in American *T. grandiflorum* and *T. erectum*.

Table 3. Comparative modes of speciation in American and Asian *Trillium* species.

	American *Trillium*	Asian *Trillium*
Population structure (diploid species)	Homogeneous in many populations	Heterogeneous in most populations
Reproductive system	Significant inbreeding, homogamy, contacted stamen and pistil, enlarged stigma	Predominantly outbreeding heterogamy, separated stamen short stigma
Frequencies of insect visits to flowers	Low, diploid species	High, mostly polyploid
Chromosome system	(2X)	(2X,3X,4X,6X,8X)
Likely speciation modes	Allopatric and parapatric speciation	Hybrid and polyploid speciation
Geographical distribution	Continental, large areas	Limited, heterogeneous montane areas
Paleogeological situation	Extensive glaciation	Extensive volcanic activity

chromosome banding pattern. *T. erectum* of the Appalachian Mountain region are fairly homogeneous within each population, but highly heterogeneous among populations. In western North America, *T. ovatum* is homogeneous in Pacific coast populations and heterogeneous in Rocky Mountain populations. On the other hand, *T. kamtschaticum* in Japan is highly heterogeneous both within and among populations, except on small islands. The American *Trillium* species rarely form hybrids even when they coexist in nature, but the Japanese *Trillium* hybridize readily, and have developed an allopolyploid mode of speciation. Why do such different modes of speciation occur in North America and Asia? American *Trillium* species are all diploid and predominantly self-pollinated, while Japanese *Trillium* species are predominantly outcrossed. The reproductive system appears to be an important factor in determining the differences in mode of speciation. Differences in amount of allopatry, extent of habitat heterogeneity, and levels of outcrossing, appear to have led to differences in modes of speciation for North American and Asian *Trillium* species.

ACKNOWLEDGMENT

It is my pleasure to present this paper in honor of Professor Herbert G. Baker, University of California, Berkeley. Almost twenty years ago I was his post-doctoral student. Having had only cytogenetics, I learned much ecology and evolution under his guidance. This paper is a combination of genetics, ecology and evolution, and was written while I was a Visiting Professor of Biology at Rhodes College in Memphis, Tennessee. I would like to acknowledge the faculty of Rhodes College for their hearty hospitality. Thanks are due also to Jane H. Bock, Meredith Lane, and Yan B. Linhart, University of Colorado at Boulder, for comments on the manuscript.

LITERATURE CITED

Darlington, C. D. and L. F. La Cour. 1940. Nucleic acid starvation of chromosomes in *Trillium*. J. Genet. 40: 185-213.

Flint, R. F. 1957. Glacial and Pleistocene Geology. Wiley, New York.

Freeman, J. D. 1975. Revision of *Trillium* subgenus *Phyllantherum* (Lilliaceae). Brittonia 27: 1-62.

Freeman, J. D. 1984. Sessile-flowered *Trillium* in Alabama. AWS Newsletter 17: 2-5.

Fukuda, I. 1961. On insects visiting *Trillium* flower. Essays and Studies. Tokyo Woman's Christian Coll. 12: 23-34.

Fukuda, I. 1962. Variation and evolution in natural population of *Trillium kamtschaticum*. Essays and Studies. Tokyo Woman's Christian Coll. 13: 91-108.

Fukuda, I. 1967. The formation of subgroups by the development of inbreeding systems in a *Trillium* population. Evolution 21: 141-147.

Fukuda, I. 1969. Chromosome compositions of natural populations in *Trillium ovatum*. Sci. Rep. Tokyo Woman's Christian Coll. 10: 110-137.

Fukuda, I. 1971. The spontaneous chromosome mutation in *Trillium ovatum*. Chromosome Information Service 12: 4-5.

Fukuda, I. 1984. Chromosome banding and biosystematics. In W.F. Grant [ed.]. Plant Biosystematics. Academic Press, New York.

Fukuda, I. 1988. Chromosome compositions of natural populations in *Trillium erectum*. Sci. Rep. Tokyo Woman's Christian Univ. 76-80: 943-961.

Fukuda, I. and R.B. Channell. 1975. Distribution and evolutionary significance of chromosome variation in *Trillium ovatum*. Evolution 29: 257-266.

Fukuda, I. and F. Grant. 1980. Chromosome variation and evolution in *Trillium grandiflorum*. Can. J. Genet. Cytol. 22: 81-91.

Fukuda, I. and Y. Kozuka. 1958. Evolution and variation in *Trillium*. A list of chromosome composition in natural populations of *Trillium kamtscaticum*. Fac. Sci. Hokkaido Univ., Sec. V 6: 273-319.

Hiraizumi, Y., T. Narise and I. Fukuda. 1961. Heterotic viability in natural populations of *Trillium kamtschaticum*. Pall. Jap. J. Genet. 36: 413-418.

Kurabayashi, M. 1960. Evolution and variation in Japanese species of *Trillium*. Evolution 12: 286-310.

Stebbins, G. L. 1974. Flowering Plants: Evolution above the Species Level. Harvard Univ. Press. Cambridge, Massachusetts.

POLYPLOID POLYMORPHISM IN GRASSES OF THE NORTH AMERICAN PRAIRIE

Kathleen H. Keeler and Binita Kwankin

INTRODUCTION

Species are generally regarded as having a single, typical and characteristic chromosome number and nuclear DNA content (Curtis and Barnes, 1985; Strickberger, 1985; Russell, 1987). Indeed, the species-specific nuclear DNA content is referred to as the constant (C) value (Cavallier-Smith, 1985; Price, 1988). However, some species vary substantially in intraspecific DNA content. In higher plants, this is often a result of intraspecific polyploidy (e.g., Lewis, 1976, 1980).

The relationship between intraspecific polyploid variation and other plant characteristics ranges from simple parallel occurrences (e.g., Porter, 1966; Fukuda, 1967) to complex character combinations (Lewis, 1976; Gould and Kapadia, 1964). Despite the breeding barriers that result from intraspecific polyploidy (Stebbins, 1971), in the cases where morphological and reproductive characters have a complex relationship to polyploidy, taxonomic recognition of polyploidy is considered impractical (Beaudry, 1963; Mosquin, 1966; Lewis, 1967, 1980). A surprising number of species fall into this category (references in Lewis, 1980; McArthur et al., 1986; Keeler et al., 1987).

The function of intraspecific and intrapopulational polyploidy is not well understood. An array of cytological traits generally follow from the presence of more DNA in the nucleus e.g., large cell volume and slower cell cycle (Stebbins, 1971; Lewis, 1980; Cavallier-Smith, 1985; Bennett, 1987; Price 1988). In addition, higher ploidy levels tend to occur in newer, more disturbed sites (Ehrendorfer, 1980; Brown and Marshall, 1981; Stebbins, 1985; Hodgson, 1987). These observations are consistent with an adaptive function for higher polyploids, if not for polymorphism.

The prairie grasses of central North America are one group in which a number of important species have been reported to vary in polyploidy (Stebbins, 1975; Sutherland, 1986). This paper investigates polyploid polymorphism in The Great Plains generally, and then focuses specifically on a single case. Similar patterns in different species and different groups suggest that functional explanations are possible for this phenomenon. First, we compared the evidence for polyploid polymorphism in three major families, the Poaceae, Asteraceae and Fabaceae, of North American grasslands. Secondly, since macrogeographic patterns must ultimately be produced by forces acting on individual plants, we determined the distribution of intraspecific polyploidy within and between sites across the range of one species, big bluestem, *Andropogon gerardii*. The results of these two investigations are aimed at determining the function, if any, of intraspecific polyploid polymorphism, and determining whether intraspecific polyploidy is an adaptive strategy of some sort, a transitory state indicating evolution or speciation in progress, or a selectively neutral condition.

SURVEY OF POLYPLOID POLYMORPHISM IN GREAT PLAINS PLANTS
Rationale
We determined the distribution of polyploid polymorphism because its widespread occurrence would suggest causation by a factor or factors shared by several plant species. While the existence of many species showing the same trend might indicate a shared adaptive response, it might also indicate shared conditions permitting tolerance of a neutral polymorphism. In any case explanations sought for a widespread characteristic are distinctly different from those for a rare phenomenon. With this in mind, we estimated the frequency of polyploidy in the three most important Great Plains plant families; grasses, composites and legumes.

Methods
Reports of intraspecific polyploid variation for Great Plains plants were sought in the literature. Frequency of chromosomal variation was estimated for the 21 grass species named by Weaver (1956, 1959) as dominant or characteristic of the major ecosystems of central North America (tallgrass prairie, midgrass prairie and shortgrass prairie and associated wet meadows, Table 1). To determine whether intraspecific chromosomal variation was characteristic, samples of two other major families in central North American prairies, composites (Asteraceae) and legumes (Fabaceae), were drawn from Lommasson (1973) (Tables 2 and 3). This method assumed that the plants portrayed in a popular book are the most conspicuous ones, and therefore some of the most common. Annual and biennial species were excluded because the important grasses (Table 1) are all perennial. The list of composites was reduced to 21 using random numbers. Since fewer than 21 legumes are portrayed in Lommasson (1973), additional

Table 1. Chromosomal Variation in Dominant Prairie Grasses. List of species based on Weaver 1956, 1959

Scientific Name	Common name	Chromosomes reported	References
Agropyron smithii (Rydb.)	western wheatgrass	56	Nielsen & Humphrey 1937, Gillette & Senn 1960, Dewey 1975
Andropogon gerardii (Vitman)	big bluestem	20, 40, 60, 70, 80, 84-86, 90	Church 1929b, 1940, Nielsen 1939, Gould 1956, 1968a,c, Löve 1964, Riley & Vogel 1982, Kwankin 1985, Keeler et al. 1987
Andropogon hallii Hack.	sand bluestem	60, 70, 100	Nielsen 1939, Church 1940, Brown 1950, Gould 1956, Reeder 1977
Andropogon scoparius Michx.	little bluestem	40	Church 1936, 1940, Nielsen 1939, Brown 1950, Gould 1956
Aristida purpurea Steud.	purple three-awn	22	Gould 1958
Bouteloua curtipendula (Michx.)Torr.	side-oats grama	20-21, 28, 35, 40-46, 48, 50, 52-56, 58-60, 62, 64, 70, 74, 76, 78, 80, 82, 84, 86-92, 94, 96, 98, 100-102, 104	Nielsen & Humphrey 1937, Fults 1942, Harlan 1949, Snyder & Harlan 1953, Freter & Brown 1955, Gould & Kapadia 1964
Bouteloua gracilis (H.B.K.) Griffiths	blue grama	20, 21, 28, 35, 40, 42, 60, 61, 77, 84	Fults 1942, Snyder & Harlan 1953

Species	Common name	Chromosome numbers	References
Bouteloua hirsuta Lag.	hairy grama	20-21, 28, 37, 40, 42, 46	Fults 1942, Gould 1958, Roy & Gould 1970, Reeder 1984
Buchlöe dactyloides (Nutt.) Englm.	buffalo grass	40, 56, 60	Nielsen 1939, Reeder 1967, 1968, 1984, Löve & Löve 1981a
Elymus canadensis L.	Canada wild rye	28, 42	Nielsen & Humphrey 1937, Nielsen 1939, Brown 1948, Hartung 1946, Bowden 1960, 1964
Koeleria cristata (Lam.) Beauv.	June grass	14, 28	Stebbins & Love 1941, Bowden 1960, Taylor & Brockman 1966
Panicum virgatum L.	switchgrass	18, 21, 25, 30, 32, 36, 54, 70, 72, 90, 102	Church 1940, Nielsen, 1944, Brown 1948, McMillian & Weiler 1959, Barnett & Carver 1967, Brunken & Estes 1975, Löve & Löve 1981b, Riley & Vogel 1982
Phalaris arundinacea L.	reed canarygrass	14, 27-31, 35, 42	Church 1929a, Nielsen & Humphrey 1937, Hanson & Hill 1953, Kadir 1974
Sorghastrum nutans Nash		40	Church 1929b, Brown 1950, Gould 1956, Celarier 1959, Bowden 1960, Löve & Löve 1981b, Riley & Vogel 1982
Spartina pectinata Link	prairie cordgrass	28, 40, 40+, 42, 84	Church 1929a, 1940, Marchant 1968, 1970, Löve & Löve 1981b

Sporobolus asper (Michx.)Kunth	tall dropseed	54, 108	Brown 1950, Riggins 1977, Davidse 1981
Sporobolus cryptandrus (Torr.) A Gray		18, 36, 38	Nielsen 1939, Brown 1950, Bowden 1960, Gould 1966, Löve & Löve 1981b
Sporobolus heterolepis (Gray) Gray	prairie dropseed	72	Nielsen 1939, Löve & Löve 1981b
Stipa comata Trin. & Rupr.	needle-and-thread	44, "44-46"	Stebbins & Löve 1941, Bowden 1960, Löve & Löve 1981b
Stipa spartea Trin	porcupine grass	44, 46	Love unpub. in Myers 1947, Löve & Löve 1981b
Stipa viridula Trin.	green needlegrass	82, 88	Johnson & Rogler 1943, Löve & Löve 1981b

species were drawn from McGregor (1986) using random numbers and excluding plants whose range within the Great Plains region is very restricted. A search for another comparable monocot family was unsuccessful as very few of the Cyperaceae of the Great Plains have reported chromosome numbers (c.f. Kolstad, 1986) and no other monocot group is among the top 20 families in the region.

Nomenclature is according to Barkley (1986), McGregor (1986) and Sutherland (1986). Polymorphism was indicated if more than 5% of the chromosome reports were of variants. Number of plants for which chromosome numbers have been determined is generally so low that the report of one variant constitutes polymorphism. Polyploid polymorphism was indicated if the variants were simple multiples of the base number (x). Base numbers were taken from Stebbins (1975, 1981, 1985) and Gould (1968b).

Results

Of the 21 prairie grass species investigated, chromosomal variants have been reported for 17 (Table 1). Of the 16 with chromosomal variation, three (*Stipa comata, S. spartea*, and *S. viridula*) are polymorphic but without polyploidy. The remaining 13 show a polyploid series among the variants. *Bouteloua curtipendula, B. gracilis*, and *Panicum virgatum* have substantial aneuploidy in addition to polyploid variation. Thus, polyploid polymorphism occurs in 67% of the dominant prairie grasses of North America.

In the Asteraceae (Table 2) 10 of 21 species (48%) have been reported to have more than one chromosome number, while for seven, multiples that might reflect polyploid polymorphism are present. It should be noted that only a few species (e.g., *Aster ericoides, Happlopappus spinulosus*) have had enough individuals counted to really establish the existence of polymorphism. In most cases only two or three plants, from different locations, have been counted (e.g., *Veronica fasciculata*, Löve and Löve, 1982). Nevertheless, it appears that 29% of the common composites are polyploid polymorphic.

For the legumes (Table 3), only *Apios americana* and *Astragalus mollissimus* have more than one reported chromosome number. *Apioes americana* is clearly aneuploid polymorphic rather than polyploid polymorphic, so that among the legumes, only *Astragalus mollissimus* can be considered polyploid polymorphic.

Thus, there are dramatic differences between the frequency of chromosome variation reported for grasses and two of the most

Table 2. Chromosomal variation in common Great Plains composites. Plants used are those in Lommasson (1973), annuals and biennials excluded, reduced to 21 randomly.

Scientific Name	Common name	Chromosomes reported	References
Antennaria neglecta Greene	pussy-toes	28	Stebbins 1932, Bayer & Stebbins 1981, Löve and Löve 1982, Bayer 1984
Aster ericoides L.	white aster	10, 20	Semple & Brouillet 1980a,b,
Aster novae-angliae L.	New England aster	8, 10, 20	Semple & Brouillet 1980a,b, Jones 1980
Cacalia plantaginea (Raf.)	Shinners Indian plantain	54	Coleman 1964, Pippen 1978
Cirsium canescens Nutt.	Platte thistle	34, 35, 36	Moore & Frankton 1963, Ownbey & Hsi 1963
Coreopsis tinctora Nutt.	plains coreopsis	12, 24, 26	Keil & Stuessy 1977, Löve and Löve 1982, Schaack 1982, Strother 1983
Crepis runcinata (James) T. & G.	hawk's beard	22	Babcock & Jenkins 1943, Tomb et al. 1978
Eupatorium perfoliatum L.	boneset	20, 58	Grant 1953, King et al. 1976, Sullivan 1976, Löve and Löve 1982

Species	Common name	Chromosome number	References
Happlopappus spinulosus (Pursch) DC.	cutleaf ironplant	8, 8 + 1, 10, 12, 16	Jackson 1957, Raven et al. 1960, De Jong and Longpre 1963, Pinkava & Keil 1977
Helenium autumnale L.	sneezeweed	32, 34, 36	Darlington & Janaki Ammal 1945, Löve and Löve 1982
Helianthus grosseserratus Martens	saw-toothed sunflower	32, 34	Cooper & Mahony 1935, Heiser & Smith 1955, Long 1961
Helianthus rigidus (Cass.) Desf.	stiff sunflower	102	Heiser & Smith 1955, Löve and Löve 1982
Helianthus tuberosus L.	jerusalem artichoke	102	Heiser & Smith 1955
Heliopsis helianthoides (L.)Sweet	ox-eye	28	Cooper & Mahony 1935, Löve and Löve 1982
Kunhia eupatorioides L.	false boneset	18	Barkley 1986
Lactuca oblongifolia Nutt.	blue lettuce	18	Barkley 1986
Ratibida pinnata (Vent.) Barnh.	gray-headed prairie cone-flower	28	Richards 1968
Rudbeckia hirta L.	black-eyed susan	38	Löve & Löve 1982
Senecio plattensis Nutt.	prairie ragwort	46, 92	Kowal 1975
Solidago gigantea Ait.	late goldenrod	18, 36, 54	Beaudry & Chabot 1959, Beaudry 1963, Melville & Morton 1982
Veronia fasciculata Michx.	ironweed	34	Löve & Löve 1982

Table 3. Chromosomal variation in common legumes of the Great Plains Plants. Plants taken from Lommasson (1973), excluding annuals and biennials, 3 plants randomly taken from *Flora of the Great Plains* to make 21.

Scientific Name	Common name	Chromosomes reported	References
Amorpha canescens Pursh	lead plant	20	Löve and Löve 1982
Apios americana Medic.	American potato bean	22, 33	Lewis et al. 1962, Seabrook & Dionne 1976
Astragalus adsurgens Pall.	standing milkvetch	32	Federov 1974, Spellenberg 1981
Astragalus canadensis L.	Canadian milkvetch	16	Vilkomerson 1943, Head 1957, Ledingham 1957, Spellenberg 1981, Löve and Löve 1982
Astragalus crassicarpus Nutt.	ground plum	22	Vilkomerson 1943, Löve and Löve 1982
Astragalus mollisimus Torr.	woolly locoweed	22, 24	Head 1957, Ledingham & Fahselt 1964, Weedin & Powell 1980, Spellenberg 1981
Astragalus racemosus Pursh	alkalai milkvetch	24	Vilkomerson 1943, Ledingham 1957
Baptisia bracteata Muhl. ex. Ell.	long-bracted wild indigo	18	McGregor 1986
Baptisia lactea (Raf.) Thieret	white wild indigo	18	McGregor 1986

Dalea purpurea Vent. (formerly *Petalostemon*)	purple prairie clover	14	Ledingham 1957, Taylor & Brockman 1966, Löve and Löve 1982, Ward 1983
Dalea villosa (Nutt.) Spreng (formerly *Petalostemon*)	silky prairie clover	14	Ledingham 1947, Löve and Löve 1982
Desmanthus illinoensis (Michx.) MacM.	prairie mimosa	28	McGregor 1986
Desmodium canadense (L.) DC.	Canadian tick clover	22	Young 1940, Rotar and Urata 1967, Löve and Löve 1982
Desmodium illinoensis Gray	Illinois tick clover	22	McGregor 1986
Glycyrrhiza lepidota Pursh	wild licorice	16	Heiser & Whitaker 1948, Ledingham 1957, Löve & Löve 1982 McGregor 1986
Lathyrus polymorphus Nutt.	hoary vetchling	14	
Lupinus plattensis S. Wats.	Platte lupine	48	Ward 1983
Psoralea argophylla Pursh	breadroot scurfpea	22	Ledingham 1957, Löve and Löve 1982
Psorealea esculenta	breadfruit scurfpea	22	Ledingham 1957, Löve & Löve 1982
Schrankia nuttallii (DC.) Standl.	sensitive brier	26	McGregor 1986
Thermopsis rhombifolia Nutt.	prairie buck bean	18	Ledingham 1975, Löve & Löve 1982 Löve & Löve 1982

abundant plant families of the Great Plains region, composites and legumes. There is significantly more polyploid polymorphism reported among common prairie grasses than among common composites (X^2 = 6.1, d.f. = 1, p<0.025) or legumes (X^2 = 17.5, d.f. = 1, p<<0.001). These grasses include the most successful plants of the region in numbers, biomass, and range. The co-occurrence of a peculiar genetic structure suggests some important shared factor. There are two basic alternative explanations. The situation must be tolerated in surviving plants, so either polyploid polymorphism is itself adaptive or it is a neutral trait unimportant to survival. Its lower frequency in the other families suggests that in them it is not selectively adaptive or tolerated as neutral. If the polymorphism is being driven by selection, it might be that the polymorphism itself is adaptive (maintained by, for example, different tolerances of the cytotypes). However, it is also possible that, although produced by selection, this polymorphism is transient, with one cytotype, presumably the higher polyploid, driving out the other in the long run.

DISTRIBUTION OF POLYPLOID POLYMORPHISM WITHIN
Andropogon gerardii
Rationale

Since the abundance of polyploid polymorphism within prairie grasses suggests a variety of explanations, we looked in detail at one species to determine the actual degree of intraspecific polyploidy and its distribution.

The obvious choice was the dominant grass of the tallgrass prairie, big bluestem (*Andropogon gerardii* Vitman). This tall (to more than 2 m.) C_4 grass ranges from the U.S. Atlantic coast to the Rocky Mountain foothills and from the Texas Gulf coast to the Canadian prairie provinces (Pohl, 1968). As the most abundant plant in North American tallgrass prairie, big bluestem formed 60-80% of the biomass on drier upland sites within the tallgrass prairie biome, over 90% in the low prairie there, and occurred in scattered stands in wet areas throughout mixed and shortgrass prairie (Weaver and Fitzpatrick, 1934). Big bluestem has been reported to have 20 to 86 chromosomes (Church, 1929b, 1940; Nielsen, 1939; Brown, 1950; Gould, 1956, 1968a,c; Löve, 1964; Riley and Vogel, 1982; Kwankin, 1985; Keeler et al., 1987), but 60 was taken as normal and the other reports considered rare variants (c.f. Pohl, 1968; Hitchcock, 1971; Gould, 1967, 1968b; Riley and Vogel,1982; Sutherland, 1986).

In our first investigation, we found individuals with 7 to 11 pg DNA/nucleus by flow cytometry, and showed nuclear DNA content to correspond to 60 to 90 chromosomes (Kwankin, 1985, Keeler et al., 1987). While a polymorphism of hexaploids, septaploids, etc. is not especially tractable, and will not readily fit theory written for diploids and tetraploids, it is not an unusual situation in the Great Plains (Table 2), where higher polyploids are relatively common.

Plants for this study were gathered in the field in September and October, by digging up one rhizome and its flowering stalk from 25 plants along a line at each site. Plants sampled were at least 3 m apart, spacing chosen to ensure each rhizome represented a different individual (Keeler, unpub. data). Plants were established in the greenhouse. Leaf samples for cell sorting used field-collected leaves when possible. The greenhouse material was kept to allow retesting of individuals as needed.

Sites

Nine-Mile Prairie. Nine-Mile Prairie is a 230 acre (93 hectare) reserve of unplowed tallgrass prairie within the city limits of Lincoln, Nebraska (Lancaster Co.). It has a history of moderate grazing by cattle, was mowed for hay in 1981-83, and burned in 1984. Annual rainfall averages approximately 684 mm (30 year mean 1950-1981, National Oceanographic and Atmospheric Commission [NOAC], 1985d). Samples were collected in fall, 1984. Sites 1-3 represent three widely-scattered transects within Nine-Mile Prairie.

Pawnee Prairie. This prairie and forest reserve is in southern Nebraska near Beatrice, Pawnee Co. It has never been plowed, but at one time was overgrazed. Since 1964, the site has been managed by Nebraska Game and Parks Commission and has been burned at 3-5 year intervals since 1977. The average annual rainfall is 766 mm (NOAC, 1985d). Plants were collected on a line in upland prairie from the northwest edge of the grassland area (Site 1), in the grassy corridor along a little-used dirt road through patches of forest (Site 2), and in the meadow by the parking area at the northwest corner of the reserve (Site 3).

Konza Prairie. Konza Research Natural Area is in Geary County, close to Manhattan, Kansas. This reserve is a National Science Foundation Long Term Ecological Research site, devoted especially to the study of fire ecology. The site has a history of grazing by cattle, and present and historical managers have used fire as a management technique. Average annual rainfall is 835 mm (NOAC 1985c). The plants were collected in October 1985 along a rocky, north-facing slope burned in 1984 (Site 1), on an exposed high area, burnt in 1984 (Site 2), in an exposed, high area rich in Indian grass, *Sorghastrum nutans*, not burned since 1980 (Site 3), and in a low wet area close to a tree-shaded stream, burned in 1983 (Site 4).

Doolittle Prairie. Doolittle Prairie, Story Co., Iowa, is an unplowed tallgrass prairie remnant managed by the Iowa Department of Natural Resources. Plants were collected in October 1987. Average annual rainfall is 983 mm. (NOAC, 1985b).

Williams Prairie. Williams Prairie, near Oxford, Iowa, (Johnson Co.) is a low wet prairie dominated by prairie cordgrass (*Spartina pectinata*) on the north and east and by big bluestem in the south and west. It was burned in spring of 1987. Rainfall at the site is approximately 915 mm per year (NOAC, 1985b). Plants were collected in August 1987.

Franzel Prairie. Franzel Prairie, Muscatine Co., Iowa, is an unplowed quarter section managed by the Muscatine Natural Resource District. Historically it was kept as a hay meadow; management began in 1986. It was burned in spring, 1987. Mean annual precipitation is 887 mm (NOAC, 1985b). Plants were collected in October 1987.

Illinois Sites. The Illinois plants were gathered along county roads in Champaign Co.: Site one mile south of Illinois Route 10, Site 2 near Bondville, Il. Mean annual rainfall is 950 mm (NOAC, 1985a). Plants were collected August 1987.

Cytological Methods

For cell sorting, fresh leaves (a minimum of 1 cm^2) were prepared according to the method of Galbraith (see Keeler et al., 1987). The technique is replicable within and between tissues of a single plant, the error between such readings being approximately that of repeated runs of the same tissue (Keeler et al. 1987). Field and greenhouse vegetative tissue are indistinguishable (Keeler, unpub. data). Differences in nuclear DNA content measured by cell sorting all corresponded to differences in chromosome number as counted by standard light microscopy (Kwankin, 1985; Keeler et al., 1987). For plants for which chromosomes were counted microscopically to date, almost all have been multiples of the base number (10); only one of the 80 individuals counted appears to be aneuploid (Kwankin, 1985).

The correlation between chromosome number and number of picograms of DNA in the nucleus was 0.97 (p<0.01). The plants at Nine-Mile Prairie which were found to have 10-11 picograms of DNA per nucleus usually had 80 chromosomes, but a few (1 in 6) had 90 chromosomes. Thus, while most of the DNA variation is explained by chromosome number, not all is. There are in fact two distinct polymorphisms, one of 7 vs 10 pg DNA per nucleus, another of chromosome number, which includes a polymorphism among high DNA individuals in having 80 or 90 chromosomes. Low DNA individuals, however, are not polymorphic for chromosome number: they have 60 chromosomes. The study to date has dealt primarily with the distribution of the population among high- and low-DNA individuals, using cell-sorting. Thus, for the purposes of this paper we will speak of high-DNA and low-DNA individuals. This obscures that fact that the DNA differences are apparently due to chromosome number, but is a precisely accurate reflection of the way the data were collected.

Analyses

The chi-square statistic (Sokal and Rohlf, 1969) was used to compare samples and prairies, using the means in Table 4. Variable numbers per sample are the results of deaths among the plants gathered. The original Nine-Mile Prairie sample was a composite of various smaller samples and so was not included in the analysis. Two sites (Sites 1 of Nine-Mile and Konza) had expectations below 5 and were excluded from the test of within-prairie uniformity. Since the prairies are significantly nonuniform without these sites, their inclusion would not affect the outcome.

Results

In the geographic survey of polyploid polymorphism in *Andropogon gerardii*, two results are apparent (Table 4). First, there is dramatic within-site variation and, second, populations with substantial frequencies of high-DNA plants occur only on the western edge of the tallgrass prairie region.

Within Konza, Nine-Mile and Pawnee prairies, the microsites were statistically significantly different from each other (Nine-Mile $X^2 = 5.52$, d.f. = 1 (Site 1 excluded), p <0.05 Pawnee $X^2 = 6.54$, d.f. = 2, p <0.05; Konza $X^2 = 25.19$, d.f. = 2 (Site 1 excluded), p <0.001). Information on environmental variation taken while the plants were being collected showed no consistent relationship to the differences; this is being pursued.

Since there is so much variation within the western sites, the eastern sites cannot be shown to differ significantly from them in frequency of higher polyploids. Nevertheless, the frequency of higher polyploids goes from 1-2 plants per site in the east (5-

Table 4. Geographic distribution of polyploid cytotypes in big bluestem as determined by flow cytometry

Location Site	DNA Content in Picograms (Percent)		Total
	ca.7 pg number (%)	ca.10 pg number (%)	
CENTRAL KANSAS			
KONZA PRAIRIE **MANHATTAN KS**			
Site 1	8(100%)	0 (0%)	8
Site 2	1 (6%)	16 (94%)	17
Site 3	11 (48%)	12 (52%)	23
Site 4	17 (89%)	2 (11%)	19
total (mean %)	37 (55%)	30 (45%)	67
EASTERN NEBRASKA			
NINE-MILE PRAIRIE **LINCOLN NE**			
Initial samples*	17 (44%)	22 (56%)	39
Site 1	8 (89%)	1 (11%)	9
Site 2	13 (65%)	7 (35%)	20
Site 3	9 (31%)	20 (69%)	29
3-site total (mean%)	30 (52%)	28 (48%)	58
PAWNEE PRAIRIE **BEATRICE NE**			
Site 1	13 (62%)	8 (38%)	21
Site 2	11 (8%)	1 (92%)	12
Site 3	9 (50%)	9 (50%)	18
total (mean %)	33 (61%)	18 (39%)	51
CENTRAL IOWA			
DOOLITTLE PRAIRIE, **STORY CO. IA**	19 (100%)		19
EASTERN IA			
FRENZEL PRAIRIE **MUSCATINE CO., IA**	16 (89%)	2 (11%)	18
WILLIAMS PRAIRIE **LINN CO., IA**	19 (95%)	1 (5%)	20
ILLINOIS			
ROADSIDE **CHAMPAIGN CO., IL**			
Site 1	19 (95%)	1 (5%)	20
Site 2	20 (1 %)	0	20
total (mean %)	39 (98%)	1	40

*Keeler et al. 1987

10%) to an average of 40% in the western part of the range of big bluestem.

DISCUSSION

Polyploid polymorphism occurs in the majority of important prairie grasses (14 of 21, Table 1). The grasses listed in Table 1 are species designated as dominants for the major grassland ecosystems of North America. Thus, these may be regarded as the most successful plants of this region, (e.g., Weaver and Fitzpatrick, 1934; Weaver, 1956, 1959; Risser et al., 1980). It appears that grasses differ from other large families (Asteraceae, Fabaceae) in being polyploid polymorphic. While these data are based on samples from families studied, rather than comprehensive lists, the patterns are so striking that the difference seems unlikely to be a sampling effect.

Polyploidy has long been recognized as a key factor in the evolution of higher plants, including grasses (e.g., Stebbins, 1971, 1975, 1981). North American grasslands are among the most recently developed ecosystems in North America; although grassy glades and parklands existed throughout the Tertiary, the extensive open prairies dominated by grasses, stressed by recurrent drought and swept by fire, have apparently only existed in their present position and composition since the last glacial retreat, perhaps 7,000 years ago (Wells, 1970; Stebbins, 1975, 1981; Axelrod, 1981; Baker and Waln, 1985). This is especially true for the big bluestem dominated tallgrass prairie (Axelrod, 1981; Baker and Waln, 1985).

Andropogon gerardii in particular must have spent the Wisconsin glacial maximum in the south and east corner of its present range, in theAtlantic Coast area shared with its close relatives. Whether big bluestem existed as a species before the Wisconsin is difficult to say. The immediate relationships of *A. gerardii* have not been determined, though it may be closely related to *A. hallii* (see below). Both stand as morphologically distinct members of their subgenus (Gould, 1956). In the absence of presumed parental species, the age and nature of the polyploid polymorphism is not easy to assess. However, in the subgenus of *Andropogon* to which *A. gerardii* and *A. hallii* belong (*Arthrolophis*), there appears to have been post-glacial range expansion northward and westward, and polyploidization during that expansion. Five species (*A. elliottii, A. floridanus, A. glomeratus, A. tracyi,* and *A. virginicus*) are reported as having 2N = 20 chromosomes (Church, 1936; Darlington and Janaki Ammal, 1945; Brown, 1950; Cave, 1963; Davidse and Pohl, 1972a,b; Moore, 1973; Fedorov, 1974; Stebbins, 1975). These five species are apparently native to the Atlantic Coastal Plain of the U.S., although several have become naturalized elsewhere. Seven species native to Florida have unknown chromosome numbers (c.f. Gould, 1956). Of the remaining three species, *Andropogon ternarius,* found on the Atlantic coastal plain from Delaware to Florida and Texas, inland to Kansas (Hitchcock 1971), is reported as having 2N = 40 chromosomes, with occasional counts of 20 and 60 (Church, 1940; Gould, 1956).

Only *Andropogon gerardii* and *A. hallii* are hexaploid (2N = 60) (Church, 1940; Gould, 1956, 1967, 1968b; Stebbins, 1975; Riley and Vogel, 1982; Kwankin, 1985; Keeler et al., 1987). *Andropogon gerardii* ranges from the Atlantic Coast in New York to the Rocky Mountain foothills in Wyoming and Colorado, from the prairie province of Canada to the Texas Gulf Coast and Florida (Hitchcock, 1971). *Andropogon hallii* is restricted to sandy soils, generally to the west of *A. gerardii,* from Canada to Texas (Hitchcock, 1971). Close similarity has long been noted between the two (e.g., Peters and Newell, 1961; Gould, 1968b; Pohl, 1968). Barnes (1986) demonstrated that the two hybridize freely in areas of natural contract. The "biological species concept" interpretation of Barnes' findings would be to consider *A. hallii* a sand ecotype of *A. gerardii.* Sutherland (1986) acknowledges the principle, but declines to merge the taxa. Since the two forms are ecologically and morphologically easily distinguished, we also retain *A. hallii* but consider it to be derived from *A. gerardii.* Chromosomal counts are consistent with this as *A. hallii* is variously reported as 2N = 60, 70 and 100 (Nielsen, 1939; Church, 1940; Brown, 1950; Gould, 1956).

One interpretation of the pattern in the subgenus *Arthrolophis* is of expansion northward and westward, perhaps coinciding with the postglacial warming, accompanied by polyploidy in the later or at least westernmost species. *Andropogon gerardii* has both the highest chromosome number in the subgenus and is among the most chromosomally variable (Table 4). If this interpretation is true, higher ploidy at the western edge of the range is the latest event in a trend of westward expansion and polyploidy, consistent with the views of Stebbins (e.g., 1975).

Whatever its history since the end of the Pleistocene, big bluestem has experienced great success. This species is particularly well adapted to fire, and will often make up virtually 100% of the biomass on a lowland tallgrass prairie after consecutive years of burning (L. Hulbert, pers. comm.). Even without annual fire, big bluestem dominated the tallgrass prairies (Weaver and Fitzpatrick, 1934; Risser et al., 1980).

Three competing explanations of polyploid polymorphism in this abundant and widespread species can be proposed. Since polyploid polymorphism is common among prairie grasses (Table 1), it is unlikely that the explanation of polyploid polymorphism for big bluestem is unique.

(1) Polyploid polymorphism might represent an adaptation fo heterogeneous environmental conditions. For example, Lewis (1976) has argued that populations polymorphic for polyploidy exploit broader ranges of environmental conditions than do individuals of any one chromosome number. This was the hypothesis that initiated this project, since polyploid polymorphism is so common within the grasses of the Great Plains (Table 1). Many cellular differences are correlated with intraspecific polyploidy, but have not been shown to explain it (Stebbins 1971, 1985; Ehrendorfer, 1980; Goldblatt, 1980; Lewis, 1980; Bennett, 1987). No simple ecological pattern was revealed based on observations of the site conditions (Konza 1 and 3 were "dry upland", 2 and 4 "low, moist"; Frenzel Prairie is quite sandy (dry), Williams Prairie is low and very wet; see Table 4).

The most polymorphic populations are on the parts of the range nearest to the ecological limits of big bluestem. Big bluestem's prehistoric western limit was the edge of the tallgrass prairie, roughly 200 km west of the Nine-Mile Prairie site. During the drought of the 1930's, the hilltop along which Site 2 at Nine-Mile was located, was dominated by short and mid grasses; big bluestem survived only on the lower slopes (Weaver 1959). The complex distribution of cytotypes at Nine-Mile prairie (Table 4) suggests both were part of the post-drought recolonization, a pattern more consistent with an adaptive explanation than one of transition or neutrality.

(2) The origin of higher ploidy levels in big bluestem populations might be part of ongoing evolution of the species, with the higher ploidy levels tending to replace the older hexaploids. Support for this model can be seen in finding more higher levels on the western edge of the range (Table 4). The within-site variation in the western sites studied complicates but does not eliminate this explanation.

Generation of higher ploidy levels in the newer areas could be produced either as part of the ongoing evolution of *Andropogon gerardii* or could signal the presence of a second, cryptic species. The existence of a cryptic second species is made less likely by the complexity of the intermingling of cytotopyes (Table 4) -- presumably a liability in a wind-pollinated species -- and the lack of distinguishable morphological characters (Kwankin,

1985; Keeler, unpubl.). What genetic data there are suggest breeding barriers, consistent with two species. The genetic data are limited but we found 1) wild-collected seed of the 60-chromosome individuals were of much higher viability than those from 80-chromosome individuals (Keeler, unpubl.) and 2) wild-collected seeds of 60-chromosome plants yielded only 7 pg DNA plants (N = 10 maternal plants, 40 seedlings), while wild-collected seeds from 80 chromosome plants segregated producing individuals ranging from 7 to 10 pg nuclear DNAs (N = 6 maternal plants, 2, 18 and 8 seedlings, with 7, 8 and 10 pg DNA respectively). The 8 pg DNA individuals were at least sometimes 70-chromosome plants (Kwankin, 1985; Keeler et al., 1987; Keeler, unpubl.). Both observations support a breeding barrier between the cytotypes and the possibility of a cryptic second species.

The sites studied were chosen to avoid the problem of hybridization with sand bluestem (*Andropogon hallii*), and all sites are more than 60 km from a known sand bluestem site. However, in the absence of direct information of *A. gerardii* on sites with *A. hallii*, a long introgressive cline cannot be ruled out.

(3) Polyploidy may be an unimportant character in these plants, since even the 7 pg DNA individuals (hexaploids) have six copies of each gene. Additional copies could represent evolutionary unimportant variation which is neutral in its impact on fitness compared to genic level adaptations to, for example, drought, frost and grazing in these same plants. Evidence for this explanation would chiefly consist of failure of all other explanations.

We cannot presently choose between these alternatives. The breeding barrier suggests this variation is not neutral. The geographic distribution is more consistent with transient polymorphism than an adaptive explanation. The history of the species suggests polyploid polymorphism is found mainly in its most recently colonized areas. This is consistent with patterns noted by Stebbins (1975, 1985) and Bennett (1987) of increased chromosome number or DNA content in newer or less stable areas, but does not explain the phenomenon. It appears that the higher polyploids are actively evolving at the western edge of the range of big bluestem.

The investigation of big bluestem was intended to elucidate the pattern both in big bluestem itself, and in other polyploid polymorphic species as well. The data on big bluestem do not lend themselves to generalizations that would explain other polyploid species.

The Great Plains grasses investigated for polyploid polymorphism actually show similar patterns, in that they do not suggest generalizations. Some geographic patterns have been seen but no functional explanations supported. In *Pancium virgatum*, examination of the polyploid polymorphism (tetraploids, hexaploids and aneuploids of the hexaploid) has indicated that chromosomal races exist. The tetraploids predominate in the southern part of the range, while hexaploid and aneuploid plants are more common in northern areas and in upland areas in the south (Nielsen, 1944, 1947; Eberhart and Newell, 1959; McMillan and Weiler, 1959; Porter, 1966; Barnett and Carver, 1967; Brunken and Estes, 1975). The reason for this pattern is not known. In *Bouteloua curtipendula*, populations having major chromosomal differences can be separated into subspecies (Gould and Kapadia, 1964). However, a substantial range of polyploid variation is found within the resultant subspecies, with some corresponding morphological variation (Gould and Kapidia, 1962, 1964; Kapadia and Gould, 1964). Again, these are patterns without known causes.

Thus, we find that many grass species show polyploid polymorphism, which sets them apart from legumes and composites, and suggests a shared cause. The cause itself is not apparent from any of the species so far examined. This suggests either that polyploid polymorphism is neutral, which its genetic consequences contradict, or a transient condition in a rapidly changing group. This explanation is at least consistent with all existing data.

SUMMARY AND CONCLUSIONS

A surprising number of plant species contain individuals of a variety of levels of polyploidy, so that the species is polymorphic for polyploidy. This paper considers possible general causes for this phenomenon in plants of the Great Plains of North America. First we compared the frequency of intraspecific polyploidy in three major plant families, Poaceae, Asteraceae, and Fabaceae. The majority of grass species (Poaceae) and the minority of composite and legume species (Asteraceae and Fabaceae) are polyploid polymorphic. The frequency of grass species with intraspecific polyploidy suggests shared causation or at least shared tolerance of this character. Having established that the situation occurs frequently in grasses, we analyzed a single species, big bluestem, *Andropogon gerardii*, looking for information on the function of polyploid polymorphism. We used flow cytometry to determine amount of nuclear DNA per cell; in big bluestem this correlates with chromosome number. Individuals with 7 pg nuclear DNA had 2N = 60 chromosomes, and were therefore hexaploid, since in *Andropogon* x = 10. Most other individuals had approximately 10 pg nuclear DNA and 80 (occasionally 90) chromosomes. Seven-pg DNA individuals were predominant everywhere, but the frequency of higher polyploids (10 pg nuclear DNA) went from virtually 0 on the eastern end of the tallgrass prairie region to 40% at the western end. Within native prairies in the west, the full range of variation existed between sites. A recent origin for polyploid polymorphism is suggested in big bluestem and probably other Great Plains grasses, and the data seem most compatible with a transitional situation, although other explanations cannot be ruled out.

ACKNOWLEDGMENTS

We thank G. Drohman for plant care, K. Harkins for the cell-sorting, and Dr. R. Morris for chromosome counting advice. The following provided invaluable assistance finding field sites: T.B. Bragg, U. Nebraska-Omaha, M.F. Willson and L.L. Getz, U. Illinois Champaign-Urbana; D. Kothenbeul, Muscatine, Iowa; L. K. Kapustka, Miami University, Oxford, Ohio; D. Glenn-Lewin, Iowa State U. We thank G.L. Stebbins, Y.B. Linhart, and an anonymous reviewer for comments on the manuscript.

LITERATURE CITED

Axelrod, D. I. 1981. Rise of the grassland biome, central North America. Bot. Rev. 51: 163-201.

Babcock, E. B., and J. A. Jenkins. 1943. Chromosomes and phylogeny in *Crepis* III. The relationships of one hundred and thirteen species. Univ. Calif. Publ. Bot. 18: 241-292.

Baker, R. G., and K .A. Waln. 1985. Quaternary pollen records from the Great Plains and central United States. In V.M. Bryant, Jr. and R.G. Halloway [eds.]. Pollen Records of Late-Quaternary North American Sediments. A.A.S.P. Foundation, New York.

Barkley, T. M. 1986. Asteraceae, Dum., the sunflower family. In Great Plains Floral Association [eds.]. Flora of the Great Plains. Univ. Kansas Press, Lawrence, KS..

Barnes, P. W. 1986. Variation in the big bluestem (*Andropogon gerardii*)-sand bluestem (*Andropogon hallii*) complex along a local dune/meadow gradient in the Nebraska Sandhills. Amer. J. Bot. 73: 172-184.

Barnett, F. L., and R. F. Carver. 1967. Meiosis and pollen stainability in switchgrass *Panicum virgatum* L. Crop Sci. 7: 301-304.

Bayer, R. J. 1984. Chromosome numbers and taxonomic notes for North American species of *Antennaria* Gaertner (Asteraceae: Inuleae). Syst. Bot. 9: 74-83.

120

Bayer, R. J., and G. L. Stebbins. 1981. Chromosome numbers of North American species of *Antennaria* Gaertner (Asteraceae: Inuleae). Amer. J. Bot. 68: 1342-1349.

Beaudry, J. R. 1963. Studies on *Solidago* L. VI. Additional chromosome numbers of taxa of the genus *Solidago.* Canad. J. Genet. Cytol. 5: 150-174.

Beaudry, J. R., and D. L. Chabot. 1959. Studies on *Solidago* L. IV. The chromosome numbers of certain taxa of the genus *Solidago.* Canad. J. Bot. 37: 209-229.

Bennett, M. D. 1987. Variation in genomic form of plants and its ecological implications. New Phytol. 106: 177-200.

Bowden, W. M. 1960. Chromosome numbers and taxonomic notes on northern grasses. III. Twenty five genera. Canad. J. Bot. 38: 541-557.

Bowden, W. M. 1964. Cytotaxonomy of the species and interspecific hybrids of the genus *Elymus* in Canada and neighboring areas. Canad. J. Bot. 42: 547-601.

Brown, A. D. H., and D. R. Marshall. 1981. Evolutionary changes accompanying colonization in plants. In G. G. E. Scudder and J. L. Reveal [eds.]. Proc. 2nd Int. Cong. Syst. and Evol. Biol. Hunt Inst. for Bot. Doc. Carnegie-Mellon Univ., PA.

Brown, W. V. 1948. A cytological study in the Gramineae. Amer. J. Bot. 35: 382-396.

Brown, W. V. 1950. A cytological study of some Texas Gramineae. Bull. Torr. Bot. Club 77: 63-76.

Brunken, J. N., and J. R. Estes. 1975. Cytological and morphological variation in *Panicum virgatum* L. Southwest Nat. 19: 379-385.

Cavallier-Smith, T. 1985. Introduction: the evolutionary significance of genome size. In T. Cavallier-Smith [ed.]. The Evolution of Genome Size. John Wiley and Sons, Chichester.

Cave, M. S. [ed.]. 1963. Index to Plant Chromosome Numbers for 1962. Univ. Carolina Press, Chapel Hill, NC.

Celarier, R. P. 1959. Cytotaxonomy of the Andropogoneae. IV. Subtribe Sorgheae. Cytologia 24: 285-303.

Church, G. L. 1929a. Meiotic phenomena in certain grasses. I. Festuceae, Aveneae, Agrostideae, Chlorideae, and Phalarideae. Bot. Gaz. 87: 608-629.

Church, G.L. 1929b. Meiotic phenomena in certain grasses. II. Paniceae and Andropogoneae. Bot. Gaz. 88: 63-94.

Church, G. L. 1936. Cytological studies in the Gramineae. Amer. J. Bot. 23: 12-16.

Church, G. L. 1940. Cytotaxonomic studies in the Gramineae *Spartina, Andropogon* and *Panicum.* Amer. J. Bot. 27: 263-271.

Coleman, J. R. 1964. Natural and artificial hybrids of *Cacalia atriplicifolia* and *C. muhlenbergii.* Rhodora 67: 55-58.

Cooper, D. C., and K. C. Mahony. 1935. Cyutological observations on certain Compositae. Amer. J. Bot. 22: 843-849.

Curtis, H., and N. S. Barnes. 1985. Invitation to Biology. 4th ed. Worth Publishers Inc., NY.

Darlington, C. D., and E. K. Janaki Ammal. 1945. Chromosome Atlas of Flowering Plants. 2nd ed. George Allen and Unwin, Ltd. London.

Davidse, G. 1981. Chromosome numbers in some miscellaneous angiosperms. Ann. Missouri. Bot. Gard. 68: 222-223.

Davidse, G., and R. W. Pohl. 1972a. Chromosome numbers and notes on some Central American grasses. Canad. J. Bot. 50: 272-283.

Davidse, G., and R. W. Pohl. 1972b. Chromosomal numbers, meiotic behavior and notes on some grasses from Central America and the West Indies. Canad. J. Bot. 50: 1441-1452.

DeJong, D. C. D., and E. K. Longpre. 1963. Chromosome studies in some Mexican Compositae. Rhodora 65: 225-242.

Dewey, D. R. 1975. The origin of *Agropyron smithii* . Amer. J. Bot. 62: 524-530.

Eberhart, S. A., and L. C. Newell. 1959. Variation in domestic collections of switchgrass *Panicum virgatum* L. Agron. J. 51: 613-616.

Ehrendorfer, F. 1980. Polyploidy and distribution. In W. H. Lewis [ed.]. Polyploidy: Biological Relevance. Plenum Press, New York.

Federov, V. 1974. Chromosome numbers of flowering plants. Reprinted by O. Koeltz Science Publishers, West Germany.

Freter, L. E., and W. V. Brown. 1955. A cytotaxonomic study of *Bouteloua curtipendula* and *B. uniflora.* Bull. Torr. Bot. Club 82: 121-130.

Fukuda, I. 1967. The biosystematics of *Achlys.* Taxon 16: 308-316.

Fults, J. L. 1942. Somatic chromosome complements in *Bouteloua.* Amer. J. Bot. 29: 45-56.

Gillett, J. M., and H. A. Senn. 1960. Cytotaxonomy and intraspecific variation of *Agropyron smithii* Rydb. Canad. J. Bot. 38: 747-760.

Goldblatt, P. 1980. Polyploidy in angiosperms: In W.H. Lewis [ed.]. Polyploidy: Biological Relevance. Plenum Press, New York.

Gould, F. W. 1956. Chromosome counts and cytotaxonomic notes on grasses in the tribe *Andropogoneae.* Amer. J. Bot. 43: 395-404.

Gould, F. W. 1958. Chromosome numbers in southwest grasses. Amer. J. Bot. 10: 757-786.

Gould, F. W. 1966. Chromosome numbers of some Mexican Grasses. Canad. J. Bot. 44: 1683-1696.

Gould, F. W. 1967. The genus *Andropogon* in the United States. Brittonia 19: 70-76.

Gould, F. W. 1968a. Chromosome numbers in Texas grasses. Canad. J. Bot. 46: 1315-1325.

Gould, F. W. 1968b. Grass Systematics. McGraw Hill, New York.

Gould, F. W. 1968c. Variability in big bluestem *Andropogon gerardi* Vitman in deKalb and Ogle counties, Illinois. Southwest. Nat. 65: 218-220.

Gould, F. W., and Z. J. Kapadia. 1962. Biosystematic studies in the *Bouteloua curtipendula* complex. I. The aneuploid rhizomatous *B. curtipendula* of Texas. Amer. J. Bot. 49: 887-892.

Gould, F. W., and Z. J. Kapadia. 1964. Biosystematic studies in the *Bouteloua curtipendula* complex. II. Taxonomy. Brittonia 16: 182-207.

Grant, W. F. 1953. A cytotaxonomic study in the genus *Eupatorium.* Amer. J. Bot. 40: 729-742.

Hanson, A. A., and H. D. Hill. 1953. The occurrence of aneuploidy in *Phalaris* spp. Bull. Torrey Bot. Club. 80: 16-20.

Harlan, J. R. 1949. Apomixis in sideoats grama. Amer. J. Bot. 36: 495-499.

Hartung, M. F. 1946. Chromosome numbers in *Poa, Agropyron*, and *Elymus.* Amer. J. Bot. 33: 516-532.

Head, S. C. 1957. Mitotic chromosome studies in the genus *Astragalus.* Madroño 14: 95-106.

Heiser, C. B., Jr., and D. M. Smith. 1955. New chromosome numbers in *Helianthus* and related genera (Compositae). Proc. Indiana Acad. Sci. 64: 250-253.

Heiser, C. B., Jr., and T. W. Whitaker. 1948. Chromosome number, polyploidy and growth habit in some California weeds. Amer. J. Bot. 35: 179-186.

Hitchcock, A. S. 1971. Manual of the Grasses of the United States. 2nd ed. A. Chase [ed.]. Dover, NY.

Hodgson, J. G. 1987. Why do so few plant species exploit productive habitats? An investigation into cytology, plant strategies and abundance within a local flora. Functional Ecol. 1: 243-250.

Hulbert, L. C. 1988. Causes of fire effects in tallgrass prairie. Ecology 69: 46-58.

Jackson, R. C. 1957. Documented chromosome numbers of plants. Madroño 14: 111-112.

Johnson, B. and G. A. Rogler. 1943. A cyto-taxonomic study of an intergeneric hybrid between *Orysopsis hymenoides* and *Stipa viridula.* Amer. J. Bot. 30: 49-56.

Jones, A. G. 1980. A classification of the New World species of *Aster* (Asteraceae). Brittonia 32: 230-239.

Kadir, Z. R. A. 1974. DNA values in genus *Phalaris* (Gramineae). Chromosoma 45: 379-386.

Kapadia, Z. J., and F. W. Gould. 1964. Biosystematic studies in the *Bouteloua curtipendula* complex III. Pollen size as related to chromosome numbers. Amer. J. Bot. 51: 166-172.

Keeler, K. H., B. Kwakin, P. W. Barnes, and D. W. Galbraith. 1987. Polyploid polymorphism in *Andropogon gerardii* Vitman (Poaceae). Genome 29: 374-379.

Keil, D. J., and T. F. Stuessy. 1977. Chromosome counts of Compositae from Mexico and the United States. Amer. J. Bot. 64: 791-798.

King, R. M., D. W. Kyhos, A. M. Powell, P. H. Raven, and H. Robinson. 1976. Chromosome numbers in Compositae. XIII. Eupatorieae. Ann. Missouri Bot. Gard. 63: 862-888.

Kolstad, D. A. 1986. Cyperaceae Juss., the sedge family. In Great Plains Flora Association, [eds.]. Flora of the Great Plains. Univ. Kansas Press, Lawrence, KS.

Kowal, R. R. 1975. Systematics of *Senecio aureus* and the allied species in the Gaspé Penninsula, Quebec. Mem. Torr. Bot. Club. 23: 1-113.

Kwankin, B. 1985. Polyploidy in *Andropogon gerardii* (Poaceae): its effect on distribution, morphology, development and DNA content. M.S. thesis. Biological Sciences, University of Nebraska-Lincoln.

Ledingham, G. F. 1957. Chromosome numbers of some Saskatchewan Leguminosae with particular reference to *Astragalus* and *Oxytropis*. Canad. J. Bot. 35: 657-666.

Ledingham, G..F., and M. D. Fahselt. 1964. Chromosome numbers of some North American species of *Astragalus* (Leguminosae). Sida 1: 313-327.

Lewis, H. 1967. The taxonomic significance of polyploidy. Taxon 16: 267-271.

Lewis, W. H. 1976. Temporal adaptation correlated with ploidy in *Claytonia virginica*. Syst. Bot. 1: 340-347.

Lewis, W. H. 1980. Polyploidy in species populations. In W.H. Lewis [ed.]. Polyploidy: Biological Relevance. Plenum Press, NY.

Lewis, W. H., A. L. Stripling, and R. G. Ross. 1962. Chromosome numbers for some angiosperms of the southern United States and Mexico. Rhodora 64: 147-161.

Lommasson, R. C. 1973. Flowering Plants of Nebraska. University of Nebraska Press, Lincoln.

Long, R. W. 1961. Biosystematics of two perennial species of *Helianthus* (Compositae). II. Natural populations and taxonomy. Brittonia 13: 129-141.

Löve, A. 1964. IOBPChromosome number reports I. Taxon 30:72.

124

Löve, A. and D. Löve. 1981a. IOBP chromosome number reports. Taxon 30: 72.

Löve, A. and D. Löve. 1981b. IOBP chromosome number reports. Taxon 30: 509-511

Löve, A. and D. Löve. 1982. IOBP chromosome number reports. Taxon 31: 344-360.

McArthur, E. D., S. C. Sanderson, and D. C. Freeman. 1986. Isozymes of an autotetraploid shrub, *Atriplex canescens* (Chenopoiaceae). Great Basin Nat. 46: 157-160.

McGregor, R. L. 1986. Fabaceae Lindl., the bean family. In Great Plain Flora Association [eds.]. Flora of the Great Plains. Univ. Kansas Press, Lawrence, KS.

McMillan, C., and J. Weiler. 1959. Cytology of *Panicum virgatum* in central North America. Amer. J. Bot. 46: 590-593.

Marchant, C. J. 1968. Evolution of *Spartina* (Graminaeae). III. Species chromosome numbers and their taxonomic significance. Bot. J. Linn. Soc. 60: 411-417.

Marchant, C. J. 1970. Chromosome pairing and fertility in *Spartina* x *caespitosa*. Canad. J. Bot. 48: 183-188.

Melville, M. R., and J. R. Morton. 1982. A biosystematic study of *Solidago canadensis* complex. I. Ontario populations. Canad. J. Bot. 60: 976-997.

Moore, R. J. 1973. Index to plant chromosome numbers 1967-1971. Dosthoek's Uitgeversmantschappij. S.V., Utrecht.

Moore, R. J., and C. Frankton. 1963. Cytotaxonomic notes on some *Cirsium* species of the western United States. Canad. J. Bot. 41: 1553-1567.

Mosquin, T. 1966. Toward a more useful taxonomy for chromosomal races. Brittonia. 18: 203-214.

Myers, M. W. 1947. Cytology and cytogenetics of forage grasses. Bot. Rev. 13: 319-422.

National Oceanographic and Atmospheric Commission. 1985a. Climatological data. Illinois. 90 (13): 1-32.

National Oceanographic and Atmospheric Commission. 1985b. Climatological data. Iowa. 96 (13): 1-31.

National Oceanographic and Atmospheric Commission. 1985c. Climatological data. Kansas. 99 (13): 1-33.

National Oceanographic and Atmospheric Commission. 1985d. Climatological data. Nebraska. 90 (13): 1-30.

Nielsen, E. L. 1939. Grass studies III. Additional somatic chromosome complements. Amer. J. Bot. 26: 366-372.

Nielsen, E. L. 1944. Analysis of variation in *Panicum virgatum*. J. Agr. Res. 69: 327-353.

Nielsen, E. L. 1947. Polyploidy and winter survival in *Panicum virgatum* L. J. Amer. Soc. Agron. 39: 822-827.

Nielsen, E. L., and L. M. Humphrey. 1937. Grass studies. I. Chromosome numbers in certain members of the tribes Festuceae, Hordeae, Aveneae, Agrostideae, Chlorideae, Phalaridaceae and Tripsaceae. Amer. J. Bot. 24: 276-279.

Ownbey, G. W., and Y-T. Hsi. 1963. Chromosome numbers in some North American species of the genus *Cirsium*. Rhodora 65: 339-354.

Peters, L. C., and L. V. Newell. 1961. Performance of hybrids between divergent types of big bluestem *Andropogon gerardi* Vitman and sand bluestem *Andropogon hallii* Hack. Crop. Sci. 1: 359-363.

Pinkava, D. J., and D. J. Keil. 1977. Chromosome counts of Compositae from the United States and Mexico. Amer. J. Bot. 64: 680-686.

Pippen, R. W. 1978. North American flora. Series 2, Part 10, pp. 151-159.

Pohl, R. W. 1968. How to Know the Grasses. [Rev. ed.] Wm. C. Brown Co. Dubuque, Iowa.

Porter, C. L. 1966. An analysis of variation between upland and lowland switchgrass *Panicum virgatum* in central Oklahoma. Ecology 47: 980-992.

Price, H. J. 1988. Plant genome size and the DNA C-value paradox. Plant Gen. Newsletter 4: 18-19, 23.

Raven, P. H., O. T. Solbrig, D. W. Kyhos, and R. Snow. 1960. Chromosome numbers in Compositae. I. Astereae. Amer. J. Bot. 47: 124-132.

Reeder, J. R. 1967. Notes on Mexican grasses. VI. Miscellaneous chromosome numbers. Bull. Torrey Bot. Club 94: 1-17.

Reeder, J. R. 1968. Notes on Mexican grasses. VII. Miscellaneous chromosome numbers 2. Bull. Torrey Bot. Club 95: 69-86.

Reeder, J. R. 1977. Chromosome numbers in western grasses. Amer. J. Bot. 64: 102-116.

Reeder, J. R. 1984. IOBP Chromosome number reports. Taxon 33: 132-133.

Richards, E. L. 1968. A monograph of the genus *Ratibida*. Rhodora 70: 348-393.

Riggins, R. 1977. A biosystematic study of the *Sporobolus asper* complex (Gramineae). Iowa State J. Sci. 1977: 287-321.

Riley, R. D., and K. P. Vogel. 1982. Chromosome numbers of released cultivars of switchgrass, indiangrass, big bluestem and sand bluestem. Crop Sci. 22: 1081-1083.

Risser, P. G., E. C. Birney, H. D. Blocker, S. W. May, W. J. Parton, and J. A. Weins. 1980. The True Prairie Ecosystem. US/IBP

126

Synthesis Series #16. Hutchinson Publishing Co. Strondsberg, PA.

Rotar, P. P., and U. Urata. 1967. Cytological studies in the genus *Desmodium*: some chromosome counts. Amer. J. Bot. 54: 1-4.

Roy, G. B., and F. W. Gould. 1970. Biosystematic investigations of *Bouteloua hirsuta* and *Bouleloua pectinata*. I. Gross morphology. Southwest. Nat. 15: 377-387.

Russell, P. J. 1987. Essential Genetics. 2nd ed. Blackwell Scientific Publishers, London.

Schaak, C. G. 1982. IOBP chromosome number reports. Taxon 31: 366-367.

Seabrook, J. A. E., and L. A. Dionne. 1976. Studies on the genus *Apios*. I. Chromosome numbers and distribution of *Apios americana* and *A. priceana*. Canad. J. Bot. 54: 2567-2572.

Semple, J. C., and L. Brouillet. 1980a. A synopsis of North American asters: the subgenera, sections and subsections of *Aster* and *Lasallea*. Amer. J. Bot. 67: 1010-1026.

Semple, J. C., and L. Brouillet. 1980b. Chromosome numbers and satellite chromosome morphology in *Aster* and *Lasallea*. Amer. J. Bot. 67: 1027-1039.

Snyder, L. A., and J. R. Harlan. 1953. A cytological study of *Bouteloua gracilis* from western Texas and eastern New Mexico. Amer. J. Bot. 40: 702-707.

Sokal, R. R., and F. J. Rohlf. 1969. Biometry. W.H. Freeman Co. San Fransisco, CA.

Spellenberg, R. 1981. Chromosome numbers and their cytotaxonomic significance for North American *Astragalus* (Fabaceae). Taxon 25: 463-476.

Stebbins, G. L. 1932. Cytology of *Antennaria*. I. Normal species. Bot. Gaz. 94: 134-151.

Stebbins, G. L. 1971. Chromosomal evolution in higher plants. Addison-Wesley. Reading, MA.

Stebbins, G. L. 1975. The role of polyploid complexes in the evolution of North American grasslands. Taxon 24: 91-106.

Stebbins, G. L. 1981. Coevolution of grasses and herbivores. Ann. Missouri Bot. Gard. 68: 75-86.

Stebbins, G. L. 1985. Polyploidy, hybridization and the invasion of new habitats. Ann. Missouri Bot. Gard. 72: 824-832.

Stebbins, G. L., and R. M. Love. 1941. A cytological study of California forage grasses. Amer. J. Bot. 28: 371-383.

Strickberger, M. W. 1985. Genetics. 3rd ed. Macmillan Publishing Co., NY.

Strother, J. L. 1983. More chromosome numbers in the Compositae. Amer. J. Bot. 70: 1217-1224.

Sullivan, V. I. 1976. Diploidy, polyploidy and agamospory among species of *Eupatorium* (Compositae). Canad. J. Bot. 54: 2907-2917.

Sutherland, D. 1986. Poaceae Barnh., the grass family. In Great Plains Flora Association [eds.]. Flora of the Great Plains. Kansas Univ. Press, Lawrence, KS.

Taylor, R. L., and R. P. Brockman. 1966. Chromosome numbers of some western Canadian plants. Canad. J. Bot. 44: 1093-1103.

Tomb, A. S., K. L. Chambers, D. W. Kyhos, A.M. Powell, and P. H. Raven. 1978. Chromosome numbers in the Compositae. XIV. Lactuceae. Amer. J. Bot. 65: 717-721.

Vilkomerson, H. 1943. Chromosomes of *Astragalus*. Bull. Torr. Bot. Club 70: 430-435.

Ward, D. W. 1983. Chromosome counts from New Mexico and southern California. Phytologia 54: 302-309.

Weaver, J. E. 1956. Grasslands of the Great Plains. Johnsen Publishing Company. Lincoln, NE.

Weaver, J. E. 1959 North American Prairie. Johnsen Publishing Company, Lincoln, NE.

Weaver, J. E., and T. J. Fitzpatrick. 1934. The prairie. Ecol. Monogr. 4: 109-295.

Weedin, J. F., and A. M. Powell. 1980. IOBP chromosome number reports. LXIX. Taxon. 29: 716-718.

Wells, P. V. 1970. Historical factors controlling vegetation patterns and floristic distributions in the Central Plains region of North America. In W. Dort, Jr., and J. K. Jones, Jr. [eds.] Pleistocene and Recent Environments of the Central Great Plains. Univ. Kansas Press Special Publ. No. 3. Lawrence, KS.

Young, J. D. 1940. Cytological investigations in *Desmodium* and *Lespedesa*. Bot. Gaz. 101: 839-850.

THE GENETIC STRUCTURE OF TROPICAL TREE POPULATIONS: ASSOCIATIONS WITH REPRODUCTIVE BIOLOGY

J.L. Hamrick and M.D. Loveless

INTRODUCTION

Genetic variation in natural plant populations is structured in time and space. Both the development and maintenance of population genetic structure are due to interactions among a complex suite of evolutionary factors: variation in the gene pool, the organization of this variation into genotypes, the spatial distribution of genotypes, the reproductive system that controls how gametes unite to form progeny, the dispersal of these progeny, random events, and the processes of growth, mortality and replacement which give rise to future populations (Burley, 1976; Clegg et al., 1978; Levin, 1978). Genetic structure, therefore, is a composite characteristic which reflects historical population processes, but it also constrains to some degree the action of those processes in present and future populations.

Tropical tree communities present an excellent system within which to examine the effects of reproductive biology on genetic organization. The biological complexity of lowland tropical forests has given rise to a diversity of ecological and life history traits. In contrast to temperate taxa, tropical trees are seldom wind-pollinated; instead most pollen vectors are insects, birds, or mammals (Frankie, 1975; Janzen, 1975; Opler et al., 1980). Seed dispersal systems are also frequently adapted to animal dispersal, although wind or gravity dispersal may be found (Foster, 1982). While temperate forests may be dominated by one or a few taxa, most tropical lowland forests are characterized by high taxonomic diversity. Even processes of growth, mortality, and replacement may differ to some degree in temperate and tropical habitats. Under conditions of localized seed dispersal, seedlings in both temperate and tropical forests will compete intraspecifically with their own cohort. But following this early stage of high mortality, interspecific competition should be relatively more important in the tropics, because of higher species diversities and less dominance (Ashton, 1969, 1973).

These features of tropical trees have led to speculations about genetic variability, structure, breeding systems, and other evolutionary processes in tropical forests. Early ecological studies (Davis and Richards, 1933; Black et al. 1950; Richards, 1952) confirmed the high species diversity and low species densities of tropical forest communities. Based on these generalizations, Corner (1954) suggested that speciation in the tropics would be favored by restricted pollen flow, with widely separated conspecifics, unsynchronized mass flowering and a homogenous environment offering little selection for genetic differentiation. Baker (1959) predicted that the density and structural complexity of the tropical forest would preclude wind pollination or even effective intraspecific pollen transfer by animal vectors, and he concluded that tropical forests should be characterized by widespread self-compatibility and autogamous mating systems.

More recently, this view of tropical forest communities and of evolution in the humid tropics has been challenged. Detailed sampling has found a significant degree of clumping between individuals of tropical tree species (Hubbell, 1979; Hubbell and Foster, 1983). As a result, pollen movement may be no more restricted by interplant distance than it is in temperate tree populations. Studies of pollinator behavior have demonstrated that at least some animal vectors are capable of flying long distances to visit conspecific trees and could be effective agents of long-distance pollen movement (Janzen, 1971; Baker, 1973; Linhart, 1973; Stiles, 1975, 1977; Webb and Bawa, 1983). Furthermore, experimental studies of tropical tree breeding systems (Bawa, 1974; Bawa and Opler, 1977; Opler and Bawa, 1978; Ashton, 1969) suggest high levels of self-incompatibility in hermaphroditic species. Tropical forest floras are also unexpectedly rich in dioecious species (Bawa 1974; Bawa and Opler, 1977). Monoecy, while not as common, is also found in some trees and shrub taxa. It appears, then, that tropical trees are predominantly outcrossed with the potential for long distance gene movement. If this is the prevailing breeding system, tropical tree populations should have relatively little spatial genetic structure (Loveless and Hamrick, 1984). To date there is little empirical evidence to test these predictions. There have been few population genetic studies of tropical trees, and those that exist were not designed to describe the distribution of genetic variation within and among populations and population subdivisions (e.g., Hamrick and Loveless, 1986).

This paper reports the results of allozyme analyses of 16 species of tropical shrubs and trees common to the moist forest of Barro Colorado Island, Republic of Panama. Levels of variation within sample sites were measured and the distribution of variation within and among sample sites was determined. Associations between assumed pollen and seed mobility and measures of intersite genetic diversity were examined. The results indicate that high levels of allozyme variation are maintained within populations of tropical trees and that among-site variation is low.

MATERIALS AND METHODS

Population collections were centered on Barro Colorado Island (BCI), an island formed in the early 1900's by the flooding of the Chagres River to form Lake Gatun during construction of the Panama Canal. BCI is located at 9° 09'N latitude 79° 51' W longitude with an elevation ranging from 25-165 m. Barro Colorado Island was chosen as a study site because a modern field station administered by the Smithsonian Tropical Research Institute is located there and because its flora has been the subject of extensive research (Croat, 1978; Leigh et al., 1982). In addition, BCI is the site of a 50 hectare mapped forest plot established by S.P. Hubbell and R.B. Foster (hereafter HF). On the HF plot every woody plant with a diameter of 1 cm or greater at 1.5m has been tagged, mapped, and identified, greatly facilitating studies of spatial structure.

Species Selection

Species were chosen for initial surveys of allozyme variation based on four criteria: (1) each species should be common on BCI and the HF plot (200 + individuals); (2) the species should represent a variety of life forms (shrubs, subcanopy trees, canopy trees); (3) they should represent a variety of pollination and seed dispersal mechanisms; and (4) information on their ecology, breeding systems etc., was available from previous studies.

Based on the results of a pilot survey of allozyme variation in 29 species on BCI (Hamrick and Loveless, 1986), 16 taxa were chosen for detailed population analyses. These 16 species represented a wide variety of breeding systems, pollination syndromes, and seed dispersal mechanisms (Table 1). A second consideration was the ease of electrophoretic analyses and the number and dependability of the genetic loci identified in the original survey.

Collection Sites

The availability of the HF plot made it possible to sample at three spatial scales. For nine of the 16 species, collections of 72

Table 1. Characteristics of 16 tropical woody species analyzed electrophoretically. Collections were made from Barro Colorado Island, Republic of Panama and other locations in Central America.

Species	Family	Growth Form	Breeding System	Pollinator	Dispersal Agent
Acalypha diversifolia	Euphorb	Shrub	Monoecious	Wind? Insects?	Explosive
Alseis blackiana	Rube	Canopy tree	Bisexual	Bees, Butterflies	Wind
Brosimum alicastrum	Morac	Canopy tree	Dioecious	Wind? Insects?	Bats, Arboreal Mammals
Cecropia insignis	Morac	Canopy tree	Dioecious	Bats? Insects? Wind?	Bats, Birds, Mammals
Dipteryx panamensis	Legume (Papil)	Canopy tree	Bisexual	Bees Mammals	Bats, Other
Erythrina costaricensis	Legume (Papil)	Understory tree	Bisexual	Hummingbirds	Birds
Gustavia superba	Lecythid	Understory tree	Bisexual	Large Bees	Mammals
Hybanthus prunifolius	Violac	Shrub	Bisexual	Small Bees	Explosive
Platypodium elegans	Legume (Papil)	Canopy tree	Bisexual	Small Bees	Wind
Poulsenia armata	Morac	Understory tree	Monoecious	Wind? Insects?	Bats? Arboreal Mammals
Psychotria horizontalis	Rube	Shrub	Bisexual, Heterostylous	Bees, Butterflies	Birds
Quararibea asterolepis	Bombac	Canopy tree	Bisexual	Bats	Mammals, Birds
Rinorea sylvatica	Violac	Shrub	Bisexual	Bees	Explosive
Sorocea affinis	Morac	Understory tree	Dioecious	Wind? Insects?	Birds
Swartzia simplex	Legume (Caesalp)	Understory tree	Bisexual	Wind? Insects?	Birds
Tachigalia versicolor	Legume (Papil)	Canopy tree	Bisexual	Bees	Wind

individuals each were obtained from four sites located approximately 300 meters from each other on the HF plot. Two additional samples of 48 individuals each were made at other locations on BCI. An attempt was made to place these sites at least 1 kilometer from the HF plot and each other. Six populations on BCI were thus available for nine species (Fig. 1). For five of the seven species, a single HF plot sample and two distant BCI samples were made, giving three populations in the survey. *Cecropia insignis* and *Dipteryx panamensis* were represented by a single population on BCI. For several species it also was possible to obtain collections from other parts of Panama or from Costa Rica. The hierarchial nature of the collection design allowed the distribution of allozyme variation to be compared at the level of the HF plot (<300m), BCI (1-2 km), and Panama (5+km).

Sampling Protocol

Small branches with attached leaves were taken from each individual. Samples were placed in plastic bags and were stored in the field in a cooler containing frozen synthetic ice packs. Within 4-6 hours of collection, samples were returned to the laboratory, and were refrigerated until processing.

Whenever possible, samples were vacuum-dried the same day as they were collected. Leaves were cut into 1-2 cm squares and placed in small gauze bags, vacuum-dried in a lyophilizer for 48 hours, and sealed in plastic bags. Leaf samples were shipped within one week to the United States for electrophoretic analysis. Long term storage was at -65 C.

Electrophoresis

Starch gel electrophoresis was performed on extracts obtained from the vacuum-dried leaf tissue. Leaves were ground in a mortar and pestle using liquid nitrogen, and the powdered plant material was slurried with a phosphate-polyvinylpyrolidone extraction buffer (Mitton et al., 1979). The crude extract was absorbed onto filter paper wicks and 24 wicks were inserted into 11% starch gels.

Depending on the species, as many as 7 buffer systems were used. Staining was carried out for up to 20 enzymes. Staining recipes are given in Soltis et al. (1983) and references therein.

Data Analyses

Banding zones with no apparent variation were classified as monomorphic. A banding zone was classified as polymorphic if the variation was consistent with Mendelian interpretations. The electrophoretic phenotypes were then treated as genetic loci.

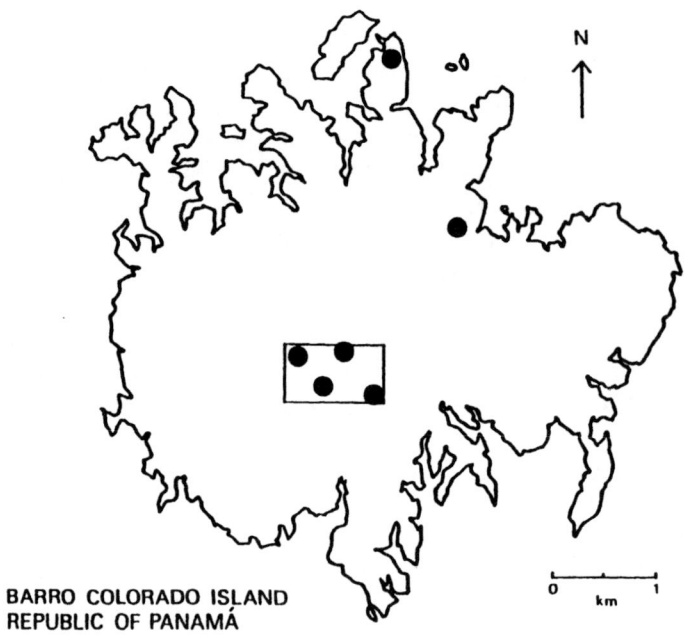

Figure 1. Collection sites on Barro Colorado Island, Republic of Panama, showing population collections for *Swartzia simplex* var. *ochnacea*. The rectangular area in the center of the island is the 50-hectare mapped plot established by S.P. Hubbell and R.B. Foster.

These data were used to calculate two commonly used measures of within population variation; the proportion of polymorphic loci (P) and the proportion of loci heterozygous per individual (H). Assuming Hardy-Weinberg conditions, H is calculated by:

$$H = 1 - \Sigma P_i^2$$

where P_i is the frequency of the ith allele at a locus.

Variation among sampling sites was calculated at two spatial levels: the HF plot and BCI. *Cecropia insignis* and *Dipteryx panamensis* were excluded from the BCI and HF plot analyses since they were represented by only one collection site on BCI. Among-site heterogeneity was measured using Nei's (1973) genic diversity statistic, G_{ST}. G_{ST} is the proportion ot the total variation (H_T) found among populations (D_{ST}), or $G_{ST} = D_{ST}/H_T$.

Levels of gene flow were estimated by two indirect methods, both based on the distribution of allele frequencies among populations. Slatkin's measure of gene flow (Slatkin, 1985; Barton and Slatkin, 1986) uses the distribution of private alleles

(alleles found in a single population) to estimate Nm, the number of migrants per generation. Slatkin (1985) has shown that the log of Nm is approximately linearly related to the log of p (1), the mean frequency of private alleles. Nm can also be estimated from the relationship $4Nm = 1/G_{ST} -1$ (Wright, 1951). The two methods are not redundant since Slatkin's measure is dependent on the distribution of rare alleles while G_{ST} values are most influenced by the distribution of common alleles.

RESULTS
Intrasite Variation
The number of loci that could be dependably resolved varied greatly among species (Table 2). *Psychotria horizontalis* and *Cecropia insignis* had the fewest loci (23 and 24, respectively), while *Dipteryx panamensis* (45) and *Erythrina costaricensis* (44) had the highest number of scoreable loci. An average of 34 loci was available for each species.

There was considerable variation among the 16 species in the proportion of polymorphic loci and mean heterozygosity (Table 2). *Tachigalia versicolor, Rinorea sylvatica, Psychotria horizontalis,* and *Cecropia insignis* had the lowest levels of within-population variation. The most variable species were *Alseis blackiana, Swartzia simplex* var. *ochnacea, Quararibea asterolepis,* and *Platypodium elegans.* Average values for percentage of polymorphic loci and mean heterozygosity are high relative to other angiosperm species (Table 3). Levels of variation estimated for these tropical tree species are similar to values obtained previously for several conifer species (Table 3). Conifers have traditionally been regarded as the most variable group of plant species (Hamrick et al., 1981).

Variation among Sites
Since off-BCI collections were available for only a few species, among-site variation was analyzed for the 14 species with three or more BCI collections. Comparisons were made among the four collection sites on the HF plot and among the three or six sites on BCI (Table 4). Values of G_{HF} and G_B were quite low in all cases indicating relatively little differentiation among sites on the HF plot or at the BCI level. All of the G_{HF} values were significantly different from zero ($P < 0.01$) although the mean G_{HF} value was only 0.048. *Erythrina costaricensis, Psychotria horizontalis,* and *Alseis blackiana* had the least genetic differentiation on the HF plot, while *Rinorea sylvatica, Hybanthus prunifolis,* and *Tachigalia versicolor* had the most differentiation.

Table 2. Levels of within population allozyme variation for 16 species of tropical trees and shrubs.

Species	Number of Populations	Number of Loci	Percent Polymorphic Loci	Mean Heterozygosity
Acalypha diversifolia	6	30	65.6	0.250
Alseis blackiana	7	28	89.3	0.340
Brosimum alicastrum	8	30	61.3	0.225
Cecropia insignis	2	24	44.7	0.146
Dipteryx panamensis	2	45	53.6	0.162
Erythrina costaricensis	6	44	68.3	0.205
Gustavia superba	4	38	65.1	0.222
Hybanthus prunifolius	7	42	66.3	0.241
Platypodium elegans	5	38	67.9	0.251
Poulsenia armata	3	33	64.2	0.250
Psychotria horizontalis	8	23	49.5	0.133
Quararibea asterolepis	3	30	65.6	0.255
Rinorea sylvatica	8	40	34.1	0.100
Sorocea affinis	5	37	73.4	0.249
Swartzia simplex	8	36	76.0	0.272
Tachigalia versicolor	6	31	29.6	0.073
MEAN VALUES	5.5	34.3	60.9	0.211

Table 3. Comparative within population allozyme variation in several plant groups.

Group	Number of Taxa	Percent Polymorphic Loci	Mean Number of Alleles	Mean Heterozygosity
Tropical Species[1]	16	60.9	-	0.211
Coniferous Trees[2]	20	67.7	2.29	0.207
Dicots[3]	74	31.2	1.49	0.113
All Plant Species[3]	113	36.8	1.69	0.141

[1] This study

[2] Hamrick et al., 1981

[3] Hamrick et al., 1979

Values of G_B were consistently higher than G_{HF} values, indicating that additional genetic variation was present in parts of the island distant from the HF plot. Mean G_B was 0.055 with *Acalypha diversifolia, Rinorea sylvatica, Hybanthus prunifolius,* and *Platypodium elegans* having the most differentiation. Populations of *Quararibea asterolepis, Gustavis superba, Psychotria horizontalis,* and *Alseis blackiana* were least differentiated from one another.

Estimates of Gene Flow

The 14 species with adequate samples were found to be relatively rich in private alleles (Table 5). All species, with the exception of *Poulsenia armata* and *Psychotria horizontalis*, had four or more private alleles. Mean frequencies of the private alleles were generally low, averaging 0.0364. This resulted in relatively high estimates of Nm. There was considerable variation for estimates of Nm among species. *Acalypha diversifolia* had the lowest estimate of Nm (0.43) while the highest value (29.94) was for *Brosimum alicastrum*. The mean value of Nm for the 14 species was 5.38. Generally, values over 1.0 indicate rather high rates of gene exchange among populations (Slatkin, 1985). The second method of estimating Nm is based on the distribution of common alleles among populations (G_B). Estimated values of Nm from this method ranged from 2.53 for *Acalypha diversifolia* to 11.11 for *Quararibea asterolepis* and *Gustavia superba*. The mean

Table 4. Distribution of allozyme variation among sample sites on the Hubbell-Foster Plot (G_{HF}) and on BCI (G_B). The predicted rank for each species indicates its potential for gene movement based on pollination and seed dispersal mechanisms, where lower numbers indicate more gene movement.

Species	G_{HF}	G_B	Predicted Rank
Acalypha diversifolia	- -	.090	1 4
Alseis blackiana	.034	.048	5
Brosimum alicastrum	.050	.055	2
Erythrina costaricensis	.029	.046	6
Gustavia superba	- -	.022	4
Hybanthus prunifolius	.065	.076	1 1
Platypodium elegans	.051	.073	8
Poulsenia armata	- -	.052	3
Psychotria horizontalis	.030	.036	9
Quararibea asterolepis	- -	.022	1
Rinorea sylvatica	.080	.077	1 2
Sorocea affinis	- -	.053	1 3
Swartzia simplex	.037	.055	1 0
Tachigalia versicolor	.059	.070	7
MEAN	.048	.055	

Table 5. Two estimates of the number of migrants per generation (Nm) based on the estimation procedures of Slatkin (1985) and Wright (1951). p(1) is the mean frequency of private alleles. G_B is the mean diversity among populations.

Species	Populations	No. of Private Alleles	Slatkin		Wright	
			p(1)	Nm	G_B	Nm
Acalypha diversifolia	3	8	0.1037	0.428	0.090	2.528
Alseis blackiana	6	9	0.0402	2.012	0.048	4.958
Brosimum alicastrum	6	4	0.0077	29.945	0.055	4.295
Erythrina costaricensis	6	6	0.0334	2.723	0.046	5.185
Gustavia superba	3	9	0.0255	4.232	0.022	11.114
Hybanthus prunifolius	6	9	0.0146	10.527	0.076	3.039
Platypodium elegans	5	10	0.0500	1.408	0.073	3.175
Poulsenia armata	3	2	0.0521	1.317	0.052	4.558
Psychotria horizontalis	6	2	0.0296	3.317	0.036	6.694
Quararibea asterolepis	3	8	0.0221	5.347	0.022	11.114
Rinorea sylvatica	7	7	0.0238	4.737	0.077	2.997
Sorocea affinis	3	13	0.0517	1.334	0.053	4.467
Swartzia simplex	8	15	0.0387	2.186	0.055	4.295
Tachigalia versicolor	6	4	0.0163	8.793	0.070	3.321

Nm value of 5.12 from this method was nearly identical to the value obtained from the Slatkin procedure.

Correlations with Predictions

Prior to the data analyses, we ranked the 14 species based on our perception of their genetic mobility (Table 4). Species with strong-flying pollinators and mobile seed dispersal agents (i.e., *Quararibea asterolepis* and *Brosimum alicastrum*) were highly ranked, while wind-pollinated species (*Acalypha diversifola* and *Sorocea affinis*) and those with explosive (and therefore short-distance) seed dispersal (*Acalypha diversifolia* and *Rinorea sylvatica*) were considered less genetically mobile. The predicted rankings were compared to the observed ranks based on G_{HF}, G_B, and N_m using a Spearman coefficient of rank correlation (Steel and Torrie, 1960). The two Nm values obtained from the Wright and Slatkin estimation procedures were also ranked and compared.

The correlation coefficient between the predicted rank and the G_{HF} values was positive and approached significance ($r_s = 0.533$; $P < 0.20$), while the observed distribution of variation on BCI (G_B) and that predicted was highly significant ($r_s = 0.679$; $P < 0.01$). The comparison between the ranked genetic diversities on the HF plot and those on BCI was also highly significant ($r_s = 0.95$; $P < 0.001$). The correlation between the Slatkin estimates of Nm and the predicted order was positive but was not significant ($r_s = 0.332$; $P < 0.30$). Estimates of Nm based on G_B gave the same highly significant rank order as the G_B values and thus had the same rank correlation coefficient. However, comparisons between the rank order of the two estimates of Nm was not significant ($r_s = 0.081$; $P < 0.50$).

DISCUSSION

There are two important conclusions to be drawn from this analysis. First, the 16 tropical taxa maintain large amounts of allozyme variation within their populations. While there is considerable variation among species, the mean values of P and H place these species among the most variable analyzed. Second, there is relatively little genetic heterogeneity among populations on BCI separated by as much as 1-2 kilometers.

The level of allozyme variation found within these populations is much higher than the mean level of variation in herbaceous angiosperm species and is similar to values obtained for predominantly outcrossed wind-pollinated conifers (Table 3). There may be several reasons why these tropical species have such high levels of variation. Earlier reviews (Hamrick et al.,

1979; Brown, 1979) suggested that outcrossed, long-lived species with high life-time fecundities tend to maintain more variation within populations. Although several of the species we analyzed are not well-studied ecologically, they would generally fall into this category. A formal estimate of the mating system of *Tachigalia versicolor* (Hamrick, Loveless and Foster, unpubl. data) indicates that this species is almost completely outcrossed. Second, while these 16 species represent a wide array of life forms and pollination and seed dispersal mechanisms (Table 2), they may not be representative of tropical woody species in general. These 16 species were chosen because they occur on BCI and the HF plot in high densities. It is possible that less common tropical woody species will contain less allozyme variation. Our data, however, do not support this argument. A rank-correlation between the frequency of adults on the HF plot and mean heterozygosity produced a correlation of 0.074 ($P<0.50$). There also may have been some bias towards more variable species when these 16 species were selected from the 29 species originally analyzed (Hamrick and Loveless, 1986).

Nevertheless, our data suggest that tropical trees maintain more variation within their populations than herbaceous angiosperms with similar pollination and seed dispersal mechanisms. This raises the question of whether woody plants in general maintain more variation within their populations (Hamrick, 1978). Evidence is accumulating which indicates that this is the case. Studies of several temperate angiosperm trees indicate that these species also maintain high levels of within-population variation (Hamrick, unpubl. data). The interaction of complex patterns of diversifying or frequency dependent selection (See Linhart, this publication) with the genetic variation introduced by high rates of gene flow could provide a partial explanation of this observation.

The second observation from this analysis is the small amount of genetic differentiation among collection sites. If most of these species are predominantly outcrossing, animal-pollinated species, we would expect G_{HF} and G_B values of 0.15 to 0.20 (Loveless and Hamrick, 1984), considerably higher than the mean values observed. Levels of diversity among collection sites on BCI separated by at least one kilometer were more similar to values found at the population subdivision level for 20 species of outcrossing plants (G_{ST} = 0.041; Loveless and Hamrick, 1984). This indicates that gene flow either via pollen or seeds must be extensive throughout BCI.

Although there is not a great deal of diversity in G values among species (0.022-0.090), there is a predictable and statistically significant pattern to the results. Species which were predicted to have high genetic mobilities (strong flying pollinators, widely dispersed seed) generally have lower G values. Species with the greatest differentiation among populations are either those with wind-pollination and explosive seeds (i.e., *Acalypha*) or those with weak-flying pollinators, patchy distributions, and weakly dispersed seeds (e.g., *Rinorea*). Baker et al. (1982) and others (Whitehead, 1969; Janzen, 1975) have speculated that wind pollination would not be very effective in tropical communities. Our results support this conclusion. Thus, while our predictions are not perfectly supported, our results indicate that knowledge of the reproductive biology of a species may allow its genetic structure to be predicted with some degree of accuracy.

The two estimates of immigration (Nm) in these tropical taxa are consistent with the conclusion that gene flow is high among populations on BCI. Predominantly selfing species have mean Nm values of 0.10 or less, whereas wind-pollinated temperate conifers have values in the range of 0.20 or higher (Hamrick, 1987; Hamrick and Griswold, 1988).

There are independent data that support the argument for high rates of gene movement among locations on BCI. In an analysis of the mating system of *Tachigalia versicolor* (Hamrick, Loveless, and Foster, unpubl. data), one cluster of six isolated adults was found to be monomorphic for an IDH locus. The frequency of the rare allele was 0.04 in the progeny of these six trees. These progeny would have had to be sired by pollen from outside the cluster. The closest flowering trees were approximately 500 meters from these six adults. Using the equation:

$$m = \frac{q_0 - q_1}{q_0 - Q}; \text{ where}$$

m = the rate of gene flow, q_0 = frequency of the rare allele in the six adults (0.0), q_1 = the frequency of the allele in the progeny (0.04), and Q = frequency of the allele in the population as a whole (0.156), we find that $m = 2.565$. This is probably a maximum estimate, since it is likely that the pollen carrying the rare allele came from trees relatively close to the cluster rather than from the island at large. If the allele frequency in the nearby trees was greater than the population as a whole, estimates of m would be reduced. Regardless of the actual value, these results demonstrate that pollen can move long distances among widely dispersed individuals in lowland tropical forests. This is consistent with the results of Webb and Bawa (1983), who monitored dyed pollen movement in two tropical species in Costa Rica.

SUMMARY AND CONCLUSIONS

Our data indicate that tropical trees maintain as much or perhaps more variation within their populations than temperate tree species. Distribution of this variation can be predicted with some accuracy if information is available on pollination and seed dispersal. The majority of the variation is found within populations, although a small, but statistically significant, amount of variation is found among populations. This supports the argument that gene flow is high among locations separated by as much as 2-3 kilometers. Certainly the picture of genetic isolation, inbreeding, and lack of variation in tropical trees which was suggested by earlier authors is not supported by our data.

ACKNOWLEDGMENTS

This research was supported by NSF grant BSR82-06946. The authors wish to thank the Smithsonian Tropical Research Institute, and S.P. Hubbell and R.B. Foster for permission to sample on the 50-hectare mapped plot. Field assistance was provided by Loretta Rosselli and Maria Felix Rios. Laboratory assistance was provided by Linda Vescio, Andrew Schnabel, and George Goldin.

LITERATURE CITED

Ashton, P. S. 1969. Speciation among tropical forest trees: some deductions in the light of recent evidence. Biol. J. Linn. Soc. 1: 155-196.

Ashton, P. S. 1973. The biological significance of complexity in lowland tropical forest. J. Indian Bot. Soc. 50A: 530-537.

Baker, H. G. 1959. Reproductive methods as factors in speciation in flowering plants. Cold Spring Harbor Symp. Quant. Biol. 24: 177-199.

Baker, H. G. 1973. Evolutionary relationships between flowering plants and animals in American and African tropical forests. In Meggers,B. J., E. S. Ayuensu, and W. D. Duckworth [eds.]. Tropical Forest Ecosystems in Africa and South America: a

Comparative Review. Smithsonian Institution Press, Washington, D. C.

Baker, H. G., K. S. Bawa, G. W. Frankie, and P. A. Opler. 1983. Reproductive biology of plants in tropical forests. In Golley, F. B. [ed.]. Tropical Rain Forest Ecosystems: Structure and Function. Elsevier Sci., Amsterdam.

Barton, N. H., and M. Slatkin. 1986. A quasi-equilibrium theory of the distribution of rare alleles in a subdivided population. Heredity 56: 409-415.

Bawa, K. S. 1974. Breeding systems of tree species of a lowland tropical community. Evolution 28: 85-92.

Bawa, K. S. and P. A. Opler. 1977. Spatial relationships between staminate and pistillate plants of dioecious tropical forest trees. Evolution 31: 64-68.

Black, G. A., T. Dobzhansky, and C. Pavan. 1950. Some attempts to estimate species diversity and population density of trees in Amazonian forests. Bot. Gaz. 111: 413-425.

Brown, A. H. D. 1979. Enzyme polymorphism in plant populations. Theor. Pop. Biol. 15: 1-42.

Burley, J. 1976. Genetic systems and genetic conservation of tropical pines. In Burley, J. and B.T. Styles [eds.]. Tropical Trees: Variation, Breeding and Conservation. Academic Press, London.

Clegg, M. T., A. L. Kahler, and R. W. Allard. 1978. Estimation of life cycle components of selection in an experimental plant population. Genetics 89: 765-792.

Corner, E. J. H. 1954. The evolution of tropical forests. In Huxley, J., A. C. Hardy, and E. B. Ford [eds.]. Evolution as a Process. 2nd ed. Humanities, New York.

Croat, T. B. 1978. Flora of Barro Colorado Island. Stanford University Press, Stanford, California.

Davis, T. A. W., and P. W. Richards. 1933. The vegetation of Moraballi Creek, British Guiana: an ecological study of a limited area of tropical rain forest. Parts I & II. J. Ecol. 21: 350-384; 22: 106-155.

Foster, R. B. 1982. The seasonal rhythm of fruitfall on Barro Colorado Island. In Leigh, E. G., Jr., A. S. Rand, and D. M. Windsor [eds.]. The Ecology of a Tropical Forest: Seasonal Rhythms and Long-Term Changes. Smithsonian Institution Press, Washington, D.C.

Frankie, G. W. 1975. Tropical forest phenology and pollinator plant coevolution. In Gilbert, L. E. and P. H. Raven [eds.]. Coevolution of Animals and Plants. Univ. of Texas Press, Austin.

Hamrick, J. L. 1978. Genetic variation and longevity. In Solbrig, O., S. Jain, G. Johnson and P. Raven [eds.]. Topics in Plant Population Biology. Columbia Univ. Press, New York.

Hamrick, J. L. 1987. Gene flow and the distribution of genetic variation in plant populations. In Urbanska, K. [ed.]. Differentiation Patterns in Higher Plants. Academic Press, London.

Hamrick, J. L., and G. B. Griswold. 1989. Association between Slatkin's measure of gene flow and the dispersal ability of plant species. Am. Nat. . In press.

Hamrick, J. L., Y. B. Linhart, and J. B. Mitton. 1979. Relationships between life history characteristics and electrophoretically detectable genetic variation in plants. Ann. Rev. Ecol. Syst. 10: 173-200.

Hamrick, J. L., and M. D. Loveless. 1986. Isozyme variation in tropical trees: procedures and preliminary results. Biotropica 18: 201-207.

Hamrick, J. L., J. B. Mitton, and Y. B. Linhart. 1981. Levels of genetic variation in trees: influence of life history characteristics. In Conkle, M. T. [ed.]. Proc. Symposium of Isozymes N. American Forest Trees and Forest Insects. Gen. Tech. Rept. PSW 48. Pacific Southwest Forest and Range Experiment Station, Berkeley, CA.

Hubbell, S. P. 1979. Tree dispersion, abundance, and diversity in a tropical dry forest. Science 203: 1299-1309.

Hubbell, S. P., and R. B. Foster. 1983. Diversity of canopy trees in a neotropical forest and implications for conservation. In Sutton, S. L., T. C. Whitmore, and A. C. Chadwick [eds.]. Tropical Rain Forest: Ecology and Management. Blackwell Scientific Publ., Oxford.

Janzen, D. H. 1971. Euglossine bees as long-distance pollinators of tropical plants. Science 171: 203-205.

Janzen, D. H. 1975. Ecology of Plants in the Tropics. Inst. Biol. Stud. Biol. No. 58. Edward Arnold, London.

Leigh, E. G., A. S. Rand, and D. M. Windsor. 1982. The Ecology of a Tropical Forest: Seasonal Rhythms and Long-Term Changes. Smithsonian Institution Press. Washington, D.C.

Levin, D. A. 1978. Some genetic consequences of being a plant. In Brussard, P. [ed.]. Ecological Genetics: the Interface. Springer Verlag, New York.

Linhart, Y. B. 1973. Ecological and behavioral determinants of pollen dispersal in hummingbird-pollinated Heliconia. Am. Nat. 107: 511-523.

Loveless, M. D., and J. L. Hamrick. 1984. Ecological determinants of genetic structure in plant populations. Ann. Rev. Ecol. Syst. 15: 65-95.

Mitton, J. B., Y. B. Linhart, K. B. Sturgeon, and J. L. Hamrick. 1979. Allozyme polymorphisms detected in mature tissues of ponderosa pine, *Pinus ponderosa* Lars. J. Hered. 70: 86-89.

Nei, M. 1973. Analysis of gene diversity in subdivided populations. Proc. Natl. Acad. Sci. USA 70: 3321-3323.

Opler, P. A., H. G. Baker, and G. W. Frankie. 1980. Plant reproductive characteristics during secondary succession in neotropical lowland forest ecosystems. Biotropica 12: 40-46.

Opler, P. A., and K. S. Bawa. 1978. Sex ratios in tropical forest trees. Evolution 32: 812-821.

Richards, P. W. 1952. The Tropical Rain Forest. Cambridge Univ. Press, Cambridge.

Slatkin, M. 1985. Rare alleles as indicators of gene flow. Evolution 39: 53-65.

Soltis, D. E., C. H. Haufler, D. C. Darrow and G. F. Gastony. 1983. Starch gel electrophoresis of ferns: a compilation of grinding buffers, gel and electrode buffers, and staining schedules. Am. Fern. J. 73: 9-27.

Steel, R. G. D., and J. H. Torrie. 1960. Principles and Procedures of Statistics. McGraw-Hill, New York.

Stiles, F. G. 1975. Ecology, flowering phenology, and hummingbird pollination of some Costa Rican *Heliconia* species. Ecology 56: 285-301.

Stiles, F.G. 1977. Coadapted competitors: the flowering seasons of hummingbird-pollinated plants in a tropical forest. Science 198: 1177-1178.

Webb, C. J., and K. S. Bawa. 1983. Pollen dispersal by hummingbirds and butterflies: a comparative study of two lowland tropical plants. Evolution 37: 1258-1270.

Whitehead, D. R. 1969. Wind pollination in the angiosperms: evolutionary and environmental considerations. Evolution 23: 28-35.

Wright, S. 1951. The genetical structure of populations. Ann. Eugenics 15: 323-354.

BREEDING SYSTEMS

Plants possess a remarkable profusion of sexual reproduction patterns. The study of these systems in their ecological and evolutionary contexts is an important component of plant evolutionary ecology. This sexual flexibility appears to be evolutionarily driven by three important constraints on plant life histories: the energetic cost of reproduction, the need to produce genetically variable progeny, and the stationary status of plants. Reproduction requires a major investment of resources because the process is energetically costly. Once committed to reproduction, this investment of resources and energy is unavailable for the growth and maintenance of vegetative tissues. Because plants are stationary, the appropriate combinations of males and females are difficult to adjust. However, these ratios are crucial if an individual is to generate progeny that are both genetically variable and reasonably well adapted to local conditions. In animals, mobility and elaborate courtships allow appropriate attraction patterns to develop between the sexes, and both male competition and female choice are important components of sexual rituals. In plants, sexual success has been achieved by going far beyond a mere two-sex system.

Because there are so many mating systems in plants, their categorization is difficult. Early studies in reproductive ecology suggested that most species were primarily cross-pollinated, an idea championed by Charles Darwin in the first (pp. 248-249) and later editions of the *Origin of Species*. Subsequent workers claimed that cross-pollination was advantageous because it helped to maintain sufficient genetic variability through repeated genetic recombination. It was argued in that context that self-pollination was an "evolutionary dead-end" because it led to a loss of genetic variability. A series of classic papers produced in the laboratories of Allard, Baker, and others in the 1960's, demonstrated that primarily self-pollinated species were remarkably common, and that there were several ways in which these autogamous species maintained large gene pools. In fact, these two alternatives, autogamy or self-pollination vs. xenogamy or cross pollination, are at opposite ends of a spectrum. Many species show the necessary developmental and genetic flexibility to not "commit" themselves to either end, and to fit "in between" in a mixed-mating system. Cruden and Lyon argue persuasively that this middle ground is advantageous under a variety of environmental conditions. They also document that a given species is not necessarily committed to a single pattern, but rather that there can be intra-specific and intra-populational variation in the relative proportions of offspring produced by cross-pollination vs. self-pollination.

Barrett has spent several years in investigations of heterostyly, one of the breeding systems that acts to promote outcrossing in higher plants. Charles Darwin was among the first to demonstrate that plants can maximize outcrossing by having distylous flowers, a condition in which some individuals have long styles and short stamens while others have short styles and long stamens. Crosses between individuals with stamens and styles of the same length often are required to produce successful seed set because the morphological manifestations of heterostyly frequently are accompanied by self-incompatibility. Heterostylous plants depend upon reliable vectors, usually insects, to transport pollen to an appropriate stigma. But dependence upon animal vectors can be a liability, and in some cases, heterostyly can break down. Barrett demonstrates that this breakdown favors the development of self pollination systems, and he puts forward possible genetic mechanisms underlying the evolutionary change from outcrossing to self-pollination. He also describes the ecological conditions, such as island colonization, under which breakdown might occur. Barrett's results underscore another feature of breeding systems in plants, that their plasticity allows for generation of intermediates and variations upon a theme.

The fate of individuals and species on islands were of interest to Darwin and Wallace and continue to provide inspiration to biologists today. Baker in 1955 pointed out that plant species which successfully colonize islands are more likely to be self-compatible than self-incompatible. Baker reasoned that the propagule which accomplishes long-distance dispersal will be much more likely to establish a new population if it can successfully reproduce without the presence of a second conspecific. This idea has provided a valuable framework for analyzing the attributes of colonizing plants on islands and elsewhere. Here, Cox presents the results of such an island analysis and finds that, in apparent contradiction to the original idea, dioecious species are quite common on certain islands. However, further analysis of his data demonstrates that many island-inhabiting dioecious species can circumvent the problem of separate sexes in several ways. They can have "leaky dioecy" whereby given individuals do not have their sexual roles absolutely fixed, but periodically produce a flower with the opposite, but appropriate, gender. Some of these species are long-lived and can wait for many decades for such "errors" to allow sexual reproduction, while others produce many-seeded fruits thereby increasing the likelihood of appropriate gender combinations reaching a distant locality together.

The papers in this section provide an introductory sample of the sorts of studies ongoing presently on plant breeding systems. In the tradition of Kölreuter, Darwin, and Guppy, Herbert Baker has been a major contributor in this field. His contributions to date are summarized in the Introduction to this volume by Stebbins.

THE EVOLUTIONARY BREAKDOWN OF HETEROSTYLY

Spencer C. H. Barrett

INTRODUCTION

The mating system is of prime importance for understanding the process of natural selection since it governs the character of genetic transmission and plays a central role in regulating the genetic structure of populations. Flowering plants display a great diversity of reproductive methods, often among closely related taxa. The variation provides suitable experimental material for population biologists interested in testing hypotheses about the evolution and adaptive significance of different mating systems. Among flowering plant families the shift from obligate outbreeding, enforced by self-incompatibility, to predominant self-fertilization is one of the major pathways of mating system evolution (Stebbins 1957, 1974; Baker, 1959a; Jain, 1976). This change in mating pattern has important systematic and ecological consequences since the evolution of selfing often initiates reproductive isolation and speciation (Baker, 1961; Barrett, 1989).

Since Darwin's original work (Darwin, 1877), heteromorphic incompatibility systems (distyly and tristyly, Fig. 1) have provided a rich source of material for mating system studies. This is because the genetic modifications affecting mating behavior are often simple, the direction of change readily interpretable and alterations in the floral polymorphisms can usually be observed without difficulty under field conditions. Modifications of heterostylous systems include the replacement of one type of outcrossing system by another such as the evolution of dioecism from distyly and the origin of distyly from tristyly. (see Ganders, 1979; Barrett, 1988a for reviews). More frequently, however, heterostylous systems break down in the direction of increased self-fertilization with the commonest pathway being the formation of self-compatible homostyles (Ernst, 1955; Baker, 1959b, 1966). Plants in these groups usually possess anthers and stigmas at the same relative position within a flower and as a result are largely autogamous (Fig. 2 d,e). Many heterostylous genera contain homostylous taxa that are small-flowered, highly self-pollinating, depauperate in isozyme variation, and adapted to

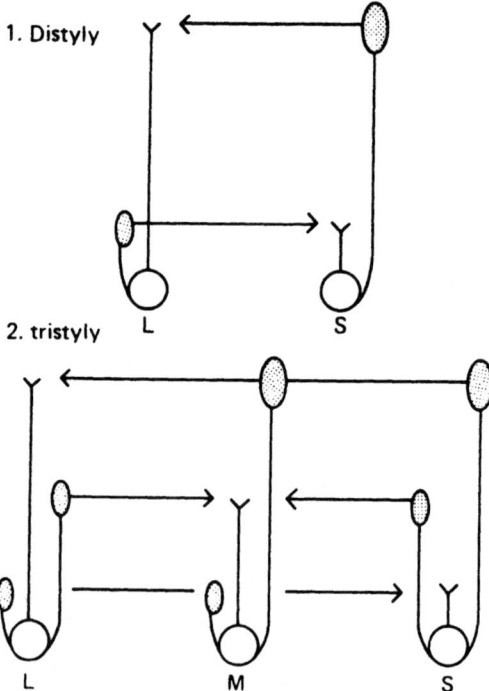

Figure 1. The heterostylous genetic polymorphisms distyly and tristyly. Distyly is controlled by a single locus with two alleles. The L morph is usually of genotype *ss* and the S morph *Ss*. In tristyly the most common mode of inheritance involves two diallelic loci (*S* and *M*) with *S* epistatic to *M*. With this genetic control the genotypes for the three morphs are: L: *ssmm*; M: *ssMm*; *ssMM*; S: *SsMm*, *Ssmm*, *SsMM*.

pioneer environments. This suggests that the shift to self-fertilization has evolved repeatedly in association with the colonization of temporary habitats. Removed from their normal pollinators and in small newly founded populations, homostylous variants would be favored over outcrossing morphs as a result of fertility selection (Baker, 1955, 1959). My own studies on two heterostylous groups, both of which are composed primarily of colonizing taxa, provide evidence in support of the role of reproductive assurance as a selective factor favoring the evolution of self-fertilization. This chapter reviews work on the evolutionary breakdown of distyly and tristyly, respectively, in the *Turnera ulmifolia* complex and in the genus *Eichhornia*. Both taxa are primarily Neotropical in distribution and the breakdown of heterostyly is associated with the colonization of geographically marginal areas.

HOMOSTYLE EVOLUTION IN THE TURNERA ULMIFOLIA COMPLEX

Turnera ulmifolia is a polymorphic complex of perennial weeds composed of diploid, tetraploid and hexaploid varieties (Shore and Barrett, 1985a). Diploid and tetraploid populations are distylous with strong self-incompatibility and a 1:1 ratio of the long- and short-styled (hereafter L and S) morphs in populations (Barrett, 1978). Dimorphism is controlled by a single gene "locus" with L plants of genotype *ss* and *ssss* and S plants of genotype *Ss* and *Ssss* in diploids and tetraploids, respectively (Shore and Barrett, 1985b, 1987). In contrast, hexaploid populations are homostylous and self-compatible. Homostyles are atypical in possessing wide variation in the relative positions of reproductive organs and hence the facility for autogamy (Fig. 2 c,d,e). The three homostylous varieties that we have studied experimentally are differentiated for morphological traits and isozyme patterns as well as being intersterile. They occur at different margins of the Neotropical range of the species complex (Argentina, Mexico, Caribbean), indicating that distyly has broken down to homostyly on at least three occasions in the complex in association with the hexaploid condition (Barrett and Shore, 1987).

Patterns of floral variation in *T. ulmifolia* are particularly complex in the Caribbean region. On large islands (e.g. Greater Antilles) populations are either tetraploid and distylous or hexaploid and homostylous, whereas on smaller islands (e.g. Bahamas) only homostyles occur. It seems likely that this distribution pattern results from cycles of island colonization and extinction. Homostyles would be favored over the self-incompatible morphs because of the capacity of single individuals to found populations (Baker, 1955; Cox, this volume). On Jamaica, populations are uniformly hexaploid and self-compatible but display a wide range of floral variation from the long homostylous phenotype (long stamens and long styles) to plants with flowers that resemble those of the typical L morph from distylous populations (Fig. 2 c,d). This variation in stigma-anther separation (herkogamy) occurs both within and among Jamaican populations and has an important influence on both the facility for spontaneous self-pollination and the outcrossing rate of maternal parents (Barrett and Shore, 1987). Individuals with well developed herkogamy experience less self-fertilization than

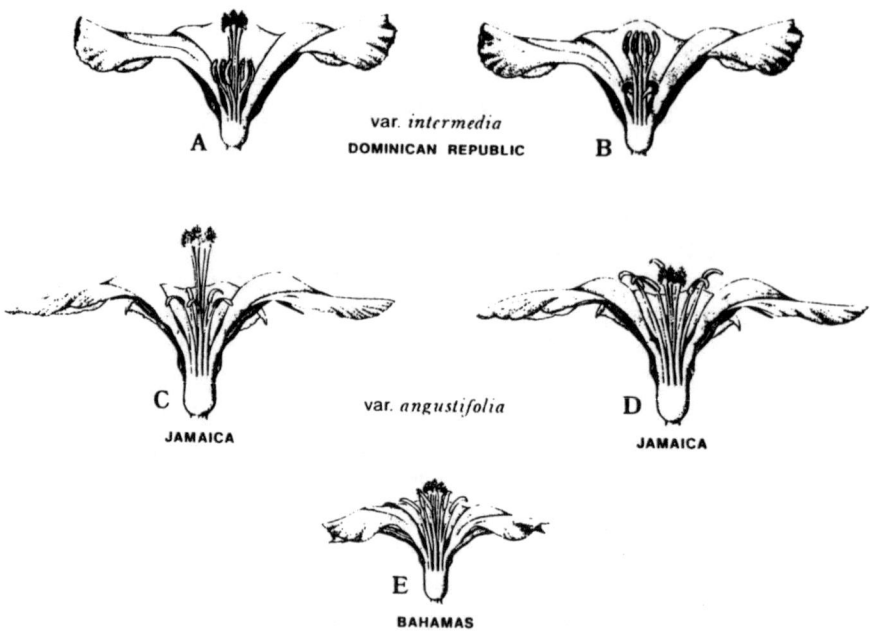

Figure 2. Floral variation in the *Turnera ulmifolia* complex. A,B) L and S morphs, respectively, of var. *intermedia* (4x) from distylous population from the Dominican Republic. C,D) Long homostyles of var. *angustifolia* (6x) from Jamaica with contrasting stigma-anther separations (herkogamy). E) Small flowered autogamous long homostyle of var. *angustifolia* from the Bahamas. After Barrett and Shore (1987).

those in which stigmas and anthers are positioned close together within a flower. The range of floral variation in *T. ulmifolia* raises questions concerning the origin and evolutionary relationships among forms within the complex. In particular, because of the atypical nature of homostyles it is pertinent to ask whether all monomorphic populations in the complex are derived from distylous ancestors as a result of crossing-over within the heterostyly supergene. An alternative possibility is that some phenotypes have originated from L plants through mutation at the SI locus. Mutations of this type occur frequently in the floral morphs of heterostylous species and have been reported in both diploid and tetraploid populations of *T. ulmifolia* (Shore and Barrett, 1986).

It is possible to distinguish between the two alternative hypotheses to account for the origin of homostylous forms in *T. ulmifolia* by assessing the fertility of controlled crosses between distylous and homostylous plants (Dowrick, 1956; Baker, 1975). If floral monomorphism has arisen by crossing-over within the distyly supergene, residual self-incompatibility in crosses with the L and S morphs is likely to be evident in the pistils and pollen of homostylous plants, irrespective of their relative stamen and style lengths. In contrast, if homostyles with well developed herkogamy are simply L plants that have lost self-incompatibility through mutation, then residual self-incompatibility reactions should be absent in crosses with the distylous morphs.

A crossing program between the distylous morphs and plants from 12 populations comprised of three hexaploid varieties of T. ulmifolia were used to distinguish between these two hypotheses (Barrett and Shore, 1987). The results provided clear evidence that the three hexaploid varieties are long homostyles, since the predicted seed set patterns from the cross-over model were revealed in all phenotypes (Fig. 3). Pollen of the S morph was compatible in crosses with homostyles, whereas pollen from the L morph was not. In reciprocals, pollen of homostyles was only compatible in crosses with the L morph. Although some homostylous populations possess "short-level" stamens, their pollen exhibits the incompatibility reaction of long-level stamens of the S morph. Formal genetic analysis of the homostyles also confirmed that they are long homostyles that have arisen through recombination in the distyly supergene (Shore and Barrett, 1985b). The finding that each homostylous variety in T. ulmifolia is a long-homostyle provides empirical support for Charlesworth and Charlesworth's theoretical model (1979a) of the breakdown of distyly. They demonstrated that if the "allele" determining the S morph is dominant, as in T. ulmifolia, long-homostyles will spread to fixation with greater probability than other self-compatible recombinant phenotypes.

Following each independent origin of long homostyly in the T. ulmifolia complex, homostyles colonized habitats not occupied by their distylous progenitors at the margins of the range of the species complex. In some populations it appears that selection pressures favoring increased outcrossing has resulted in the re-establishment of herkogamy. This may be more readily achieved in T. ulmifolia through selection on polygenic variation for reproductive organ length than by the de novo development of alternative outcrossing systems based on self-incompatibility or dioecism. It is remarkable that despite the shift to hexaploidy and considerable morphological and genetic divergence from their distylous ancestors, homostylous forms in T. ulmifolia still retain their ancestral incompatibility behavior, although it serves no apparent function and cannot influence the mating patterns of

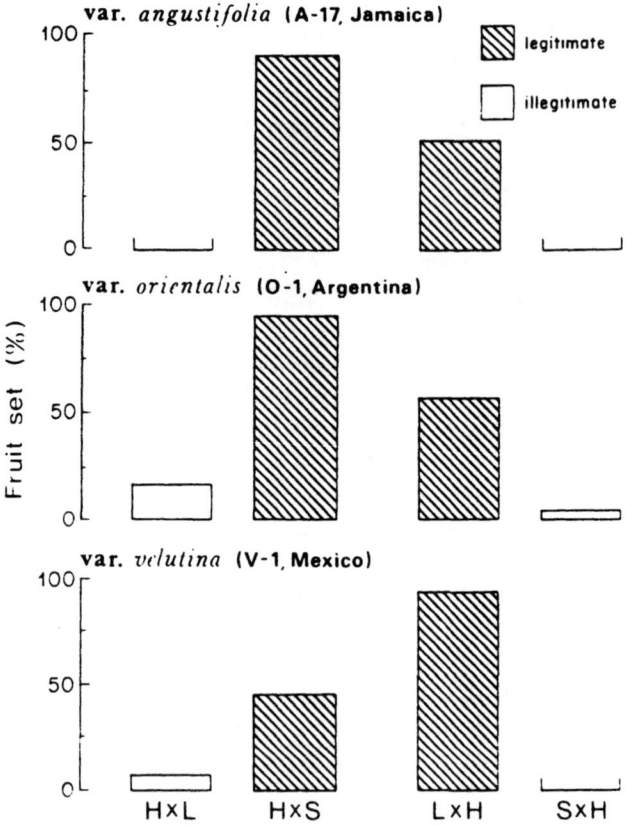

Figure 3. Compatibility relationships between distylous and homostylous varieties within the *Turnera ulmifolia* complex. Histograms illustrate the percentage fruit obtained in crosses between the long- and short-styled morphs (L, S, respectively) and three homostylous (H) varieties. In a distylous population, legitimate crosses are compatible whereas illegitimate crosses are incompatible. For further details see Barrett and Shore (1987).

these self-compatible plants. The evolutionary lability of mating systems in the *T. ulmifola* complex illustrated in Figure 4 suggests that the evolution of selfing from outcrossing is not necessarily a unidirectional change as has been frequently postulated in other plant groups.

In *Primula* (Ernst, 1955), *Armeria* (Baker, 1966), *Linum* (Baker, 1975), and *Turnera*, the breakdown of dimorphic incompatibility is the result of recombination in the supergene that controls distyly. However, this is not the only genetic mechanism by which incompatibility can be modified in heterostylous groups. A number of taxa are known in which the

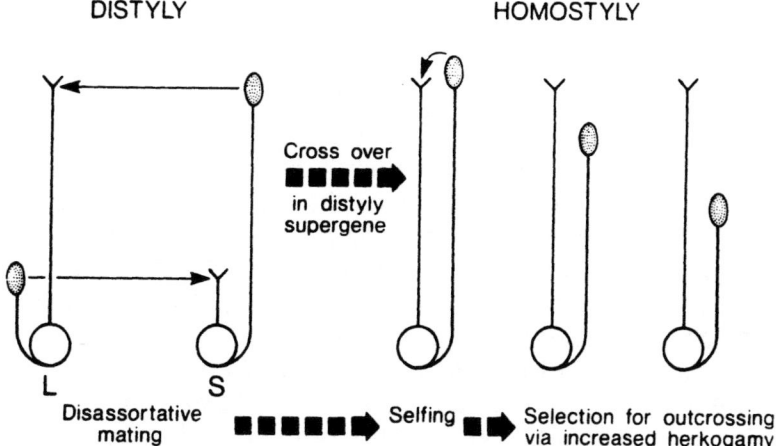

DISTYLY   HOMOSTYLY

Figure 4. Model of the evolutionary relationships among distylous and homostylous forms in the *Turnera ulmifolia* complex. After Barrett and Shore (1987).

style morphs are highly self-compatible but the stamen-style polymorphism is unmodified. Since it is unlikely that heterostyly can evolve prior to the development of an incompatibility system (Charlesworth and Charlesworth, 1979b; but see Lloyd and Webb, 1989) for an alternative view) it seems more probable that these taxa have secondarily lost their incompatibility systems through the accumulation of self-compatibility mutations at the incompatibility locus. The occurrence of populations of heterostylous species that maintain significant genetic variation for incompatibility expression (e.g., Barrett and Anderson, 1985) supports this view.

SEMI-HOMOSTYLE EVOLUTION IN EICHHORNIA

Among the three tristylous families (Lythraceae, Oxalidaceae and Pontederiaceae) there is good evidence of the repeated breakdown of trimorphic incompatibility to give rise to predominantly self-fertilizing populations (Charlesworth, 1979; Barrett, 1988a). The commonest pathway involves relaxation and eventual loss of the incompatibility system followed by modifications in the relative position of style and stamen heights. The resulting phenotypes, known as semi-homostyles, have one set of anthers adjacent to the stigma and, as a result, are largely self-pollinating. In contrast to the distylous groups discussed above, where a quantum change in incompatibility system and mating behavior occurs through recombination, semi-homostyle formation in tristylous species usually develops in stages with the mating system evolving gradually towards increased levels of self-fertilization. The evolutionary breakdown of tristyly is particularly evident in *Eichhornia*, a small monocotyledonous genus of freshwater aquatics that inhabit Neotropical lakes, marshes, and seasonal pools. Our studies of this group have attempted to integrate information from ecology, genetics, and development in an effort to understand the processes responsible for the repeated breakdown of this complex genetic polymorphism (Barrett, 1988b). Since, with the exception of the notorious aquatic weed, *E. crassipes* (water hyacinth), the group is poorly known, much of our early work attempted to document the taxonomic distribution of breeding systems and chromosome numbers in the genus, and to determine the geographical distribution and ecological preferences of individual taxa. This information has been used to interpret both the phylogenetic relationships and biogeographical history of the group (Eckenwalder and Barrett, 1986).

Of the eight species of *Eichhornia*, three possess large showy flowers and are primarily tristylous, and the remaining five are small-flowered and largely self-pollinating. The selfers exhibit phenotypes which suggest that they are semi-homostylous derivatives of tristylous ancestors. Evidence to support this interpretation is based on the occurrence in populations of the semi-homostylous taxa, of segregation of residual tristylic characters involving weak heteromorphisms of pollen size, style length, and style coloration. Studies of each tristylous species have revealed that floral trimorphism has become modified in the direction of increased selfing with semi-homostylous populations occurring at the margins of their respective Neotropical distributions (Barrett, 1978, 1979, 1985a). Recent work has focussed on *E. paniculata* since populations of this species display a wide spectrum of floral modifications ranging from complete tristyly to semi-homostyly (Fig. 5). Unlike its tristylous congeners (*E. crassipes* and *E. azurea*) the species does not exhibit extensive clonal propagation, and is a short-lived perennial or annual diploid which regenerates exclusively by seed. These features simplify experimental studies and enable genetic changes to be detected more readily in natural populations.

Eichhornia paniculata has a markedly disjunct distribution, with the major centers of occurrence in N.E. Brazil and the Caribbean (Cuba and Jamaica). Surveys of style morph frequency and studies of the ecological genetics of populations in the two regions have provided useful insights into the breakdown process. Figure 6 illustrates stages in the breakdown of tristyly to semi-

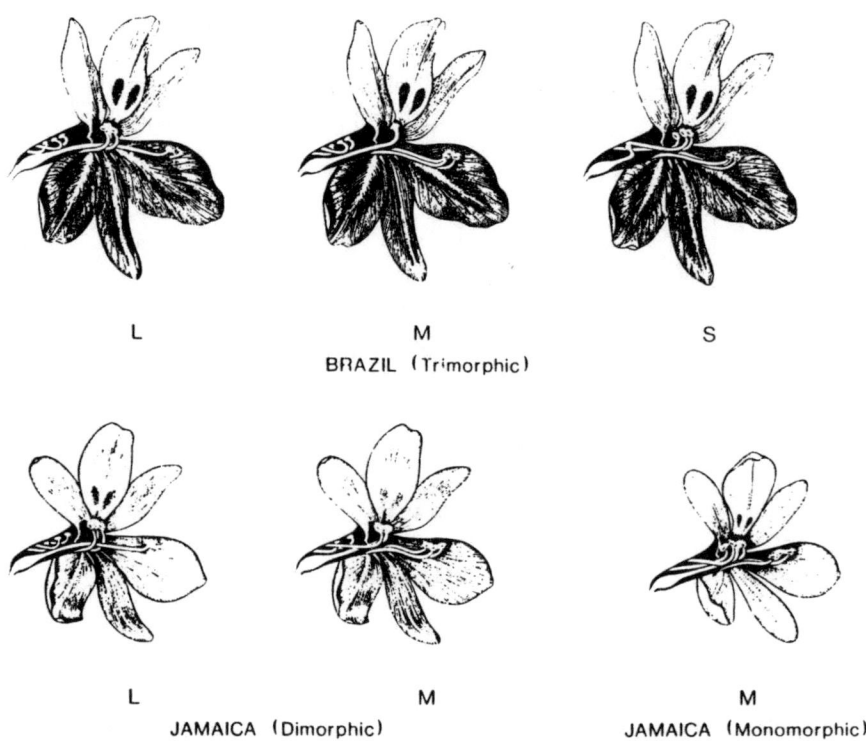

L   M   S

BRAZIL (Trimorphic)

L   M   M

JAMAICA (Dimorphic)   JAMAICA (Monomorphic)

Figure 5. The evolutionary breakdown of tristyly to semi-homostyly in *Eichhornia paniculata*. The change from outcrossing to selfing is accompanied by a reduction in the size and conspicuousness of flowers. Genetic modifications in stamen position that cause self-pollination occur in the M morph of dimorphic and monomorphic populations.

homostyly. The model is based on observations of the patterns of style morph distribution in natural populations and studies of the mating systems and reproductive ecology of populations (Barrett, 1985b; Glover and Barrett, 1986; Barrett et al., 1989). Among populations that we have surveyed in Brazil, 58 were trimorphic, with the S morph under-represented, 21 were dimorphic, composed of the L and M morph, and 5 were fixed for self-pollinating variants of the M morph. On Jamaica the S morph is absent from the island, all plants of the M morph are self-pollinating variants, and the L morph occurs at low frequency in a small number of populations (Fig. 7).

The evolution of selfing in *E. paniculata* involves two key stages. The first involves loss of the *S* allele, and hence the S

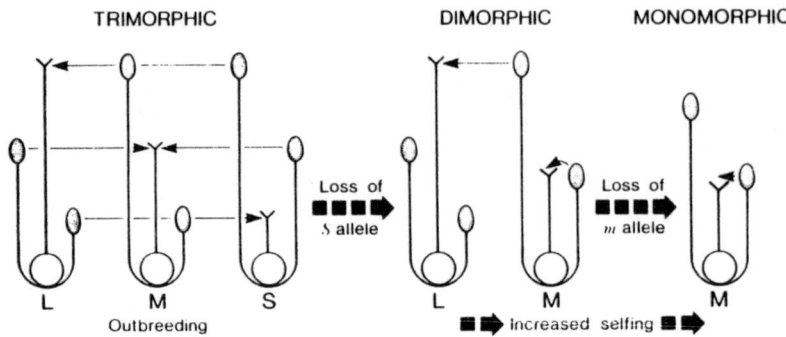

Figure 6. Model of the breakdown of tristyly to semi-homostyly in *Eichhornia paniculata*. Arrows indicate the predominant matings. Note modifications in the short-stamen position of the M morph in dimorphis and monomorphic populations. After Barrett (1985b).

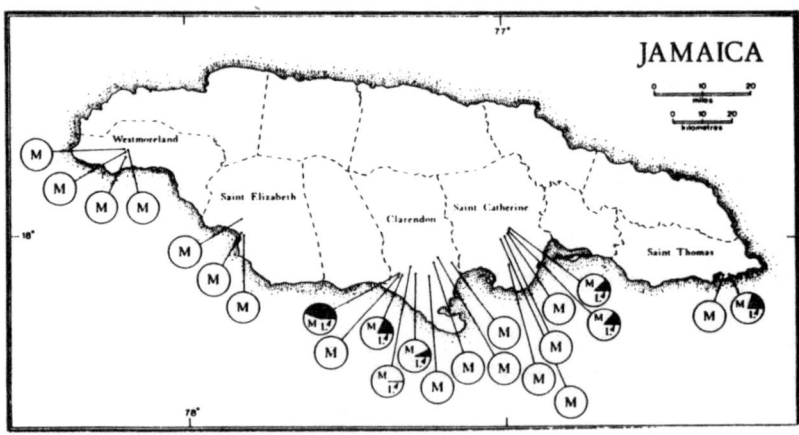

Figure 7. Pattern of style morph distribution in *Eichhornia paniculata* populations from Jamaica. Surveys were conducted in January 1979, 1984, 1987. The figure illustrates all populations that were located during these time periods. All plants of the M morph in Jamaica exhibit modified short stamens. After Barrett, 1988b.

morph; and the second, the loss of the *m* allele and thus the L morph. The most likely explanation for loss of the S morph from populations of *E. paniculata* involves the effect of stochastic influences on population size as a result of bottlenecks and random dispersal. Since the species is highly self-compatible, polymorphic populations can arise from selfing and segregation of genotypes heterozygous at the *S* and *M* loci. Since the dominant *S* allele is carried only by the S morph, separate introductions of this morph are necessary for it to become established in populations. In contrast, the *m* allele can be carried by all three morphs and the *M* allele by the M and S morphs. Computer simulation studies by Heuch (1980) on the effects of random fluctuations of population size in tristylous systems confirm that the S morph is most often lost from populations. Survey data of style morph distribution in the related *E. crassipes* indicate a similar pattern, with the S morph absent from many parts of the New World range (Barrett and Forno, 1982), as well as from the entire adventive Old World range (Barrett, 1977). In the latter case, genotypes of the S morph were presumably not represented among the clones transported to the Old World by humans.

Evidence to support the bottleneck hypothesis comes from surveys of the size, density and distribution of *E. paniculata* populations in N.E. Brazil (Barrett et al., 1989). Trimorphic populations are often large in size, dense, and the majority that we have located are concentrated in two major geographical areas. In contrast, dimorphic and monomorphic populations are often composed of a small number of scattered individuals and most are geographically isolated from the main population centers. This pattern suggests that colonizing episodes, perhaps to ecologically marginal sites, have led to loss of the S morph from populations. Studies of the demography and population genetics of tristylous and non-tristylous populations would be valuable in assessing the bottleneck hypothesis.

Two contrasting but not mutually exclusive selective mechanisms can explain the loss of the L morph from populations of *E. paniculata*. The first involves reproductive assurance favoring the M morph, the second automatic selection of the M morph through mating asymmetries between the morphs. Both are initiated by genetic modifications of the M morph which enable it to self-pollinate. These changes appear to be quite widespread in nature as virtually all non-tristylous populations that we have observed contain selfing variants of the M morph. The modifications involve different degrees of elongation of the three filaments of the short-stamen level. The initial step, which is commonly observed in dimorphic populations, involves a single stamen that elongates to a position equivalent to that of the mid-level stigma (Fig. 6). This results in autonomous self-pollination of the flower. Subsequent modifications, which are largely restricted to monomorphic populations, involve elongation of the two remaining stamens of the short level. These finally take up a position in the mid region of the flower resulting in full semi-homostyly.

Modified M plants of *E. paniculata* rarely occur in tristylous populations and the genes that modify stamen position have no significant phenotypic effects when transferred to the L and S morphs. As a result, in dimorphic populations plants of the M morph frequently display genetic modifications of stamen position, whereas the L morph remains unmodified. This difference has a profound effect on the mating system of populations. Unlike tristylous populations where each morph is highly outcrossed, in dimorphic populations the M morph experiences a high level of self-fertilization whereas the L morph remains largely outcrossing (Fig. 8). With this mating asymmetry among the morphs and no major fitness differences between progeny arising from them, the M morph will replace the L morph. This is because genes that cause an increased rate of self-fertilization have an automatic advantage, since the maternal parent will transmit genes via both pollen and ovules to selfed progeny and thus evade the "cost of meiosis." It should be noted that while we have verified that the maternal outcrossing rates of the L and M morphs are consistent with the model presented in (Fig. 6), we have yet to measure the male fertility of the two morphs to confirm that the M morph has higher male fitness. Experiments with electrophoretic markers similar to those undertaken by Schoen and Clegg (1985) are in progress to enable us to do this. Populations of *E. paniculata* on the island of Jamaica are short-lived as a result of frequent droughts and human disturbance. As a consequence, it seems probable that reproductive assurance may be a more important selective factor in determining the relative frequency of style morphs on the island than the mating asymmetry hypothesis discussed above. The large number of monomorphic M populations on the island (Fig. 7) most likely results from repeated colonizing events and periods of low density. These conditions would favor establishment of self-pollinating variants of the M morph over the non-autogamous L morph. Erratic pollinator service and the absence of specialized long-tongued bees on Jamaica prevent the normal functioning of heterostyly and result in a considerable fertility advantage to the M morph in comparison with the L morph in dimorphic populations (Fig. 9). In contrast, in tristylous populations from

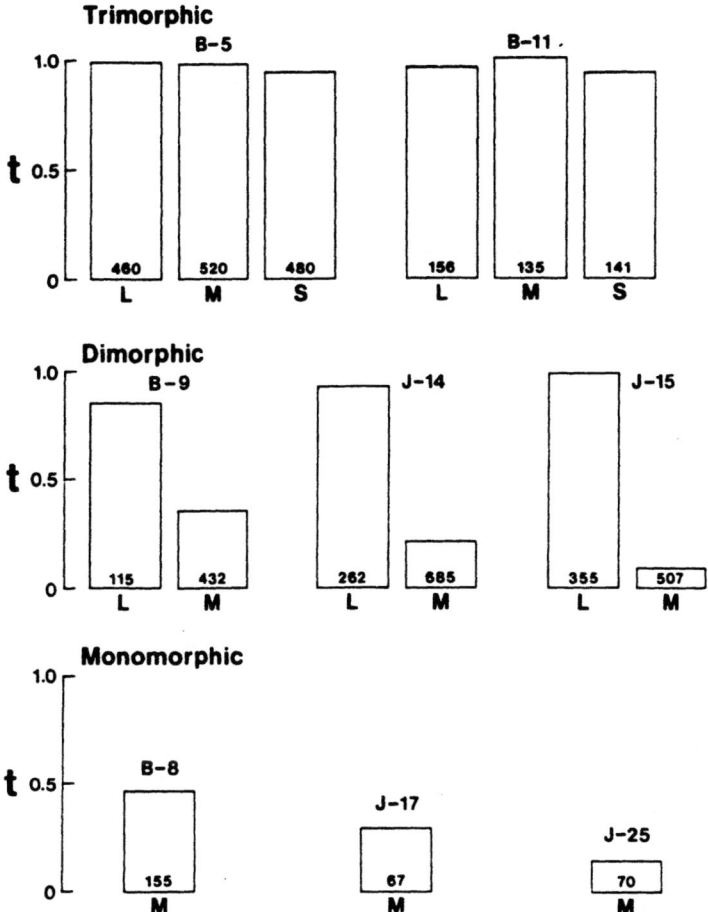

Figure 8. Multilocus estimates of outcrossing rate (t) of the floral morphs in trimorphic, dimorphic and monomorphic populations of *Eichhornia paniculata* from N.E. Brazil (B) and Jamaica (J). Outcrossing rates were estimated by the method of Ritland and Jain (1981) using isozyme loci. The number of progeny assayed per morph are indicated. Within trimorphic populations there were no significant differences between the outcrossing rates of floral morphs whereas in dimorphic populations the morphs differed significantly (P<<0.001) in outcrossing rate.

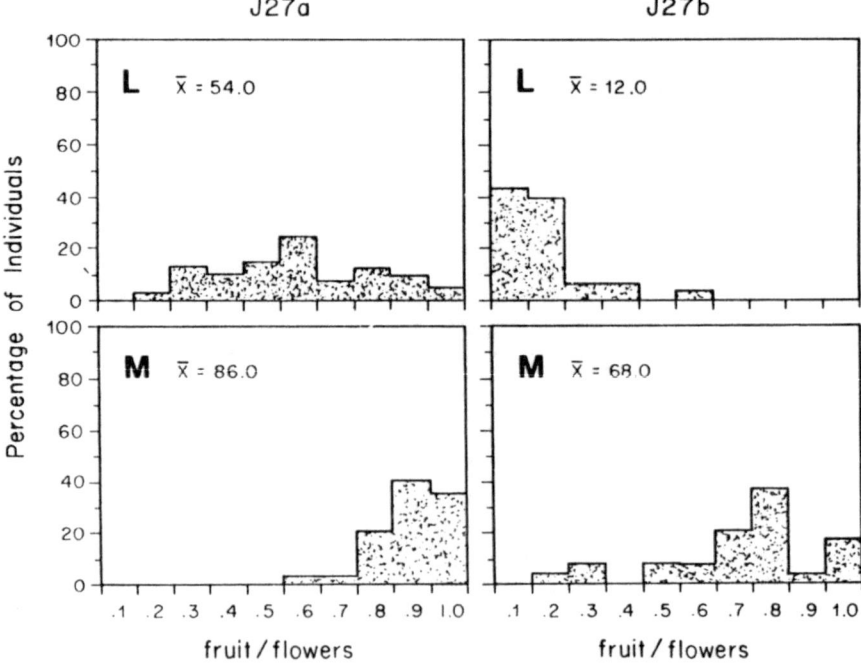

Figure 9. Fruit set of floral morphs in two dimorphic populations of *Eichhornia paniculata* from Jamaica. All plants of the M morph in both populations were selfing variants with a high capacity for autonomous self-pollination. The L morph is non-autogamous because of well-developed herkogamy.

N.E. Brazil specialized pollinators are abundant, fruit set is generally high, and estimates of disassortative mating indicate that most matings occur between the floral morphs (Barrett et al., 1987, 1989)

Investigations of the mating systems and genetic structure of *E. paniculata* populations from N.E. Brazil and Jamaica point to the importance of both founder effects and increased selfing on genetic diversity (Glover and Barrett, 1987; B. C. Husband and S. C. H. Barrett, unpubl. data). A survey of 21 allozyme loci in 11 populations revealed that Jamaican populations were genetically depauperate in comparison with those from Brazil. The low levels of allozyme polymorphism on Jamaica probably result from a restricted number of long-distance dispersal events to the island. The predominance of self-pollinating variants of the M morph on Jamaica suggests that they were favored during establishment after long-distance dispersal. Among the 11 populations surveyed, outcrossing rates were significantly correlated with the number of polymorphic loci, alleles per locus, observed heterozygosity, and genetic diversity of populations. These results indicate the important influence of both the mating system and founder events on levels of genetic variation in plant populations.

CONCLUSION

Experimental studies on the evolutionary genetics of two ecologically distinct heterostylous groups indicate that the evolution of self-fertilization from outcrossing is associated with colonizing episodes and periods of low population density. The geographical distribution of floral forms in both groups provide evidence in support of Baker's Law. In *Turnera ulmifolia* self-compatible homostyles occur at the margins of the range of the complex and have successfully established on many small islands in the Caribbean where self-incompatible forms are absent. In *Eichhornia paniculata* most populations in N.E. Brazil are tristylous and likely outcrossing, whereas on the island of Jamaica autogamous semi-homostylous variants of the M morph predominate. Although in this species all populations, regardless of floral morphology, are highly self-compatible, the capacity for autonomous self-pollination in homostyles must have provided a selective advantage in establishment and subsequent spread on the island following long-distance dispersal. Baker's Law contrasts the colonizing ability of self-compatible and self-incompatible plants. Self-compatible species populations that are highly autogamous are likely to be favored in establishment after long-distance dispersal.

SUMMARY

The evolution of predominant selfing from obligate outcrossing is one of the major pathways of mating system evolution in flowering plants. Heterostylous genetic polymorphisms (distyly and tristyly) provide a rich source of material in which to test hypotheses concerned with the evolution of self-fertilization. Studies of two Neotropical heterostylous groups (*Turnera ulmifolia* complex and *Eichhornia* spp.) provide evidence that the breakdown of heterostyly to homostyly is associated with colonizing episodes and periods of low population density. The self-pollinating habit of homostyles provides reproductive assurance under conditions of uncertain pollinator service. In *T. ulmifolia* homostyles arise in one step by recombination in the supergene that controls distyly. Crossing studies indicate that this has occurred on at least three occasions in the species complex in association with changes in ploidal level. Selection for increased outcrossing appears to have favored the evolution of herkogamy in homostylous populations of *T. ulmifolia*. The breakdown of tristyly in *Eichhornia paniculata* occurs in distinct stages, with the mating system evolving gradually towards increased levels of self-fertilization. Stochastic influences on population size and the spread of mating system modifier genes result in changes from floral trimorphism through dimorphism to monomorphism. Polymorphic sexual systems, such as heterostyly, provide excellent material for empirical studies of mating system evolution.

ACKNOWLEDGMENTS

My research on *Turnera* and *Eichhornia* began while I was a graduate student at Berkeley and was stimulated by Herbert Baker's classic studies on heterostylous breeding systems. I thank both him and Irene Baker for the support, encouragement and freedom they gave me during this period. I also thank Deborah Glover, Brian Husband, Martin Morgan, Jennifer Richards and Joel Shore for advice and assistance, Elizabeth Campolin for drawing the figures, and the Natural Sciences and Engineering Research Council of Canada for financial support.

LITERATURE CITED

Baker, H. G. 1955. Self-compatibility and establishment after "long-distance" dispersal. Evolution 9: 347-348.

Baker, H. G. 1959a. Reproductive methods as factors in speciation in flowering plants. Cold Spring Harb. Symp. in Quant. Biol. 24: 177-191.

Baker, H. G. 1959b. The contribution of auteocological and genecological studies to our knowledge of the past migration of plants. Amer. Natur. 93: 255-272.

Baker, H. G. 1961. Rapid speciation in relation to changes in the breeding systems of plants. In Recent Advances in Botany, University Toronto Press, Toronto, Ontario, Canada.

Baker, H. G. 1966. The evolution, functioning and breakdown of heteromorphic incompatibility systems. I. Plumbaginaceae. Evolution 20: 349-368.

Baker, H. G. 1975. Sporophyte-gametophyte interaction in *Linum* and other genera with heteromorphic incompatibility. In D.L. Mulcahy [ed.]. Gamete Competition in Plants and Animals. North-Holland, Amsterdam, Netherlands.

Barrett, S. C. H. 1977. Tristyly in *Eichhornia crassipes* (Mart.) Solms. (Water Hyacinth). Biotropica 9: 230-238.

Barrett, S. C. H. 1978a. Heterostyly in a tropical weed: the reproductive biology of the *Turnera ulmifolia* complex. Can. J. Bot. 56: 1713-1725.

Barrett, S. C. H. 1978b. Floral biology of *Eichhornia azurea* (Sw.)Kunth. (Pontederiaceae). Aquat. Bot. 5: 217-228.

Barrett, S. C. H. 1979. The evolutionary breakdown of tristyly in *Eichornia crassipes* (Mart.) Solms. (Water Hyacinth) Evol. 33: 499-510.

Barrett, S. C. H. 1985a. Floral trimosphism and monomorphism in continental and island populations of *Eichhornia paniculata* (Spreng.) Solms. (Pontederiaceae) Biol. J. Linn. Soc. 25: 41-60.

Barrett, S. C. H. 1985b. Ecological genetics of breakdown in tristyly. In J. Jaeck and J.W. Woldendorp [eds.]. Structure and Functioning of Plant Populations II: Phenotypic and Genotypic Variation in Plant Populations. North-Holland, Amsterdam, Netherlands.

Barrett, S. C. H. 1988a. The evolution, maintenance and loss of self-incompatibility systems, In J. and L. Lovett Doust [eds.]. Reproductive Strategies of Plants: Patterns and Strategies. Oxford University Press, Oxford.

Barrett, S. C. H. 1988b. Evolution of breeding systems in *Eichhornia* (Pontederiaceae) : a review. Ann. Missouri. Botan. Gard.75: 741-760.

Barrett, S. C. H. 1989. Mating system evolution and speciation in heterostylous plants, In D. Otte and J. Endler [eds.]. Speciation and Its Consequences. Sinauer Associates, Sunderland, MA (in press).

Barrett, S. C. H., and J. M. Anderson. 1985. Variation in expression of trimorphic incompatibility in *Pontederia cordata* L. (Pontederiaceae). Theor. Appl. Genet. 70: 355-362.

Barrett, S. C. H., and I. W. Forno. 1982. Style morph distribution in New World populations of *Eichhornia crassips* (Mart.) Solms-Laubach (water hyacinth). Aquat. Bot. 13: 299-306.

Barrett, S. C. H., and J. S. Shore. 1987. Variation and evolution of breeding systems in the *Turnera ulmifolia* L. complex (Turneraceae). Evol. 41: 340-354.

Barrett, S. C. H., A. H. D. Brown, and J. S. Shore. 1987. Disassortative mating in tristylous *Eichhornia paniculata* (Spreng.) Solms. (Pontederiaceae). Heredity 58: 49-55.

168

Barrett, S. C. H., M. T. Morgan, and B. C. Husband. 1989. Dissolution of a complex genetic polymorphism: the evolution of self-fertilization in tristylous *Eichhornia paniculata* (Pontederiaecae). Evolution (in press).

Charlesworth, B., and D. Charlesworth. 1979a. The maintenance and breakdown of distyly. Amer. Nat. 114: 449-513.

Charlesworth, D., and B. Charlesworth. 1979b. A model for the evolution of heterostyly. Amer. Nat. 114: 467-498.

Charlesworth, D. 1979. The evolution and breakdown of tristyly. Evolution 33: 489-498.

Darwin, C. D. 1877. The Different Forms of Flowers on Plants of the Same Species. John Murray, London.

Dowrick, V. P. J. 1956. Heterostyly and homostyly in *Primula obconica*. Heredity 10: 219-236.

Eckenwalder, J.E., and S.C.H. Barrett. 1986. Phylogenetic systematics of Pontederiaceae. Syst. Bot. 11: 373-391.

Ernst, A. 1955. Self-fertility in monomorphic Primulas. Genetica 27: 391-448.

Ganders F. R. 1979. The biology of heterostyly. New Zeal. J. Bot. 17: 607-635.

Glover, D. E., and S. C. H. Barrett. 1986. Variation in the mating system of *Eichhornia paniculata* (Spreng.) Solms. (Pontederiaceae). Evol. 40: 1122-1131.

Glover, D. E., and S. C. H. Barrett. 1987. Genetic variation in continental and island populations of *Eichhornia paniculata* (Pontederiaceae). Heredity 59: 7-17..

Heuch, I. 1980. Loss of incompatibility types in finite populations of the tristylous plant, *Lythrum salicaria*. Hereditas 92: 53-57.

Jain, S. K. 1976. The evolution of inbreeding in plants. Ann. Rev. Ecol. Syst. 7: 469-495.

Lloyd, D. G. and C. J. Webb. 1989. The evolution of heterostyly. In M. Nicholls [ed.]. The Evolution and Adaptive Significance of Heterostyly. Springer-Verlag, Berlin (in press).

Ritland, K., and S. K. Jain. 1981. A model for the estimation of outcrossing rate and gene frequencies using n independent loci. Heredity 47: 35-52.

Schoen, D. J., and M. T. Clegg. 1985. The influence of flower color on outcrossing rate and male reproductive success in *Ipomoea purpurea*. Evol. 39: 1242-1249.

Shore, J. S., and S. C. H. Barrett. 1985a. Morphological differentiation and crossability in the *Turnera ulmifolia* complex (Turneraceae). Syst. Bot. 10: 308-321.

Shore, J. S., and S. C. H. Barrett. 1985b. Genetics if distyly and homostyly in the *Turnera ulmifolia* complex (Turneraceae). Heredity 55: 167-174.

Shore, J. S., and S. C. H. Barrett. 1986. Genetic modifications of dimorphic incompatibility in the *Turnera ulmifolia* complex (Turneraceae). Can. J. Genet. Cytol. 28: 796-807.

Shore, J. S., and S. C. H. Barrett. 1987. Inheritance of floral and isozyme polymorphisms in *Turnera ulmifolia* L. (Turneraceae). J. Heredity 78: 44-48.

Stebbins, G. L. 1957. Self fertilization and population variability in the higher plants. Amer. Nat. 91: 337-354.

Stebbins, G. L. 1974. Flowering Plant Evolution above the Species Level. Belknap, Cambridge, MA.

FACULTATIVE XENOGAMY: EXAMINATION OF A MIXED MATING SYSTEM

Robert William Cruden and David L. Lyon

INTRODUCTION

Although various workers have indicated that plant mating systems be considered in the context of a trichotomy, i.e., predominantly outbred, predominantly inbred, and mixed (e.g., Grant, 1958, Baker, 1959), mating systems are frequently presented and/or interpreted in the context of a dichotomy with xenogamous, outbreeding species, on one branch and autogamous, inbreeding species oh the other (e.g., Feagri and Pijl, 1978; Lande and Schemske, 1985; Richards, 1986). With the exception of species that produce both chasmogamous and cleistogamous flowers, species with mixed mating systems such as facultative xenogamy (Cruden, 1977), have received minimal attention (e.g. Grant, 1958, Wyatt, 1983; Richards, 1986). This paradigm reflects at least in part the fact that most of the interest has been focused at the two ends of the spectrum, i.e., either on primarily xenogamous or primarily autogamous species. Favorite topics include (1) a long and continuing interest in adaptations that facilitate xenogamy and outbreeding (e.g., Darwin, 1895; Stebbins, 1950; Richards, 1986); (2) the recent interest in the evolution of weedy and/or colonizing species, thus the focus on autogamy and inbreeding (e.g., Baker 1955; Stebbins, 1957; Baker and Stebbins, 1965; Ornduff, 1969; Baker, 1974; Jain, 1976; Lloyd, 1979); (3) models that show genes for self-pollination should spread rapidly following their introduction into a population of outbreeding plants (e.g., Wells, 1979); and (4) experimental protocols involving the exclusion of pollinators that distinguished quickly and accurately between species that can and those that can not self-pollinate.

There is a consensus that most, if not all, facultatively autogamous species are derived from xenogamous species (e.g., Stebbins, 1957; Ornduff, 1969; Richards, 1986) and that a "facultatively self-fertilizing" stage may be a necessary step in the evolution of autogamy (Stebbins, 1957). However, it does not follow that an autogamous mating system is the inevitable evolutionary outcome once self-pollination can occur without significant inbreeding depression (Lloyd, 1979; Holsinger, 1986). Schemske and Lande (1985) argued that selection might

favor intermediate mating systems that were able to track
environmental variation, and suggested that plants that produce
both chasmogamous and cleistogamous flowers are explicable in
such a context.

Here, we examine facultative xenogamy, a mixed mating
system of strictly chasmogamous, perfect-flowered plants.
Cruden (1977) observed that facultatively xenogamous species
grow in habitats or flower during periods when pollinator activity
is unreliable. He suggested that the flowers of such species should
be cross-pollinated if pollinators were present but would self-
pollinate in their absence. It follows that the fruit set of
facultatively xenogamous species should be high relative to that of
xenogamous species because pollination success of the former is
less subject to environmental conditions that decrease pollinator
activity and thereby reduce the fruit set of xenogamous species.

We have three general objectives: 1) to demonstrate that the
flowers of facultatively xenogamous species can function as Cruden
(1977) predicted, 2) to show that facultative xenogamy is
different from both xenogamy and facultative autogamy, as well as
other mixed mating systems, and 3) indicate when and/or where
facultative xenogamy is adaptive. We will conclude, based on the
available data, that facultative xenogamy is a unique and
evolutionarily stable mating system rather than an intermediate
stage in the evolution of autogamy.

Specifically, we present data from a number of species that
directly test Cruden's (1977) hypothesis that the flowers of
facultatively xenogamous species may be cross-pollinated if
pollinators are present and self-pollinate in their absence. We
examine differences among mating systems, including 1) pollen-
ovule rations (P/O's), 2) outbreeding rates, and 3) the
relationship between P/O and pollen grain size. We review and
update the data on P/O's because Cruden (1977) showed that P/O's
differ significantly among mating systems. We examine
outbreeding rates because those of facultatively xenogamous
species should be intermediate between those of xenogamous and
facultatively autogamous species. In addition, Schmeske and Lande
(1985) argued that the distribution of breeding systems should be
bimodal, a conclusion that Waller (1986) has questioned. The
relationship between P/O and pollen grain size is of interest
because in xenogamous plants, P/O is negatively related to pollen
grain size and the ratio of the stigmatic area to the pollen-bearing
area of the pollinator (Cruden and Miller-Ward, 1981). If the
negative relationship between P/O and pollen grain size reflects
vector mediated pollination then it should occur among
facultatively xenogamous species. However, in facultatively

autogamous species the evolution of floral traits is not constrained by the pollinator-flower interaction; thus a negative relationship is not expected. There is an alternative hypothesis: P/O and pollen number are essentially reciprocals and a negative relationship is predicated regardless of mating system. Finally, we examine the balance between xenogamy and autogamy in facultatively xenogamous species and discuss habitats in which facultative xenogamy is adaptive. We also compare facultative xenogamy to two other mixed-mating systems. One, which we term "morphological" is found in species with two types of flowers: cleistogamous, self-pollinated ones and chasmogamous, cross-pollinated ones. The other, which we term "genetic" exists in species whose populations are mixtures of individuals some of which can self-pollinate and others cannot, the ability being genetically determined.

MATERIALS AND METHODS

Most of the species we studied (Tables 1-3, 7) were selected because they were available locally and casual examination indicated they were either facultatively autogamous or facultatively xenogamous. The mating systems of each species were defined on the basis of an out-crossing index (Cruden, 1977) and whether there was a reasonable period of time during which cross-pollination could occur, i.e., whether self-pollination occurred at the time of flower opening or soon thereafter or at the time of flower closure. The species in Tables 4-6 were assigned to their respective mating system based on descriptions of their floral behavior, pollination biology, etc.

We added an emasculation treatment to the standard pollinator exclusion protocol to determine if pollen is effectively moved by a vector. Our experimental design included three treatments: 1) open-pollinated flowers, 2) open-pollinated, emasculated flowers, and 3) flowers of caged plants. Data from an emasculation treatment must be interpreted carefully. First, it does not distinguish between xenogamy and geitonogamy. Second, if pollinators, such as bees, are visiting the flowers primarily for pollen, they may recognize and avoid emasculated flowers, thus providing underestimates of vector-mediated pollination. Third, the stigma and style may function abnormally (Young, 1982; also see Estes and Thorp, 1975). Of these potential problems we encountered only one, i.e., small bees remained on emasculated flowers of *Linum ulcatum* for shorter periods of time then they did on unemasculated flowers.

Pollen-ovule ratios were determined from material preserved in alcohol. The number of pollen grains in one anther

Table 1. Characteristics of facultatively xenogamous species. The species are arranged with lowest P/O first.

Species	Pollen/Ovule Ratio $\bar{X}$±S.E.	Pollen grain Volume (μm^3) $\bar{X}$±S.E.	Pollen vectors	Annual or Perennial	Breeding System Characteristics[a]
Geranium carolinianum L.[c]	22±1	144,627±11,202		An	H, SC, A[c]
Linum sulcatum Riddell[d]	36±2	206,499±7,134	bees, flies	An	H, SC, A[c]
Veronica wormskjoldii Roem. & Schult.[e]	48±3	5,373±250		An	
Epilobium sp.[f]	49±6	27,165±1,868		An?	
Malva neglecta Wallr.[g]	138±5	152,880±4,529		An	Sc, A[c]
Oxalis stricta L.[d]	138±8	12,251±422	bees	An?	H, Sc, A[c]
Gentianella amarella (L.) Boern.[h]	212±17	14,436±504		An	H
Circaea alpina L. spp. alpina[e]	392±19	2,482±93		Pr	
Polygonum pennsylvanicum L.[i]	394±20	69,075	bees	An	H
Polygonum virginianum L.[g]	406±17	35,131±2,215	bees	Pr	H, SC, A[c]
Abutilon theophrasti Medic.[g]	508±23	34,800±818		An	Pg, SC A[c]
Solanum nigrum L.[g]	617±30	1,825±49		An	Pg, Sc, A[c]

Phyla lanceolata (Michx.) Greene[d]	797±24	10,490±515		Pr	
Adenocaulon bicolor Hook.[e]	895[b]	9,856±548	flies, wasps	Pr	Pg, SC
Agrimonia pubescens Wallr.[d]	2,008±266	16,080±710	flies	Pr	H, Sc, A[c]
Ellisia nyctelea L.[g]	2,182±178	3,852±239	bees, flies	An	H, SC, A[c]
Lappulla redowskii (Hornem.) Greene[d]	2,228±67	661±38		An	
Polanisia dodecandra (L.) DC.[j]	2,465±204	2,862±128		An	SC, A[c]
Tiarella unifoliata Hook.[e]	5,964±345	1,617,±108	flies	Pr	Pg

a. H = homogamous, Pg = protogynous, SC = self-compatible, A[c] = autogamous on closing.
b. From Cruden (1977).
c. Florida: Leon County, Tall Timbers Research Station.
d. Iowa: Dickinson County, Cayler Prairie.
e. Montana: Glacier National Park.
f. Wyoming: Albany County; 7-8 mi. W Centennial.
g. Iowa: Johnson County, Iowa City.
h. Colorado: Clear Creek County, Loveland Pass.
i. Iowa: Louisa County, Cone Marsh.
j. Iowa: Cherokee County, 1-2 mi. N Cherokee.

was multiplied by the number of anthers and divided by the number of ovules in the ovary. The sample size for all P/O counts is 10.

Pollen grain volumes were calculated from measurements obtained from stained pollen grains. The pollen from a minimum of one flower from each of five individuals was used and means are based on 25 haphazardly selected grains. The pollen grains were dissected from the anthers in 95% ethanol, washed in ethanol, stained in basic fuchsin in 95% ethanol, washed in absolute alcohol, dehydrated in an alcohol-xylene series and mounted in piccolyte. The critical measurements, either the diameter or the long and short axes, were made with an ocular micrometer at 40X. The data were analyzed with a variety of statistical procedures, including one-way ANOVA and Student-Newman-Keuls test, the Wilcoxon two sample test (Sokal and Rohlf, 1981), and regression analysis (Ryan *et al.*, 1985). The test for skewness is from Sokal and Rohlf (1981). The parametric analyses were performed on log or arcsine transformed data. Means of untransformed data plus or minus one standard error are given in the text and tables.

RESULTS
Field Studies
The perennial species occurred in late successional seres; among annual species, some were found in late and others in early successional seres (Tables 1-3). The flowers of these species (Table 1) were either protogynous or homogamous, self-compatible, and had delayed autogamy, i.e., they self-pollinated during flower closure. Many produced nectar and most were visited by small solitary bees and/or flies.

In *Linum sulcatum* Riddell, *Polygonum virginianum* L., *Mirabilis nyctaginea* (Michx.) MacM., and *Alisma plantago-aquatica* L. there was no pollination prior to the first visits by insects (Table 2). The deposition of equivalent numbers of pollen grains on the stigmas of emasculated and open pollinated flowers of three species demonstrated that pollinators (small bees, flies, and moths) moved pollen between flowers. The few pollen grains remaining in the flowers at the time of abscission or closure (Table 2) probably contributed little to the pollination of those flowers. Flowers of caged plants self-pollinated (Table 2). The stigmas of both open-pollinated and self-pollinated flowers generally received sufficient numbers of pollen grains to result in fruit set.

Data from other species are consistent with the above pattern. The stigmas of 34 flowers from 14 individuals of

Table 2. Stigmatic pollen loads and residual pollen in four facultatively xenogamous species.

	Linum sulcatum[a]		Polygonum virginianum[b]		Mirabilis nyctaginea[c]		Alisma plantago-aquatica[d]	
	n	$\bar{X}$±S.E.	n	$\bar{X}$±S.E.	n	$\bar{X}$±S.E.	n	$\bar{X}$±S.E.
Pollen grains per stigma(s)								
at flower opening	50	0	10	0	80	0	50	0
open pollinated flowers	22	34.0±2.5	20	3.2±0.3	30	17.4±2.7	34	0.0±0.9
emasculated flowers	14	17.1±3.6[e]	22	3.2±0.4	17	14.7±2.2	24	7.2±0.6
flower of caged plants		[f]	21	3.5±0.6	20	11.4±1.6	14	14.1±1.5
Pollen grains per flower								
at time of opening	10	355±16	10	406±17	9	278±13	10	6492±391
at time of closing	17	9.8±2.0	21	5.6±1.4	20	17.0±2.1	22	218±30
Number ovules per flower	10		1		1		10	17.60.3

a. Iowa: Dickinson Co., Cayler Prairie.
b. Iowa: Johnson Co., Iowa City.
c. Dickinson Co., Iowa Lakeside Laboratory.
d. Iowa: Dickinson Co., 2 mi W and 1 mi N Milford.
e. Bees stayed a shorter time on emasculated flowers.
f. The stigmas of all 27 flowers (9 plants) received pollen grains during flower closure.

Agrimonia pubescens (Wallr.) received no pollen on opening and prior to selfing, the stigmas of the upper and lower pistils of 55 flowers had received 5.0 ± 0.7 and 4.9 ± 0.8 pollen grains, respectively, approximately 5 pollen grains per ovule. Likewise, in *Scilla sibirica* (Andr.) the stigmas received no pollen during flower opening and little pollen was left in the anthers of flowers visited by *Apis mellifera* (L.).

Additional evidence that facultatively xenogamous species are effectively pollinated by flower visitors and self-pollinate in their absence is found in a comparison of fruit and seed set among open pollinated, emasculated, and caged flowers (Table 3). In *Scilla sibirica* fruit set was equivalent in open-pollinated and emasculated flowers but caged (c) flowers set significantly fewer seeds than open-pollinated (o) or emasculated (e) flowers ($F_{2,60}$ = 10.73; for c-e Least Significant Range [LSR] = 119.5 < 171.1, $p < .01$; for c-o: LSR = 92.9 < 115.8, $p < .01$). In *Calylophus serrulatus* (Nutt.) Raven, open-pollinated flowers set significantly more fruits than caged flowers ($t = 6.75$, $p < 0.001$) and open-pollinated and emasculated flowers set more seed than caged flowers ($F_{2,146}$ = 9.91, $P < .001$; for o-c: LSR = 3.44 << 40.0 and e-c: LSR = 5.91 << 40.8, $p << .01$ for both). The difference in seed set between open-pollinated and emasculated flowers is not significant (LSF = 18.57 > 0.8). An analysis of the fruit set data for *C. serrulatus* from caged plants indicates no skewness ($g_1 = 0.23$; $t_s = 0.23$) but does show a platykuritic distribution ($g_2 = -1.58$; $t_s = 2.72$; $p < .01$). The distribution appears to be bimodal and plants at the study site had either large or small flowers (Cruden, personal observation). There may be a greater propensity for autogamy in the smaller flowered plants.

Pollen-Ovule Ratio

Most of the known P/O's for species with one of the three mating systems are included in Fig. 1. Wind-pollinated species and those with pollinia or polyads were excluded from the analysis. Although their ranges overlap considerably, there are significant differences among the three mating systems ($F_{2,218}$ = 296.59; $p << .001$; for X-FX: LSR = 2.71 < 6.8; X-FA: LSR = 3.00 < 13.6; and Fx-FA: LSR = 2.91 < 6.8, for all $p < .01$).

If the extreme data for facultative xenogamy (1 species) and xenogamy (6 species) are omitted from the analyses, the mean P/O for each mating system is nearly the same as Cruden (1977) reported: facultative autogamy - 200 ± 27 (n = 67), facultative xenogamy - 1008 ± 191 (n = 61); xenogamy - 5739 ± 915 (n = 86). With the outliers included, mean P/O for facultative

Table 3. Fruit and seed set in two facultatively xenogamous species.

		Calylophus serrulatus[a]		Scilla sibirica[b]	
		n	$\overline{X}\pm$S.E.	n	$\overline{X}\pm$S.E.
% Fruit Set -	Open Pollinated	56	90.4±1.3	23	100
	Emasculated	26	88	15	100
	Caged	15	64.1±4.2	51	96
% Seed Set -	Open Pollinated	44	26.9±1.4	23	75.4±3.5
	Emasculated	15	26.7±1.9	15	82.1±6.3
	Caged	90	21.0±0.7	25	58.7±3.5

a. Iowa: Dickinson County, Cayler Prairie. n for *C. serrulatus* fruit set in open and caged plants are stems, thus it is a x of x's; all other n's are flowers. The low seed set reflects, at least in part, translocation heterozygosity that reduces pollen and ovule fertility by half (Towner, 1977).

b. Iowa: Johnson County, Iowa City.

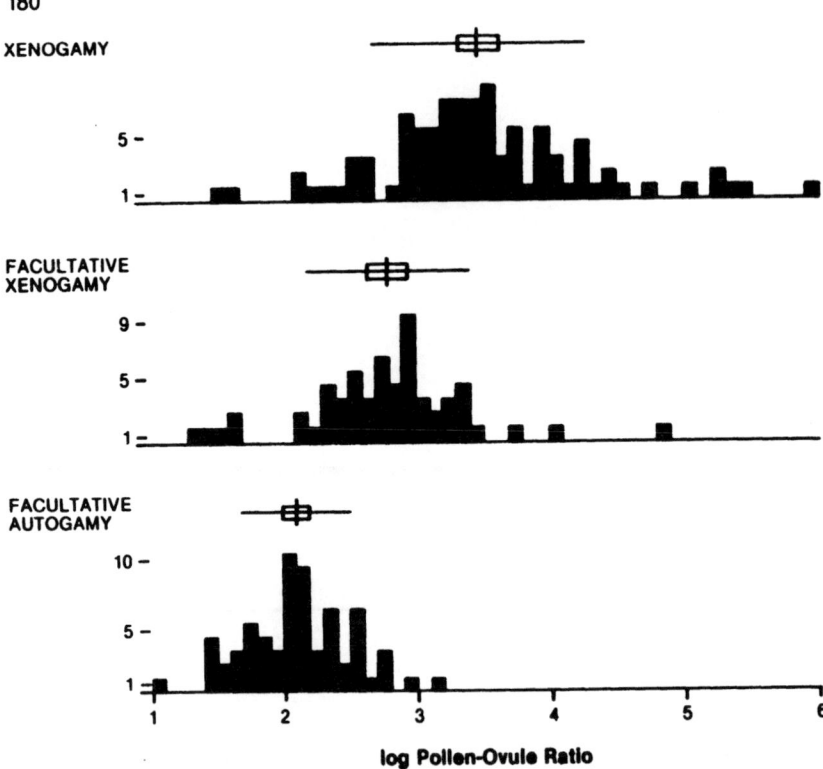

XENOGAMY

FACULTATIVE XENOGAMY

FACULTATIVE AUTOGAMY

log Pollen-Ovule Ratio

Figure 1. Summary of pollen-ovule ratios. The vertical bar indicates the mean, the solid bar, 2 standard errors, and the horizontal bar the standard deviation. The data were obtained from Adams, et al. (1981), Anderson and Beare (1983), Armstrong and Drummond (1986), Arnold (1982), Breckenridge and Miller (1982), Casper (1983), Cruden (1977), Cruden and Jensen (1979), Cruden and Miller-Ward (1981), Dulberger (1981), Hannan (1981), Harder, et al. (1985), Kubitzki and Kurz (1984), Mackiernan and Norman (1979), Pellmyr (1986a, b), Pellmyr and Patt (1986), Philbrick (1984), Schlising, et al. (1980), Schoen (1977), Short (1981), Spira (1980), Stucky and Beckman (1981), Thomas and Murray (1981), Tomlinson et al. (1979), Uno (1982) and Tables 3 and 4. Four species in Cruden (1977) have been reassigned based on new information, i.e., *Monarda fistulosa* L. (facultative xenogamy to xenogamy) (Cruden et al., 1984), *Capsella bursapastoris* (L.) Medic (facultative xenogamy to facultative autogamy), *Cryptotaenia canadensis* (L.) DC. and *Sanguinaria canadensis* L. (xenogamy to facultative xenogamy).

xenogamy is 2232 ± 1239 (n = 62) and that for xenogamy is 24, 138 ± 9518 (n = 92).

Outbreeding vs Inbreeding

There is a strong relationship between mating system and breeding system (Fig. 2). Mean outbreeding rates are generally above 80% in xenogamous species (Table 4) and below 10% in facultatively autogamous species (Table 5). Mean outbreeding rates in facultatively xenogamous taxa are intermediate, have a wide range (Table 6), and are significantly different from both facultatively autogamous (t_s = 5.71, n = 25, 20, p < .001) and xenogamous species (t_s = 5.01, n = 25, 15, p < .001). The inclusive data set shows no evidence of skewness (t_s = 0.55, p > .50) but is platykurtic (t_s = -2.64, p < .05). The results are equivalent when all the species with mixed mating systems (Table 6) and species, rather than subspecies (as above), are used. The outbreeding rates of species with mixed mating systems exceed those of facultatively autogamous species (t_s 5.81, n = 27, 20, p < .001) and are lower than those of xenogamous species (t_s = 4.86, n = 27, 15, p < .001). The test for skewness (t_s = 0.48, p > .50) is not significant but that for kurtosis is significant: (t_s = -4.89, p<0.01). Further, when the data for facultatively autogamous species are excluded, the remaining data show no evidence of skewness (t_s = -0.31, p > .50) and are uniformly distributed (X^2 = 8.22, p > .50).

Pollen-Ovule Ratio and Pollen Volume

The relationship between pollen-ovule ratio and pollen volume varies among mating systems. Among xenogamous and facultatively xenogamous species there is a significant, negative relationship between the two (Figs. 3a, 3b). The relationship between P/O and pollen volume is not significant in facultatively autogamous species (Fig. 3c, Table 7).

To test if this relationship differed among mating systems both the slopes (Fig. 3d) and y-intercepts were compared with a t-test (Ryan et al., 1985). The slopes for xenogamous and facultatively xenogamous species are not significantly different (t = 1.776, p > .05) but their y-intercepts are significantly different (t = 2.905, p < .01). In contrast, both the slopes and y-intercepts for xenogamous and facultatively autogamous species are significantly different (t = 3.586, p < .005; t = 5.810, p < .001). Although the slopes for facultatively xenogamous and facultatively autogamous species are not significantly different (t

Table 4. Mean percent outbreeding in xenogamous species.

	Percent Outbreeding	Source
Boraginaceae		
Amsinckia spectabilis F. & M.	50	Ganders et al. 1985
Compositae		
Borrichia frutescens (L.) DC.	113	Antlfinger 1982
Helianthus annuus L.	76	Ellstrand et al. 1978
Gramineae		
Aegilops speltoides Tausch.	85	Zohary and Imber 1963
Lolium perenne L.	100	Kannenberg and Allard 1967
Leguminosae		
Pithecellobium pedicellare Benth.	95	O'Malley and Bawa 1987
Limnanthaceae		
Limnanthes alba Hartw.	99	Arroyo 1975
	80	Ritland and Jain 1981
Onagraceae		
Clarkia unguiculata Lindl.	100	Vasek 1965
Oenothera organensis Munz	88	Levin et al. 1979
Proteaceae		
Banksia attenuata R. Br.	110	Scott 1980
B. menziesii R. Br.	104	Scott 1980
Rosaceae		
Rubus idaeus L.	95-100	Crane; in Haskell 1954
Scrophulariaceae		
Collinsia heterophylla Buist ex Grah.	89	Weil and Allard 1964
Mimulus guttatus DC.	88	McNair in Richards 1986
Turneraceae		
Turnera ulmifolia L. var. *elegans* (Otto) Urb.	100-114	Barrett and Shore 1987

Table 5. Percent outbreeding in facultatively autogamous species.

	x̄	Range	Source
Boraginaceae			
Amsinckia spectabilis F. & M.	3.0		Ganders et al. 1985
Caryophyllaceae			
Spergula arvensis L.		0 - 3	New 1959
Spergularia marina (L.) Griseb.	1.5	1 - 2	Sterek & Dijkhuizen 1972
Compositae			
Senecio vulgaris L. (rayless)	<1		Marshall & Abbott 1982
			Hull 1974
Convolvulaceae			
Ipomoea hederaceae (L.) Jacq.	7		Ennos 1981
Gramineae			
Avena barbata Brot.	2.6	1.3-4.5	Marshall & Allard 1970
Avena fatua L.		1 - 12	Imam & Allard 1965
Bromus mollis L.	4.8	2.1-9.5	Jain et al. 1970
Elymus canadensis L.	6.9	0 - 31	Sanders & Hamrick 1980
Festuca microstachys Nutt.		0-0.001	Kannenberg & Allard 1967
Hordeum jubatum L.	2.0	1 - 3	Babbel & Wain 1977
Hordeum spontaneum Koch.	1.6	0-9.6	Brown et al. 1978
Hordeum vulgare L.	1.5	1 - 2	Allard et al. 1968
Triticum dicoccoides Koern.	3.0		Nevo et al. 1982
Leguminosae			
Lupinus bicolor Lindl.	4.1	0 - 29	Harding et al. 1974
Lupinus pachylobus Greene	0	0	Harding et al. 1974
Lupinus polycarpus Greene	1	0 - 1	Harding et al. 1974
Primulaceae			
Primula vulgaris Huds.		5 - 10	Crosby 1959
Solanaceae			
Lycopersicon pimpinellifolium (Jusl.) Mill.	8.8	0 - 23	Rick et al. 1977
Valerianaceae			
Plectritis brachystemon F. & M.		0-3.3	Ganders et al. 1977a

Table 6. Percent outbreeding in species with mixed mating systems. Mean and/or ranges are given. Where both are provided, the mean is in parentheses.

	Percent Outbreeding	Source
<u>Morphological</u>		
Caryophyllaceae		
Spergularia media (L.) C. Presl	8-15	Sterek and Dijkhuizen 1972
Cruciferae		
Cheiranthes cheiri L.	73.2	Bateman 1956
Leguminosae		
Lupinus affinus J.F. Agardh.	13-50	Harding et al. 1974
Lupinus albus L.	8-9, 25	Green et al. 1980; Faluyi and Williams 1981
Lupinus nanus Dougl.		
ssp. *apricus* (Greene) D. Dunn	5-97 (36)	Harding et al. 1974
spp. *menkerae* (C.P. Sm.) D. Dunn	0-65 (30)	Harding et al. 1974
spp. *latifolius* (Benth.) D. Dunn	33-100 (72)	Harding et al. 1974
spp. *vallicola* (C.P. Sm.) D. Dunn	24-88 (53)	Harding et al. 1974
spp. *nanus*	50-83 (66)	Harding et al. 1974
Lupinus pilosus Murr.	30, 51, 60	Horovitz and Harding 1983
Lupinus succulentus Dougl. ex Koch	14-97	Harding and Barnes 1977
Limnanthaceae		
Limnanthes floccosa Howell		
spp. *californica* Arroyo	74	Arroyo 1975
spp. *grandiflora* Arroyo	79	Arroyo 1975
spp. *pumila* (Howell) Arroyo	79	Arroyo 1975
spp. *bellingerina* (M.E. Peck) Arroyo	31	Arroyo 1975
spp. *floccosa*	5-35	Arroyo 1975
Onagraceae		
Clarkia exilis Lewis and Vasek	43-45	Vasek 1964, 1967
	37, 44, 89	Vasek and Harding, 1976
Clarkia tembloriensis Vasek	8.4-83 (55)	Vasek and Harding 1976
Papaveraceae		
Papaver dubium L.	20-30	Humphreys and Gale 1974
Polemoniaceae		
Gilia achilleifolia Benth	15-96 (57)	Schoen 1982b
Phlox cuspidata Scheele	22	Levin 1978

Rosaceae

Rubus "Merton thornless"	17.4	Haskell 1954

Solanaceae

Lycopersicon pimpinellifolium (Jusl.) Mill	1.7, 14-40	Rick et al. 1977

Turneraceae

Turnera ulminifolia L. var. *angustifolia* (Mill.) DC.	4-79 (19)	Barrett and Shore 1987

Valerianaceae

Plectritis congesta (Lindl.) D.C.	48-80	Ganders et al. 1977b; Carey 1983

<u>Morphological</u>

Compositae

Senecio vulgaris L. (with ray flowers)	22	Campbell and Abbott 1976

<u>Genetic</u>

Boraginaceae

Amsinciia spectabilis F. & M.	26	Ganders et al. 1985
Borago officinalis L.	79	Crowe 1971

Convolvulaceae

Ipomoea purpurea (L.) Roth	70	Ennos 1981

Leguminoseae

Vicia faba L.	30-70	Holden and Bond 1980
	15-46	Porceddu et al. 1980

Myrtaceae[a]

Eucalyptus delegatensis R.T. Bak.	77	Moran and Brown 1980
E. obliqua L'Heirt.	76	Brown, et al. 1975
E. pauciflora Sieb. ex Spreng.	63	Phillips and Brown 1977
E. stoatei C.A. Gardner	82	Hooper and Moran 1981

Solanaceae

Nicotiana rustica L.	0-70 (18)	Breese 1959

---

a. The species in *Eucaluptus* are included here based on discussion of their mating systems (Brown et al., 1975; Eldridge, 1976). The conclusion that these species have mixed mating systems is based on the assumption that they are ±SC. However, there is evidence of inbreeding depression, certation in favor of cross-pollen (Eldridge, 1976), and SI (Krug and Silveira Alves, 1949). There is an alternative hypothesis: *Eucalyptus* are xenogamous and the reduced outbreeding rates reflect mating between related individuals.

Table 7. Pollen-ovule ratios and pollen grain volume of facultatively autogamous species.

	Pollen-Oule Ration (X±S.E.)	Pollen Volume (μm^3) X±S.E.
Comastoma tenellum (Rottb.) Toyokun.[a]	25±2	12,039±1,273
Veronica arvensis L.[b]	30±1	5,643±357
Spergularia rubra (L.) J & C. Presl[c]	47±2	1,527±145
Oxalis corniculata L.[b]	49±4	18,209±685
Cerastium vulgatum L.[b]	50±4	15,456±924
Stellaria media (L.) Cyrillo[d]	58±5	8,687±396
Malva rotundifolia L.[e]	62±6	204,335±9,086
Matricaria matricarioides (Less.) Porter[f]	89±6	3,343±168
Polygonum caespitosum Blume[g]	126±7	42,269±2,707
Verbena bracteata Leg. & Radr.[d]	152±9	8,710±1,216
Lycopus virginicus L.[b]	162±10	9,307±482
Polygala sanguinea L.[h]	194±13	11,422±956
Cuscuta indecora Choisy[i]	332±22	3,474±206
Galium aparine L.[b]	333±19	4,594±340
Capsella bursa-pastoris (L.) Medic.[b]	354±24	2,004±84
Lycopus americanus Muhl.[b]	397±14	6,754±373
Polygonum aviculare L.[b]	572±52	6,804±317
Lepidium virginicum L.[b]	616±42	2,213±107
Bidens bipinnata L.[g]	891±138	6,229±270
Polygonum convolvulus L.[d]	1,355±97	6,537±309

a. Colorado: Clear Creek County, Loveland Pass.
b. Iowa: Johnson County, Iowa City.
c. Oregon: Multnomah-Washington County line, Springville Rd.
d. Iowa: Dickinson County, Iowa Lakeside Laboratory.
e. Minnesota: Jackson County, Sioux Valley.
f. Wyoming: Bighorn County, Rt. 16, Willow Park Picnic Area, 45 mi. SW Buffalo.
g. Missouri: St. Louis County, St. Louis.
h. Iowa: Linn County, Matsell Bridge County Conservation Area.
i. Iowa: Cherokee County, 1-2 mi N Cherokee.

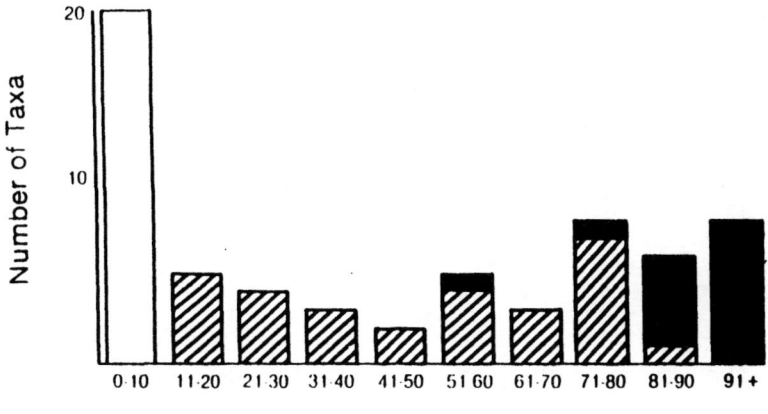

Percent Outbreeding

Figure 2. Percent outbreeding in angiosperms with different mating systems. Open bar = facultatively autogamous species, hatched bar = facultatively xenogamous species, solid bar = xenogamous species. The data are in Tables 4-6.

= 1.776, p > .05), the y=intercepts are significantly different (t = 2.383, p < .05).

DISCUSSION
Facultative Xenogamy

Our first objective was to test Cruden's (1977) hypothesis that the flowers of facultatively xenogamous species may be cross-pollinated if pollinators are present but self-pollinate in their absence. With the use of open-pollinated, emasculated flowers we showed that pollinators effectively moved pollen between flowers, and the high fruit and seed set of caged plants showed that flowers self-pollinated in the absence of pollinators.

Facultative xenogamous species exhibit a number of adaptations that facilitate cross-pollination, including showy corollas, nectar production, and delayed self-pollination. The flowers may be homogamous, protogynous (this study), or protandrous, for example in *Limnanthes* (Arroyo, 1973b), *Lobelia* (Young, 1982), *Pyrrhopappus* (Barber and Estes, 1978), and *Panax* (Schlessman, 1985). Flower size is highly variable and reflects the plants' pollinators. Some species have relatively large flowers, *e.g.*, *Quamoclit coccinea* (L.) Moench (hummingbird-pollinated; Cruden et al., 1983), *Leonotis*

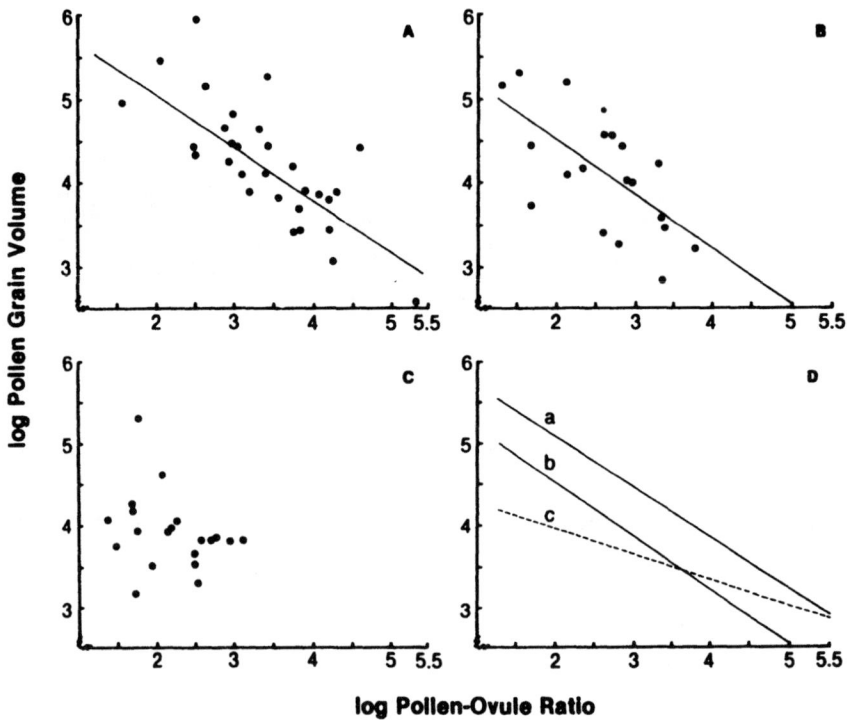

Figure 3. Relationship between pollen-ovule ratio and pollen grain volume. Among xenogamous (A) and facultatively xenogamous species (B) there is a significant and negative relationship between pollen-ovule ratio and pollen grain volume (y = 6.98 - 0.840x, r = .753, n = 30, p < .01; and y = 5.26 - 0.651x, r = .674, n = 19, p < .01, respectively). The relationship among facultatively autogamous species (C) is not significant (r = 0.321, n = 20, p > .05). The volumes for xenogamous species were calculated from data in Cruden and Miller-Ward (1981), Casper (1983), Dulberger (1981), Dulberger and Ornduff (1980), and Uno (1982). Unpublished data for four species (Cruden, 1989) are also included. The data for facultatively xenogamous and facultatively autogamous species are in Tables 1 and 7, respectively. Differences between the regression lines and their y-intercepts (D) are discussed in the text.

nepetaefolia R. Br. and *Lobelia* spp. (pollinated by sunbirds and/or other perching birds; Cruden, 1976; Young, 1982), *Mirabilis jalapa* L. (hawkmoth-pollinated; Martínez del Río and Búrquez, 1986), *M. nyctaginea* (pollinated by moths, small bumblebees, and solitary bees; Cruden, 1973), *Hepatica americana* (DC.) Ker (Motten, 1982), various subspecies of *Limnanthes floccosa* Howell (Arroyo, 1973b), and *Sanguinaria canadensis* (Lyon and Cruden, pers. obs.) that are visited and pollinated by larger *Andrena*, and *Lamium amplexicaule* L. (Knuth, 1909) and *Scilla sibirica* that are visited by *Apis mellifera.* Smaller flowered species, including most of the species we studied (Tables 1 and 2), were generally visited by small, solitary bees and/or flies.

The flowers of the species we studied were regularly visited by bees and/or other insects (Cruden, 1973; Cruden and Lyon, personal observation). These visitors transferred sufficient pollen to account for the quite high levels of fruit and seed set that are typical of facultatively xenogamous species (Cruden and Westley, unpublished), for example, *Pyrrhopappus carolinianus* (96%; Estes and Thorp, 1975), *P. geiseri* (88%; Barber and Estes, 1978), *Melilotus* spp. (88-96%; Sano, 1977), *Lamium purpureum* (95%; Macior, 1978), *Sanguinaria canadensis* (81-96%; Schemske et al., 1978; Motten, 1986), and *Jeffersonia diphylla* (L.) Pers. (90%; Smith et al., 1986). The insects left so little pollen in the flowers of most of the species we studied that little if any seed set could be attributed to self-pollination in visited flowers. Thus, self-pollination was restricted primarily to flowers that were missed by flower visitors or times when there was little or no pollinator activity.

Delayed autogamy is characteristic of facultatively xenogamous species whether their flowers are homogamous or dichogamous. In the absence of pollinators, the flowers of most of the species we studied self-pollinated as did those of other facultatively xenogamous species. Self-pollination produced full sets of seed in *Clarkia exilis* (Vasek, 1964), *Cheiranthus cheiri* (Bateman, 1956), *Hepatica americana* (Motten, 1982), *Jeffersonia diphylla* (L.)Pers. (Smith et al., 1986), *Leonotis nepetaefolia* (Cruden, 1976), *Panax quinquefolium* (Schlessman, 1985), *Tribulus terrestris* L. (Subba Reddi et al., 1981), and *Sanguinaria canadensis* (Lyon pers. obs.). However, seed set was reduced in the absence of pollinators in *Plectritis congesta* (Ganders et al., 1977b), *Lupinus albus* (Faluyi and Williams, 1981), *L. nanus* (Harding et al., 1974), *Papaver* (Rogers, 1969), and *Scilla sibirica*, and in *Adenocaulon bicolor* Hook. Few seed were set in the absence of pollinators (Cruden, personal observation).

Balance between Outbreeding and Inbreeding

In facultatively xenogamous species, the balance between xenogamy and autogamy is primarily a function of pollinator activity, and presumably the outbreeding rate reflects that balance. In both *Lupinus pilosus* (Horovitz and Harding, 1983) and *L. albus* (Green et al. 1980) outbreeding rates were positively related to pollinator number and activity. Low (<20%) levels of outbreeding in early flowering populations of *L. nanus* ssp. *apricus* compared to later flowering populations (>50%), and the striking between-year differences at various sites (e.g., 33 and 68% at Mace 2; and 5 and 45% at Spanish Flat 1) (Harding et al., 1974) were consistent with differences in pollinator activity. Arroyo (1973b) suggested the larger flowered subspecies of *Limnanthes floccosa* (*i.e., californica, grandiflora,* and *pumila*) are largely self-pollinating only in years with little pollinator activity. In commercial fields of *Phaseolus lunatus* L. the level of self-pollination varied from year to year and between sites (Harding & Tucker, 1964), a pattern that is consistent with variation in pollinator activity. Finally, in *Mirabilis jalapa*, pollinator activity was temperature-dependent and the number of stigmas receiving cross-pollen and the number of pollen grains per stigma were positively correlated with temperature (Martínez del Río and Búrquez, 1986).

The outbreeding rate also can reflect the number of open flowers, pollinator behavior, and the balance between xenogamy and geitonogamy. In essence, xenogamy will be highest in species with few open flowers, high pollen carry-over, and pollinators that visit few flowers per plant. In some species subject to geitonogamy the outbreeding rate may be increased by cryptic self-incompatibility. For example, in *Cheiranthus cheiri* L., if both cross- and self-pollen reach a stigma at the same time, cross-pollen tubes are favored (Bateman, 1956; see also Cruden et al., 1984).

Data from some facultatively xenogamous species indicate their mating and breeding systems are adapted to local conditions, in particular, to levels of pollinator activity. In *Limnanthes floccosa* outbreeding in the large flowered subspecies (Table 6) greatly exceeds that of the small flowered subspecies *floccosa* and *bellingeriana* (Arroyo, 1973a), which flower earlier in the year and occur in ecologically marginal habitats, such as dry stony flats, roadsides, etc. (Arroyo, 1973b), where pollinators may be scarce. Schoen (1982a, b) presented evidence that the amount of self-pollination varied among populations of *Gilia achilleaefolia* Benth. and that there was a strong positive correlation between amount of outbreeding and mean androecium weight. In contrast, Rick et al. (1977) found no obvious ecological differences that were associated with facultatively xenogamous and facultatively autogamous populations of *Lycopersicon pimpinellifolium*.

Mating Systems Compared

Our second objective was to show that facultative xenogamy is a unique mating system. The available data are consistent with that idea and also support a trichotomous view of mating systems for two reasons. (1) The very nature of facultative xenogamy distinguishes it from both xenogamy and facultative autogamy. Xenogamous species are not adapted for self-pollination and rarely do, whereas facultatively autogamous species are not adapted for cross-pollination and rarely cross-pollinate. In contrast, facultatively xenogamous species are adapted for cross-pollination but can self-pollinate easily in the absence of pollinators. (2) The pattern of resource allocation to sexual function differs significantly among mating systems (Cruden and Lyon, 1985). Although the flowers of the facultatively xenogamous species we studied (Cruden and Lyon, 1985) were intermediate in size and differed significantly from both xenogamous and facultatively autogamous species on a dry weight basis, the pattern of relative investment in flower parts was not intermediate (Table 8). The percent dry weight biomass of sepals and petals of facultatively xenogamous and facultatively autogamous species were equivalent and differed significantly from those of xenogamous species. In contrast, the percent dry weight biomass of the stamens of xenogamous and facultatively autogamous species were significantly different and those of facultatively xenogamous species were intermediate. Finally, the relative biomass of the pistils of xenogamous and facultatively xenogamous species were significantly lower than those of facultatively autogamous species. The ratio of dry weight stamens to dry weight pistils was essentially the same in xenogamous and facultatively xenogamous species. The latter observation is consistent with facultatively xenogamous species being adapted for cross-pollination, and the uniqueness of the pattern of relative biomass allocation supports the notion that facultative xenogamy is a distinct mating system.

Other data also are consistent with facultative xenogamy being a discrete mating system. These include differences in P/O's, outbreeding rates, and the relationship between P/O and pollen grain size. Although there is considerable overlap of their ranges, P/O's differ significantly among mating systems (Fig. 1). In addition, there is a strong association between mating system and breeding system. Mean outbreeding rates are generally above

Table 8. Percent dry weight biomass of flower parts from plants with different mating systems. Percents based on data in Cruden and Lyon (1985, Table 2). Percents within a row with same superscript are not significantly different.

	Facultative Autogamy	Facultative Xenogamy	Xenogamy
Calyx	48.4[a]	42.4[a]	23.2[b]
Corolla	21.5[a]	24.8[a]	36.6[b]
Stamens	11.9[a]	20.8[ab]	26.0[b]
Pistil	18.2[a]	12.0[b]	14.2[b]

80% in xenogamous species and below 10% in facultatively autogamous species (Fig. 2). Outbreeding rates in facultative xenogamous species are intermediate and have a wide range.

Our examination of outbreeding rates addresses an additional question. Schemske and Lande (1985) presented evidence consistent with their hypothesis that outbreeding rates should have a biomodal distribution. This conclusion has been questioned. Aide (1986) concluded that the bimodality in Schemske and Lande's (1985) data was a consequence of the wind-pollinated species in their sample, and that there was no evidence of bimodality among the animal-pollinated species. Likewise, we found no evidence of biomodality in our data set even though it contained a large number of facultatively autogamous species all with outbreeding rates of less than 10%, as well as wind-pollinated grasses. (We did not include the gymnosperms included in Schemske and Lande's [1985] sample). Our data, which include more facultatively xenogamous taxa than did Schemske and Lande (1985), are neither biomodal nor significantly skewed but are platykurtic, and the outbreeding rates of facultatively xenogamous and xenogamous species are evenly distributed.

Waller (1986) questioned Schemske and Lande's (1985) conclusion on genetic grounds. He suggested that if the "...variation did not have a genetic basis, Lande and Schemske's (1985) model would not apply." The variation in outbreeding rates in the facultatively xenogamous species considered here is easily explained by variable levels of pollinator activity, i.e., it appears to have a nongenetic basis. However, there also are species with genetically controlled mixed mating systems and variable outbreeding rates. This suggests that mixed mating systems, regardless of how outbreeding is regulated, are evolutionarily stable. Finally, the question may be moot, as the bimodality observed in outbreeding rates (Aide, 1986; Schemske and Lande, 1985) may simply be a sampling artifact. If only xenogamous and autogamous species are examined, a bimodal distribution of outbreeding rates is inevitable.

The negative relationship between P/O and pollen grain size bears on several questions. First, the differences among y-intercepts are consistent with a trichotomous view of mating systems. Second, the negative slopes among xenogamous and facultatively xenogamous species are consistent with their being animal pollinated. We suggest that the negative relationship reflects, at least in part, the pollen-bearing area of the pollinator which limits the amount of pollen exposed to the stigmas. Given that constraint and variation in pollen grain size, a negative relationship is inevitable. In essence, the pollen-bearing area accommodates a limited amount of pollen, which can be divided into a variable number of pollen grains. In facultatively autogamous species, because the amount of pollen exposed to the stigmas is not constrained by the pollen-bearing-area, pollen grain number and size were free to evolve independently, and a significant relationship between the two traits is not expected. Third, the absence of the relationship among facultatively autogamous species is inconsistent with the hypothesis that P/O's and pollen grain size are simply reciprocals. Further, if P/O's and pollen grain size were simply reciprocals the relationship should be linear with a slope of -1. The slopes of the relationship among xenogamous (m = .840) and facultatively xenogamous (m = .651) species are inconsistent with that assumption. The observed slopes are consistent with an alternative hypothesis. Because a greater percentage of large pollen grains are transferred to stigmas relative to small grains (Cruden, 1989), fewer large pollen grains need be produced and the relationship between P/O and pollen grain volume will be described by a slope greater than -1. Given the relatively small sample sizes, the small difference between the slope for xenogamous species and -1, and an alternative hypothesis (Charnov, 1982, p. 264), our hypothesis merits further testing.

Finally, we offer a word of caution. Although facultatively xenogamous species are adapted for cross-pollination and are quite different from facultatively autogamous species in virtually all respects, they might, nonetheless, be classified as facultatively autogamous on the basis of pollinator exclusion tests. For example, our data shoqw that both *Calylophus serrulatus* and *Sanguinaria canadensis* (Lyon and Cruden, personal observation) are facultatively xenogamous. In contrast, other populations of these species were described as autogamous (Towner, 1977, p.

61; Schemske et al., 1978), a conclusion that is consistent with the pollinator exclusion protocol.

Mixed Mating Systems Compared

We recognize two classes of mixed mating systems that are quite different from facultative xenogamy. The first ("morphological") includes plants that produce two types of flowers and the second ("genetic") includes populations comprised of different kinds of plants. In the first group are CH/CL species, i.e., those that produce both chasmogamous (cross-pollinating) and cleistogamous (self-pollinating) flowers, for example, *Impatiens* (Schemske, 1978; Waller, 1980), *Lithospermum caroliniense* (Walt.) MacM. (Levin, 1968), *Myosurus* (Stone, 1959), *Viola, Specularia perfoliata* (L.) A. DC, *Stipa spartea* Trin. (Cruden, pers. obs..) and many others (Uphof, 1938). Also in this group are gynomonoecious species, for example, *Lasthenia minor* (DC.) Ornduff ssp. *maritima* (Gray) Ornduff (Ornduff, 1966) and *Senecio vulgaris* (Marshall and Abbott, 1982). In these Compositae the ray-flowers are pistillate and only cross-pollinate, and the perfect disk-flowers generally self-pollinate.

In such "morphological" systems the outbreeding rate is first a function of the number of flowers that can cross-pollinate. That number may reflect a plant's genotype and/or its environment (Clay, 1982, 1983b; Wilken 1982). The maximum outbreeding rate is a function of the number of genetically determined cross-pollinating flowers. That number may reflect local adaptation to more or less predictable environmental factors, for example, amounts of precipitation and numbers of pollinators (Campbell and Abbott, 1976; Hull, 1974; Wilken, 1982). Because a plant may produce fewer cross-pollinating flowers than maximum in response to less than optimal environmental conditions, the actual outbreeding rate is a function of the number of cross-pollinating flowers that are actually produced, and subsequent conditions that affect pollinator numbers and/or activity.

Within a population the balance between xenogamy and autogamy may reflect variable and unpredictable differences in local environmental conditions, such as available water and/or light, grazing, and nutrient levels (Darwin, 1895; Uphof, 1938; Harlan, 1945; Brown, 1952; James, 1965; Langer and Wilson, 1965; Schemske, 1978; Waller, 1980; Clay, 1983b). Not surprisingly, in CH/CL species those individuals growing in good sites are larger and produce both more flowers and a greater percentage of CH flowers than plants in poor sites (Schemske, 1978; Waller, 1980; Wilken, 1982). It follows that outbreeding rates will also reflect the immediate environment of the individual. These species tend to live in a relatively variable world, and are adapted to the unpredictable nature of the local habitat (Brown, 1952; Schemske, 1978; Wilken, 1982; Clay, 1983b).

In "genetic" systems, populations are comprised of different types of individuals, some of which are strictly outbred and others that have flowers that can self-pollinate. In these species outbreeding is a function of the numbers of "outbreeding" flowers or individuals, whose ability to cross-pollinate or outbreed is genetically regulated (Jain, 1976). Maximum outbreeding is thus a function of the frequency of alleles that determine self-incompatibility and/or cross-pollination plus pollinator activity. Included here are species such as *Borage officinalis* (Crowe, 1971) and *Vicia faba* (Holden and Bond, 1960; also see Lord and Heslop-Harrison, 1986)). A variation on this theme is the segregation of alleles for timing of anther dehiscence and stigma exsertion, which affect the level of outbreeding in *Nicotiana rustica* (Breese, 1959). Outbreeding in *Ipomoea purpurea* is positively related to the distance between the stigmas and anthers (Ennos, 1981), as is cross-pollination in *Erythronium grandiflorum* (Thomson and Stratton, 1985).

Facultative xenogamy differs fundamentally from these other mixed mating systems. The flowers of facultatively xenogamous species are morphologically and functionally equivalent and the balance between xenogamy and autogamy is primarily a function of pollinator activity. In contrast, in other mixed mating systems, the balance between xenogamy and autogamy is a function of a combination of the relative number of flowers and plants adapted for cross-pollination, and outbreeding and pollinator activity (also see Sun and Ganders, 1988).

Where and When Facultative Xenogamy Is Adaptive

The species we studied fall into six groups. The first three include species of late successional communities that flower when and/or where pollinator activity may be limited. Woodland species, such as *Hepatica americana* (Motten, 1982), *Sanguinaria canadensis* Schemske, 1978; Lyon, unpublished), *Jeffersonia diphylla* (Smith et al., 1986), *Scilla sibirica*, and *Lamium amplexicaule*, flower in the spring when pollinator activity is unreliable due to cool and/or rainy weather. The second group includes summer flowering species also in woods and forests, such as *Polygonum virginianum, Circaea lutetiana* L. ssp. *canadensis* (L.) Asch. and Magnus, *Cryptotaenia canadensis* (L.) DC., *Phryma leptostachya* L., *Adenocaulon bicolor* and *Tiarella unifoliata*

Hooker. In such habitats pollinator activity occurs mostly in sun spots, thus limiting the number of visits a flower is likely to receive (Beattie, 1971; Cruden, pers. obs.) and increasing the likelihood of a flower being missed. A third group of species occurs in habitats typified by unpredictable weather conditions, for example, alpine fell-fields and sub-alpine meadows where pollinator activity may be affected by strong winds, cool to cold temperature, rain, and/or clouds. The P/O's of several alpine species are consistent with their being facultatively xenogamous, for example, *Ranunculus pedatifidus* Smith (P/O = 404.1 ± 30.5) and *Claytonia megarhiza* (Gray) Parry (P/O = 542.1 ± 37.9) (Cruden unpubl.).

In such late successional communities, inbred as well as outbred progeny may contribute to individual fitness. In only a few species was there evidence of inbreeding depression, which suggests that in many, if not most facultatively xenogamous species, inbred offspring contribute to individual fitness, perhaps by increasing adaptation to local conditions (Jain 1976). However, in *Lobelia* there was evidence of outbred seedlings enjoying a selective advantage over inbred seedlings (Young, 1982). Likewise, in CH/CL species the evidence for outbreeding superiority is mixed. There is evidence that the CH seeds of some species enjoy a selective advantage (Waller, 1982; Mitchell-Olds and Waller, 1985) but the reverse is true in others (Clay, 1983a; Schoen, 1984).

The other three groups of facultatively xenogamous species occur in disturbed sites. First, some species are adapted for colonizing disturbances in relatively stable ecosystems where the resultant populations may persist for numbers of years. Initially autogamy must be adaptive (Baker, 1955; Stebbins, 1957; Jain, 1976) and subsequently xenogamy may provide the heterosis that contributes to individual fitness over ecological time as well as the genetic variation that is necessary for evolutionary change as the site changes and the genotypes of the colonizers no longer enjoy a selective advantage (Beckman and Mitton, 1984). Such facultatively xenogamous species do not enjoy the best of two reproductive worlds; rather their mating and breeding systems reflect the fact that they live in two ecological worlds. Included here are species such as *Spergularia media* (Sterek and Dijkhuizen, 1972), *Agrimonia pubescens*, *Polygonum pennsylvanicum*, and *Phyla lanceolata*. A second group includes fugitive species, such as *Mirabilis hirstuta* (Pursh) MacM. (Platt, 1976) and *Verbena stricta* Vent. (Cruden, 1977) that colonize badger mounds and other disturbances in prairies (Platt,

1975). The third group includes a small number of species, such as *Abutilon theophrasti, Ellisia nyctelea, Lappulla redowskii,* and *Solanum nigrum,* that are successful in weedy and/or highly disturbed sites where pollinators are frequently rare or lacking and autogamy is highly adaptive.

SUMMARY AND CONCLUSIONS

Facultative xenogamy is a mixed mating system in which the balance between self- and cross-pollination is a function of pollinator activity. Facultatively xenogamous species are self-compatible, adapted for cross-pollination, have delayed autogamy, occur primarily in climax or other stable communities, and flower when pollinator activity may be low or unreliable. In most species both fruit and seed set are high.

Facultative xenogamy is a unique mating system. It is similar to xenogamy in various respects but is quite different from facultative autogamy. The flowers of facultatively xenogamous species are adapted for cross-pollination, in contrast to autogamous species, and self-pollinate in the absence of the pollinators in contrast to the flowers of xenogamous species. The pattern of resource allocation to flower parts is unique. A negative relationship between pollen-ovule ratio and pollen grain size is found among facultatively xenogamous and xenogamous species but not among facultatively autogamous species. In addition, the pollen-ovule ratios and outbreeding rates of facultatively xenogamous species are intermediate between those of xenogamous and facultatively autogamous species.

Facultative xenogamy is different from other mixed mating systems. Most species with other mixed mating systems fit into one of two categories. One, which we term "morphological," consists of plants that produce morphologically different flowers that are either cross- or self-pollinating. The other, which we term "genetic," consists of populations with a mix of outbred individuals and others with flowers that can self-pollinate. In these other mixed mating systems, outbreeding reflects the relative numbers of flowers or plants adapted for cross-pollination, in addition to pollinator activity. In contrast, populations of facultatively xenogamous species are characterized by a single flower type and the level of pollinator activity is the primary determinant of the balance between xenogamy and autogamy, hence outbreeding rate.

We know where to expect facultative xenogamy, but we do not fully understand its adaptive nature. We can only speculate as to the adaptive nature of it. It is clear that xenogamy increases fecundity, but does it also contribute to or increase adaptation to local conditions (Jain, 1976)? In short, facultative xenogamy is relatively well-understood functionally and ecologically, but the various selective pressures that maintain it are poorly known, and it merits the scrutiny that CH/CL systems have received.

ACKNOWLEDGEMENTS

This work was supported by scholarships from the University House of the University of Iowa. Our thanks to Jay Semel and Gretchen Miller for making our tenure at University House both enjoyable and productive.

LITERATURE CITED

Adams, D. E., W. E. Perkins, and J. R. Estes. 1981. Pollination systems in *Paspalum dilatatum* Poir. (Poaceae). Amer. J. Bot. 68: 389-394.

Aide, T. M. 1986. The influence of wind and animal pollination on variation in outcrossing rates. Evolution 40:434-435.

Allard, R. W., S. K. Jain, and P. L. Workman. 1968. The genetics of inbreeding populations. Adv. Genet. 14: 55-131.

Anderson, R. C., and M. H. Beare. 1983. Breeding system and pollination ecology of *Trientalis borealis* (Primulaceae). Amer. J. Bot. 70: 408-415.

Antlfinger, A. E. 1982. Genetic neighborhood structure of the salt marsh composite, *Borrichia frutescens*. J. Hered. 73: 128-132.

Armstrong, J. E., and B. A. Drummond, III. 1986. Floral biology of *Myristica fragrans* Houtt. (Myristicaceae), the nutmeg of commerce. Biotropica 18: 32-38.

Arnold, R. M. 1982. Pollination, predation and seed set in *Linaria vulgaris* (Scrophulariaceae). Amer. Midl. Nat. 107: 360-369.

Arroyo, M. T. Kalin. 1973a. Chaisma frequency evidence on the evolution of autogamy in *Limnanthes floccosa* (Limnanthaceae). Evolution 27: 679-688.

Arroyo, M. T. Kalin. 1973b. A taximetric study of intraspecific variation in autogamous *Limnanthes floccosa* (Limnanthaceae). Brittonia 25: 177-191.

Arroyo, M. T. Kalin. 1975. Electrophoretic studies of genetic variation in natural populations of allogamous *Limnanthes alba* and autogamous *Limnanthes floccosa* (Limnanthaceae). Heredity 35: 153-164.

Babbel, G. R., and R. P. Wain. 1977. Genetic structure of *Hordeum jubatum*. I. Outcrossing rates and heterozygosity levels. Can. J. Genet. Cytol. 19: 143-152.

Baker, H. G. 1955. Self-compatibility and establishment after "long-distance" dispersal. Evolution 9: 347-349.

Baker, H. G. 1959. Reproductive methods as factors in speciation in flowering plants. Cold Spring Harbor Symp. Quant. Biol. 24: 177-191.

Baker, H. G. 1974. The evolution of weeds. Ann. Rev. Ecol. Syst. 5: 1-24.

Baker, H. G., and G. L. Stebbins [eds.]. 1965. The Genetics of Colonizing Species. Academic Press, New York.

Barber, S. C., and J. R. Estes. 1978. Comparative pollination ecology of *Pyrrhopappus geiseri* and *Pyrrhopappus carolinianus*. Amer. J. Bot. 65: 562-566.

Barrett, S. C. H., and J. S. Shore. 1987. Variation and evolution of breeding systems in the *Turnera ulmifolia* L. complex (Turneraceae). Evolution 41: 340-352.

Bateman, A. J. 1956. Cryptic self-incompatibility in the wallflower: *Cheiranthus cheiri* L. Heredity 10: 257-261.

Beattie, A. J. 1971. Itinerant pollinators in a forest. Madroño 21: 120-124.

Beckman, J. S., and J. B. Mitton. 1984. Peroxidase allozyme differentiation among successional stands of ponderosa pine. Amer. Midl. Nat. 112: 43-49.

Breckenridge, F. G. and J. M. Miller. 1982. Pollinator biology, distribution, and chemotaxonomy of the *Echinocereus enneacanthus* complex (Cactaceae). Syst. Bot. 7: 365-378.

Breese, E. L. 1959. Selection for differing degrees of out-breeding in *Nicotiana rustica*. Ann. Bot. 23: 331-344.

Brown, A. H. D., A. C. Matheson, and K. G. Eldridge. 1975. Estimation of the mating system of *Eucalyptus obliqua* L'Herit. by using allozyme polymorphisms. Aust. J. Bot. 23: 931-949.

Brown, A. H. D., D. Zohary, and E. Nevo. 1978. Outcrossing rates and heterozygosity in natural populations of *Hordeum spontaneum* Koch in Israel. Heredity 41: 49-62.

Brown, W. V. 1952. The relation of soil moisture to cleistogamy in *Stipa leucotricha*. Bot. Gaz. 113: 438-444.

Campbell, J. M., and R. J. Abbott. 1976. Variability of outcrossing frequency in *Senecio vulgaris* L. Heredity 36: 267-274.

Carey, K. 1983. Breeding system, genetic variability, and response to selection in *Plectritis* (Valerianaceae). Evolution 37: 947-956.

Casper, B. B. 1983. The efficiency of pollen transfer and rates of embryo initiation in *Cryptantha* (Boraginaceae). Oecologia (Berl.) 9: 262-268.

Charnov, E. L. 1982. The Theory of Sex Allocation. Princeton University Press, Princeton, NJ.

Clay, K. 1982. Environmental and genetic determinants of cleistogamy in a natural population of the grass *Danthonia spicata*. Evolution 36: 734-741.

Clay, K. 1983a. The differential establishment of seedlings from chasmogamous and cleistogamous flowers in natural populations of the grass *Danthonia spicata* (L.) Beauv. Oecologia (Berl.) 57: 183-188.

Clay, K. 1983b. Variation in the degree of cleistogamy within and among species of the grass *Danthonia*. Amer. J. Bot. 70: 835-843.

Crosby, J. L. 1959. Outcrossing in homostyle primroses. Heredity 13: 127-131.

Crowe, L. K. 1971. The polygenic control of outbreeding in *Borago officinalis*. Heredity 27: 111-118.

Cruden, R. W. 1973. Reproductive biology of weedy and cultivated *Mirabilis* (Nyctaginaceae). Amer. J. Bot. 60: 802-809.

Cruden, R. W. 1976. Intraspecific variation in pollen-ovule ratios and nectar secretion - preliminary evidence of ecotypic adaptation. Ann. Missouri Bot. Gard. 63: 277-289.

Cruden, R. W. 1977. Pollen-ovule ratios: A conservative indicator of breeding systems in flowering plants. Evolution 31: 32-46.

Cruden, R. W. 1989. Pollen-ovule ratios, pollination efficiency, and sex allocation: Examination of alternative hypotheses. In S. Buchmann, [ed.]. Experimental Studies in Pollination and Pollinator Foraging Efficiency. Univ. of Arizona Press, Tucson. (In press).

Cruden, R. W., and K. Jensen. 1979. Viscin threads, pollination efficiency and low pollen-ovule ratios. Amer. J. Bot. 66: 875-879.

Cruden, R. W., and D. L. Lyon. 1985. Patterns of biomass allocation to male and female functions in plants with different mating systems. Oecologia (Berl.) 66: 299-306.

Cruden, R. W., and S. Miller-Ward. 1981. Pollen-ovule ratio, pollen size, and the ratio of stigmatic area to the pollen-bearing area of the pollinator: an hypothesis. Evolution 35:964-974.

Cruden, R. W., S. M. Hermann, and S. Peterson. 1983. Patterns of nectar production and plant-pollinator coevolution. In B. Bentley and T. Elias, [eds.]. The Biology of Nectaries. Columbia Univ. Press, New York.

Cruden, R. W., L. Hermanutz, and J. Shuttleworth. 1984. The pollination biology and breeding system of *Monarda fistulosa* (Labiatae). Oecologia (Berl.) 64: 104-110.

Darwin, C. 1895. The effects of cross and self fertilization in the vegetable kingdom. D. Appleton and Company, New York.

Dulberger, R. 1981. The floral biology of *Cassia didymobotrya* and *C. auriculata* (Caesalpiniaceae). Amer. J. Bot. 68: 1350-1360.

Dulberger, R., and R. Ornduff. 1980. Floral morphology and reproductive biology of four species of *Cyanella* [Tecophilaeaceae]. New Phytol. 86: 45-56.

Eldgridge, K. G. 1976. Breeding systems, variation and genetic improvement of tropical Eucalypts. In J. Burley and B. T. Styles [eds.]. Tropical Trees: Variation, Breeding, and Conservation. .(Printed for the Linnaean Soc., London). Academic Press, London.

Elstrand, N. C., A. M. Torres, and D. L. Levin. 1978. Density and the rate of apparent outcrossing in *Helianthus anuus* (Asteraceae). Syst. Bot. 3: 403-407.

Ennos, R. A. 1981. Quantitative studies of the mating system in two sympatric species of *Ipomoea* (Convolvulaceae). Genetica 57: 93-98.

Estes, J. R., and R. W. Thorp. 1975. Pollination ecology of *Pyrrhopappus carolinianus* (Compositae). Amer. J. Bot. 62: 148-159.

Feagri, K., and L. van der Pijl. 1978. The Principles of Pollination Ecology. Third ed. Pergamon Press, Oxford.

Faluyi, M. A., and M. Williams. 1981. Studies on the breeding system in lupin species: a) Self- and cross-compatibility in three European lupin species, b) percentage outcrossing in *Lupinus albus*. Z. Pflanzensucht. 87: 233-239.

Ganders, F. R., K. Carey, and A. J. F. Griffiths. 1977a. Outcrossing rates in natural populations of *Plectritis brachystemon* (Valerianaceae). Canad. J. Bot. 55: 2070-2074.

Ganders, F. R., K. Carey, and A. J. F. Griffiths. 1977b. Natural selection for a fruit dimorphism in *Plectritis congesta* (Valerianaceae). Evolution 31: 873-881.

Ganders, F. R., S. K. Denny, and D. Tsai. 1985. Breeding systems and genetic variation in *Amsinckia spectabilis* (Boraginaceae). Canad. J. Bot. 63: 533-538.

Grant, V. 1958. The regulation of recombination in plants. Cold Spring Harbor Symposia on Quantitative Biology 23: 337-363.

Green, A. G., A. H. D. Brown, and R. N. Oram. 1980. Determination of outcrossing rate in a breeding population of *Lupinus albus* L. (White Lupine). Z. Pflanzenzucht. 84: 181-191.

Hannan, G. L. 1981. Flower color polymorphism and pollination biology of *Platystemon californicus* Benth (Papaveraceae). Amer. J. Bot. 68: 233-243.

Harder, L. D., J. D. Thomson, M. B. Cruzan, and R. S. Unmasch. 1985. Sexual reproduction and variation in floral morpholgy in an ephemeral vernal lily, *Erythronium americanum*. Oecologia (Berl.) 67: 286-291.

Harding, J., and K. Barnes. 1977. Genetics of *Lupinus* . X. Genetic variability, heterozygosity and outcrossing in colonial populations of *Lupinus succulentus*. Evolution 31: 247-255.

Harding, J., and C. L. Tucker. 1964. Quantitative studies on mating systems 1. Evidence for the non-randomness of outcrossing in *Phaseolus lunatus*. Heredity 19: 369-381.

Harding, J., C. B. Mankinen, and M. H. Elliott. 1974. Genetics of *Lupinus* VII. Outcrossing, autofertility, and variability in natural populations of the *nanus* group. Taxon 23: 729-738.

Harlan, J. R. 1945. Cleistogamy and chasmogamy in *Bromus carinatus* Hook. and Arn. Amer. J. Bot. 32: 66-72.

Haskell, G. 1954. The genetic detection of natural crossing in blackberry. Genetica 27: 162-167.

Holden, J. H. W., and D. A. Bond. 1960. Studies on the breeding systems of the field bean, *Vicia faba* (L.). Heredity 15: 175-192.

Holsinger, K. E. 1986. Dispersal and plant mating systems: the evolution of self-fertilization in subdivided populations. Evolution 40: 405-413.

Hooper, S. D., and G. F. Moran. 1981. Bird pollination and the mating system of *Eucalyptus stoatei*. Aust. J. Bot. 29: 625-638.

Horovitz, A., and J. Harding. 1983. Genetics of *Lupinus* XII. The mating system of *Lupinus pilosus*. Bot. Gaz. 144: 276-279.

Hull, P. 1974. Self-fertilization and the distribution of the radiate form of *Senecio vulgaris* L. in central Scotland. Watsonia 10: 69-75.

Humphreys, M. O., and J. S. Gale. 1974. Variations in wild populations of *Papaver dubium* Vill. The mating system. Heredity 33: 33-41.

Imam, A. G., and R. W. Allard. 1965. Population studies in predominantly self-pollinated species. VI. Genetic variability between and within populations of wild oats from differing habitats in California. Genetics 51: 49-62.

Jain, S. K. 1976. The evolution of inbreeding in plants. Ann. Rev. Ecol. Syst. 7: 469-495.

Jain, S. K., D. R. Marshall, and K. Wu. 1970. Genetic variability in natural populations of softchess (*Bromus mollis* L.). Evolution 24: 649-659.

James, S. H. 1965. Complex hybridity in *Isotoma petraea* 1. The occurrence of interchange heterozygosity, autogamy and a balanced lethal system. Heredity 20: 341-353.

Kannenburg, L. W., and R. W. Allard. 1967. Population studies in predominantly self-pollinated species. VIII. Genetic variability in the *Festuca microstachys* complex. Evolution 21: 227-240.

Knuth, P. 1909. Handbook of Flower Pollination. Vol. III. J.R. Ainsworth Davis, (Trans.). Clarendon Press, Oxford.

Krug, C. A., and A. Silveira Alves. 1949. *Eucalyptus* improvement. Part I. J. Hered. 40: 133-139.

Kubitzki, K., and H. Kurz. 1984. Synchronized dichogamy and dioecy in neotropical Lauraceae. Pl. Syst. Evol. 147: 253-266.

Lande, R., and D. S. W. Schemske. 1985. The evolution of self-fertilization and inbreeding depression in plants. I. Genetic models. Evolution 39: 24-40.

Langer, R. H. M., and D. Wilson. 1965. Environmental control of cleistogamy in prairie grass (*Bromus unioloides* H.B.K.). New Phytologist 64: 80-85.

Levin, D. A. 1968. The breeding system of *Lithospermum caroliniense*: adaptation and counteradaptation. Amer. Nat. 102: 427-441.

Levin, D. A. 1978. Genetic variation in annual *Phlox*: self-compatible versus self-incompatible species. Evolution 32: 245-263.

Levin, D. A., K. Ritter, and N. C. Ellstrand. 1979. Protein polymorphism in the narrow endemic *Oenothera organensis*. Evolution 33: 534-542.

Lord, E. M., and Y. Heslop-Harrison. 1986. Pollen-stigma interaction in the Leguminosae: stigma organization and the breeding system in *Vicia faba* L. Annl. Bot. 54: 827-836.

Lloyd, D. G. 1979. Some reproductive factors affecting the selection of self-fertilization in plants. Amer. Nat. 113: 67-79.

Macior, L. W. 1978. Pollination ecology of vernal angiosperms. Oikos 30: 452-460.

Mackiernan, J. M., and E. M. Norman. 1979. Reproductive biology of *Nemastylis floridana* Small (Iridaceae). Florida Scientist 42: 229-236.

Marshall, D. F., and R. J. Abbott. 1982. Polymorphism for outcrossing frequency at the ray floret locus in *Senecio vulgaris* L. I. Evidence. Heredity 48: 227-235.

Marshall, D. R., and R. W. Allard. 1970. Maintenance of isozyme polymorphism in natural populations of *Avena barbata*. Genetics 66: 393-399.

Martínez del Río, C., and A. Búrquez. 1986. Nectar production and temperature dependent pollination of *Mirabilis jalapa* L. Biotropica 18: 28-31.

Mitchell-Olds, T., and D. M. Waller. 1985. Relative performance of selfed and outcrossed progeny in *Impatiens capensis*. Evolution 39: 533-544.

Moran, G. F., and A. H. D. Brown. 1980. Temporal heterogeneity of outcrossing rates in alpine ash (*Eucalyptus delegatensis* R.T. Bak.). Theoret. Appl. Genet. 57: 101-105.

Motten, A. F. 1982. Autogamy and competition for pollinators in *Hepatica americana* (Ranunculaceae). Amer. J. Bot. 69: 1296-1305.

Motten, A. F. 1986. Pollination ecology of the spring wild flower community of a temperate deciduous forest. Ecol. Monogr. 65: 21-42.

Nevo, E., E. Golenberg, and E. Beiles. 1982. Genetic diversity and environmental associations of wild wheat, *Triticum dicoccoides*, in Israel. Theoret. Appl. Genet. 62: 241-254.

New, J. K. 1959. A population study of *Spergularia arvensis*. II. Genetics and breeding behaviour. Ann. Bot. n.s. 23: 23-33.

O'Malley, D. M., and K. S. Bawa. 1987. Mating system of a tropical forest tree species. Amer. J. Bot. 74: 1143-1149.

Ornduff, R. 1966. A biosystematic survey of the goldfield genus *Lasthenia* (Compositae: Helenieae). Univ. Calif. Publ. Bot. 40: 1-92.

Ornduff, R. 1969. Reproductive biology in relation to systematics. Taxon 18: 121-144.

Pellmyr, O. 1986a. Pollination ecology of two nectariferous *Cimicifuga* sp. (Ranunculaceae) and the evolution of andromonoecy. Nordic J. Bot. 6: 129-138.

Pellymr, O. 1986b. The population ecology of two nectarless *Cimicifuga* sp. (Ranunculaceae) in North America. Nordic J. Bot. 6: 713-723.

Pellymr, O., and J. M. Patt. 1986. Function of olfactory and visual stimuli in pollination of *Lysichiton americanum* (Araceae) by a staphylinid beetle. Madroño 33: 47-54.

Philbrick, C. T. 1984. Aspects of floral biology, breeding system, and seed and seedling biology in *Podostemum ceratophyllum* (Podostemaceae). Syst. Bot. 9: 166-174.

Phillips, M. A., and A. H. D. Brown. 1977. Mating systems and hybridity in *Eucalyptus pauciflora*. Aust. J. Biol. Sci. 30: 337-344.

Platt, W. J. 1975. The colonization and formation of equilibrium plant species associations on badger disturbances in a tall-grass prairie. Ecol. Monogr. 45: 285-305.

Platt, W. J. 1976. The natural history of a fugitive prairie plant (*Mirabilis hirsuta* (Pursh) MacM.). Oecologia (Berlin) 22: 399-409.

Porceddu, E., L. M. Monti, L. Frusciante, and N. Volpe. 1980. Analysis of cross-pollination in *Vicia faba* L. Z. Pflanzenzucht. 84: 313-322.

Richards, A. J. 1986. Plant Breeding Systems. George Allen & Unwin, London.

Rick, C. M., J. F. Fobes, and M. Holle. 1977. Genetic variation in *Lycopersicon pimpinellifolium*: Evidence of evolutionary change in the mating systems. Pl. Syst. Evol. 127: 139-170.

Ritland, K., and S. Jain. 1981. A model for the estimation of outcrossing rate and gene frequencies using n independent loci. Heredity 47: 35-52.

Rogers, S. 1969. Studies on British poppies I. Some observations on the reproductive biology of the British species of *Papaver*. Watsonia 7: 55-63.

Ryan, B. F., B. L. Joiner, and T. A. Ryan, Jr. 1985. Minitab handbook. 2nd ed. Duxbury Press, Boston.

Sanders, T. B., and J. L. Hamrick. 1980. Variation in the breeding system of *Elymus canadensis*. Evolution 34: 117-122.

Sano, Y. 1977. The pollination systems of *Melilotus* species. Oecologia Planatarum 12: 383-394.

Schemske, D. W. 1978. Evolution of reproductive characteristics in *Impatiens* (Balsaminaceae): the significance of cleistogamy and chasmogamy. Ecology 59: 596-613.

Schemske, D. W., and R. Lande. 1985. The evolution of self-fertilization and inbreeding depression in plants. II. Empirical observations. Evolution 39: 41-52.

Schemske, D. W., M. F. Willson, M. N. Melampy, L. J. Miller, L. Verner, K. M. Schemske, and L. B. Best. 1978. Flowering ecology of some spring woodland herbs. Ecology 59: 351-366.

Schlising, R. A., D. H. Ikeda, and S. C. Morey. 1980. Reproduction in a Great Basin evening primrose *Camissonia tanacetifolia* (Onagraceae). Bot. Gaz. 141: 290-293.

Schlessman, M. 1985. Floral biology of the American ginseng (*Panax quinquifolium*). Bull. Torrey Bot. Club 112: 129-133.

Schoen, D. J. 1977. Morphological, phenological, and pollen-distribution evidence of autogamy and xenogamy in *Gilia achilleifolia* (Polemoniaceae). Syst. Bot. 2: 280-286.

Schoen, D. J. 1982a. Male reproductive effort and breeding system in an hermaphroditic plant. Oecologia (Berl.) 53: 255-257.

Schoen, D. J. 1982b. The breeding system of *Gilia achilleifolia*: Variation in floral characteristics and outcrossing rate. Evolution 36: 352-360.

Schoen, D. J. 1984. Cleistogamy in *Microlaena polynoda* (Gramineae): An examination of some model predictions. Amer. J. Bot. 71: 711-719.

Scott, J. A. 1980. Estimation of the outcrossing rate for *Banksia attenuata* R. Br. and *Banksia menziesii* R. Br. (Proteaceae). Aust. J. Bot. 28: 52-59.

Short, P. S. 1981. Pollen-ovule ratios, breeding systems and distribution patterns of some Australian Gnaphaliinae (Compositae: Inuleae). Muelleria 4: 395-417.

206

Smith, B. H., M. L. Ronsheim, and K. R. Swartz. 1986. Reproductive ecology of *Jeffersonia diphylla* (Berberidaceae). Amer. J. Bot. 73: 1416-1426.

Sokal, R. R., and F. J. Rohlf. 1981. Biometry, 2nd ed. W.H. Freeman and Company, San Francisco.

Spira, T. P. 1980. Floral parameters, breeding system and pollinator type in *Trichostema* (Labiatae). Amer. J. Bot. 67: 278-284.

Stebbins, G. L. 1950. Variation and Evolution in Plants. Columbia Univ. Press, New York.

Stebbins, G. L. 1957. Selffertilization and population variability in the higher plants. Amer. Nat. 41: 337-354.

Sterek, A. A., and L. Dijkhuizen. 1972. The relation between the genetic determination and ecological significance of the seedwing in *Spergularia media* and *S. marina*. Acta Bot. Neerl. 21: 481-490.

Stone, D. E. 1959. A unique balanced breeding system in the vernal pool mouse tails. Evolution 13: 151-174.

Stuckey, J. M., and R. L. Beckmann. 1982. Pollinator biology, self-incompatibility, and sterility in *Ipomoea pandurata* (L.) G.F.W. Meyer (Convolvulaceae). Amer. J. Bot. 69: 1022-1031.

Subba Reddi, C., E. U. B. Reddi, and N. S. Reddi. 1981. Breeding structure and pollination ecology of *Tribulus terrestris*. Proc. Indian Nat. Sci. Acad. B47: 185-193.

Sun, M., and F. R. Ganders. 1988. Mixed mating systems in Hawaiian *Bidens* (Asteraceae). Evolution 42: 516-527.

Thomas, S. M., and B. G. Murray. 1981. Breeding systems and hybridization in *Petrorhagia* sect. *Kohlrauschia* (Caryophyllaceae). Pl. Syst. Evol. 139: 77-94.

Thomson, J. D., and D. A. Stratton. 1985. Floral morphology and cross-pollination in *Erythronium grandiflorum* (Liliaceae). Amer. J. Bot. 72: 433-437.

Tomlinson, P. B., R. B. Primack, and J. S. Bunt. 1979. Preliminary observations on floral biology in mangrove (Rhizophoraceae). Biotropica 11: 256-277.

Towner, H. F. 1977. The biosystematics of *Calylophus* (Onagraceae). Ann. Missouri Bot. Gard. 64: 48-120.

Uno, G. E. 1982. Comparative reproductive biology of hermaphroditic and male-sterile *Iris douglasiana* Herb. (Iridaceae). Amer. J. Bot. 69: 818-823.

Uphof, J. C. Th. 1938. Cleistogamic flowers. Bot. Rev. 4: 21-49.

Vasek, F. C. 1964. Outcrossing in natural populations. I. The Breckenridge Mountain population of *Clarkia exilis*. Evolution 18: 213-218.

Vasek, F. C. 1965. Outcrossing in natural populations. II. *Clarkia unguiculata*. Evolution 19: 152-156.

Vasek, F. C. 1967. Outcrossing in natural populations. III. The Deer Creek population of *Clarkia exilis*. Evolution 21: 241-248.

Vasek, F. C., and J. Harding. 1976. Outcrossing in natural populations. V. Analysis of outcrossing, inbreeding, and selection in *Clarkia exilis* and *Clarkia tembloriensis*. Evolution 30: 403-411.

Waller, D. M. 1980. Environmental determinants of outcrossing in *Impatiens capensis* (Balsaminaceae). Evolution 34: 747-761.

Waller, D. M. 1984. Differences in fitness between seedlings derived from cleistogamous and chasmogamous flowers in *Impatiens capensis* (Balsaminaceae). Evolution 38: 427-440.

Waller, D. M. 1986. Is there disruptive selection for self-fertilization? Amer. Nat. 128: 421-426.

Wells, H. 1979. Self-fertilization: advantageous or deleterious? Evolution 33: 252-255.

Weil, J., and R.W. Allard. 1964. The mating system and genetic variability in natural populations of *Collinsia heterophylla*. Evolution 18: 515-525.

Wilken, D. H. 1982. The balance between chasmogamy and cleistogamy in *Collomia grandiflora* (Polemoniaceae). Amer. J. Bot. 69: 1326-1333.

Wyatt, R. 1983. Pollinator-plant interactions and the evolution of breeding systems. In L. Real [ed.]. Pollination Biology. Academic Press, Orlando, FL.

Young, T. P. 1982. Bird visitation, seed-set, and germination rates in two species of *Lobelia* on Mount Kenya. Ecology 63: 1983-1986.

Zohary, D., and D. Imber. 1963. Genetic dimorphism in fruit types in *Aegilops speltoides*. Heredity 18: 223-231.

BAKER'S LAW: PLANT BREEDING SYSTEMS AND ISLAND COLONIZATION

Paul Alan Cox

INTRODUCTION

In 1955 a seminal paper entitled "Self-compatibility and establishment after 'long distance' dispersal" appeared in the journal *Evolution*. It was written by Herbert G. Baker, a young botanist at the University of Ghana, who articulated a novel but strongly intuitive concept of island colonization. He suggested (p. 348) that "with self-compatible individuals a single propagule is sufficient to start a sexually-reproducing colony" on an island or new habitat. With self-incompatible taxa, however, Baker argued that establishment of a new breeding population required, at the minimum, the arrival of two propagules at roughly the same time and same place, as well as the previous establishment of the appropriate pollinator. Based on these considerations, he proposed a concept with Ledyard Stebbins later termed "Baker's Law" (Stebbins, 1957). Simply stated, Baker's Law proposes that "self-compatible rather than self-incompatible taxa will be favored in establishment after long-distance dispersal" (Baker, 1967).

Since ecology has few concepts that have been sufficiently demonstrated to merit the appellation of "law", it may be argued that Stebbins was premature in elevating Baker's concept to the status of a law. Stebbins argued, however, that the correlation of self-compatibility and successful island colonization

> occurs so widely and has such significance for studies of the origin and migration of genera of flowering plants, including some animals (Baker, 1955) that it deserves recognition as Baker's Law. (Stebbins, 1957, p. 349).

The two criteria used by Stebbins, explanatory power and significance for future studies, should be joined by a third , originality, to justify terming this concept Baker's Law. Even though previous workers, such as Henslow (Baker, 1965) may have guessed the link between breeding systems and colonization ability, the correlation was first documented and most clearly explained by Baker. Even the writings of Darwin, which, if read carefully, can be found to contain many of the ideas currently being investigated in plant evolutionary ecology, are silent on this topic. In considering plant colonization ability, earlier workers tended to emphasize the importance of propagule structure, abiotic and biotic dispersal mechanisms, and historical processes but gave little attention to the significance of plant breeding systems. Thus Ridley's encyclopedic treatise *The Dispersal of Plants Throughout the World* (1930) failed to discuss any possible effects of plant breeding systems on island colonization. The topic was also ignored in the earlier work on plant dispersal by Guppy, who may, however, have had an inkling of the idea when he suggested that the limited range of the genus *Sararanga* (Pandanaceae) might be due to its "dioecious habit" (Guppy, 1906, p. 156).

It therefore appears that the three criteria of explanatory power, significance in influencing future studies, and originality are satisfied and that Baker's Law is properly termed. The purpose of this essay is to discuss the relevance of Baker's Law to plant ecology today and to point out fruitful areas for future research in this topic.

SELF-COMPATIBILITY

A plant is said to be self-compatible when under field conditions the deposition of its pollen on its own stigmas will result in successful fertilization and seed set. Self-compatible plants are usually thought of as having monomorphic or hermaphroditic flowers, but even plants possessing dimorphic flowers such a monoecious individuals or even sexually inconsistent individuals of dioecious species can be said to be self-compatible if they potentially can set seed through self-fertilization.

Baker's law predicts that self-compatible taxa will be favored in establishment after long-distance dispersal particularly if they do not require the services of a pollinator to effect self-fertilization. The breeding systems of many strand plants in island habitats conform to this prediction. Their independence of a particular pollinator facilitates island colonization regardless of the island's faunal composition.

Convolvulus siculus (Convolvulaceae), a strand plant found throughout the Mediterranean, is a good example of such a self-pollinating species. Initially its bifurcate stigma is protected against receipt of its own pollen and subsequent self-pollination by the extrorse dehiscence of the anthers, although geitonogamous pollinations (pollinations with pollen from other flowers on the same plant) are possible. However, when the anthers eventually shrivel they place pollen on the erect stigma. Kerner von Marilaun (1895, p. 333) argued that differing angles of the stigma lobes predisposed one lobe to self-pollination and the other lobe to cross pollination. The superiority of this type of breeding system in long-distance dispersal is clear since the establishment

of a single propagule is sufficient to initiate the spread of a
sexually reproducing population.

DICHOGAMY, SELF-INCOMPATIBILITY AND HETEROSTYLY

Baker's law predicts that self-incompatible taxa will be
disadvantaged in island colonization. Establishment of self-
incompatible taxa on islands requires the advent of at least two
propagules arriving at about the same time and growing in
relative proximity to each other. In addition for species other
than those with abiotic pollination systems the previous presence
on the island of any necessary pollinator is also required.
However, Baker's Law predicts that self-incompatible taxa will
themselves vary in their success in colonization in relationship to
the severity of their outbreeding mechanism.

Take for example the case of dichogamy, or the temporal
separation of male and female functions. In a protogynous island
species with single inflorescence such as *Joinvillea
gaudichaudiana* (Joinvilleaceae) the stigmatic lobes are exposed
and fertilized before the anthers are exerted and dehisce. An
example of the opposite case of dichogamy, namely protandry, in
island plants occurs in *Scaevola taccada* (Goodeniaceae) where the
anthers dehisce in bud and the pollen is presented by the indusium
(cup-like structure) surrounding the stigma. For such
dichogamous breeding systems, the possibility of self-
fertilizations, including geitonogamous fertilizations, seems low.
However, the nearby successful immigration of another propagule
is likely to produce a breeding population since dichogamous
systems are generally synchronized within but not between
individuals. Even a single large rhizomatous individual is
unlikely to have male and female stages completely synchronized
and thus may be self-fertilized if the pollinator is present.

In multiple allele self-incompatibility systems, the
probability of the second propagule established resulting in an
initial seed crop is also high, particularly since selection favors
the rapid fixation of new allelic variants in such systems. This is
because a plant possessing a new self-incompatibility allele will
be initially capable of fertilizing the entire population. The allele
will thus rapidly increase in frequency until it reaches
equilibrium. Therefore the low probability of two propagules
having the same self-incompatibility alleles (and hence being
incompatible) in inversely proportion to the number of self-
incompatibility alleles (assuming the source population is in
genetic equilibrium). Numerical simulations potentially could
provide interesting insights into the effects of sampling error on
this process.

In contrast, the probability of the second propagule being of the appropriate sex in a dioecious system is only 50% if the source population has a Fischerian sex ratio. However, the apparent advantage of self-incompatibility systems over dioecism in long-distance dispersal, disappears when one considers the breeding potential of the first generation of seeds. Since in a dioecious population, any male X female cross is legitimate, we would expect roughly 50% of the F_2 to produce seed. However, since the offspring of the first two initial immigrants of a population with gametophytic self-incompatibility will, as a result of their relatedness, share self-incompatibility alleles, the seed-fertility of the F_2 is like to be very low. It is for this reason that Anderson and Stebbins (1984) proposed that dioecism should be a more likely outbreeding system in small founder populations. It is interesting in this regard that Rick (1966) found an absence of self-incompatibility in Galapagos representatives of Solanaceae that have strong self-incompatibility systems on the mainland. This trend towards self-compatibility in island species has recently been further confirmed by McMullen (1987) who studied the breeding systems 52 species of angiosperms in the Galapagos. Of these, 51 species proved to be hermaphroditic, and all the species (40) that yielded conclusive results in the pollination experiments proved to be self-compatible. The breeding systems of the remaining 11 species have yet to be determined.

A Mediterranean example can be found in the genus *Nigella* (Ranunculaceae). The self-incompatible species *Nigella arvensis* occurs throughout Turkey and Greece, while the related self-compatible species *Nigella doefleri* is confined to the small islands of the Cyclades (Richards, 1986; Strid, 1970). Perhaps, through a similar mechanism, self-compatible individuals are more likely to establish small founder populations at the margins of a species' geographical range.

More investigations of the breeding systems of island taxa are needed to determine the incidence and prevalence of self-compatibility in island taxa. Grant and Grant (1981), for example, found self-incompatibility in *Opuntia* in the Galapagos. As predicted by Baker's Law, such incidents should be relatively infrequent. Not only should self-incompatible species be disadvantaged in establishment after long-distance dispersal, but the genetic complexities of multiple allele self-incompatibility systems makes them unlikely to evolve in self-compatible taxa after establishment (Baker, 1959). In this light the recent work of Carr *et al.* (1986) demonstrating the presence of a strong self-incompatibility system in *Argyroxiphium sanwicense, Wilkesia gymnoxiphium,* and several species of the genus *Dabautia* is of tremendous interest since these three gerera probably descended from a single ancestral immigrant to Hawaii. This case may therefore likely represent autochthonous evolution of self-incompatibility after island colonization.

Heterostyly is apparently very rare in oceanic floras (Baker, 1967; Carlquist, 1974) and is strikingly absent in genera such as *Psychotria* and *Plumbago* that have heterostylous species in continental areas. Baker's Law suggests that losses of heterostyly in such taxa will precede rather than occur after successful long-distance dispersal. This appears to have occurred in the *Turnera ulmifolia* complex (Turneraceae) that is usually composed of self-incompatible distylous diploids and tetraploids, except for self-compatible homostylous hexaploid populations at the edge of its range (Barrett and Shore, 1987; Barrett, this volume).

DIOECISM

In the formulation of Baker's Law, Baker (1965, 1967) specifically suggested that dioecious taxa (species with male and female flowers confined to different plants) should be disadvantaged in establishment after long-distance dispersal. Partially in response to a paper by Bawa (1982) that suggested that "dioecious taxa may have been disproportionately successful in colonizing the (Hawaiian) islands", an extensive comparison was made of the relative representation of dioecious taxa in floras of oceanic island and comparable mainland floras (Baker and Cox, 1984). This analysis of 22 island floras indicated that percentages of dioecious species in the islands generally do not differ from those in latitudinally comparable mainland floras. This analysis also documented strong correlations between percentages of dioecious species and the maximum height and proximity to the equator of the various islands.

Thus, dioecism does not appear to be as much of an impediment to island colonization as may have been predicted by Baker's law. Obviously some of the problem may be explained by the evolution of dioecism in hermaphroditic taxa subsequent to island colonization and establishment. However, close examination (Bawa, 1982) of the Hawaiian flora reveals that of 14 genera that are dioecious in Hawaii but hermaphroditic elsewhere, 10 have dioecious representatives in locations other than Hawaii. Of course, dioecism could have evolved several times in a genus; a clear evolutionary picture of such events can be provided only by reference to appropriate phylogenetic analysis (Cox, 1989).

Nevertheless it is clearly untenable at this point to ascribe the high incidence of dioecism in tropical oceanic high islands merely to the agency of such autochthonous breeding system evolution. Ecological and other biological solutions that mitigate the effects of Baker's Law on the dispersal and establishment of dioecious taxa in oceanic islands must be sought. Several such possibilities are discussed below.

1. Colonization Saturation.

Extensive numerical simulations (Cox, 1985) predict an exponentially inverse relationship between number of years until establishment of a dioecious population and the mean annual number of propagules hitting an island. For example, in a computer simulation of a very small island an average of 322 years were required for successful population establishment at a mean immigration flux of 0.2 seeds per year. At a mean annual immigration flux of 1.8 seeds, an average of only 35 years were required for population establishment (Cox, 1985). These results suggest that the disadvantages of dioecism in island colonization could be greatly reduced if the numbers of propagules dispersed were increased. In this regard, most of the dioecious species on oceanic islands (as well as mainland areas) are large shrubs or trees or plants of large clone sizes (such as seagrasses) capable of producing many propagules. These habits also correlate with perennation strategies as noted below.

2. Perennation and Long Life-Spans.

a. Perennation from sexual propagules. Numerical simulations also show the salubrious effects of long life-span on colonization ability of dioecious plants. In model islands with fixed mean annual immigration flux (1.8 propagules per year), an average of 35 years are required for the establishment of dioecious taxa having life-spans of 50 to 500 years. However, in the simulations, plants of 5 year life-spans required an average of 179 years and dioecious annuals required an average of 1918 years for successful establishment of breeding populations (Cox, 1985). These predictions are consistent with the striking absence of dioecious annuals from oceanic island floras and indeed with the general paucity of dioecious annuals in continental floras. Dioecism in annuals, unless ameliorated by some other factor, such as apomixis, may impose severe constraints on populations spread through seed dispersal.

Unlike dioecious annuals, dioecious perennial plants can "sit and wait" on an island for the immigration of the opposite sex. For example, in 1900 a single female individual of *Baccharis viminea*

(Compositae) was found in San Clemente Island (Trask, 1904), and it could still be seen in 1963 (Raven, 1963). It would be interesting to see if its many years of solitude has since been rewarded by the advent of a male individual.

b. Perennation from asexual propagules. Dispersal of asexual propagules such as bulbils, rhizomes, buds, and branches with subsequent establishment and vegetative growth is another possible mechanism by which the deleterious effects of dioecism on colonization ability are negated. It is likely that many dioecious seagrass species colonize new habitats through such vegetative propagation as I have found numerous bits of rhizomes and attached foliage of species such as *Syringodium isoetifolium* (Crymodoceaceae), *Halodule pinnifolia* (Cymodoceaceae), and *Halophila ovalis* (Hydrocharitaceae) washed ashore on island beaches throughout the western South Pacific. Presumably some pieces root in shallow water and establish new populations through vegetative growth. Such mechanisms are not limited to tropical islands; *Elodea canadensis* (Hydrocharitaceae) which is indigenous to North America, first was recorded in Europe in a pond at Dunse Castle, Berwick, England in 1842 with a possible second introduction being made in Foxton, Leicestershire in 1847 in a shipment of American timber (Ridley, 1930). Even though only a pistillate individual was present, it rapidly spread vegetatively throughout England, choking rivers and streams, and succeeded in reaching Ireland by 1886. The rapid and immense spread of the species throughout Europe has been primarily through vegetative means, even though staminate plants were later introduced.

Asexual propagation is also very common in dioecious bryophytes (During and van Tooren, 1987). Gametophytes produce propagules from their rhizoids, leaf lamina, or even specialized gemmae cups. However, since asexual propagation is such a common feature in the bryophytes (During, 1979) it is still unclear if this feature correlates more closely with dioecism than monoecism (Brent Mischler, pers. comm.). In any case, though, the possibility of asexual propagation allows dioecious species as well as self-incompatible species to have a much better chance of successful establishment after long-distance dispersal.

3. Multi-Seeded Propagules

a. Abiotically dispersed assemblages. By having two or more seeds dispersed in a single unit, a dioecious plant can greatly increase the likelihood of population establishment. The probability that a multi-seeded propagule from such a dioecious taxon will carry at least one male and one female seed, and hence potentially be capable of establishing a breeding population in a single immigration event is equal to:

$$p = \sum_{i=2}^{6} p(i) \; (0.5)^{(i-2)}$$

where p(i) is the probability that the syncarp will produce i seedlings, and r is the sex ratio of seeds, usually taken to be 0.5.

In some species, great numbers of seeds are dispersed together as a unit. This phenomenon, termed synaptospermy (Murbeck, 1919; van der Pijl, 1969) occurs in some dioecious genera such as *Spinacia* (Chenopodiaceae) where, in wild species, the entire infructescence is dispersed as a single glomerule, or *Cotula* (Compositae) where the achenes are dispersed individually in monoecious species, but entire fruiting heads are the units of dispersal in dioecious species (Lloyd, 1972). Such a mechanism is clearly advantageous in establishment after long-distance dispersal since it ensures that entire breeding populations, rather than single individuals are produced. Neither are such mechanisms limited to angiosperms; Schuster (1966) notes that the four spores resulting from meiosis tend to be dispersed as a unit in dioecious bryophytes. Because each tetrad contains spores that will germinate into male gametrophytes and spores that will produce female gametophytes, the two sexes of gametophytes will likely develop in proximity to each other.

b. Biotically-dispersed assemblages. The endozootic dispersal of small seeded dioecious plants to islands by birds and (in tropical regions) bats, serves to concentrate seeds in their feces, thus resulting in the establishment of entire breeding populations rather than single isolated individuals. Dioecious species of the genus *Ficus*, which range from small shrubs to large Banyan-like trees in the South Pacific are good examples of this strategy. The numerous tiny seeds of *Ficus* species are frequently dispersed in the Pacific by flying foxes of the genus *Pteropus* which regularly consume over 2.5 times their body weight of fruit per night (the equivalent of an adult human male eating 10 bushels of fruit in a single sitting), and which can be seen flying between islands. Thus several thousand *Ficus* seeds may be defecated or spit out in a single location.

Bawa (1980) and Givnish (1980) have both shown correlations between fleshy fruits and dioecism. They suggest that competition for resources between male and female functions coupled with selection for large fruit displays by frugivores may drive the evolution of dioecism in taxa with fleshy fruits. Perhaps a more parsimonious (but non-exclusive) explanation is that dioecious plants with fleshy fruits are favored in establishment after long-distance dispersal because of the benefits of multi-seeded dispersal (Baker and Cox, 1984).

4. Leaky Dioecy

Occasional sexual inconstancy in dioecious taxa has long been known but seldom commented upon (see Table IV in Lloyd and Bawa, 1984). The infrequent production of a staminate flower by an otherwise pistillate individual, a pistillate flower by an otherwise staminate individual or a hermaphroditic flower by either sex - a phenomenon termed "leaky dioecy" (Baker and Cox, 1984) - may appear at first to be of little ecological consequence, but may actually be of profound significance in the establishment of dioecious taxa after long-distance dispersal. As early as 1895, von Marilaun published accounts of leaky dioecy in *Urtica dioica* (Urticaceae), *Salix sp.* (Salicaceae), *Vitis cordata* (Vitaceae), *Mercurialis annua* (Euphorbiaceae), and in *Lychnis diurna* and *L. vespertina* (Caryophyllaceae). Baker and Cox (1984) list 21 taxa of dioecious island plants in which leaky dioecy has been found to occur, and many more cases undoubtedly could be found in the literature. Leaky dioecy permits a single individual of a dioecious taxon to set seed. Even if leakage in the dioecious breeding system is a relatively infrequent event, even if it occurs but once in the lifetime of a single individual dioecious colonist, it can result in the establishment of a sexually reproducing population. In these cases, the dictates of Baker's Law are not abrogated but are rather fulfilled - individuals with leaky dioecy are likely to be favored in establishment after long distance dispersal precisely because they are self-compatible. There are costs, however. Seeds from self-fertilizing individuals of normally dioecious *Salix* species germinate poorly, possibly as a result of chromosomal aberrations (T. Elmqvist, pers. comm.).

Given the modular nature of sexuality in angiosperms (Cox, 1988) and their resultant sexual plasticity, it is likely that leaky dioecy will emerge as one of the major ways that island taxa overcome the deleterious effects of a dioecious breeding system in long-distance dispersal.

5. Apomixis

The occurrence of apomixis in dioecious taxa has been little investigated even though the phenomenon of apomixis was first

discovered in a dioecious species. Five seeds of the dioecious
Alchornea ilicifoila (Euphorbiaceae) from Australia were planted
at Kew. There, J. Smith (1841), found that each seed grew into a
pistillate individual, but all plants still set viable seed in the
absence of males. For this Smith deduced the reality of apomixis
and dealt a death blow to the theory of the homunculus in the
sperm (Baker, 1979). Perhaps modern plant evolutionary
ecologists need to be more diligent in studying the living
collections of botanical gardens (Cox, Wallace, and Baker, 1984).

The striking lack of subsequent studies on apomixis in
dioecious species probably is an artifact of the manner in which
apoximis is usually discovered - via bagging studies - as very few
investigators have actually bagged pistillate individuals of
dioecious plants. Still there are some indications in the literature
concerning the presence of apomixis among dioecious taxa. Barlow
et al. (1978) suggested apomixis as a possible explanation for the
highly female-skewed sex ratios in *Viscum album* (Loranthaceae)
in Romania, and seed has been found to develop in pistillate
cultures of the dioecious seagrass *Halophila stipulacea*
(Hydrocharitaceae) where no male flowers were observed
(McMillan, 1980; Cook, 1982). I am currently studying
reproductive ecology of a variety of seagrasses to determine if
apomixis occurs in other dioecious taxa. The well established case
of apomixis in *Pandanus tectorius* will be discussed in the next
section. Apomixis in self-incompatible taxa is also a topic worthy
of further investigation.

BREEDING SYSTEMS AND COLONIZATION ABILITY IN PANDANACEAE

The Pandanaceae, a paleotropical family of arborescent
monocotyledons, provide a good test of Baker's Law since the three
genera differ in range, number of species, and reproductive
ecologies. All however, are dioecious.

The smallest of the three genera, *Sararanga*, consists of only
two species distributed along the edge of the Tethys geosyncline in
the Solomon Islands, New Guinea, the Admiralty Islands, New
Britain, and the Philippines. Even within its range, *Sararanga*
populations are very patchily distributed and somewhat difficult
to find. Given its bird-dispersed fruits, "it is not at first sight
easy to understand why its distribution should be so limited,
unless this is connected with its dioecious habit" (Guppy, 1906,
p. 156). In my study of the reproductive ecology of *Sararanga* in
Guadalcanal, Tulagi, and Nggela Islands of the Solomon Islands, I
surveyed several hundred individuals but could find no evidence of
leaky dioecy, or apomixis (Cox, 1989). *Sararanga sinuosa* also

appears to require pollination by specialized insects. It is not rhizomatous, and its phyllatoactic divergence between adult and juvenile shoots (Cox, 1989) and resultant determination of the meristems suggest that *Sararanga* may be incapable of vegetative propagation. The only feature which might mitigate the deleterious effect of dioecism on *Sararanga* is its fleshy fruits. Therefore Baker's Law would predict *Sararanga* to be a poor candidate for establishment after long-distance dispersal. This prediction is verified by its current distribution along the edge of the Tethys geosyncline. Thus the current distribution of *Sararanga* may be due more to vicariant events than long-distance dispersal.

The next largest genus of the Pandanaceae is *Freycinetia*, a group of about 200 species of lianas ranging from Tahiti to Sri Lanka on an east-west axis and from the uplands of Vietnam to New Zealand on a north-south axis. Like *Sararanga* , *Freycinetia* species also have fleshy fruits that are dispersed by both birds and flying foxes. It has, however, some additional features in its reproductive system that, under Baker's Law, explain its far greater success at island colonization. First, unlike *Sararanga, Freycinetia* species do not require a specialist pollinator, but can be pollinated by vertebrates ranging from the white-eyes (*Zosterops japonica*) that are 11 cm. in length to very large flying foxes with wingspans exceeding 1.5 m. (Cox, 1982, 1983, 1984). It was probably this flexibility that allowed *Freycinetia arborea* in Hawaii to survive extinction of its indigenous pollinators (Cox, 1983). Secondly, *Freycinetia* species are capable of vegetative growth, with *F. cammiana* being propagated from cuttings for the horticultural trade. Thirdly, *Freycinetia* species have leaky dioecy and are self-compatible (Cox, 1981; Cox et al., 1984; Poppendick, 1987). The combination of these three features with abiotically dispersed as well as biotically-dispersed multi-seeded proagules (bird and bat feces) has led to a spectacular success for the genus *Freycinetia* in island colonization, despite its dioecious condition. Its species are found today on virtually every high island in the tropical Pacific west of Tahiti.

An even more spectacular success in island colonization, however, has been achieved by the third genus of the Pandanaceae, the genus *Pandanus*. The genus *Pandanus* is composed of nearly 600 species of trees ranging from the Society Islands to west Africa on an east-west axis, and from the foothills of the Himalayas to Australia on a north-south axis. The genus *Pandanus* as represented by the type species *Pandanus tectorius* has a range of features that, as predicted by Baker's Law, have led to its prominence as one of the most successful species in long-distance dispersal despite its dioecious condition. First of all, like *Freycinetia, P. tectorius* does not require a specialist pollinator, and indeed does not require a pollinator at all. Extensive field experiments and wind-tunnel studies have shown *Pandanus tectorius* to be wind-pollinated, with optimal air velocities for pollination occurring at approximately 1 meter per second (the speed of a barely noticeable breeze) (Cox, 1985). Secondly *Pandanus tectorius* produces multi-seeded propagules that are both abiotically and biotically dispersed. Birds, bats, and even crabs are attracted to the sweet, colorful syncarps (Ridley, 1930; Lee, pers. comm.) which also float and are dispersed by sea. An analysis of seedling production in Maui showed that any given syncarp of *P. tectorius* has a 55.1% probability of producing at least 1 male and 1 female seedling. Thus there is a good chance that a single immigration event could result in the establishment of a breeding population. Thirdly, bagging studies and electrophoresis show that *P. tectorius* is facultatively apomictic (Cox, 1985); that is to say that apomictic seed is produced in the absence of pollination. Thus a single female individual can fill an island with her apomictic descendants long before the first male arrives. This facultative system of apomixis has, through genetic drift, undoubtedly led to rapid speciation and produced many of the headaches confronting *Pandanus* taxonomists. It is clear, however, with abiotically or biotically dispersed, multi-seeded sturdy fruits (dissection by electric bandsaw being the preferred way of getting endosperm samples for electrophoresis), wind pollination, apomixis, and long-life spans, that the disadvantages of a dioecious breeding system in long-distance dispersal have all but been overcome in *Pandanus tectorius*.

The three genera of the Pandanaceae provide an instructive example of Baker's Law in action. *Saranga* appears to be unable to overcome the consequences of its dioecious breeding system and may be on its way to extinction, while *Freycinetia* has enjoyed moderate success principally through the aegis of leaky dioecy, while the combination of anemophily, apomixis, and multi-seeded propagules have led to the spectacular success of *Pandanus* in island colonization.

CONCLUSIONS

Baker's Law, as originally formulated, dictated that self-compatible taxa have the advantage over other taxa in establishment after long-distance dispersal. These considerations seemed to suggest that self-incompatible taxa, and particularly dioecious taxa, should be under-represented in island floras as compared to mainland floras. This particular prediction is not supported by an analysis of island floras: dioecious taxa seem to fare far better in long-distance dispersal than predicted by Baker's Law. Closer examination reveals, however, that such taxa have a variety of mechanisms to overcome their dioecious handicap ranging from perennation to apomixis. Thus Baker's Law is not disproven, but as predicted by Stebbins (1957) has become a powerful tool in structuring studies concerning island colonization. A particular point that emerges from such considerations is the need to consider island colonization on a species by species basis, and particularly the need to carefully examine island species for self-incompatibility systems.

SUMMARY

In a 1955 paper Herbert Baker proposed that self-compatible taxa will be more likely to succeed in establishment in long-distance dispersal than self-incompatible taxa. Stebbins termed this concept Baker's Law. He was justified in doing so because the concept met the criteria of (1) being original with Baker, (2) explaining widely occurring patterns and correlations, and (3) being useful in guiding future studies. An example of the utility of Baker's Law is the way in which it helps to structure the study of the colonization of islands by dioecious plants. Dioecious plants have a variety of mechanisms that serve to ameliorate their disadvantages in long-distance dispersal, including colonization saturation, prennation, multiple-seeded propagules, leaky dioecy, and apomixis. These features are discussed for the Pandanaceae, a family of three genera of dioecious monocotyledons that vary in their success in island colonization.

ACKNOWLEDGMENTS

I thank Herbert and Irene Baker for being such wise and kind mentors; I hope Herbert's influence can be seen in the useful parts of this paper while I must claim sole responsibility for the foolish parts. This study was supported by an NSF Presidential Young Investigator Award BSR-8452090.

LITERATURE CITED

Anderson, G. J. and G..L. Stebbins. 1984. Dioecy versus gametophytic self-incompatibility: a test. Amer Nat. 124: 423-428.

Baker, H. G. 1955. Self-compatibility and establishment after "long-distance" dispersal. Evolution 9: 347-348.

222

Baker, H. G. 1959. Reproductive methods as factors in speciation of flowering plants. Cold Spring Harbor Symp. Quant. Biol. 24: 177-191.

Baker, H. G. 1965. Charles Darwin and the perennial facts: a controversy and its implication. Huntia 2: 141-161.

Baker, H. G. 1967. Support for Baker's Law as a rule. Evolution 21: 853-856.

Baker, H. G. 1979. Anthecology: Old Testment, New Testament, Apocrypha. New Zealand J. of Bot. 17: 431-440.

Baker, H. G. and P. A. Cox. 1984. Further thoughts on islands and dioecism. Ann. of the Missouri Bot. Gard. 71: 230-239.

Barlow, B. A., D. Wiens, C. Wiens, W. H. Busby, and C. Brighton. 1978. Permanent translocation heterozygosity in *Viscum album* and *V. cruciatum*, sex assoication, balanced lethals, sex ratios. Heredity 40: 33-38.

Barrett, S. C. H. and J. S. Shore. 1987. Variation and evolution of breeding systems in the *Turnera ulmifolia* L. complex (Turneraceae). Evolution 41: 340-354.

Bawa, K. S. 1980. Evolution of dioecy in flowering plants. Ann. Riv. Ecol. and Syst. 11: 15-40.

Bawa, K. S. 1982. Outcrossing and the indicence of dioecism in island floras. Amer. Nat. 119: 866-871.

Carlquist, S. 1974. Island Biology. Columbia University Press, New York.

Carr, G. D., E. A. Powell, and D. W. Kyhos. 1986. Self-incompatability in the Hawaiian Madiinae (compositae): an exception to Baker's Rule. Evolution 40: 430-434.

Cook, C. D. K. 1982. Pollination mechanisms in the Hydrocharitaceae. In Studies on Aquatic Vascular Plants. J. J. Symoens, S. S. Hooper, and P. Compere, [eds.]. R. Bota. Soc. Belg. Brussels, Belgium.

Cox, P. A. 1981. Bisexuality in the Pandanaceae: new findings in the genus *Freycinetia*. Biotropica 13: 195-198.

Cox, P. A. 1982. Vertebrate pollination and the maintenance of dioecism in *Freycinetia*. Amer. Nat. 120: 65-80.

Cox, P. A. 1983. Extinction of the Hawaiian avifauna resulted in a change of pollinators for the ieie, *Freycinetia arborea*. Oikos 41: 195-199.

Cox, P. A. 1984. Chiropterophily and ornithophily in *Freycinetia* in Samoa. Plant Evol. and Syst. 144: 277-290.

Cox, P. A. 1985. Islands and dioecism: insights from the reproductive ecolgy of *Pandanus tectorius* in Polynesia. In J. White [ed.], Studies on Plant Demography: A Festschrift for John L. Harper, pp. 359-372. Academic Press, London.

Cox, P. A. 1988. Monomorphic and dimorphic sexual starategies: a modular approach. In J. Lovett Doust and L. Lovett Doust [eds.] Plant Reproductive Ecology. Oxford University Press, Oxford.

Cox, P. A. 1989. Breeding systems and the evolution of Pandanaceae. Ann. of the Missouri Botan. Gard. In press.

Cox, P. A., B. Wallace, and I. Baker. 1984. Monoecism in the genus *Freycinetia*. Biotropica 16: 313-314.

During H. J. 1979. Life strategies of bryophytes: a preliminary review. Linbergia 5: 2-18.

During H. J., and B. F. Van Tooren. 1987. Recent developments in bryophyte population ecology. Trends in Ecol. and Evol. Biol. 2: 89-93.

Givnish, T. J. 1980. Ecological constraints on teh evolution of breeding systems in seed plants: dioecy and dispersal in gymnosperms. Evolution 34: 959-972.

Grant, B. R., and P. R. Grant. 1981. Exploitation of *Opuntia* cactus by birds on the Galapagos. Oecologia 49: 179-187.

Guppy, H. B. 1906. Observations of a naturalist in the Pacific between 1896 and 1899. Vol. II. Plant-Dispersal. Macmillan, London.

Kerner von Marilaun, A. 1895. The Natural History of Plants. Vol. II. F. W. Oliver [trans.]. Gersham, London.

Lloyd, D. G. 1972. A revision of the New Zealand subantarctic and South American species of *Cotula* Section *Leptinella*. New Zaland J. Bot. 10: 277-372.

Lloyd, D. G., and K. S. Bawa. 1984. Modification of the gender of seed plants in varying conditions. Evolutionary Biol. 17: 255-338.

McMillan, C. 1980. Flowering under controlled conditions by *Cymodocea seratula, Halophila stipulacea, Syringodium isoetifolium, Postera capensis* and *Thalassica hemprichii* from Kenya. Aquatic Bot. 8: 323-336.

McMullen, C. K. 1987. Breeding systems of selected Galapagos Islands angiosperms. Amer. J. Bot. 74: 1694-1705.

Murbeck, S. 1919. Beitrage zur Biologe der Wustenpflanzen, I. Lunds Univ. Arsskr. 15 (10).

Pijl, L., van der 1969. Principles of dispersal in higher plants. Springer-verlag, Berlin.

Poppendick, H. 1987. Monoecy and sex changes in *Freycinetia* (Pandanaceae). Annals of the Missouri Bot. Gard. 74: 314-320.

Raven, P. H. 1963. A flora of San Clemente Island, California. Aliso 5: 289-347.

Richards, A. J. 1986. Plant Greeding Systems. George Allen & Unwin, London.

Rick, C. M. 1966. Some plant-animal relationships in the Galapagos Islands. In R. I. Bowman [ed.] The Galapagos: Proceedings of the Galapagos International Scientific Project. Univ. of California Press, Berkeley.

Ridley, H. N. 1930. The dispersal of plants throughout the world. L. Reeve, Kent, England.

Schuster, R. M. 1966. The Hepaticae and Anthocerotae of North America. Vol. 1. Columbia University Press, New York.

Smith, J. 1841. Notice of a plant which produces perfect seeds without any apparent action of pollen. Trans. of the Linnean Soc., London. 18: 509-512.

Strid, A. 1970. Studies in the Aegean flora. XVI. Biosystematics of the *Nigella arvensis* complex. Opera Botanica 28: 1-169.

Stebbins, G. L. 1957. Self-fertilization and population variability in the higher plants. Amer. Nat. 41: 337-354.

Trask, B. 1904. Flora of San Clemente Island. Bull. So. Calif. Acad. Sci. 3: 76-78, 90-95.

REPRODUCTIVE BIOLOGY

Sexual reproduction in plants includes pollination, seed set, and much more. When plants produce flowers, these flowers can attract a vast array of animals, some of whom can serve as pollinators, though others are simply consumers that exploit the flowers. Such exploitation is costly to the plants. Flower colors and odors are well known signals used by plants to attract specific pollinators, and to deter unwanted visitors. Thanks primarily to the efforts of Herbert and Irene Baker, in recent years students of pollination have recognized that nectars are not just sugary liquids which provide temporary sustenance to any visitor. In fact, they are important components of the total flowering "signal" sent to potential pollinators. Nectars are complex mixtures of carbohydrates, amino acids, proteins, antioxidants, and sometimes toxic substances. The appropriate ingredients in these mixtures can therefore serve to attract one type of visitor and discourage visits by others. In a similar vein, pollen grains are not merely gametophytes. Pollen grains are covered by a complex coat called the pollenkitt, and Dobson reviews what is known about the structure and possible functions of this covering. Just as nectar composition in various species appears to have been evolutionarily adjusted to attract and provide essential nutrients to specific pollinator groups, so it is that pollenkitt shows interspecific variability which suggests a variety of adaptively useful functions.

Once pollination and fertilization are successful, a plant faces the challenge of converting embryos into mature seeds. The complexities of this conversion are illustrated in several essays. Reproduction is energetically expensive, and for this reason, plant size is often a good predictor of reproductive output. However, Fone demonstrates for *Collinsonia verticillata* the extent of pollination is, in fact, of much greater importance than various elements of vegetative growth in determining seed production. In a detailed analysis of the relationships among plant parts, Fone also demonstrates the usefulness of path analysis in elucidating these relationships. Resources available to single plants are finite, and patterns of allocation indicate that when resources are allocated to one plant part or activity, they are unavailable to another. In *Collinsonia*, there is a negative correlation between allocation to leaves and to seeds. Clearly, plant growth and maintenance may decrease when seeds are produced.

In a study of two grass species, *Poa annua* and *Poa pratensis*, Wagner finds that overall reproductive effort is not correlated with plant size in any consistent fashion, and that it varies from year to year. These two studies involve species with very different architectures, examined under different ecological conditions and using different methodologies. Yet they arrive at similar conclusions, and serve as useful warnings that no single factor is identifiable as a reliable predictor of reproductive output.

The reproductive success of individual plants is not dependent solely on the energy available to them. Seeds consist of concentrated packages of highly attractive nutrients, and many animals utilize seeds as a primary or sole source of food. These animals can have detrimental effects on reproduction. In a study of *Hibiscus moscheutos*, Spira demonstrates that seed-eating insects can be the primary determinants of the total reproductive output of a plant.

Rainfall and its timing play critical roles in the reproduction and population dynamics of certain species such as desert annuals, plants of buffalo wallows in the Great Plains (Uno, this volume), and vernal pools of the Central Valley of California. Schlising documents the latter with *Sidalcea hirsuta* whose seed production varied as much as 100-fold among years. Variation in ecological conditions among years is to be expected under any circumstances, and can affect plant growth and reproduction. Wagner, Spira, and Schlising illustrate the extent to which experimental results can vary among years. And more importantly, the year to year differences give a strong warning against publishing ecological and evolutionary studies of plant reproduction based upon data from a single season.

POLLENKITT IN PLANT REPRODUCTION

Heidi E. M. Dobson

INTRODUCTION
        Pollenkitt forms an oily, colorless to yellow-orange coating on pollen grains of angiosperms and is especially abundant in animal-pollinated species (Hesse, 1980, 1984). It is derived from contents of the anther's tapetal cells and, just prior to anther dehiscence, becomes deposited on the surface and in the structural exinous cavities of the pollen grain (Heslop-Harrison, 1968; Dickinson, 1973; Keijzer, 1987). Evidence from several studies shows pollenkitt to contain a variety of oily constituents in addition to the more obvious carotenoid pigments (see Stanley and Linskens, 1974; Dobson, 1985). These other components include principally glycerides, sterols, and various terpenoid, aromatic, and aliphatic compounds; also present are flavonoids, phospholipids, enzymes, glycoproteins, glycolipids, carboxylated polysaccharides, phenolics, and, in special cases, cytoplasmic remains of the tapetum (Nilsson et al., 1957; Heslop-Harrison, 1968; Wittgenstein and Sawicki, 1970; Dickinson, 1973; Clarke et al., 1979; Roberts et al., 1979; Mattsson, 1983; Bernhardt, 1984; Klungness and Peng, 1984; Dobson et al., 1987; Dobson, 1988).
        The earliest references to pollenkitt date from the general period of Sprengel's landmark book on pollination in 1793. One of the first mentions was made by Kölreuter (1761), who suggested that pollenkitt serves in fertilization by being transmitted along with stigmatic substances to the ovules. In the mid 1800s, discussion arose about the production site of pollenkitt oil (Fritzsche, 1837; von Mohl, 1852), a question not resolved until a century later (Mühlethaler, 1955; Pankow, 1958). Also deliberated were the functional aspects of pollenkitt (Meyen, cited by Troll, 1928; Fischer, cited by Wodehouse, 1935). Following this, Kerner (1898) provided detailed descriptions of pollenkitt from several plant species, which he accompanied with a more thorough treatment of the possible functions. In particular, he pointed to pollenkitt's capacity to bind pollen grains to each other, to the anther wall (prior to pollination), and to pollinators. The importance of the oil's adhesive qualities, and thus its role in enhancing pollen dispersal by flower visitors, became increasingly recognized. This is reflected in Knoll (1930) naming it "Pollenkitt" (pollen cement). As interest increased in pollination ecology in the 1900s, the roles for pollenkitt received

further consideration (Troll, 1928; Wodehouse, 1935).
Especially insightful were speculations by Knoll (1930), who, in addition to suggesting that pollenkitt (carotenoid) pigments visually attracted flower visitors (a point made already by earlier workers), also proposed that constituents of pollenkitt might provide nutrition and olfactory attractants to pollinators.

Several functions of pollenkitt are currently acknowledged (Stanley and Linskens, 1974; Dobson, 1985, in press). Some relate to the survival of the male gametophyte, with pollenkitt providing protection against 1) damage by ultraviolet light, 2) excessive moisture loss, and 3) microbial attack. Others pertain to the successful transport of pollen from anther to stigma. Pollenkitt may facilitate pollinator service by providing 1) visual attraction, 2) olfactory attraction, and 3) nutrition (all three reviewed in greater detail by Dobson, in press), as well as by increasing pollen adhesion. Finally, pollenkitt may play an important part in pollen-pistil interactions that are prerequisites for fertilization. These diverse roles are not mutually exclusive, and it seems likely that in most cases more than one may apply. Each of these roles is briefly reviewed below.

ULTRAVIOLET RADIATION

Pollen pigments, flavonoids and carotenoids, have been suggested to protect pollen by filtering ultraviolet light (Asbeck, 1955; Stanley and Linskens, 1974). The amount of exposure to these damaging wavelengths depends on both geographic parameters (e.g., latitude, elevation) and floral morphology (e.g., position of dehisced anthers within the corolla) (Flint and Caldwell, 1983). Some flowers have poricidally dehiscent anthers, where pollen is contained within the anther sacs until shaken out by wind or pollinators. Here,the anther walls appear to act as protective screens and are bright yellow in contrast to the light colored pollen (Buchmann, 1983).

Flavonoids (water-soluble) are common in pollen grains, but occasionally are present also in pollenkitt (Wiermann and Vieth, 1983). They show high UV absorbance (McClure, 1975; Harborne, 1976); their actual role in shielding pollen against damaging radiation damage, however, is not completely clear (Coe et al., 1981).

Carotenoids, less widespread as pollen pigments than flavonoids, are found primarily in pollenkitt (Pankow, 1958; Heslop-Harrison, 1968; Wittgenstein and Sawicki, 1970; Dobson, 1988). One of their functions may be to serve as photoprotectors (Burnett, 1976). Carotenoid composition of pollen changes with elevation, suggesting that carotenoids differ in their adaptive value (Neamtu and Bodea, 1970).

DESSICATION

By acting as a seal on the grain's surface, pollenkitt may be important in protecting pollen from a decrease in viability due to water loss (Heslop-Harrison, 1979a, 1979b; Hesse, 1980). Pollen types differing in the amount of pollenkitt can show correspondingly striking contrasts in viability. Thus, grass pollen, which is covered by a very thin lipid seal, loses viability within 1-2 hours after exposure to air, whereas pollen with a thick oil coat tends to retain viability for a much longer time (Knox, 1984). Other factors besides pollenkitt, however, may also be important, such as properties of the plasma membranes (Heslop-Harrison, 1979c).

MICROBIAL ATTACK

Pollenkitt contains a variety of chemicals (Dobson et al., 1987; Dobson, 1988) that are typical of essential oils (Harborne and Turner, 1984), and indirect evidence suggests that some of these might offer defense against fungi and bacteria. Indeed, many constituents of essential oils, including volatiles from leaves and flowers, exhibit antimicrobial activity (Maruzella, 1962; Morris et al., 1979; Toth et al., 1987). Investigations are needed that focus on the possible role of pollenkitt in deterring or mitigating attack by microbes on the male gametrophyte, a function already proposed for the pollen exine (Chaloner, 1976).

VISUAL ATTRACTION

Pollen displays a wide range in colors owing to the presence of flavonoids and carotenoids (Stanley and Linskens, 1974). Flavonoids contained in the cytoplasm and wall produce the non-yellow as well as many of the light yellow colorations of pollen, whereas many yellow-orange hues are due mainly to carotenoids (carotenes and xanthophylls) located in pollenkitt (Heslop-Harrison, 1968; Wiermann and Weinert, 1969; Goodwin, 1976; Wiermann, 1981; Dobson, 1988). These pigmentations can contribute to visual floral patterns by enhancing contrasts between different floral parts or between flowers and their background (Vogel, 1978; Douglas, 1983; Penny, 1983; Linsey and Bell, 1985). Conversely, removal of pollen by flower-visiting animals may serve as an avoidance signal to subsequent pollen-seeking visitors. The modified key floral patterns represent a form of "post-pollination" color change (Gori, 1983).

Pollen coloration may serve in attracting pollinators.

Indeed, in plants having different colors of pollen in feeding versus fertilizing anthers, the fodder pollen tends to be more brightly pigmented (often yellow) than the fertilizing pollen (Vogel, 1978; Mori et al., 1980; Muller, 1981; Pacini and Bellani, 1986). Studies on honey bees, however, provide equivocal evidence for the importance of pollen color per se in the actual selection of flowers and pollen by insects (Levin and Bohart, 1955; Lepage and Boch, 1968; Boch, 1982). Other investigations indicate that pigments responsible for pollen coloration may also serve functions associated with pollen survival (Mori et al., 1980). Nevertheless, the more frequent occurrence of carotenoids, and thus pollenkitt pigments, in entomophilous than anemophilous plants, as reported in surveys by Lubliner-Mianowska (1955) and Wiermann and Vieth (1983), does intimate that pollenkitt pigmentation may represent an adaptation to pollination by animals, most probably as an attractant.

OLFACTORY ATTRACTION

Previous suggestions that volatiles responsible for the distinctive odors of different pollen types arise from pollenkitt (Knoll, 1930; Porsch, 1956; Faegri and van der Pijl, 1979; Buchmann, 1983) have been confirmed by recent chemical studies. In a preliminary survey of 60 angiosperm species in California (Dobson, 1988), thin-layer chromatographic analysis of pollenkitt revealed a diversity of chemical classes, many typical of essential oils (e.g., esters, alcohols, acids, aldehydes, ketones, and hydrocarbons, of aliphatic, terpenoid, and aromatic derivation). Furthermore, these chemicals were present in species-specific mixtures. Closer examination and comparison of the aromas of pollen (collected by adsorption-desorption) and of pollenkitt extracts in *Rosa rugosa* (Rosaceae), both analyzed by gas chromatography and mass spectrometry, indicate that major chemicals found in headspace of pollen grains are also present in pollenkitt (Fig. 1; see also Dobson et al., 1987). Esters (aliphatic, terpenoid, or aromatic) were the major volatiles identified in pollenkitt, along with alcohols, ketones, and aldehydes in smaller quantities. Differences between pollen and pollenkitt in the relative abundance of individual substances may in large part be due to the different natures of the samples, i.e., head space versus extract. In an earlier study of pollen volatiles in *Vitis* (Vitaceae), Egorov and Egofarova (1971) reported primarily terpenes.

Given that pollen does release volatiles, the question arises as to whether these originate from pollenkitt (i.e., are tapetal) or

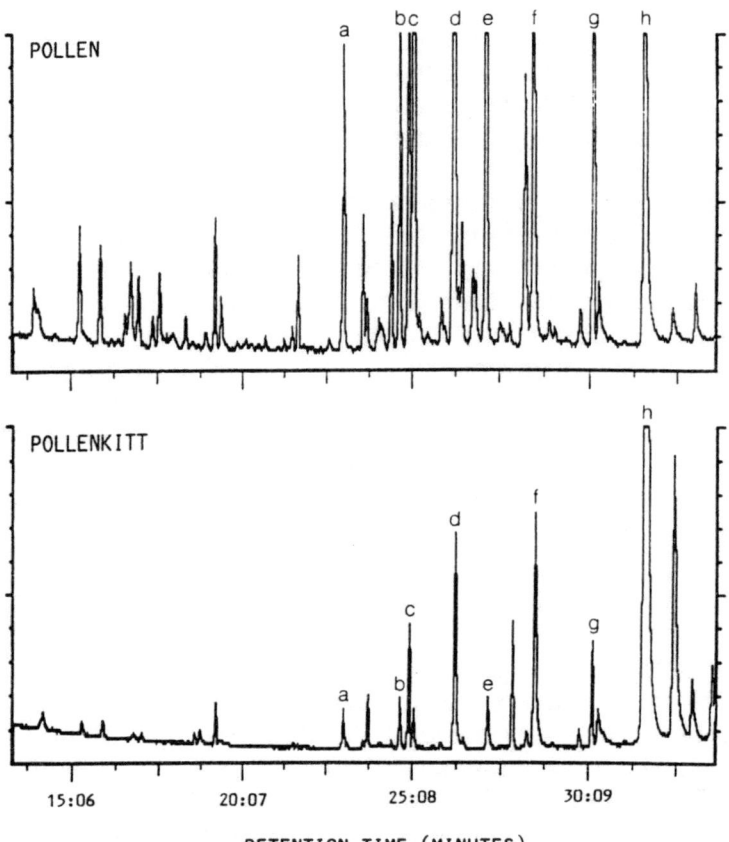

RETENTION TIME (MINUTES)

Figure 1. Gas chromatograms of volatiles detected in pollen (headspace) and pollenkitt (extract) of *Rosa rugosa* L.. GC-MS analyses were performed by J. Bergström (see Dobson *et al.*, 1987) Most of the major substances in pollen were also found in pollenkitt (*i.e.* a = citronellyl acetate; b = geranial; c = geranly acetate; d = 2-tridecanone; e = geranyl acetone; f = tetradecanal; g=methyeugenol; h = tetradecyl acetate.

whether they represent chemicals from other parts of the flower that become adsorbed onto the pollenkitt subsequent to anther dehiscence and pollen exposure. When pollen of *Papaver rhoeas* (Papaveraceae) and *Dactylis glomerata* (Gramineae) was placed in separate glass petri dishes, each surrounded with 22 cut flowers of *Rosa rugosa* (stems in water), and allowed 24 hours to pick up fragrance components from the flowers, analysis of the pollen volatiles showed undetectable or only trace amounts of *R. rugosa* chemicals (Dobson and Bergström, unpubl.). This suggests that volatiles from petals and other fragrance-releasing floral organs may contribute only little to pollen odor. Additional experiments carried out over fewer hours are needed to verify that low detection levels are not caused by volatilization of odorous pollen substances over the sampling period; also, pollen types that have differing quantities of pollenkitt need to be compared in order to establish the extent to which oil abundance influences adsorption of foreign chemicals.

Interspecific comparisons show that the volatile profiles of pollen (which comprise chemicals similar to those reported in floral fragrances by Williams (1983) vary widely, both quantitatively and quantitatively (Dobson and Bergström, in prep.). Among the dozen species analyzed to date, only trace amounts of volatiles were detected in the four wind-pollinated species. In contrast, volatiles in the eight animal-pollinated species were much more prominant. These data suggest that pollen - and pollenkitt - odor may indeed represent an adaptation to pollination by animals, serving as an olfactory attractant to flower visitors. The line between wind- and animal-pollinated species is not a sharp one, however, and some anemophilous pollens have been described as having distinctive odors (Porsch, 1956).

The actual role of pollenkitt as an olfactory attractant of animal pollinators is suggested by several studies, although it still remains to be conclusively established under field conditions. Among bees, which are an important group of pollinators, honey bees show increased attraction to otherwise unattractive materials when these are supplemented with pollen extracts containing pollenkitt (Louveaux, 1959; Hügel, 1962; Taber, 1963; Lepage and Boch, 1968; Hohmann, 1970). Furthermore, honey bees can be trained to distinguish among different plant species on the basis of pollen odor alone (von Frisch, 1923; von Aufsess, 1960). Behavioral tests on pollen-specific solitary bees indicate that, when offered a choice of either pollen or pollenkitt from four plant species, individuals of Colletes fulgidus (Colletidae) exhibited a significant feeding preference for their host-plant (Dobson, 1987). Similar experiments on other bees indicate that their ability to discriminate between plants by pollen odor alone varies interspecifically, some species relying more on visual cues than others (Dobson, unpubl.; Dobson and Tengö, in prep.). Pollen odor also has been inferred in the attraction of flower-feeding beetles to their host plants (Charpentier, 1985; Andersen and Metcalf, 1987). Neurophysiological studies using electro-antennogram tests are currently being applied to several species of bees and pollen-feeding beetles to evaluate the ability of their antennal sensilla to respond to different pollen odors (Dobson and Ågren, in prep.).

ANIMAL NUTRITION

Pollen is eaten by a variety of animals, including insects, birds, and mammals (Faegri and van der Pijl, 1979; Dobson, 1985). The digestion of pollen has been investigated in several species (Haslett, 1983; Klungness and Peng, 1984; Turner, 1984; Peng et al., 1985), but information on the utilization of distinct pollen components is limited primarily to protein constituents (Turner 1984; Schmidt and Buchmann, 1985; Schmidt et al., 1987; Smith and Green, 1987). Pollenkitt has received only cursory mention.

Non-polar fractions of pollen have been shown to contain chemicals that are phagostimulatory to honey bees within the hive (Robinson and Nation, 1968; Doull, 1974; Schmidt, 1985) and pollenkitt, as part of this fraction, may be implied. Furthermore, the frequent detection of glycerides, esters, fatty acids, and sterols in pollenkitt (Roberts et al., 1979; Dobson, in press), compounds also typical for pollen reserves (Stanley and Linskens, 1974) and more abundant as such than starch in species pollinated by pollen-feeding insects (Baker and Baker, 1979, 1983), suggests that pollenkitt may provide some nutrition to flower visitors. Certain constituents, particularly terpenoids, might also be used as precursors to some of the diverse chemicals found in bee secretions (Francke et al., 1984; Duffield et al., 1984).

Feces of pollen-feeding flies and bees typically contain a yellow, pollenkitt-like, material (Knoll, 1930; Dobson, 1985, in press; Nowicke and Meselson, 1984). Histochemical studies of pollen digestion in adult honey bees (Klungness and Peng, 1984; Peng et al., 1985) indicate that some yellow pigments, as well as other uncharacterized pollenkitt components, pass through the bee gut unabsorbed. Parallel studies on larvae of solitary bees (Dobson and Peng, in prep.) are of particular interest, since pollen is the main food of larvae and possible sequestration or utilization of pollenkitt components would seem most likely to occur at this feeding stage.

POLLEN ADHESION

Characteristics of both the exine (spines, threads, irregular surface patterns) and the pollenkitt (consistency, quantity, distribution) are critical in determining pollen adhesiveness (Sprengel, 1793; Kerner, 1898; Knoll, 1930; Skvarla et al., 1978; Hesse, 1981a, 1986). The adaptive significance of exine and pollenkitt features in pollination is plainly shown by comparing zoophilous and anemophilous plants. Animal-pollinated species typically have pollen with pronounced exine scupturing and/or evident pollenkitt in contrast to wind-pollinated species, in which pollen grains are smooth, powdery, and thus non-clumping, thereby facilitating dispersal in air currents. In the case of zoophily, the need for any one of these different adhesive features depends also on the body-surface characteristics of the pollinator (e.g., hair types, scales, feathers, cuticular aspects) (Gregory, 1964; Roberts and Vallespir, 1978; Thorp, 1979; Hesse and Waha, 1983; Hemsley and Ferguson, 1985) and possibly on the electrostatics of plant and animal tissue (Erickson and Buchmann, 1983).

Variation in pollenkitt characters and/or exine ornamentation that is associated with different pollinator agents (and thus differing needs for pollen adhesion) have been documented at the family, genus, and species levels (Wodehouse, 1935; Grasshoff and Beaman, 1970; Page, 1978; Skvarla et al., 1978; Hesse, 1980; Adams et al., 1981; Ferguson and Skvarla, 1982; Stelleman, 1983; Hemsley and Ferguson, 1985; Ferguson and Pearce, 1986; Grayum, 1986). Likewise, decrease in pollen adhesiveness in association with autogamy has been noted for several taxa (Grasshoff and Beaman, 1970; McNeill and Crompton, 1978; Skvarla et al., 1978; Prentice et al., 1984; Lord and Eckard, 1984). In some plant groups, however, exine and pollenkitt traits may show little relationship with pollination mode (Kress, 1986), or alternatively, pollen attachment to pollinators may be achieved by other means, such as non-exinous threads (Hesse and Waha, 1983; Waha, 1984; Hesse, 1986) or non-pollenkitt, sticky substances (Vogel, 1981; Grayum, 1986).

The most significant studies of correlations between pollenkitt and pollination mode have been made by Hesse (1980, 1981b), who documented numerous examples of variation in pollenkitt between closely related entomophilous and anemophilous species in several families. Increased pollen stickiness (pollenkitt) in response to insect pollination has also been noted for some species within the predominantly wind-pollinated genus *Paspalum* (Gramineae) (Adams et al., 1981). Within the species *Plantago lanceolata* (Plantaginaceae), Stelleman (1983) observed interpopulational differences in pollenkitt occurrence that are related to habitat and, perhaps, to pollination. Plants growing in open areas have powdery pollen that is dispersed primarily by wind, while plants in closed habitats, where pollen dispersal requires greater reliance on insects, have "gluey" pollen grains.

Pollenkitt characteristics are closely tied to, and

determined by, the tapetum and the exine. Whereas the quantity and consistency of pollenkitt depends on the tapetal material released into the anther loculus, exine morphology (e.g., closed versus open tectum) influences the distribution of pollenkitt on the grain surface (Hesse, 1980, 1981a, 1981b). Thus, the degree of stickiness conferred by pollenkitt fo pollen grains appears to be associated with two main factors: 1) pollenkitt consistency (i.e., homogeneity and electron density), which may reflect chemical composition, and 2) pollenkitt distribution. In entomophilous species, pollenkitt tends to be located mostly on the exine surface as a homogeneous, electron-dense layer, and the exine cavities are almost empty. Pollen of anemophilous species presents a situation opposite to this, while plants pollinated by both wind and insects show intermediate pollenkitt features.

Several species showing pollen polymorphism in association with heterostyly or, especially, with heteranthery (feeding versus fertilizing pollen) offer examples of contrasting pollen adhesiveness. In heterostylous species of *Waltheria* (Sterculiaceae), the striking dimorphism in exine ornamentation of the pollen types supposedly reflects differences in adhesion essential for pollination (Kohler, 1976); a similar case is discussed by Baker (1956) in his delightful detective work on *Rudgea jasminioides* (Rubiaceae). In heterantherous species, the anthers and pollen are separated by functions related strictly to reproduction (fertilization), on the one hand, and to pollinator reward (feeding), on the other. This offers excellent opportunities for elucidating the adaptive significance of pollenkitt. Detailed studies are few. In *Couroupita guianensis* (Lecythidaceae), the fertilizing pollen is covered with pollenkitt droplets, but the fodder pollen is not, suggesting possible adaptation for increasing pollen attachment to pollinators or stigmatic surfaces (Mori et al., 1980; Ormond et al., 1981). In *Lagerstroemia* species (Lythraceae), it is the fodder pollen that has greater quantities of pollenkitt (Muller, 1981; Pacini and Bellani, 1986); perhaps in these cases the oily coat serves mainly to enhance pollinator attraction. Undoubtedly, more than one selective force influences pollenkitt deposition on pollen grains.

POLLEN-PISTIL INTERACTIONS

Following pollination, a pollen grain must undergo adhesion, hydration, germination, and pollen-tube growth to fertilize the ovules. Completion of these steps depends on complex structural and chemical interactions between pollen and pistil, and substances located within the pollen wall can be decisive

(Howlett et al., 1979; Roberts et al., 1979; Shivanna, 1982; Dumas et al., 1984; Knox, 1984; Dickinson, 1985).

Adhesion of pollen to stigma involves chemicals originating from both pollen and pistil. Pollenkitt constituents, especially lipoproteins, pigments, and carbohydrate-containing compounds, may be important, particularly in species with dry stigmas (Clarke et al., 1979; Shivanna, 1982; Knox, 1984; Dumas et al., 1984). In some heterostylous plants, both exine morphology and pollenkitt affect adhesion and hydration (Baker, 1966; Dulberger, 1975, 1987; Ghosh and Shivanna, 1982; Mattsson, 1983). In the case of *Armeria maritima* (Plumbaginaceae), the two morphologically dissimilar pollen types differ in quantity of exine oils, suggesting that the variation serves partly to store - and later release - differing amounts of (chemically different?) pollenkitt involved in pollen-stigma adhesion (Mattsson, 1983). Pollenkitt droplets have also been shown to differentially influence pollen-stigma interactions of the pollen morphs in *A. maritima* (Mattsson, 1983) and *Primula* (Primulaceae) (Shivanna et al., 1983) by modulating water exchange between pollen and ambient air.

Proteinaceous diffusates from the pollen wall (e.g., chemicals with allergenic, antigenic, and enzymatic activity) have been suspected to act as recognition factors during pollen germination and pollen-tube growth in both intraspecific and interspecific incompatibility reactions (Howlett et al., 1979; Knox, 1984). This activity, however, is associated primarily with the intine (inner wall of pollen grain) rather than exine (Knox and Heslop-Harrison, 1970). By sealing the pollen surface, pollenkitt may influence the rate at which these intine proteins are discharged (Howlett et al., 1973; Heslop-Harrison et al., 1973).

Evidence that the pollenkitt plays an active role in incompatibility systems is ambiguous (Dumas, 1977; Roggen, 1974; Roberts et al., 1979; Mattsson, 1983; Knox, 1984). In some species, pollenkitt-containing exine diffusates from compatible pollen have been effective in overcoming the various physiological barriers associated with either interspecific or sporophytic-intraspecific (i.e., 'dry' stigma) incompatibility (Shivanna, 1982; Ghosh and Shivanna, 1984). Sporophytic incompatibility systems appear to require exines that allow storage of tapetum-derived materials; in pollen with reduced exine, such as in monocotyledons, sporophytic systems are generally unknown and the barriers occur at the gametophytic level (Heslop-Harrison, 1976).

The evolutionary development of baculate exines with

internal cavities, supposedly a response to stress on the pollen wall from volume changes due to uptake or loss of water (harmomegathy) (Walker, 1976), is thought to have pre-adapted the exine for storage of oily pollen-coat material, and later, of sporophytic incompatibility substances as well (Heslop-Harrison, 1975). Indeed, primitive angiosperm species with baculate exines generally have gametophytic incompatibility systems and their exine cavities are filled with tapetal lipids. In summary, reticulate and perforate exines, which have cavities containing tapetal material that readily diffuses out onto moist stigmatic surfaces, tend to be associated with sporophytic incompatibility. In contrast, imperforate and microperforate exines occur in species with gametophytic incompatibility systems (Zavada, 1984).

CONCLUSIONS

Factors influencing the evolution and production of pollenkitt are clearly of diverse nature and are related to physiological parameters of plant reproduction as well as to ecological parameters of pollen dispersal. Close examination of how pollenkitt varies in association with environmental conditions and plant reproductive biology will provide insight into the evolutionary pressures underlying the differential occurrences and diverse properties of pollenkitt within and among plant species.

ACKNOWLEDGMENTS

I am very grateful to Herbert and Irene Baker for sharing their broad knowledge and keen interest in pollination ecology, and especially their enthusiasm for pollen biology, during my undergraduate and graduate years in close association with them. I also thank Lennart Ågren and anonymous reviewers for helpful comments on the manuscript.

LITERATURE CITED

Adams, D. E., W. E. Perkins, and J. R. Estes. 1981. Pollination systems in *Paspalum dilatatum* Poir. (Poaceae): an example of insect pollination in a temperate grass. Amer. J. Bot. 68: 388-394.

Anderson, J. F., and R. L. Metcalf. 1987. Factors influencing distribution of *Diabrotica* spp. in blossoms of cultivated *Cucurbita* spp.. J. Chem. Ecol. 13: 681-699.

Asbeck, F. 1955. Fluoreszierender Blütenstaub. Naturwissenschaften 42: 632.

Aufsess, A. von. 1960. Geruchliche Nahorientierung der Biene bei entomophilen and ornithophilen Blüten. Z. Vergl. Physiol. 43: 469-498.

Baker, H. G. 1956. Pollen dimorphism in the Rubiaceae. Evolution 10: 23-31.

Baker, H. G. 1966. The evolution, functioning and breakdown of heteromorphic incompatibility systems. I. The Plumbaginaceae. Evolution 20: 349-368.

Baker, H. G., and I Baker. 1979. Starch in angiosperm pollen grains and its evolutionary significance. Amer. J. Bot. 66: 591-600.

Baker, H. G., and I. Baker. 1983. Some evolutionary and taxonomic implications of variation in the chemical reserves of pollen. In D. L. Mulcahy and E. Ottaviano [eds.]. Pollen: Biology and Implications for Plant Breeding. Elsevier, New York. pp. 43-52.

Bernhardt, P. 1984. The pollination biology of *Hibbertia stricta* (Dilleniaceae). Pl. Syst. Evol. 147: 267-277.

Boch, R. 1982. Relative atttractiveness of different pollens to honeybees when foraging in a flight room and when fed in the hive. J. Apic. Res. 21: 104-106.

Buchmann, S. L. 1983. Buzz pollination in angiosperms. In Jones, C. E., and R. J. Little [eds.]. Handbook of Experimental Pollination Biology. Sci. Acad. Press, New York.

Burnett, J. H. 1976. Functions of carotenoids other than in photosynthesis. In T. W. Goodwin [ed.]. Chemistry of Plant Pigments, 2nd ed. Acad. Press, London.

Chaloner, W. G. 1976. The evolution of adaptive features in fossil exines. In Ferguson, I. K., and J. Muller [eds.]. The Evolutionary Significance of the Exine. Academic Press, London.

Charpentier, R. 1985. Host-plant selection by the pollen beetle *Meligethes aeneus*. Entomol. Exp. Appl. 38: 277-285.

Clarke, A., P. Gleeson, S. Harrison, and R. B. Knox. 1979. Pollen-stigma interactions: identification and characterization of surface components with recognition potential. Proc. Nat. Acad. Sci. USA 76: 3358-3362.

Coe, E. H., S. M. McCormick, and S. A. Modena. 1981. White pollen in maize. J. Heredity 72: 318-320.

Dickinson, H. G. 1973. The role of plastids in the formation of pollen grain coatings. Cytobios 8: 25-40.

Dickinson, H. G. 1985. The cytophysiological basis of the sporophytically controlled self-incompatibility mechanism operating in *Brassica*. In Sussex, I., A. Ellingboe, M. Crouch, and R. Malmberg [eds.]. Plant Cell/Cell Interactions. Cold Spring Harbor Lab., New York.

Dobson, H. E. M. 1985. Pollen and Pollen-Coat Lipids: Chemical Survey and Role in Pollen Selection by Solitary Bees. PhD Dissertation, Dept. of Botany. Univ. California, Berkeley.

Dobson, H. E. M. 1987. Role of flower and pollen aromas in host-plant recognition by solitary bees. Oecologia 72: 618-623.

Dobson, H. E. M. 1988. Survey of pollen and pollenkitt lipids: chemical cues to flower visitors? Amer. J. Bot. 75: 170-182.

Dobson, H. E. M. In press. Possible roles of pollenkitt in flower selection by bees: an overview. In S. L. Buchmann [ed.]. Experimental Studies in Pollination and Pollinator Foraging Efficiency. Univ. Arizona Press, Tucson.

Dobson, H. E. M., J. Bergström, G. Bergström, and I. Groth. 1987. Pollen and flower volatiles in two *Rosa* species. Phytochemistry 26: 3171-3173.

Douglas, S. 1983. Floral color patterns and pollinator attraction in a bog habitat. Can. J. Bot. 61: 3494-3501.

Doull, K. M. 1974. Effects of attractants and phagostimulants in pollen and pollen supplement on the feeding behavior of honeybees in the hive. J. Apic. Res. 13: 47-54.

Duffield, R. M., J. W. Wheeler, and G. C. Eickwort. 1984. Sociochemicals of bees. In W. J. Bell and R. T. Cardé [eds.]. Chemical Ecology of Insects. Chapman and Hall, London.

Dulberger, R. 1975. Intermorph structural differences between stigmatic papillae and pollen grains in relation to incompatibility in Plumbaginaceae. Proc. R. Soc. London B 188: 257-274.

Dulberger, R. 1987. Fine structure and cytochemistry of the stigma surface and incompatibility in some distylous *Linum* species. Annals Bot. 59: 203-217.

Dumas, C. 1977. Lipochemistry of the progamic stage of a self-incompatible species: neutral lipids and fatty acids of the secretory stigma during its glandular activity, and of the solid style, the ovary and the anther of *Forsythia intermedia* Zab.. Planta 137: 177-184.

Dumas, C., R. B. Knox, and T. Gaude. 1984. Pollen-pistil recognition: new concepts from electron microscopy and cytochemistry. Intern. Rev. Cytol. 90: 239-272.

Egorov, I. A. and R. K. H. Egofarova. 1971. Investigations of essential oils of grape pollen. (In Russian). Dokl. Akad. Nauk. SSSR Ser. Biol. 199: 1439-1442.

Erickson, E. H. and S. L. Buchmann. 1983. Electrostatics and pollination. In C. E. Jones and R. J. Little [eds.]. Handbook of Experimental Pollination Biology. Academic Press, New York.

Faegri, K. and L. van der Pijl. 1979. The Principles of Pollination Ecology. 3rd ed. Pergamon Press, Oxford.

Ferguson, I. K. and J. J. Skvarla. 1982. Pollen morphology in relation to pollinators in Papilionoideae (Leguminosae). Bot J. Linn. Soc. London. 84: 183-193.

Ferguson, I. K. and K. J. Pearce. 1986. Obsaervations on the pollen morphology of the genus *Bauhinia* L. (Leguminosae: Caesalpinioideae) in the neotropics. In S. Blackmore and I. K. Ferguson [eds.]. Pollen and Spores. Academic Press, London.

Flint, S. D., and M. M. Caldwell. 1983. Influence of floral optical properties on the ultraviolet radiation environment of pollen. Amer. J. Bot. 70: 1416-1419.

Francke, W., W. Schröder, G. Bergström, and J. Tengö. 1984. Esters in the volatile secretions of bees. Nova Acta Reg. Soc. Sci. Upsaliensis ser. V:C, 3: 127-136.

Frisch, K. von. 1923. Über die "Sprache" der Bienen. Zool. Jahrb. Allg. Zool. Physiol. 40: 1-186.

Fritzsche, J. 1837. Über den Pollen. Mem. Acad. Imp. Sci. St. Petersbourg.

Ghosh, S., and K. R. Shivanna. 1982. Studies on pollen-pistil interaction in *Linum grandiflorum.* Phytomorph. 32: 385-395.

Ghosh, S., and K. R. Shivanna. 1984. Structure and cytochemistry of the stigma and pollen-pistil interaction in *Zephyranthes.* Annals Bot. 53: 91-105.

Goodwin, T. W. 1976. Distribution of carotenoids. In T. W. Goodwin [ed.]. Chemistry and Biochemistry of Plant Pigments. 2nd ed. Academic Press, New York.

Gori, D. F. 1983. Post-pollination phenomena and adaptive floral changes. In E. C. Jones and R. J. Little [eds.]. Handbook of Experimental Pollination Biology. Sci. Acad. Press, New York.

Grasshoff, J. L., and J. H. Beaman. 1970. Studies in *Eupatorium* (Compositae), III. Apparent wind pollination. Brittonia 22: 77-84.

Grayum, M. H. 1986. Correlations between pollination biology and pollen morphology in the Araceae, with some implications for angiosperm evolution. In S. Blackmore and I. K. Ferguson [eds.]. Pollen and Spores. Academic Press, London.

Gregory, D. P. 1964. Hawkmoth pollination in the genus *Oenothera.* Aliso 5: 385-419.

Harborne, J. B. 1976. Functions of flavonoids in plants. In T. W. Goodwin (ed.). Chemistry and Biochemistry of Plant Pigments. 2nd ed. Academic Press, London..

Harborne, J. B., and B. L. Turner. 1984. Plant Chemosystematics. Academic Press, London.

Haslett, J. R. 1983. A photographic account of pollen digestion by adult hoverflies. Physiol. Entomol. 8: 167-171.

Hemsley, A. J., and I. K. Ferguson. 1985. Pollen morphology of the genus Erythrina (Leguminosae: Papilionoideae) in relation to floral structure and pollinators. Annals Missouri Bot. Gar. 72: 570-590.

Heslop-Harrison, J. 1968. Tapetal origin of pollen-coat substances in Lilium. New Phytol. 67: 779-786.

Heslop-Harrison, J. 1975. The physiology of the pollen grain surface. Proc. R. Soc. London B 190: 275-299.

Heslop-Harrison, J. 1976. The adaptive significance of the exine. In I. K. Ferguson and J. Muller [eds.]. The Evolutionary Significance of the Exine. Academic Press, London.

Heslop-Harrison, J. 1979a. An interpretation of the hydrodynamics of pollen. Amer. J. Bot. 66: 737-743.

Heslop-Harrison, J. 1979b. Pollen walls as adaptive systems. Annals Missouri Bot. Gar. 66: 813-829.

Heslop-Harrison, J. 1979c. Aspects of the structure, cytochemistry and germination of the pollen of rye (Secale cereale L.). Annals Bot. 44 suppl.: 1-47.

Heslop-Harrison, J., Y Heslop-Harrison, R. B. Knox, and B. Howlett. 1973. Pollen-wall proteins: 'gametophytic' and 'sporophytic' fractions in the pollen walls of the Malvaceae. Annals of Bot. 37: 403-412.

Hesse, M. 1980. Pollenkitt in relation to pollination ecology. Calcutta Univ. Res. J. 1: 29-33.

Hesse, M. 1981a. Pollenkitt and viscin threads: their role in cementing pollen grains. Grana 20: 145-267.

Hesse, M. 1981b. The fine structure of the exine in relation to the stickiness of angiosperm pollen. Rev. Palaeobot. Palynol. 35: 81-92.

Hesse, M. 1984. Pollenkitt is lacking in Gnetatae: Ephedra and Welwitschia further proof for its restriction to the angiosperms. Rev. Palaeobot. Palynol. 35: 81-92.

Hesse, M. 1986. Nature, form and function of pollen-connecting threads in angiosperms. In S. Blackmore and I. K. Ferguson [eds.]. Pollen and Spores. Academic Press, London.

Hesse, M., and M. Waha. 1983. The fine structure of pollen wall in Strelitzia reginae (Musaceae). Pl. Syst. Evol. 141: 285-298.

Hohmann, H. 1970. Über die Wirkung von Pollenextrakten und Duftstoffen auf das Sammel- und Werbeverhalten Höselnder Bienen (*Apis mellifera* L.). Apidologie 1: 157-178.

Howlett, B. J., R. B. Knox, and J. Heslop-Harrison. 1973. Pollen-wall proteins: release of allergen antigen E from intine and exine sites in pollen grains of ragweed and *Cosmos*. J. of Cell Sci. 13: 603-619.

Howlett, B. J., H. I. M. V. Vithanage, and R. B. Knox. 1979. Pollen antigens, allergens and enzymes. Curr. Adv. Pl. Sci. 35: 1-17.

Hügel, M. F. 1962. Etude de quelques constituents du pollen. Ann. Abeille 5: 97-133.

Keijzer, C. J. 1987. The process of anther dehiscence and polllen dispersal II. New Phytol. 105: 499-507.

Kerner, A. von Marilaun. 1898. The Natural History of Plants. Vol. II (trans. by F. W. Oliver). Blackie and Son. London.

Klungness, L. M., and Y. Peng. 1984. A histochemical study of pollen digestion in the alimentary canal of honeybees (*Apis mellifera* L.). J. Insect Physiol. 30: 511-521.

Knoll, F. 1930. Über Pollenkitt und Bestaubungsart. Zeit. Bot. 23: 609-675.

Knox, R. B. 1984. Pollen-pistil interactions. In H. F. Linskens and J. Heslop-Harrison [eds.]. Cellular interactions. Enclycl. Pl. Physiol. 17. Springer-Verlag, New York.

Knox, R. B., and J. Heslop-Harrison. 1970. Pollen-wall proteins: localization and enzymic activity. J. Cell Sci. 6: 1-27.

Kohler, E. 1976. Pollen dimorphism and heterostyly in the genus *Waltheria* L. (Sterculiaceae). In I. K. Ferguson and J. Muller [eds.]. The Evolutionary Significance of the Exine. Linn. Soc. Symp. Ser. 1: 147-161.

Kölreuter, D. J. G. 1761. Vorlaufige Nachricht von Einigen das Geschlecht der Pflanzen Betreffenden Versuchen und Beobachtungen. Verlag Wilhelm Engelmann, Leipzig.

Kress, W. J. 1986. Exineless pollen structure and pollination systems of tropical *Heliconia* (Heliconiaceae). In S. Blackmore and I. K. Ferguson [eds.]. Pollen and Spores. Academic Press, London.

Lepage, M., and R. Boch. 1968. Pollen lipids attractive to honeybees. Lipids 3: 530-534.

Levin, D. A., and G. E. Bohart. 1955. Selection of pollens by honey bees. Amer. Bee J. 95: 392-393, 402.

Lindsey, A. H., and C. R. Bell. 1985. Reproductive biology of Apiaceae. II. Cryptic specialization and floral evolution in *Thaspium* and *Zizia*. Amer. J. Bot. 72: 231-247.

Lord, E. M., and K. J. Eckard. 1984. Incompatibility between the dimorphic flowers of *Collomia grandiflora*, a cleistogamous species. Science 223: 695-696.

Louveaux, J. 1959. Recherches sur la récolte du pollen par les abeilles (*Apis mellifera* L.). Ann. Abeille 2: 13-111.

Lubliner-Mianowska, K. 1955. The pigments of pollen grains. (In Polish). Acta. Soc. Bot. Pol. 24: 609-617.

Maruzella, J. C. 1962. The germicidal properties of perfume oils and perfumery chemicals. Amer. Perf. Cosm. 77, Sect. 2:167-170.

Mattsson, O. 1983. The significance of exine oils in the initial interaction between pollen and stigma in *Armeria maritima*. In D. L. Mulcahy and E. Ottaviano [eds.]. Pollen : Biology and implications for Plant Breeding. Elsevier, New York.

McClure, J. W. 1975. Physiology and functions of flavonoids. In J. B. Harborne, T. J. Mabry, and H. Mabry [eds.]. The Flavonoids. Academic Press, New York.

McNeill, J., and C. W. Crompton. 1978. Pollen dimorphism in *Silene alba* (Caryophyllaceae). Can. J. Bot. 56: 1280-1286.

Mohl, H. von. 1852. Principles of the Anatomy and Physiology of the Vegetable Cell (translation by A. Henfrey). John van Voorst, Paternoster Row, London.

Mori, S. A., J. E. Orchard, and G. T. Prance. 1980. Intrafloral pollen differentiation in the New World Lecythindaceae, subfamily Lecythidoideae. Science 209: 400-403.

Morris, J. A., A. Khettry, and E. W. Seitz. 1979. Antimicrobial activity of aroma chemicals and essential oils. J. Amer. Oil Chem. 56: 595-603.

Mühlethaler, K. 1955. Die Struktur einiger Pollenmembranen. *Planta* 46: 1-13.

Muller, J. 1981. 1981. Exine architecture and function in some Lythraceae and Sonneratiaceae. Rev. Palaeobot. Palynol. 35: 93-123.

Neamtu, G., and C. Bodea. 1970. Pigmentii carotinoidici din plante recoltate din regiuni montane si subalpine. Stud. Cercet. Biochim. *13: 307-311.*

Nilsson, M., R. Hyhage, and E. von Sydow. 1957. Constituents of pollen II. Long-chain hydrocarbons and alcohols. Acta Chem. Scand. 11: 634-639.

Nowicke, J. W., and M. Meselson. 1984. Yellow rain - a palynological analysis. Nature 309: 205-206.

Ormond, W. T., M. C. B. Pinheiro, and A. R. C. de Castells. 1981. A contribution to the floral biology and reproductive system of *Couroupita quianensis* Aubl. (Lecithidaceae). Annals Missouri Bot. Gard. 68: 514-523.

Pacini, E., and L. M. Bellana. 1986. *Lagerstroemia indica* L. pollen: form and function. 1986. In S. Blackmore and I. K. Ferguson [eds.]. Pollen and Spores. Academic Press, London.

Page, J. S. 1978. A scanning electron microscope survey of grass pollen. Kew Bull. 32: 313-319.

Pankow, H. 1958. Über den Pollenkitt bei *Galanthus nivalis* L. Flora 146: 240-253.

Peng, Y.-S., M. E. Nasr, J. M. Marston, and Y. Fang. 1985. The digestion of dandelion pollen by adult worker honeybees. Physiol. Entomol. 10: 75-82.

Penny, J. H. J. 1983. Nectar guide colour contrast: a possible relationship with pollination strategy. New Phytol. 95: 707-721.

Porsch, O. 1956. Windpollen und Blumeninsekt. Osterr. Bot. Z. 103: 1-18.

Prentice, H. C., O. Mastenbroeck, W. Berendsen, and P. Hogeweg. 1984. Geographic variation in the pollen of *Silene latifolia* (*S. alba, S. pratensis*): a quantitative morphological analysis of population data. Can. J. Bot. 62: 1259-1267.

Roberts, I., A. D. Stead, and H. G. Dickenson. 1979. No fundamental change in lipids of the pollen grain coating of *Brassica oleracea* following either self- or cross-pollinations. Incompatibility Newsl. 11: 77-79.

Roberts, R. B., and S. Vallespir. 1978. Speciatization of hairs bearing pollen and oil on the legs of bees. Annals Entolmol. Soc. Am. 71: 619-627.

Robinson, F. A., and J. L. Nation. 1968. Substances that attract caged honeybee colonies to consume pollen supplements and substitutes. J. Apic. Res. 7: 83-88.

Roggen, H. 1974. Pollen washing influences (in)compatibility in *Brassica oleracea* varieties. In H. F. Linskens [ed.]. Fertilization in Higher Plants. North-Holland Publ., Amsterdam.

Schmidt, J. O. 1985. Phagostimulants in pollen. J. Apic. Res. 25: 107-114.

Schmidt, J. O., and S. L. Buchmann. 1985. Polen digestion and nitrogen utilization by *Apis mellifera* L.. Comp. Biochem. Physiol. A 82: 499-503.

Schmidt, J. O., S. C. Thoenes, and M. D. Levin. 1987. Survival of honey bes, *Apis mellifera*, fed various pollen sources. Ann. Entomol. Soc. Am. 80: 176-183.

Shivanna, K. R. 1982. Pollen-pistil interaction and control of fertilization. In B. M. Johri [ed.]. Experimental Embryology of Vascular Plants. Springer-Verlag, New York.

Shivanna, K. R., J. Heslop-Harrison, and Y. Heslop-Harrison. 1983. Heterostyly in *Primula.* 3. Pollen water economy: a factor in the intramorph- incompatibility response. Protoplasma 117: 175-184.

Skvarla, J. J., P. H. Raven, W. F. Chissoe, and M. Sharp. 1978. An ultrastructural study of viscin threads in Onagraceae pollen. Pollen et Spores 20: 5-143.

Smith, A. P., and S. W. Green. 1987. Nitrogen requirements of the sugar glider (*Petaurus breviceps*), an omnivorous marsupial, on a honey-pollen diet. Physiol. Zool. 60: 82-92.

Sprengel, C. K. 1793. Das entdeckte Geheimnis der Natur in Bau und in der Befruchtung der Blumen. Friedrich Vieweg, Berlin.

Stanley, R. G., and H. F. Linskens. 1974. Pollen. Springer-Verlag, New York.

Stelleman, P. 1983. Pollination strategy in *Plantago lanceolata* L.. In D. L. Mulcahy and E. Ottaviano [eds.]. Pollen: Biology and Implications for Plant Breeding. Elsevier, New York.

Taber, S. 1963. Why bees collect pollen. Rep.-Abstr. XIX, the International Beekeeping Congr. Prague. p. 114.

Thorp, R. W. 1979. Structural, behavioral, and physiological adaptations of bees (Apoidea) for collecting pollen. Annals Missouri Bot. Gard. 66: 788-812.

Toth, G., E. Lemberkovics, and J. Kutasi-Szabo. 1987. The volatile components of some Hungarian honeys and their antimicrobial effects. Amer. Bee J. 127: 496-497.

Troll, W. 1928. Über Antherenbau, Pollen und Pollination von *Galanthus* L.. Flora 123: 321-343.

Turner, V. 1984. *Banksia* pollen as a source of protein in the diet of two Australian marsupials *Cercartetus nanus* and *Tarsipes rostratus.* Oikos 43: 53-61.

Vogel, S. 1978. Evolutionary shifts from reward to deception in pollen flowers. In A. J. Richards [ed.]. The Pollination of Flowers by Insects. Linn. Soc. Symp. Ser. 6: 89-96. Academic Press,

Vogel, S. 1981. Die Klebstoffhaare an den Antheren von *Cyclanthera pedata (Cucurbitaceae).* Pl. Syst. Evol. 137: 291-316.

Waha, M. 1984. Zur Ultrastruktur und Funktion pollen vergindender Fäden bei Ericaceae und andereen Angiospermemfamilien. Pl. Syst. Evol. 147: 189-203.

Walker, J. W. 1976. Evolutionary significance of the exine in the pollen of primitive angiosperms. In I. K. Ferguson and J. Muller [eds]. The Evolutionary Significance of the Exine. Linn. Soc. Symp. Ser. 1:251-308.

Wiermann, R. 1981. "Yellow flavonols" as components of pollen pigmentation. Z. Naturforsch. 36a: 204-206.

Wiermann, R., and K. Vieth. 1983. Outer pollen wall, an important accumulation site for flavonoids. Protoplasma 118: 230-233.

Wiermann, R. and H. Weinart. 1969. Unterwuchungen zum Phenylpropanstoff-wechsel des Pollens. Z. Pflanzenphysiol. 61: 173-183.

Williams, N. H. 1983. Floral fragrance as cues in animal behavior. In E. C. Jones and R. J. Little [eds.]. Handbook of Experimental Pollination Biology. Sci. Acad. Press, New York.

Wittgenstein, E., and E. Sawicki. 1970. Analysis of the non-polar fraction of giant ragweed pollen: carotenoids. Mikrochim. Acta: 765-783.

Wodehouse, R. P. 1935. Pollen Grains. McGraw-Hill, New York.

Zavada, M. S. 1984. The relation between pollen exine sculpturing and self-incompatibility mechanisms. Pl. Syst. Evol. 147: 63-78.

REPRODUCTIVE BIOLOGY OF *HIBISCUS MOSCHEUTOS* (MALVACEAE)

Timothy Spira

INTRODUCTION

Hibiscus moscheutos L. ssp. *moscheutos* Clausen (Malvaceae), the swamp rose mallow (hereafter referred to as *Hibiscus*), grows in freshwater and brackish marshes and swamps along the east coast of the United States from Massachusetts to Texas and less frequently along inland waterways to Ohio and Indiana (Beal 1977). Individual plants (genets) consist of few to many upright stems, 1-2 m tall, which emerge each year from a perennial rootstalk. In Maryland, vegetative shoots emerge in April, flowering begins in late July, reaches a peak in August, and typically ends in early September. The flowers are large (10-15 cm across), with pink or white petals, which frequently have a conspicuous red nectar guide at their base. Anthesis occurs in early morning and flowers generally close by late afternoon or evening. While individual flowers generally last a single day, unpollinated flowers may remain open for two or more days (Skutch and Burwell, 1928; pers. obs.). Fruits (capsules) dehisce in late summer or fall, releasing spherical, hard-coated seeds (<3 mm) that float and thus may be water dispersed (Cahoon and Stevenson, 1986).

Although the flowers of *Hibiscus* are among the largest in North America, surprisingly little is known about its reproductive biology. The purpose of this study was to provide information on its breeding system, pollination biology, and reproductive efficiency. In addition, I investigated seed loss due to pre-dispersal seed predators.

STUDY AREA

The research was conducted in 1985 and 1986 at the Smithsonian Environmental Research Center on the inner coastal plain of Chesapeake Bay approximately 15 km S of Annapolis, Anne Arundel County, Maryland (38°53'20"N,76°33'20"W). The population studied was located in a freshwater wetland dominated by *Hibiscus* and *Saururus cernuus*.

METHODS
Breeding System and Reproductive Efficiency

Newly opened flowers were hand-pollinated at regular intervals along transects early in the morning prior to pollinator activity. Dehiscing anthers were brushed against stigma lobes of the same flower in 20 self-pollinated flowers while dehiscing anthers from plants 25-100 m away from maternal plants were used as the pollen source in 36 hand cross-pollinated flowers.

In each treatment, pollen was liberally applied to stigmas after which flowers were tagged and enclosed in a cheesecoth "bag" to prevent additional pollinations. To test for apomixis, newly opened unpollinated flowers were also tagged and bagged. Bags were removed after flower senescence and flowers were scored for fruit and seed set 3-4 weeks later. Flowers that set fruit had enlarged ovaries (at least 1.5 cm long) and contained developing seeds. In contrast, ovaries of flowers that failed to set fruit had abscised (leaving a naked pedicel) or were small, senescent, and lacked developing seeds. Seed set was determined by distinguishing enlarged ovules (potential seeds) from aborted (flat, shrivelled, often discolored) ovules with the aid of a dissecting microscope.

Fruit and seed set from open-pollinated flowers sampled along the same transects described above served as a control. Additional data on fruit and seed set from open-pollinated flowers were obtained by sampling every flower on 4-5 ramets of 10-15 clones during the 1985 and 1986 seasons.

Pollination Biology

To assess the potential for stigmas to contact anthers and thereby effect self-pollination, the distance between the uppermost anthers and the stigmas was measured in flowers (N = 31) sampled at regular intervals along transects. The composition and relative frequency of flower visitors was determined by observing individual flowers for 15 minute intervals (N = 75) at various times of day throughout the 1986 flowering season. Insects foraging on *Hibiscus* flowers were also observed and collected in 1985.

Seed Predation

To estimate pre-dispersal seed predation, I compared the number of developing seeds within immature fruits (prior to seed destruction) with the number of intact seeds within mature fruits. Immature fruits ($\leq$ 1.5 cm long but still green) were sufficiently developed that developing seeds were distinguishable from aborted ovules. Mature fruits were collected just prior to capsule dehiscence. Thirty immature and 30 mature fruits were sampled at regular intervals along transects in both 1985 and 1986.

Intact seeds within mature fruits harvested in 1985 were pooled and seeds (N = 96) were randomly subsampled and tested for viability. Seeds were sliced longitudinally and submerged in a 0.15% tetrazolium solution in foil-wrapped petri dishes for approximately 24 hr. Embryos which stained pink to red were considered viable (Iseley, 1952).

RESULTS AND DISCUSSION
Breeding System
Fruit set was similar among treatments ranging from 58.6% in open-pollinated to 66.7% in cross-pollinated flowers (Fig. 1). Seed set within developed fruits ranged from 82.6 to 91.1% (Fig. 1). There was no significant difference between self- and cross-pollinated flowers (T-test; p>0.05) indicating that *Hibiscus* flowers are clearly self-compatible. Seed set in open- and cross-pollinated flowers were also not significantly different (T-test; p > 0.5), whereas open-pollinated flowers had significantly greater seed set than self-pollinated flowers (T-test; p < 0.5). The fact that fruit and seed set in open-pollinated flowers was generally similar to hand-pollinated flowers suggests sexual reproduction was not pollen-limited.

Pollination Biology
At anthesis, most anthers had dehisced and the stigma lobes appeared to be receptive to pollen. However, automatic self-pollination was prevented by the stigmas protruding well beyond the uppermost anthers (anther-stigma distance, $\bar{x} \pm$ S.E. = 3.2 $\pm$ 0.006; Range = 2.7 - 4.1 cm; N = 31). Wind pollination was unlikely as the pollen grains were sticky and tended to clump together. Furthermore, flowers covered with a single layer of cheesecloth (porous to pollen but not insect pollinators) accumulated few to no pollen grains on stigmas and failed to mature fruits indicating that flowers were not apomictic and that a vector other than wind was needed for successful pollination.

Spatial separation of anthers from stigmas (herkogamy) effectively prevents self-pollination in this self-compatible species. It does not, however, prevent pollination between flowers on the same plant (geitonogamy). Even though *Hibiscus* clones are multi-stemmed (15 to 25 stems per genet is common), the number of open flowers each day is generally less than 5. Most stems do not produce an open flower on a given day and those that

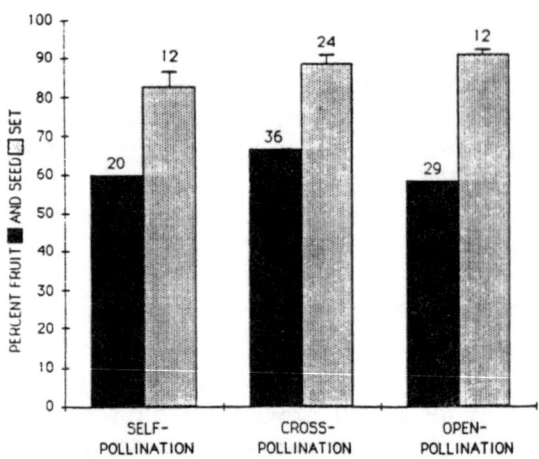

Figure 1. Fruit and seed set of *Hibiscus* flowers in response to pollination treatment. Self- and cross-pollinated flowers were hand-pollinated and flower visitors were excluded. Open-pollinated flowers were exposed to pollinators. Sample sizes for fruit and seed set are shown above bars as is one standard error for the seed set data.

do tend to have only one or two open flowers. The relatively small number of open flowers per genet at a given time should decrease geitonogamous pollinations and promote outcrossing in this species.

Hibiscus flowers were actively visited and pollinated by bees, particularly *Ptilothrix bombiformis* (Cresson), a large (12-18 mm) non-social anthophorid bee which made 94% of the observed visits (N = 323) to flowers in 1986. Individualized flowers averaged 4.3 visits per 15 min. period in 1986 (Spira, unpubl.). Other flower visitors included *Bombus pennsylvanicus* (Deeger), a small as yet unidentified bee, and several moths and butterflies. Aside from *P. bombiformis*, only *B. pennsylvanicus* appeared to be an effective pollinator.

Several other investigators have found *Ptilothrix bombiformis* to be the most abundant visitor to *Hibiscus* flowers (Robertson 1888, 1893, 1925; Weiss and Dickerson, 1919; Michener, 1947; Blanchard, 1976). The distribution of *P. bombiformis* extends throughout the range of *Hibiscus* and much of the activity of adult bees is centered around *Hibiscus* flowers (Blanchard, 1976). Apparently, the only pollen used by this bee comes from several species of *Hibiscus* (Blanchard, 1976, Rust, 1980). Adults emerge about the time *Hibiscus* begins flowering and are no longer active by the end of the flowering season (Blanchard, 1976; personal observation). Males forage for nectar, search for mates, and sleep in *Hibiscus* at night and during inclement weather (Blanchard, 1976; personal observation). Females collect pollen and drink nectar from flowers. Nests are formed in the ground (often near large strands of *Hibiscus*) and each is provisioned with *Hibiscus* pollen on which a single egg is laid (Blanchard, 1976; Rust, 1980).

Individuals of *P. bombiformis* and *B. pennsylvanicus* collectively foraged for nectar on 87% of the flowers visited, 6% of the flowers were visited for pollen, and both nectar and pollen were collected from 7% of the flowers visited. As female bees foraged for pollen, their ventral surfaces generally became covered with it. Also, bees foraging for nectar frequently crawled over the anthers to reach the nectaries at the base of the flower. Consequently, bees foraging for pollen and/or nectar tend to accumulate large amounts of pollen on their ventral surfaces.

Most bee visits, however, did not result in pollination as the visitor failed to contact any of the flower's stigmas. Only 27% of flower foraging *P. bombiformis* and *B. pennsylvanicus* (N = 238) appeared to contact a stigma while foraging for nectar or pollen. Flower visiting bees which contacted a stigma did so by using the stigma lobes as a landing platform prior to foraging for pollen and/or nectar or by inadvertently brushing against a stigma while flying into or out of the flower. When they did contact the stigma, they generally deposited large amounts of pollen on the stigmas (up to 889 grains; unpublished data). An average of almost 1900 pollen grains were counted on stigmas of one day old flowers sampled during the 1985 season (unpublished data). Since the average number of ovules per flower is 138, 14 times as many pollen grains reached stigmas as there were ovules in ovaries.

Reproductive Efficiency

The proportion of open-pollinated flowers that developed into fruit was only 52.6% in 1985 and 2.4% in 1986. The difference between years is highly significant (T-test, $p < 0.01$). In addition to substantial between-year variation, there was also large between-plant variation. Of the 15 plants sampled in 1985, fruit set ranged from 11 to 79%. In 1986, fruit set ranged from 0 to 7% and flowers on 5 of the 10 plants sampled failed mature any fruits. Many factors may limit fruit set including limited resources (e.g., light, water, photosynthate, mineral nutrients), inclement weather (e.g., frost), and herbivores and pathogens (Stephenson, 1981 and references therein). Although experiments were not conducted to determine what might limit fruit set in *Hibiscus*, one likely influence was a severe regional drought during the spring and summer of 1986. The soil surface was unusually dry during the summer and *Hibiscus* plants exhibited symptoms of water stress such as drooping (rather than upright) stems and wilted leaves. These characteristics were most pronounced in July and August when plants were flowering and setting fruit.

The proportion of ovules that set seed within developed fruit was significantly greater (T-test; $p < 0.01$) in 1985 than 1986 with values ranging from 91.6 ± 0.8 percent in 1985 to 83.7 ± 2.8 the following year (Fig. 2). Even though fruit set was extremely low in 1986, seed set within developed fruits remained quite high.

Seed Predation

The number of seeds in immature versus mature fruits is shown in Fig. 3. The reduction in seed number reflects the destruction of seeds by predispersal seed predators. In 1985, the $\overline{X}$ number of seeds in immature fruits was 126.2 ± 3.0, while in mature fruits, there were only 59.1 ± 5.5 seeds. Based on these values, approximately 53% of the potential seeds within fruits were consumed by predators. In the following year, there were 108 ± 4.3 seeds per young fruit and only 11.7 ± 1.9 seeds per mature fruit, indicating that approximately 89% of the potential seeds within fruits were destroyed. Viability (tetrazolium) tests on intact seeds sampled from mature fruits in 1985 indicated that 87.5% (N = 96) were viable.

Substantial predation on developing *Hibiscus* seeds has been noted by other investigators, including Cushman (1911), Blanchard (1976), and Cahoon and Stevenson (1986). The two primary seed predators in this study were the bruchid beetles *Athaeus hibisci* Oliver (Bruchidae) and the weevil *Conotrachelus fissunguis* Leconte (Curculionidae). Bruchid larvae burrow through the ovary wall and enter a seed where they eventually pupate (Weiss and Dickerson, 1919). While infested seeds appear to develop normally, their entire contents are consumed such that only a hollow seedcoat remains after the adult emerges (Blanchard, 1976; Cahoon and Stevenson, 1986). Adult weevils lay their eggs within developing fruits and the larvae feed on developing seeds (Schoof, 1942; Weiss and Dickerson, 1919).

SUMMARY AND CONCLUSIONS

The reproductive biology of *Hibiscus moscheutos* (Malvaceae) was studied in a marsh near Annapolis, Maryland for two years. The unusually large flowers (10-15 cm across) of

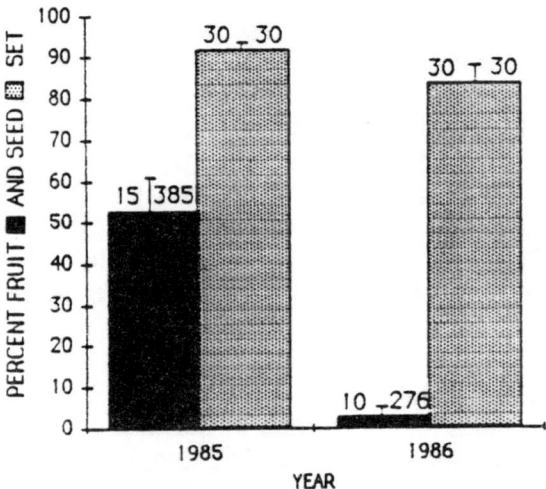

Figure 2. Fruit and seed set in 1985 and 1986. Numbers to left of standard error bars reflect the number of plants sampled and numbers to right of bars refer to the number of flowers sampled.

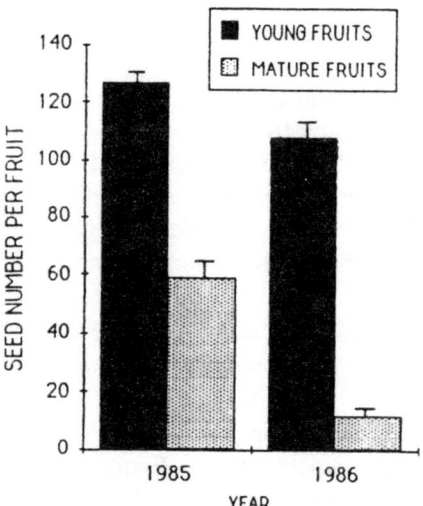

Figure 3. Pre-dispersal seed predation in 1985 and 1986 was estimated by comparing the number of seeds in immature fruits (prior to seed destruction) with the number of seeds in mature fruits prior to seed dispersal. Standard error bars shown. N = 30 fruits per treatment.

this species were self-compatible but spatial separation of anthers and stigmas (herkogamy) prevented self-pollination within flowers. The primary pollinator was the solitary bee, *Ptilothrix bombiformis*, which actively collected both nectar and pollen from flowers. In spite of large stigmatic pollen loads, fruit set was only 52.6% and 2.4% in successive years indicating that factors other than pollen generally limit fruit set. Seed set within developed fruits was high (>83%); however, the larvae of bruchid beetles and weevils reduced the number of developing seeds by 53.2% and 89.2% in the two years studied. Thus, pre-dispersal seed predators dramatically reduced reproductive output in this species.

ACKNOWLEDGEMENTS

I thank Jen Leak for assistance in the field and laboratory and Jane Bock, Yan Linhart, Dennis Whigham, Lisa Wagner and an anonymous reviewer for their helpful comments on an earlier draft of the manuscript. Insects were kindly identified by O.J. Blanchard (bruchid beetles and weevils) and Robin Thorpe (bumble bees). Financial support was provided by the Visiting Scientist Program of the Smithsonian Institution, the Whitehall Foundation, and by the Faculty Research Committee of Georgia Southern College.

LITERATURE CITED

Beal, E. O. 1977. A Manual of Marsh and Aquatic Vascular Plants of North Carolina with Habitat Data. North Carolina Agricultural Experimental Station Tech. Bull. No. 247.

Blanchard, O. J., Jr. 1976. A Revision of Species Segregated from *Hibiscus* Sect. *Trionum* (*Medicus*) deCandolle *sensu lato* (Malvaceae). PhD Dissertation, Cornell Unviersity. Ithaca, New York.

Cahoon, D. R., and J. C. Stevenson. 1986. Production, predation, and decomposition in a low salinity *Hibiscus* marsh. Ecology 67: 1341-1350.

Cushman, R. A. 1911. Notes on the host plants and parasites of some North American Bruchidae. J. Econ. Entomol. 4: 489-510.

Iseley, D. 1952. Employment of tetrazolium chloride for determining viability of small grain seed. Proc. Assoc. Off. Seed Anal. 42: 143-153.

Michener, C. D. 1947. Bees of a limited area in southern Mississippi (Hymenoptera: Apoidea). Amer. Midl. Nat. 38: 443-455.

Robertson, C. 1888. Zygomorphy and its causes. II. Botan. Gaz. (Crawfordsville) 13: 203-208.

Robertson, C. 1893. Flowers and insects. XI. Botan. Gaz.
    (Crawfordsville) 18: 267-274.
Robertson, C. 1925. Habits of the *Hibiscus* bee, *Emphor bombiformis..*
    Psyche 32: 278-282.
Rust, R. W. 1980. The biology of *Ptilothrix bombiformis*
    (Hymenoptera: Anthophoridae). J. Kansas Entomol. Soc. 53:
    427-436.
Schoof, H. F. 1942. The genus *Conotrachelus* DeJean (Coleoptera,
    Curculionidae) in the North/Central United States. Illinois Biol.
    Monogr. 19: 1-170.
Skutch, A. F., and R. L. Burwell, Jr. 1928. The period of anthesis in
    Hibiscus. Torreya 28: 1-5.
Stephenson, A. G. 1981. Flower and fruit abortion: proximate causes
    and ultimate functions. Ann. Rev. Ecol. Syst. 12: 253-281.
Weiss, H. B., and E. L. Dickerson. 1919. Insects of the swamp rose
    mallow, *Hibiscus moscheutos* L., in New Jersey. J. New York
    Entomol. Soc. 27: 39-68.

RELATIONSHIPS BETWEEN COMPONENTS OF PLANT FORM AND SEED OUTPUT IN *COLLINSONIA VERTICILLATA* (LAMIACEAE)

Allan L. Fone

INTRODUCTION

Pollination is necessary for fruit and seed production in most flowering plants, but once it has occurred, a host of other factors act to determine eventual seed output, among them plant size and form. Demographic studies of plants have shown that larger size usually results in greater fecundity (Baskin and Baskin, 1979; van der Meijden and van der Waals-Kooi, 1979; Solbrig, 1981; Bierzychudek, 1982; Kachi and Hirose, 1985; Stanton, 1985). This is not surprising, since the number of floral meristems should increase with plant size, resulting in an increase in potential fruit and seed production. However, as described below, the effects of plant form on fruit and seed production may be more complex than that implied by the relationship between size and fecundity.

Since plants may be viewed as modular organisms (Harper, 1981; White, 1984; Watkinson and White, 1985), plant growth and form can be described in terms of numbers, sizes, and types of parts that make up the plant body (e.g., leaves, flowers, and shoots). The overall relationship of plant size and form to seed production can be analyzed by examining the effects of each type of component. In this context, traits such as the numbers of leaves and flowers per plant, and leaf size, are often referred to as yield components. A particular yield component may affect seed and fruit output not only directly but also indirectly through correlations with other yield components (Dewey and Lu, 1959). If fecundity increases with size, then components of plant form might be expected to have positive effects on fruit and seed production. However, relationships between yield and its components or among yield components themselves are not necessarily positive (Dewey and Lu, 1959; Eaton and Kyte, 1978; Primack and Antonovics, 1981; Lovett Doust et al.,1983; Kiang and Chiang, this volume).

There have been many analyses of effects of yield components on seed and fruit production for plants of agronomic or horticultural importance (Dewey and Lu, 1959; Eaton and Kyte, 1978; Jolliffe et al., 1982; Lovette Doust et al., 1983). However, the effects of yield components on seed production have rarely been estimated for plants from natural populations. A study of two species of *Plantago* (Maddox and Antonovics, 1983) is probably the best example of such an analysis. Other studies on natural populations have focused on either the genetic basis of size and yield components or their degree of plasticity (Primack and Antonovics, 1981; Solbrig, 1981; Mitchell-Olds, 1986; Marshall et al., 1986).

The present paper describes a study of the relationships between plant form and seed and fruit production in *Collinsonia verticillata* Baldwin ex Ell. (Lamiaceae), a perennial herb that grows in rich woods in the southeastern United States. Several vegetative and reproductive traits were used to describe variation in plant form. The direct and indirect effects of these traits on seed and fruit production per plant under natural conditions were estimated using multiple regression and path analysis (Li, 1975; Sokal and Rohlf, 1981). To provide a basis for judging the overall importance of plant size and form to seed and fruit production in *C. verticillata*, pollination and seed abortion also were studied.

METHODS
Plant Form and Seed Output

Collinsonia verticillata is not highly variable in general growth form, but individuals vary in size and number of parts. Plants typically produce one aerial shoot per year from an underground rhizome. Shoots that flower usually have two pairs of leaves and an unbranched, leafless inflorescence axis, with flowers in whorls (verticels) of six at the nodes of the inflorescence. A flower can produce a maximum of four seeds, since each ovary has only four ovules.

In spring 1986, 70 florwering individuals at Oconee Station Cove, South Carolina, were sampled to determine the numbers of fruits and seeds matured per plant, plant height (to the highest leaf node), the sum of all leaf lengths (a measure of total leaf area), inflorescence length, number of nodes per inflorescence, and number of flowers per plant. Forty-seven plants produced mature seeds, and only these plants were used to analyze relationships between plant form and seed output. In traditional yield component analysis (Eaton and Kyte, 1978; Jolliffe et al., 1982; Lovett Doust et al., 1983), data on plant form and reproduction would be used to construct a series of multiplicative yield components whose product equals plant yield (e.g., plant height, leaf length/plant height, inflorescence height/leaf length, nodes/inflorescence height, fruits/flowers, and seeds/fruits). However, I used only the original variables for this study, because ratios are relatively inaccurate and provide no information on the form of the relationship between the original variables (Sokal and Rohlf, 1981).

Several potentially important plant traits could not be measured, because this would have required destructive sampling. However, preliminary work showed that the sum of all leaf lengths was correlated strongly with traits such as total leaf area ($r = 0.92$, $n = 138$), rhizome mass ($r = 0.86$, $n = 37$), total leaf mass ($r = 0.92$, $n = 37$), and stem mass ($r = 0.92$, $n = 37$). In the present study, I measured a randomly chosen leaf from each leaf pair and doubled its length to estimate total leaf length for each leaf pair. Lengths of single leaves were added without modification.

I used multiple regression to estimate the relationships of plant height, leaf length, inflorescence height, number of nodes per inflorescence, number of flowers per plant, and number of fruits per plant to seed output. Only plants that produced mature seeds were included in this analysis. The use of multiple regression assumes that the response variable (in this case, seeds per plant) is directly affected by each regressor variable (the remaining plant attributes) and that correlations exist among all regressor variables (Li, 1975; Sokal and Rohlf, 1981). Each regression coefficient (b) estimates a direct effect. The magnitudes of direct effects can be compared using standardized regression coefficients (b'), which are calculated by multiplying the usual regression coefficient by the ratio of the standard deviation of the response variable to the standard deviation of the regressor variable. The multiple regression also assumes that all possible indirect effects occur between the response and regressor variables. Indirect effects are most important when there are strong correlations among variables (Li, 1975). The models of direct and indirect effect used here in the multiple regression and in the path analysis described below do not necessarily represent actual direct and indirect causal effects, since only descriptive data were collected and not all relevant variables were investigated.

The multiple regression of seed output on the remaining plant traits represents one possible hypothesis about how plant traits affect seed set. It is also possible that some traits have no effects or only indirect effects on seed production. I investigated a series of path models that represented alternatives to the multiple regression model. Different combinations of traits were included in these models, and the relationships among the traits were varied. Path analysis (Li, 1975) was used to estimate values for direct effects, which are called path coefficients, and to calculate the extent to which each model explained variation in number of seeds per plant (R^2). Path coefficients are mathematically equivalent to standardized regression coefficients. The path model presented in the results was selected for its combination of few variables, few direct and indirect effects, and high R^2.

Pollination Rates and Fruit-Set

Natural pollination levels in *C. verticillata* were studied by estimating stigmatic pollen loads. Styles were collected from three flowers on each of 20 plants. These were stained with lactophenol in aniline blue, and pollen grains on the stigmatic surfaces were counted using a compound microscope.

To test if pollination effectiveness could be limited by self-incompatibility, I performed an experiment with three pollination treatments: artificial self-pollination, artificial cross-pollination, and natural pollination (control). Six flowers on each of 10 plants were allocated to the experiment, with two flowers per treatment. Artificially pollinated flowers were caged to exclude pollinators. In the absence of artificial pollination, caged flowers did not set fruits. Pollen for crosses came from newly dehisced anthers on plants that were at least 60 m away from the experimental plants. To ensure the possibility of full seed-set, the amount of pollen placed on the stigmatic surfaces of each style exceeded the number of ovules per flower.

Patterns of Seed-Set and Seed Abortion

To determine if seed abortion rates increase with the number of seeds initiated per flower, numbers of seeds initiated and matured per flower were scored for flowers on 48 plants (a subset of the 70 plants originally sampled). In *C. verticillata*, the ovary has four lobes, corresponding to the four ovules. If an ovary lobe had enlarged, it was counted as a developing seed. This procedure may have underestimated seed initiation rates if some fertilized ovules were aborted before ovary lobe expansion was obvious.

To determine if the chance of an ovule becoming a seed was independent of the fates of other ovules in the same flower, a frequency distribution for the number of mature seeds per flower was constructed using flowers from all 70 plants originally sampled. This frequency distirubtion was compared to an expected distribution that was calculated with the assumption that all ovules had an equal chance of becoming seeds. G-tests were used to test for significant differences between actual and expected frequency distributions of mature seeds per flower, and also to test for significant differences in rates of seed abortion among flowers that initiated different numbers of seeds.

RESULTS
Plant Form and Seed Output

Not surprisingly, the average numbers of seeds and fruits produced per plant was submaximal for *C. verticillata* (Table 1). In general, coefficients of variation for reproductive traits were larger than those for vegetative traits (Table 1). This was especially true for seed and fruit output per plant.

Results of the regression analysis can be depicted in a path diagram (Fig. 1), where single-headed arrows are the regression coefficients (b), which estimate direct effects. Double-headed arrows are correlations among plant traits treated as regressors. Finally, U_s represents the portion of variation in seed number that is not taken into account by any of the plant traits treated as regressor variables. U_s can be estimated indirectly at $1 - R^2$.

The regression analysis showed that fruit production and leaf length had significant positive effects on seed production (Fig. 1). The direct effects of fruits per plant on seeds per plant (b = 1.5) indicates that a mature fruit usually had only one or two seeds. Some inflorescence characteristics had negative direct effects on seed production, but these were nonsignificant. The standardized regression coefficients show that fruits per plant had a greater effect on seeds per plant than any other variable, including leaf length.

There were significant positive correlations among reproductive traits, but vegetative traits were not significantly correlated with reproductive traits (Fig. 1). Thus, indirect effects on seed production would be expected to be weakest for vegetative traits and strongest for inflorescence traits (inflorescence length, nodes, and flowers). For example, flowers per plant had a relatively strong effect on seeds per plant through its correlation with fruits per plant. An interesting result was that negative correlations only occurred between vegetative and reproductive traits (e.g., between leaf size and number of flowers).

Table 1. Means ($\bar{x}$), standard deviations (SD), and coefficients of variation (CV) for vegetative and reproductive characteristics measured on plants of *Collinsonia verticillata* (n = 47, except for inflorescence length, where n = 45). F-tests were used to compare CVs (critical values of F = 1.63 to 1.64, P = 0.05).

	x	SD	CV(%)
Height (cm)	18.6	2.6	14
Sum of leaf lengths (cm)	51.3	12.0	23
Inflorescence length (cm)	12.5	3.2	26[a]
Nodes per inflorescence	4.8	1.6	34[a]
Flowers per plant	16.9	7.4	44[ab]
Fruits per plant	5.0	3.6	72[ab]
Seeds per plant	7.1	5.7	80[ab]

[a]CV significantly greater than CV for plant height
[b]CV significantly greater than CV for leaf length

• • • • •

Table 2. Observed and expected frequency distributions for number of mature seeds per flower in *Collinsonia verticillata*.

Number of mature seeds	Number of flowers	
	Observed	Expected
0	956	900.0
1	172	262.7
2	47	28.8
3	12	1.4
4	6	0.03

Test for differences between distributions: G = 69.2, 1 d.f., P < 0.001. Flowers with 2 or more seeds were pooled to avoid expected frequencies that were too low for the G-test.

• • • • •

Table 3. Seed abortion rates for flowers of *Collinsonia verticillata*.

Number of seeds initiated	Number of flowers	Probability of seed aborting	Probability of flower aborting all seeds
1	179	0.16	0.16
2	55	0.16	0.07
3	16	0.29	0.12
4	17	0.41	0.06

Test for differences in seed abortion rates: G = 18.6, 3 d.f., P < 0.001; for probability of aborting all seeds: G = 3.4, 2 d.f., P > 0.05 (three- and four-seeded flowers pooled to avoid low frequencies).

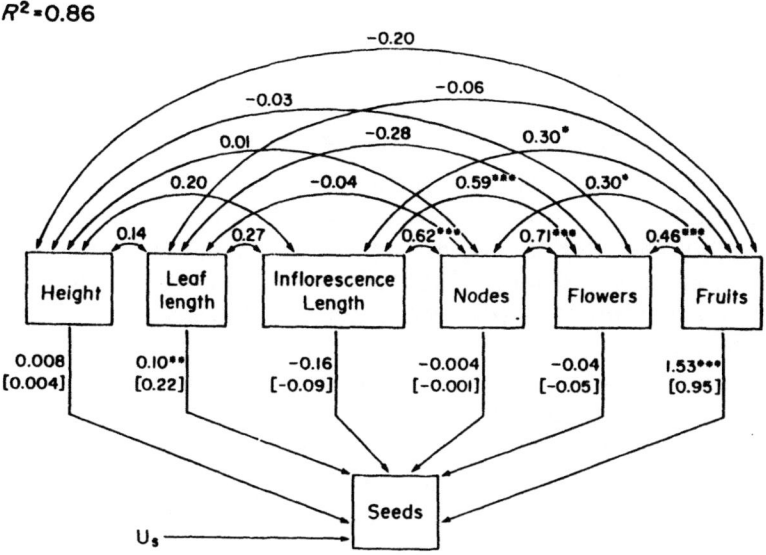

Figure 1. Multiple regression of number of seeds per plant on components of plant form for *Collinsonia verticillata*. Regressor variables were plant height, sum of leaf lengths, inflorescence length, nodes per inflorescence, flowers per plant, and fruits per plant. Double-headed arrows indicate correlation coefficients. Single-headed arrows indicate regression coefficients, with standardized coefficients in brackets. U_S represents unexplained variation for seeds per plant. Significance levels: *, **, *** = $P < 0.05$, 0.01, 0.001, respectively.

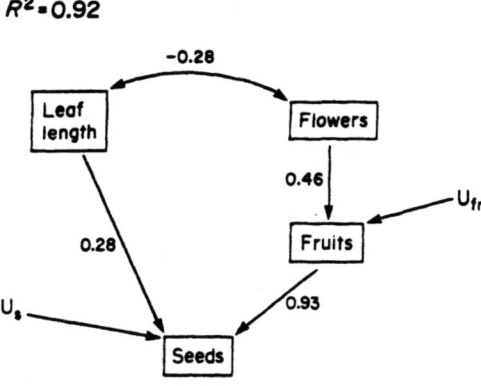

Figure 2. Path analysis of seed and fruit production for *Collinsonia verticillata*. Numbers next to single-headed arrows are path coefficients, which are equivalent to standardized regression coefficients. U_{fr} represents unexplained variation for number of fruits per plant. Remaining symbols as in Fig. 1.

Path analysis of data for *C. verticillata* led to a model with fewer variables and a simpler network of direct and indirect effects, compared to the full multiple regression model (Fig. 2). However, this path model explained a larger percentage of variation in number of seeds per plant than did the multiple regression model (R^2 = 92% vs. 86%). The path model retained the strong positive direct effects of leaf size and fruit number on seed number, and the weak negative correlation between leaf size and number of flowers. The path coefficient for the effect of leaf size on number of seeds was larger than the corresponding coefficient in the multiple regression. The path coefficient for this effect (0.45) is equivalent to a regression coefficient of 0.35, indicating that fruit-set per plants was about 35%. Only 21% of the variation in number of fruits per plant was determined by number of flowers, with the remainder unexplained by any other factor in the path diagram.

Pollination Rates and Fruit-Set

Sixty percent of all sampled styles carried pollen grains. The mean number of pollen grains per style was 5.5 + 1.3 ($\bar{x}$ + S.E.). Of these pollen grains, only 1.9 $\pm$ 0.4 had germinated, while 2.7 $\pm$ 0.9 were ungerminated and 0.9 $\pm$ 0.2 were empty. These results indicate that flowers should regularly produce fewer than four seeds. However, the mean number of germinated pollen grains may have been underestimated, because some of the pollen grains that were classified as empty actually may have been germinated grains for which pollen tubes could not be discerned. Also, pollen grains may have fallen off styles during sampling.

In the artificial pollination experiment, the percentage of flowers setting fruits was 5% for self-pollinations, 10% for the cross-pollinations, and 5% for natural pollinations. These fruit-set rations (one or two flowers out of 20 flowers per treatment) were too low to test for significant differences among treatments using a G - test. However, the formation of fruits from some self-pollinated flowers indicates that *C. verticillata* is at least partially self-compatible.

Seed-Set and Seed Abortion

Most flowers that set fruits produced only one seed (Table 2). Analysis of frequencies of mature seeds per flowers showed that one-seeded flowers were less common than expected, while flowers with zero or two or more seeds were more common than expected, assuming all ovules had an equal chance of becoming seeds (Table 2). Analysis of seed-abortion rates for flowers with different seed-initiation rates showed that the frequency of aborted seeds increased with the number of developing seeds per flower (Table 3). The frequencies of flowers aborting all seeds did not differ for flowers that initiated different numbers of seeds (Table 3).

DISCUSSION

In *Collinsonia verticillata*, the number of seeds produced per plant was mainly influenced by factors controlling the number of flowers per plant that set fruits. Leaf size and number of fruits had significant direct effects on seed production, but direct effects for other yield components were nonsignificant. Much of the effect of increased plant size on seed production would be expected to occur through indirect effects that involve positive correlations among traits. In the multiple regression, inflorescence traits had relatively strong indirect effects on seed yield, by virtue of significant positive correlations with numbers of fruits. On the other hand, indirect effects of vegetative traits on seed yield were small. The absence of significant correlations between vegetative and reproductive traits indicates that plants with larger vegetative parts did not always produce inflorescences with more flowers, which suggests that developmental constraints linking vegetative and reproductive growth were weak. This notion is supported by the coefficients of variation for different plant characteristics, which show that the development of inflorescence traits was generally more variable than that of vegetative traits.

Most of the variation in number of fruits per plant was determined by factors not under plant control. In the path model, these factors were represented by the unexplained variation in number of fruits per plant (U_{fr}). The low stigmatic pollen loads and the high percentage of flowers that failed to produce seeds strongly suggest that pollination rates limited fruit-set per plant. In addition, the artificial pollination experiments showed that pollination does not guarantee fruit-set, a contention also supported by the difference between the percentage of styles bearing pollen (60%) and the percentage of flowers setting fruit per plant (35%). The presence of an incompatibility system could not be ruled out by the artificial pollination experiment. In addition, pollination effectiveness could be limited by short duration of stigmatic receptivity or pollen that rapidly loses viability. Currently, the factors that determine whether pollination leads to fruit-set in *C. verticillata* are unknown.

In the multiple regression and path models for *C. verticillata*, number of fruits per plant accounted for most of the variation in seed output per plant. In large part this was because most flowers that set fruit had only a single seed. Although submaximal seed-set in these flowers could be attributed to low stigmatic pollen loads or to seed abortion, evidence from the artificial pollination experiment and the seed abortion study suggests that other factors also were involved in determining the number of mature seeds per flower. In the pollination experiment, pollen loads in excess of ovule number failed to elicit maximum seed-set, and artificially pollinated flowers were not more successful at setting seeds than were the naturally pollinated controls. As discussed above, there seem to have been factors limiting pollination effectiveness. In the seed abortion study, rates of seed abortion increased with the number of seeds initiated per flower, so that one might expect to find an increase in the relative frequency of single-seeded flowers and a decrease in the relative frequency of flowers with two or more seeds. However, under natural conditions, there was an excess of flowers with two or more mature seeds and a deficit of single-seeded flowers. Thus, the expected effect of seed abortion, if present at all, was masked by the effects of other factors.

Two important factors that often limit seed and fruit production per plant are resources (Willson and Price, 1980; Wyatt, 1980; Lee and Bazzaz, 1982, 1986; Stephenson, 1984; Delph, 1986) and low pollination rates (Willson and Schemske, 1980; Bierzychudek, 1981; Bertin, 1982; McCall and Primack, 1985). Although seed output per plant in *C. verticillata* was probably limited by low pollination rates, data on seed abortion indicate that resource limitation may have affected seed-set at the level of the individual flower. In addition, the significant direct effect of leaf size on seed output per plant suggests that seed production was partly limited by resources available from leaves. Similar effects of leaf size on seed production were reported by Maddox and Antonovics (1983) in their analysis of yield components in *Plantago*.

In the path model for *C. verticillata*, leaf size had a weak negative indirect effect on seed production through number of flowers and fruits per plant. This effect was due to alow negative correlation between leaf size and number of flowers, which was not quite significant. Negative relationships among traits that affect yield may be due to competition between sinks for limited resources (Eaton and Kyte, 1978; Primack and Antonovics, 1981; Lovett Doust, et al., 1983), or due to developmental constraints

(Watson and Casper, 1984). In either case, the negative correlation between leaf size and flower number in *C. verticillata* suggests that increased vegetative growth occurred partially at the expense of reproductive growth.

In *C. verticillata*, plant traits other than leaf size and flower number exerted relatively minor effects on seed and fruit production. This result would be expected in cases where fruit-set per plant is limited by low pollination rates. In situations where pollination rates do not limit fruit production, the morphological and physiological features that control resource allocation would be expected to be more important. However, even if fruit-set per plant is pollen-limited, plant form could affect fruit and seed production on a smaller scale if the allocation of resources to developing fruits and seeds within a plant is restricted by morphological factors (e.g., Watson and Casper, 1984; Lee and Bazzaz, 1986). This might help explain observed patterns of seed abortion and seed-set per flower in *C. verticillata*.

The effects of plant form on reproductive success may be analyzed with other methods such as yield component analysis and related procedures (e.g., Jolliffe et al., 1982; Lovett Doust et al., 1983; Eaton and Kyte, 1978). Yield component analysis assumes that the form of a mature plant may be broken down into a series of components that represents a growth sequence leading to fruit and seed production. Thus, leaf size, inflorescence length, nodes per inflorescence, flowers per plant, and fruits per plant could be viewed as a developmental series in which the formation of certain components precedes other components, i.e., leaves before inflorescence. However, when shoots of *C. verticillata* first appear in spring, the unexpanded leaves, inflorescence axis, and flower buds are already formed and expand almost simultaneously. Shoot form probably is determined early in development, and a sequential view does not seem quite appropriate. Therefore, in the regression and path analyses, I assumed that traits other than the numbers of fruits and seeds per plant were correlated characters reflecting the age, size, or allometry of each plant.

In general, the analysis of components of seed yield in *C. verticillata* showed that plant reproductive fitness was determined by a network of contingencies involving plant growth, pollination rates, leaf size, and apparent constraints on the number of seeds per flower. A major strength of multiple regression and path analysis is that relationships among many plant traits can be studied simultaneously. The effects of plant form on seed output in *Collinsonia verticillata* were fairly easily to estimate because plants of this species have relatively simple gross structure. However, multiple regression and path analysis can be applied to more complicated plant forms. Path analysis is more flexible than multiple regression because the network of relationships among plant attributes can vbe varied, allowing the investigation of different models (Li, 1975; Sokal and Rohlf, 1981; Maddox and Antonovics, 1983).

SUMMARY AND CONCLUSIONS

The influence of plant size and form on seed production was investigated for *Collinsonia verticillata*, a woodland perennial native to the southeastern United States. Plants were characterized in terms of leaf size, inflorescence length, nodes per inflorescence, and the number of flowers and fruits produced. Multiple regression showed that increased number of fruits per plant and greater leaf size were associated with greater seed output per plant. The significant relationship between leaf size and seed production suggested that larger leaves were more effective at providing resources to developing seeds. Although inflorescence traits had relatively strong positive indirect effects on seed production, the indirect effects of vegetative traits on fruit and seed production were small because correlations between vegetative and reproductive traits were weak. Thus, plants that had larger vegetative parts did not necessarily produce larger inflorescences.

Path analysis was used to construct a more realistic network of direct and indirect effects than that upon which the multiple regression was based. Results of the path analysis showed that variation in seed production in *C. verticillata* could largely be explained by a simple model that took into account effects of leaf size, number of flowers, and number of fruits. Leaf size had a minor adverse effect on seed production through its negative correlation with number of flowers. This negative correlation may mean that increased vegetative growth occurred partially at the expense of reproductive growth.

Although some plant traits had significant effects on seed output per plant, pollination appeared to be more important than plant form in determining seed yield in *C. verticillata*. Variation in the percentage of flowers setting fruit per plant was probably controlled by pollination rates, and seed-set per plant essentially paralleled fruit-set because flowers that set fruit usually had only one seed. Seed-set per flower was probably limited by a combination of factors, including low stigmatic pollen loads and seed abortion.

This study of seed production in *C. verticillata* shows that the effects of plant form on reproduction are not necessarily as simple as one might expect, given the generalization that fecundity increases with size. Plant traits may have negative as well as positive effects on seed production, and these effects may be either direct or indirect. In general, relationships among plant traits probably reflect underlying developmental or physiological processes. However, in most situations, the causal relationships linking plant characteristics, as well as environmental factors, to seed output will not be fully known, and many models of direct and indirect causal effects will be possible. Path analysis should prove to be more useful than multiple regression for analyzing the factors that affect seed production, because it can be used to construct and evaluate a wider range of models.

ACKNOWLEDGMENTS

This research was supported by a grant-in-aid from Highlands Biological Foundation, Highlands, North Carolina. Thanks to R. Wyatt for much advice and for suggesting that *Collinsonia verticillata* might be an interesting to study. Comments on the early manuscript were made by S. Dewey and C. DePamphilis.

LITERATURE CITED

Baskin, J. M., and C. M. Baskin. 1979. Studies on the autecology and population biology of the weedy monocarpic perennial, *Pastinaca sativa*. J. Ecology 67: 601-610.

Bertin, R. I. 1982. Floral biology, hummingbird pollination, and fruit production of trumpet creeper (*Campsis radicans*, Bignoniaceae). Amer. J. . Bot. 69: 122-134.

Bierzychudek, P. 1981. Pollinator limitation of plant reproductive effort. Amer. Nat. 117: 838-840.

Bierzychudek, P. 1982. The demography of jack-in-the-pulpit, a forest perennial that changes sex. Ecol. Monogr. 52: 335-351.

Delph, L. F. 1986. Factors regulating fruit and seed production in the desert annual *Lesquerella gordonii*. Oecologia 69: 471-476.

Dewey, D. R., and K. H. Lu. 1959. Correlation and path coefficient analysis of components of crested wheatgrass seed production. Agron. J. 51: 515-518.

Eaton, G. W., and T. R. Kyte. 1978. Yield component analysis in the cranberry. Journal of the Amer. Hort. Soc. 103: 578-583.

Harper, J. L. 1981. The concept of population in modular organisms. In R.M. May [ed.] Theological Ecology: Principles and Applications, 2nd ed. Blackwell Scientific Publications, Oxford.

Jolliffe, P. A., G. W. Eaton, and J. Lovett Doust. 1982. Sequential analysis of plant growth. New Phytol. 92: 287-296.

Kachi, N., and T. Hirose. 1985. Population dynamics of *Oenothera glazioviana* in a sand-dune system with special reference to the adaptive significance of size-dependent reproduction. J. Ecol. 73: 887-903.

Lee, T. D., and T. A. Bazzaz. 1982. Regulation of fruit and seed production in an annual legume: *Cassia fasciculata*. Ecology 63: 1363-1373.

Lee, T. D., and T. A. Bazzaz. 1986. Maternal regulation of fecundity: non-random ovule abortion in *Cassia fasciculata* Michx. Oecologia 68: 459-465.

Li, C. C. 1975. Path Analysis: a Primer. Boxwood Press, Pacific Grove, California.

Lovett Doust, J., L. Lovett Doust, and G.W. Eaton. 1983. Sequential yield component analysis and models of growth in bush beans (*Phaseolus vulgaris* L.). American Journal of Botany 70: 1063-1070.

Maddox, G. D., and J. Antonovics. 1983. Experimental ecological genetics in *Plantago*: a structural equation approach to fitness components in *P. aristata* and *P. patagonica*. Ecology 64: 1092-1099.

Marshall, D. L., D. A. Levin, and N. L. Fowler. 1986. Plasticity of yield components in response to stress in *Sesbania macrocarpa* and *Sesbania vesicaria* (Leguminosae). Amer. Nat. 127: 508-521.

McCall, C., and R. B. Primack. 1985. Effects of pollen and nitrogen availability on reproduction in a woodland herb, *Lysimachia quadrifolia*. Oecologia 67: 403-410.

Meijden, E. van der, and R. E. van der Waals-Kooi. 1979. The population ecology of *Senecio jacobaea* in a sand dune system. I. Reproductive strategy and the biennial habit. J. Ecology 67: 131-153.

Mitchell-Olds, T. 1986. Quantitative genetics of survival and growth in *Impatiens capensis*. Evolution 40: 107-116.

Primack, R. B., and J. Antonovics. 1981. Experimental ecological genetics in *Plantago*. V. Components of seed yield in the ribwort plantain *Plantago lanceolata* L. Evolution 35: 1069-1079.

Solbrig, O. T. 1981. Studies on the population biology of the genus *Viola*. II. The effect of plant size on fitness in *Viola sororia*. Evolution 35: 1080-1093.

Sokal, R. R., and F. J. Rohlf. 1981. Biometry. W.H. Freeman, San Fransisco.

Stanton, M. L. 1985. Seed size and emergence time within a stand of wild radish (*Raphanus raphanistrum* L.): the establishment of a fitness hierarchy. Oecologia 67: 524-531.

Stephenson, A. G. 1984. The regulation of maternal investment in an indeterminate flowering plant (*Lotus corniculatus*). Ecology 65: 113-121.

Watkinson, A. R., and J. White. 1985. Some life-history consequences of modular construction in plants. Philos. Trans. Royal Soc. London B 313: 31-51.

Watson, M. A., and B. B. Casper. 1984. Morphogenetic constraints on patterns of carbon distribution in plants. Ann. Rev. Ecol. Syst. 15: 233-258.

White, J. 1984. Plant metamerism. In R. Dirzo and J. Sarukhan [eds.]. Perspectives on Plant Population Ecology. Sinauer Associates, Sunderland, Massachusetts.

Willson, M. F., and P. W. Price. 1980. Resource limitation of fruit and seed production in some *Asclepias* species. Canad. J. Bot. 58: 2229-2233.

Willson, M. F., and D. W. Schemske. 1980. Pollinator limitation, fruit production, and floral display in pawpaw (*Asimina triloba*). Bull. Torrey Bot. Club 107: 401-408.

Wyatt, R. 1980. The reproductive biology of *Asclepias tuberosa*. I. Flower number, arrangement, and fruit-set. New Phytol. 85: 119-131.

SIZE DEPENDENT REPRODUCTION IN *POA ANNUA* AND *P. PRATENSIS*

Lisa K. Wagner

INTRODUCTION

Variation in plant reproductive effort has been related to a variety of factors, including life history strategies (Harper, 1967; Pitelka, 1977; Primack, 1979), successional stage of habitat (Abrahamson, 1979; Stewart and Thompson, 1980; Weaver and Cavers, 1980), and site-to site variation (Hume and Cavers, 1983; Quinn and Hodgkinson, 1984). Within-species differences in reproductive allocation among sites and between years have also been reported for a number of species (Jolls, 1980; Soule and Werner, 1981; Willson, 1985; Quinn and Hodgkinson, 1984; Whigham, 1984) and various hypotheses proposed to account for this variability (Evenson, 1983; Hancock and Pritts, 1987).

Reproductive output, measured as numbers of flowers, fruits, and/or seeds, is correlated with plant size in many species (Harper, 1977). Similarly, if total reproductive biomass (RWT - the biomass in fruits, seeds and accompanying reproductive tissues) is regressed against total biomass, or total vegetative biomass excluding reproductive biomass (TVEGWT), relatively high r^2 values are obtained, suggesting that a linear size-dependent relationship exists between reproductive weight and plant size (Weiner, 1988). It has been suggested that proportional reproductive effort (RE) - the ratio or proportion of reproductive biomass to vegetative biomass - may also be size-dependent (Samson and Werk, 1986; Weiner, 1988) and that size-dependent effects may be responsible for much of the observed variability of RE in natural populations. These potential size-dependent relationships are summarized in Figure 1.

If the relationship between total reproductive weight (RWT) and total vegetative weight (TVEGWT) is linear, then, above some threshold size for initial reproduction (V_o), there may be a simple allometric relationship between plant size and seed reproduction (Fig. 1a: lines A, B, C) determined by developmental, structural and physiological constraints (Samson and Werk, 1986). As a consequence of this linear relationship, proportional RE (RWT/TVEGWT) may also vary with plant size, depending in the value of the y-intercept and the slope of the regression line (Fig. 1b: lines A', B', C'). For example, if the intersection with the y axis is positive (Fig. 1a: line C), a monotonically decreasing function of RE versus TVEGWT results (Fig. 1b: line C'); if the y-intercept is negative (Fig. 1a: line B), a monotonically increasing function of RE versus TVEGWT is the result (Fig. 1b: line B'). If the y-intercept is zero, a flat function of RE versus TVEGWT would result (Fig. 1b: A'). If a species is characterized by a non-zero y-intercept, RE may then vary among populations simply as a result of mean differences in plant size (Samson and Werk, 1986).

There have been few studies of the relationship between RWT, TVEGWT, and RE in natural populations. In this paper, I discuss the relationship of plant size to absolute and proportional reproductive allocation patterns in two weedy *Poa* species (*P. annua* and *P. pratensis*) and examine whether a size-dependent linear reproduction model (Weiner, 1988) is compatible with patterns of reproduction and growth in these species.

METHODS

I studied *Poa annua*, a cosmopolitan weedy annual grass, and *P. pratensis*, a widespread rhizomatous perennial, in two sites at the Smithsonian Environmental Research Center in Edgewater, Maryland. *Poa annua* was sampled in an unmowed coarse lawn near the Center's dock and along a little-used dirt road adjacent to a cultivated field. *P. pratensis* was studied in the same lawn where *P. annua* was sampled and in an old field near the roadside population of *P. annua*.

To determine patterns of reproductive effort, ten to fifteen randomly selected plants in each population were excavated at reproductive maturity in 1984 and 1985. Individual genets were collected for *P. annua*. For *P. pratensis*, distinct clones were collected and divided into separate ramets. Each ramet was a cluster of tillers sharing a common root mass. Roots of each sample were rinsed in a dispersing solution (Malone, 1967) to remove soil and organic debris. Samples were then dried to constant weight at 50° C. Individual tillers were divided into reproductive and vegetative fractions prior to weighing. Dry weight was used as a measure of energy allocation (Hickman and Pitelka, 1975), although how to best measure reproductive effort remains a matter of some debate (Thompson and Stewart, 1981; van Andel and Jager, 1981; Abrahamson and Caswell, 1982; Bazzaz and Reekie, 1985). The terms biomass and weight refer to dry weight of plant parts.

Total reproductive weight (RWT) for each genet of *P. annua* and ramet of *P. pratensis* was calculated by summing individual

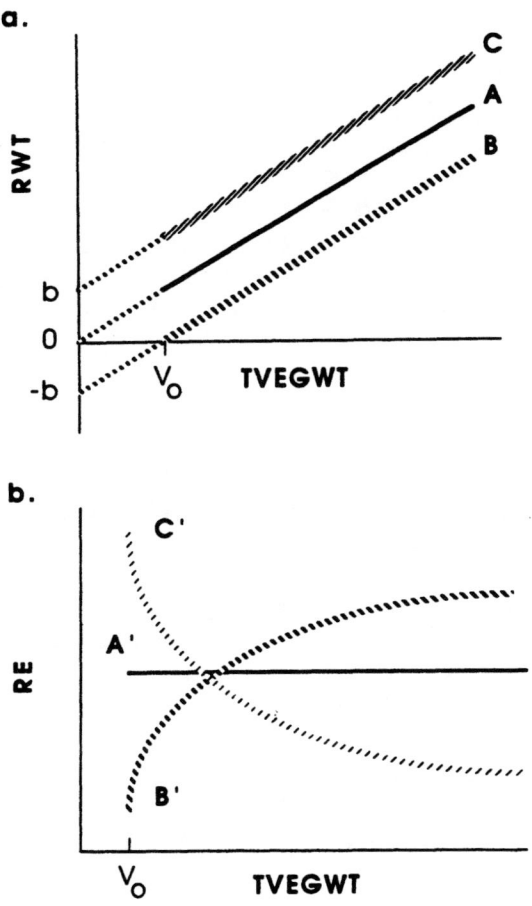

Figure 1. Graphical model of reproductive allocation in plants (redrawn from Samson and Werk [1986] by permission). a) Reproductive biomass (RWT) as a linear function of total vegetative biomass (TVEGWT). V_o is the minimum plant size for reproduction. b) Proportional reproductive allocation (RE - the ratio of RWT to TVEGWT) as a function of TVEGWT for lines shown in (a). Lines A, A', B, B', and C, C' represent possible relationships between plant size and seed reproduction.

tiller data for each sample. RWT included seeds, surrounding glumes, and all supporting structures above the base of the panicle. Total vegetative weight (TVEGWT) was the summed biomass of stems, leaves, and roots for each individual and was equivalent to total plant biomass minus reproductive biomass. Reproductive effort (RE) per individual was calculated as the ratio of RWT to TVEGWT.

Reproductive weight was plotted against total vegetative weight and linear regression analysis (GLM procedure, SAS) carried out to determine the degree of linearity in the relationship between individual plant size. TVEGWT was used rather than total plant weight to reduce the possibility of auto-correlation (Samson and Werk, 1986). RE (RWT/TVEGWT) was plotted against plant size to determine whether RE was size-dependent in these populations.

RESULTS

Reproductive weight (RWT) was strongly correlated with plant size (TVEGWT) in both species (Figs. 2a and 3a). In *P. annua* , the relationship between RWT and TVEGWT was linear for both populations (Table 1) and much of the variation of RWT could be explained by variation in TVEGWT. Regression lines were similar in both years for the roadside population, whereas in 1984, RWT in the lawn population increased more slowly with increasing TVEGWT resulting in a smaller mean RE than in 1985 (Fig. 2b; Table 1). Y-intercepts were not significantly different from zero in either population.

RWT was also linearly related to TVEGWT in both populations of *P. pratensis* (Fig. 3a). Regression parameters for the lawn population were similar in both years (Table 1). In contrast, the old field population exhibited a much lower rate of allocation to reproduction in 1985 than in 1984 (Table 1; Fig. 3a). Y-intercepts of regression lines for the lawn population of *P. pratensis* were not significantly different from zero ($p < 0.01$) in either year. In the old field population , however, Y-intercepts differed significantly from zero with a negative value in 1984 (-0.018 ± 0.006; $p < 0.01$) and a positive value in 1985 (0.039 ± 0.013; $p < 0.01$).

Proportional reproductive effort (RE) was not related to plant size in *P. annua* (Fib. 2b). In both years, plants in lawn and roadside populations allocated biomass to reproduction independently of plant size and RE was essentially a flat function of TVEGWT (linear regressions were not significant [$p > 0.05$], with r^2 values ranging form 0.02 to 0.13) as would be predicted by Y-

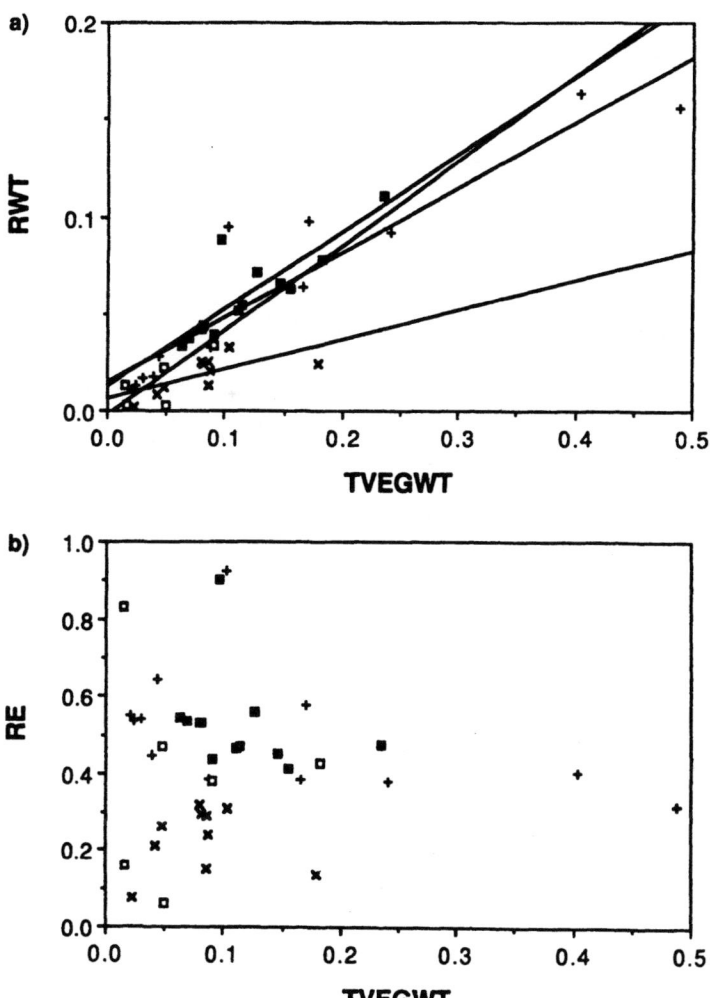

Figure 2. a) RWT (g) as a function of TVEGWT (g) for lawn and roadside populations of *P. annua in 1984 and 1985.* Symbols represent lawn, 1984 (x): lawn, 1985 (+); roadside, 1984 (open squares); roadside, 1985 (closed squares). Lines correspond to the best-fit regression line for each population sample. b) RE (=RWT/TVEGWT) plotted against TVEGWT for lawn and roadside populations of *P. annua.* Symbols as in (a).

278

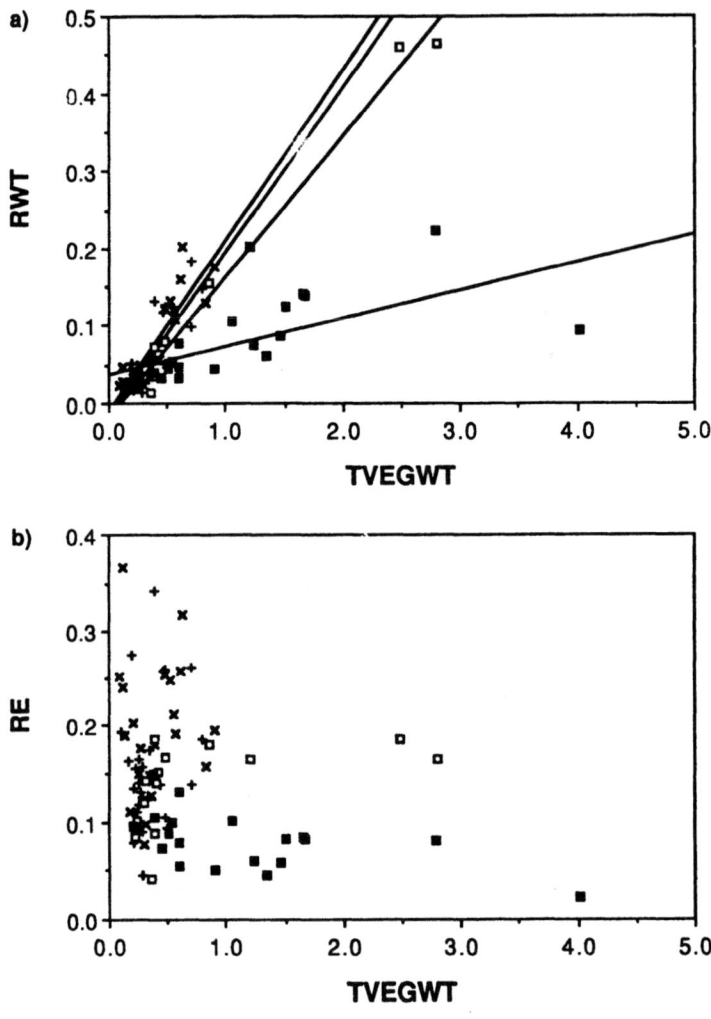

Figure 3. a)RWT (g) as a function of TVEGWT (g) for lawn and old field populations of *P. pratensis* in 1984 and 1985. Symbols represent lawn, 1984 (x); lawn, (+); old field, 1984 (open squares); old field, 1985 (closed squares). b) RE plotted against TVEGWT for lawn and old field populations of *P. pratensis*. Symbols as in (a).

Table 1. Regression parameters and mean values of RE for populations of *Poa annua* and *P. pratensis* at the Smithsonian Environmental Research Center, Edgewater, Maryland.

Species	Site	Year	RWT = mTVEGWT + b[#]				
			m	b	n	r^2	RE ± s.e.
Poa annua	lawn	1984	.152	.006	10	.47	.229 ± .084
		1985	.333	.015	12	.88	.508 ± .047
	roadside	1984	.437	-.002	7	.89	.408 ± .036
		1985	.399	.013	11	.69	.525 ± .040
P. pratensis	lawn	1984	.220	-.009	22	.79	.172 ± .058
		1985	.213	-.017	29	.67	.155 ± .067
	old field	1984	.182	-.018	15	.99	.120 ± .046
		1985	.035	.039	20	.45	.083 ± .031

[#]RWT = reproductive biomass (fruits, surrounding glumes, and rachis);TVEGWT = vegetative biomass (stems, leaves, and roots); RE = RWT/TVEGWT. Linear regression coefficients were significant at P < 0.01 or less and all slopes were significantly different from zero.

intercepts of zero. Values of RE within *P. annua* populations were highly variable (Fig. 2b). Mean values for RE in the roadside population were not significantly different in 1984 and 1985 whereas RE in the lawn population differed between years (*t*-test significant at p.0.01), probably due to some seed dispersal prior to sampling in 1984. In 1985, RE in the lawn population was not significantly different than either year in the roadside population.

Poa pratensis populations exhibited a more complicated pattern of RE in relationship to plant size (Fig. 3b). RE was not related to TVEGWT in the lawn population in either year (linear regressions were not significant; $r^2 = 0.04$ [1984] and $r^2 = 0.06$ [1985]). In the old field population, on the other hand, RE and TVEGWT were positively correlated in 1984 (linear regression significant at p<0.05; $r^2 = 0.26$), whereas in 1985, the relationship was negative (linear regression significant at p<0.05; $r^2 = 0.30$). Mean RE in the lawn population was not significantly different between years. In both 1984 and 1985, however, mean RE in the lawn population was greater than the old field population (*t*-test significant at p<0.01).

DISCUSSION

How well does a linear size-dependent reproduction model fit *Poa* species? If RWT is plotted against TVEGWT, strongly linear relationships characterize both populations of *P. annua* and *P. pratensis* and a linear size-dependent model of plant reproduction (Samson and Werk, 1986; Weiner, 1988) fits the available data quite well. In general, the relatively high r^2 values indicate that a significant fraction of the variation in RWT can be attributed to variation in TVEGWT. *P. annua* populations have Y-intercepts not significantly different from zero (Samson and Werk's A category, Fig. 1a), resulting in little variation in RE with size (A', Fig. 1b).

Strongly modular plant architecture of the kind exhibited by *P. annua* and many other annual grasses, in which reproductive shoots, or tillers, are successively produced, is ideally suited to size-dependent reproductive behavior; as Weiner (1988) pointed out "a linear size-dependent reproduction model is consistent with the notion that plant growth is modular...if all modules are reproductive, a simple proportion of biomass would be expected to be in reproductive tissues".

In *P. annua*, reproduction commences when the plants are very small, which results in Y-intercepts close to zero. New tillers are produced for as long as conditions are favorable and are almost always reproductive. Each tiller is largely self-supporting and independent during the time of grain maturation; radioactive carbon studies in this species (Ong and Marshall, 1975) showed that reproductive photosynthesis by the inflorescence accounts for almost 75% of the assimilate needed to develop seeds, with the remainder of photosynthate coming form the first leaf. Following seed dispersal, considerable photosynthate is exported from tillers, demonstrating that tillers are capable of re-integration.

A lack of size-dependence in RE also characterized the lawn population of *P. pratensis* (Fig. 3b) and the data fell into Samson and Werk's A category. In both years of study, mean RE in the lawn population was a flat function of TVEGWT. In the old field population of *P. pratensis*, however, RE was size-dependent and the observed trends in RE fit those predicted for negative and positive Y-intercepts based on Samson and Werk's model. In 1984, RE increased and leveled off with increasing plant size (Fig. 3b), consistent with expectations based on a negative Y-intercept (lines B and B', Fig. 1). In 1985, RE decreased weakly with increasing plant size (Fig. 3b), again corresponding to predicted values.

The strongly linear relationship between RWT and TVEGWT in *P. pratensis* is also related to its modular growth form. *P. pratensis* is characterized by physiological independence of tillers when plants are undisturbed, similar to *P. annua* (Nyahoza et al., 1973, 1974), such that reproductive output is largely dependent on photosynthate produced by that tiller. Reproductive tillers, however, are produced in a single reproductive period in spring in contrast to *P. annua*'s extended period of reproductive tiller production (Wagner, unpub. data).

The differences between the lawn population and the old field population in size-dependence of RE are somewhat difficult to interpret. In perennials, potentially competing demands for photosynthate between vegetative growth, sexual reproduction, and storage requirements may obscure the relationship between reproductive weight and vegetative weight. Physiological constraints in the movement of photo-assimilate may also play a role (Watson, 1986). Furthermore, variations in root biomass related to clone size and potential difficulties in obtaining an adequate sample of root biomass in perennials may also introduce variation in RE.

Variation in size-dependence of RE may also be related to the degree of openness of the site and the growth form of the plants. The lawn population is normally mowed during growth periods and occasionally trampled and driven over, and consequently remains more open than the old field population. Mean ramet size was considerably smaller in the lawn population with generally less allocation to stem and leaf biomass and greater allocation to reproduction than in the old field population (Wagner, unpub. data). The old field population occurs in a more competitive environment, which is reflected in greater allocation to vegetative biomass and increased tiller height. Potentially greater plant age in the old field population may also affect values of reproductive effort.

This study supports the need for consideration of potential size-dependence in studies of RE in plants. Whether or not a species shows size-dependence in RE, information about the relationship of absolute and relative reproductive effort to plant size is necessary to evaluate the reproductive pattern of a species. Such information provides a basis for further study of the ecological, morphological, and physiological factors that affect reproductive effort in plants.

SUMMARY

Variation in reproductive effort (RE) was investigated in relation to plant size in two populations of *Poa annua* , an annual, and *Poa pratensis,* a rhizomatous perennial, for two successive years. In both species, reproductive weight (RWT) was linearly related to vegetative weight (TWEGWT), with r^2 values ranging form 0.45 - 0.99 (p<0.01). Variation in plant size largely explained variation in reproductive weight. However, RE - the ratio of reproductive biomass to vegetative biomass - was not related to size in *P. annua* and in a lawn population of *P. pratensis.* An old field population of *P. pratensis* exhibited size - dependence of RE, increasing with size in 1984 and decreasing with size in 1985. The linear relationship between reproductive weight and vegetative weight probably reflects modular growth patterns of *Poa* species; each tiller is effectively an independent module that produces most of the assimilate required for seed ripening, much of it from the inflorescence itself.

ACKNOWLEDGEMENTS

This work was carried out with support from a Smithsonian Institution Post-Doctoral Fellowship, the Smithsonian Institution Office of Fellowships and Grants' Visiting Scientist Program, and the Faculty Research Committee of Georgia Southern College. I especially thank Dennis Whigham who provided valuable advice and made me welcome in his lab. Jake Weiner first suggested the line of analysis described in this paper and pointed out its significance.

LITERATURE CITED

Abrahamson, W. G. 1979. Patterns of resource allocation in wildflower populations of fields and woods. Amer. J. Bot. 66a: 71-79.

Abrahamson, W. G. and H. Caswell. 1982. On the comparative allocation of biomass, energy, and nutrients in plants. Ecology 63: 982-991.

Bazzaz, F. R., and E. G. Reekie. 1985. The meaning and measurement of reproductive effort in plants. In J. White [ed.], Studies on Plant Demography: a Festschrift for John L. Harper. Academic Press, London.

Evenson, W. E. 1983. Experimental studies of reproductive energy allocation in plants. In C. E. Jones and R. L. Little [eds.], Handbook of Experimental Pollination Biology. Scientific and Academic Editions, New York.

Hancock, J. F., and M. P. Pritts. 1987. Does reproductive effort vary across different life forms and seral environments? A review of the literature. Bull. Torrey Bot. Club 1987: 53-59.

Harper, J. L. 1967. A Darwinian approach to plant ecology. J. Ecol. 55: 247-270.

Harper, J. L. 1977. Population Biology of Plants. Academic Press, London.

Hickman, J. C., and L. F. Pitelka. 1975. Dry weight indicates energy allocation in ecological strategy analysis of plants. Oecologica (Berl.) 21: 117-121.

Hume , L., and P. B. Cavers. 1983. Resource allocation and reproductive and life-history strategies in widespread populations of *Rumex crispus*. Can. J Bot. 61: 1276-1282.

Jolls, C. L. 1980. Phenotypic patterns of variation in biomass allocation in *Sedum lanceolatum* Torr. at four elevational sites in the Front Range, Rocky Mountains, Colorado. Bull. Torrey Bot. Club 107: 65-70.

Malone, C. R. 1967. A rapid method for the enumeration of viable seeds in the soil. Weeds 15: 381-382.

Nyahoza, F., C. Marshall, and G. R. Sagar. 1973. The inter-relationship between tiller and rhizomes of *Poa pratensis* L. - an autoradiographic study. Weed Research 13: 304-309.

Nyahoza, F., C. Marshall, and G. R. Sagar. 1974. Assimilate distribution in *Poa pratensis* L. - a quantitative study. Weed Research 13: 251-256.

Ong, C. K., and C. Marshall. 1975. Assimilate distribution in *Poa annua* L. Ann. Bot. 39: 413-421.

Pitelka, L. F. 1977. Energy allocation in annual and perennial lupines (*Lupinus*: Leguminosae). Ecology 58: 1055-1065.

Primack, R. B. 1979. Reproductive effort in annual and perennial species of *Plantago* (Plantaginaceae). Amer. Nat. 114: 51-62.

Quinn, J. A., and K. C. Hodgkinson. 1984. Plasticity and population differences in reproductive characters and resource allocation in *Danthonia caespitosa* (Gramineae). Bull. Torrey Bot. Club 111: 19-27.

Samson, D. A., and K. S. Werk. 1986. Size-dependent effects in the analysis of reproductive effort in plants. Amer. Nat. 127: 667-680.

Soule, J. D., and P. A. Werner. 1981. Patterns of resource allocation in plants, with special reference to *Potentilla recta* L. Bull. Torrey Bot. Club 108: 311-319.

Stewart, A. J. A., and K. Thompson. 1982. Reproductive strategies of six herbaceous perennial species in relation to a successional sequence. Oecologia (Berl.) 52: 269-272.

Thompson, K., and A. J. A. Stewart. 1981. The measurement and meaning of reproductive effort in plants. Amer. Nat. 177: 205-211.

Van Andel, J. and J. C. Jager. 1981. Analysis of growth and nutrition of six plant species of woodland clearings. J. Ecol. 69: 871-882.

Watson, M. A. 1986. Integrated physiological units in plants. Trends in Evol. and Ecol. 1: 119-123.

Weaver, S. E., and P. B. Cavers. 1980. Reproductive effort in two perennial weed species in different habitats. J. of Appl. Ecol. 17: 505-513.

Weiner, J. 1988. The influences of competition on plant reproduction. In J. Lovett Doust and L. Lovett Doust [eds.], Plant Reproductive Ecology: Patterns and Strategies. Oxford University Press, New York.

Whigham, D. F. 1984. The effect of competition and nutrient availability on the growth and reproduction of *Ipomoea hederacea* in an abandoned old field. J. Ecol. 72: 721-730.

Willson, M. F. 1985. Plant Reproductive Ecology. Wiley, New York.

YEARLY FLUCTUATIONS IN A VERNAL POOL ANNUAL, *SIDALCEA HIRSUTA*

Robert A. Schlising

INTRODUCTION

Vascular plants that grow in vernal pools are receiving attention and study as they become rarer due to continued destruction of their unique habitat. Vernal pools are not restricted to California, but in North America they are found mainly in this state (Thorne, 1984; Moran, 1984; Zedler, 1987). The typical vernal pool occurs at a low elevation in that part of California with a Mediterranean climate. Here the winter rain accumulates in depressions, and often stands for months due to an impervious layer of rock or cemented soil materials, and to conditions conducive to low evaporation. Hot and dry conditions, beginning in late spring, evaporate the water and turn the pools into severely xeric habitats. Thus the plant and animal life of vernal pools must cope with extremes ranging from standing water to intense drought.

Vernal pools are sometimes, and perhaps more appropriately, referred to as "temporary annual pools" or "periodic environments" (Alexander, 1976) because usually they are filled with water from late fall until spring. It is their flora, mostly dicotyledonous annuals, which is predominantly vernal. Most of the species flower with the spring dry-down, and set seed as the summer dry season begins. This often produces rings of different colors as different species, at slightly different depths in the pool, flower in response to the drying. Such pools have been extirpated widely as the lowland areas of California have been converted to crop fields and urban areas.

Complexes of these pools still in existence sometimes show inter-connections--especially during periods of "high water," when precipitation has been heavy. But frequently vernal pools occur as discrete, island-like habitats, disjunct from one another by meters or kilometers. This habitat then, in addition to being alternately aquatic and xeric, is peculiar in terms of its overall patchiness within an area (often grassland).

The disjunct nature of vernal pools provides the arena for studying evolutionary and population phenomena. For example, phytogeography, species diversity, and even pools as islands have been examined (Holland and Jain, 1981, 1984; Jain, 1976; Ebert and Balko, 1984). Aspects of the sharp environmental gradient from pool to non-pool have been studied (e.g., Linhart, 1972, 1976, 1988; Zedler, 1984), as have aspects of vegetational distinctness of adjacent pool and non-pool (e.g., Holland and Jain, 1977; Rosario and Lathrop, 1984; Schlising and Sanders, 1982). Distinct assemblages of co-occurring species are well documented in vernal pool floras (Lathrop and Thorne, 1983; Lin, 1970; Schlising and Sanders, 1983). Systematic treatments and ecological studies for plant groups restricted to, or well-represented in, vernal pools have provided information on the plants' biology and evolution (e.g., McLaughlin, 1974 on *Callitriche*; Weiler, 1962, and Martin and Lathrop, 1986, on *Downingia*: Ornduff, 1966, 1976 on *Lasthenia*; Scheidlinger, 1984, on *Pogogyne*: Griggs, 1980, 1981, on *Orcuttia*; and numerous studies by Jain and co-workers on *Limnanthes*, with a good summary in Jain, 1984). Many of these authors, and still others (e.g., Stebbins, 1976) indicate that vernal pools provide good biological field-laboratories for studying evolution or dynamics of plant populations because of the isolated or island-like nature of many vernal pools.

The present study in the northern Sacramento Valley was conceived in the spring of 1982 after finding a population of *Sidalcea hirsuta* Gray (Malvaceae) contained within one vernal pool and disjunct from the next pool population by 1.5 km (Broyles, 1983). *Sidalcea hirsuta*, the vernal pool checker, is one of only five annual species in this mainly western-American genus. This species occurs in California in the Central Valley and in valleys of the North Coast Ranges (Moore, 1971), mainly in deeper pools. The plants produce showy, magenta-pink flowers 3-4 cm across, and at flowering and fruiting can range from about 20 to 100 cm tall. The fruit separates into usually 7-9 one-seeded carpels (="seeds"). There is no information on the biology of *S. hirsuta*, and the study on *S. calycosa* in shallow pools of a burned grassland near Chico, California (Hunter, 1986), is the only ecological study reported for any of the annual species of *Sidalcea*. Therefore, a several-year study of *S. hirsuta* was initiated by asking apparently simple questions: 1) Does population size fluctuate between years? 2) Does the number of fruits produced per plant, a feature related to plant size, vary from year to year? and 3) Does seed set vary by fruit and by total seed crop through several years?

STUDY AREA AND METHODS

The study site was on the Vina Plains Preserve, just north of Singer Creek in southernmost Tehama County, about 27 km north of Chico, California. The area consists of annual grassland, with scattered populations of a perennial bunchgrass, *Stipa pulchra*. This 616 ha preserve contains about 10 large vernal pools and numerous small pools and drainage-ways. A hard volcanic substrate maintains standing water in the pools, although deposits of Anita clay and other soils overlay this hardpan in several of the pools (Broyles, 1983, 1987). Water stands in some of the deeper pools from fall or winter until as late as June in some years.

This region, in the northern Sacramento Valley, has a well-developed Mediterranean climate. Complete climatic data are available for Orland (Glenn County), about 25 km by air to the southwest of the preserve and at about the same elevation (66 to 77 m) (U.S. Dept. Commerce, 1982-1986). The highest (30-year) monthly mean temperature is 25.9°C in July, the lowest 7°C in January. Precipitation comes as rain, mostly from October through April, and the mean yearly rainfall (1951-1980) is 505 mm.

Sidalcea hirsuta was studied in an oval-shaped pool about 228 m long by 140 m at its widest, in the central, ungrazed portion of the preserve. The *Sidalcea* population occurred in a crescent-shaped area along the southwestern part of the pool each of the four years (1983-1986) of this study. Boundaries of the population were quite discrete, since the plants rarely grow above the pool margin in the surrounding grassland, and do not occur in deeper parts of the pool.

Each year, the length and maximum width of the population was measured, and all the plants were counted. Plants were then sampled (systematic-randomly) by using a grid system that spanned the vast majority of the population, omitting a few plants at the north and south "tails."

Nearest individuals to crossings of metric tapes in the grid were sampled by counting fruits per plant. Ten fruits per plant (or varying numbers up to 10 in some instances) were collected and stored individually in gelatin capsules for later counts of seeds per fruit. Plants and fruits were counted, and fruit collected, in late May or early June (depending on the onset of the dry summer each year), as soon as all plants were totally past flowering and bore mature and dried seeds. In 1987, a fruit from near the base of the inflorescence (a "proximal" fruit) and also a fruit from near the tip of a branch on the same plant (a "distal" fruit) were collected from 50 plants for comparison.

Statistics were calculated by hand, or with a computer using SPSS. Fruits per plant and seeds per fruit were compared for the four years using ANOVA and Tukey's paired comparisons (Neder, et al., 1985).

RESULTS

Population size varied dramatically among years (Table 1), from 1,389 in 1983 to a low of 652 in 1984, to highs of 21,113 in 1985 and 16,695 in 1986. Mean fruits per plant also varied among years, from a high of 259.5 fruits per plant in 1983, to a low of 9.2 in the third season. High standard deviations indicate great variability in fruits per plant, but mean fruits per plant sampled in 1983 is significantly different from means the other years ($p < 0.01$). Seeds per fruit also were variable, and mean seeds per fruit in 1983 and 1986 samples were significantly higher than those in 1984 and 1985 ($p < 0.01$). Total seeds produced per population are estimated (not accounting for seeds removed and lost due to the sampling), and show dramatic fluctuations from year to year.

Mean number of fruits per plant is shown in relation to precipitation for each year in Fig. 1. The rainfall increments are shown from July through June, since this reflects the Mediterranean-climatic seasons to which these annuals are adapted (i.e., seed germination occurs with the fall rains, and the plants produce seeds and then die in May or June as the summer drought begins). In addition to monthly increments of rainfall, Fig. 1 shows cumulative rainfall for each of the four growing seasons. In 1982-83, when rainfall was the highest of the four years, *Sidalcea* plants were large, and numbers of fruits were the highest of the four years. The 1983-84 and 1984-85 growing seasons had rainfall below normal, which was unevenly distributed during the growing seasons relative to 1982-83. In both 1984 and 1985 the plants were small, and many bore only one or two mature fruits (compared with a low of 38 fruits in the 1983 sample). Although plants were small in both 1984 and 1985, there were over 30 times more individuals in 1985 than in the year before. Finally, in 1985-86, high total precipitation was associated with a large population size and a high mean number of fruits per plant (Table 1). Clearly, population size and fruits per plant are not related.

Since numbers of fruits per plant (which is an indication of plant size in *Sidalcea hirsuta*) appears related to seasonal rainfall, fruits per plant were regressed on total rainfall (Fig. 2). The regression, using log-transformed data, was performed using Tellagraf, by Computer Associates, on the Prime Computer, and both a linear and the quadratic regression shown in Fig. 2 indicate significant relationships ($p < 0.001$). All counts of fruits

Table 1. Population and seed crop parameters for *Sidalcea hirsuta*, Vina Plains Preserve, Tehama County, California.

Year	Total Individuals	Fruits per plant $\overline{x} \pm SD$	(n)	Seeds per fruit $\overline{x} \pm SD$	(n)	Total seed crop (millions of seeds)
1983	1,389	259.5 ± 204.4*	(30)	7.7 ± 0.4**	(30)	2.77
1984	652	11.7 ± 25.0	(30)	6.4 ± 1.1	(30)	0.05
1985	21,113	9.2 ± 14.6	(30)	6.5 ± 0.8	(30)	1.26
1986	16,695	43.4 ± 48.6*	(32)	7.4 ± 0.6**	(32)	5.35

* significantly different from other 3 years ($p < 0.01$); other means not significantly different from each other
** significantly different from other 2 years ($p < 0.01$)

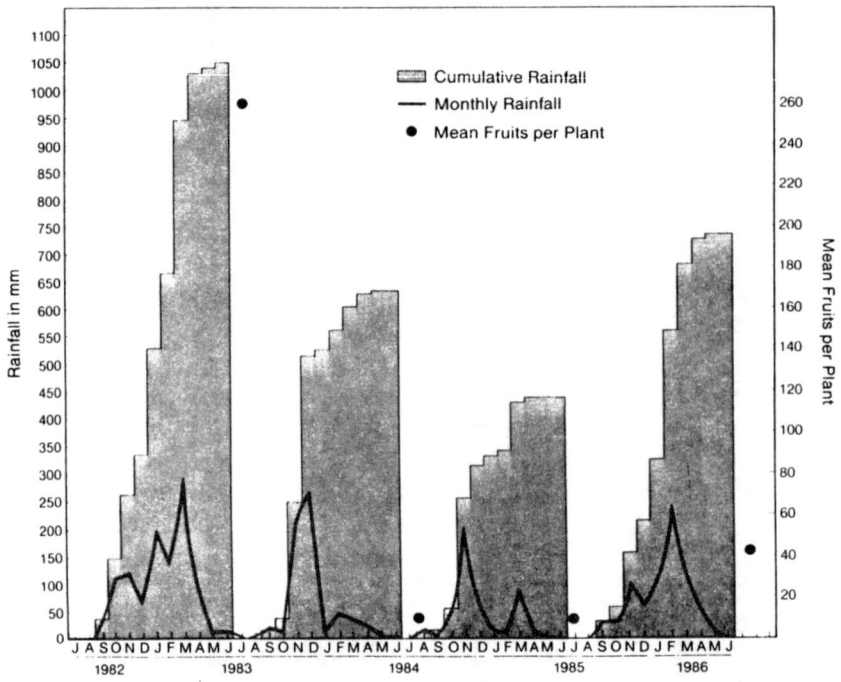

Figure 1. Mean fruits per plant in *Sidalcea hirsuta*, shown in relation to cumulative and monthly rainfall through four growing seasons at Vina Plains Preserve.

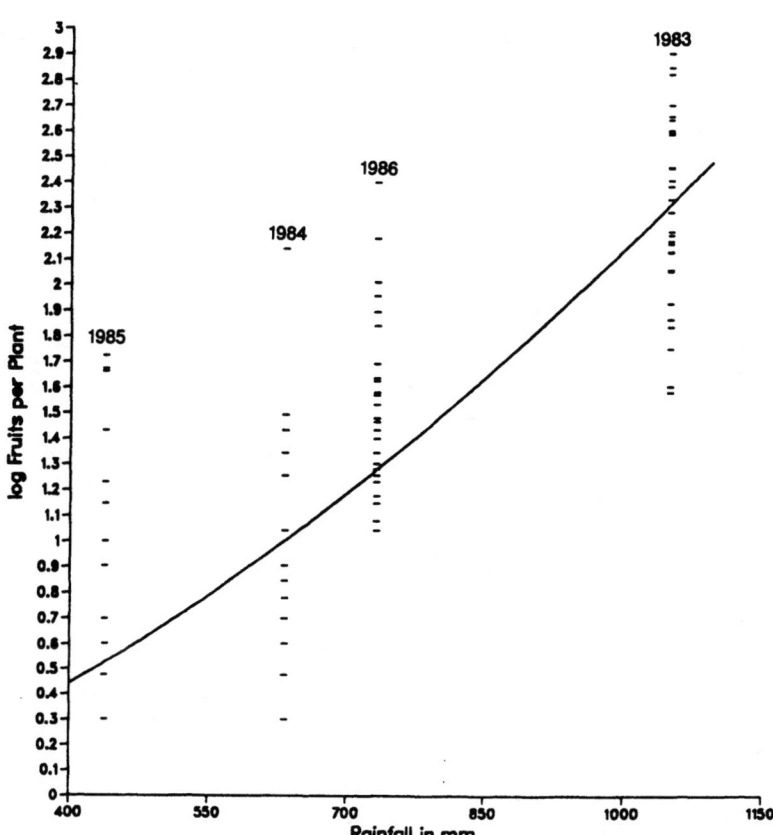

Figure 2. Regression showing relationship of fruits per plant (log-transformed) in *Sidalcea hirsuta* to cumulative rainfall for four growing seasons at Vina Plains Preserve. n = 30 or 32 for each year, as in Table 1: r^2 = 0.663, F = 116.92.

and seeds for 1983-1986 appearing in Table 1 and Figs. 1 and 2 are based on sample sizes of 30 and 32. Since the total number of plants in the population was dramatically higher in 1985 and 1986, about 300 additional fruit and seed samples were made in these two years. Means of these more expansive samplings were closely similar to, or lower than, the means presented (Table 1). Thus, in 1985, for fruits per plant: N = 30, $\bar{x}$ = 9.2 ± 14.6 (Table 1) versus n = 300, $\bar{X}$ = 7.8 ± 14.8. In 1986, for fruits per plant: n = 32, $\bar{x}$ = 43.3 ± 48.6 versus n = 306, $\bar{X}$ = 38.4 ± 48.2.

Similarly, seeds per fruit were sampled in 1985: n = 30, $\bar{X}$ = 6.5 ± 0.8 (Table 1) versus n = 300, $\bar{X}$ = 6.68 ± 1.3. In 1986, for seeds per fruit: n = 32, $\bar{x}$ = 7.4 ± 0.6 versus n = 306, $\bar{X}$ = 7.4 ± 0.7.

Fruits collected from proximal (lower and interior) and distal (branch-tip) portions of inflorescences differed significantly ($p < 0.0001$; t = 3.5863) in numbers of seeds per fruit (proximal: n = 50, $\bar{x}$ = 7.4 ± 0.6 versus distal: n = 50, $\bar{x}$ = 6.7 ± 1.2 seeds per fruit).

Total length of the *Sidalcea* population (length of a line through the plants tangent to the pool margin) varied little through the four years: 1983, 45 m; 1984, 38 m; 1985, 47 m; and 1986, 47 m.

OBSERVATIONS AND DISCUSSION
Amount and Timing of Precipitation

The data indicate that the amount of precipitation greatly affects the growth and fruiting of *Sidalcea hirsuta* at the Vina Plains Preserve. Many studies on vernal pools, or on vernal pool plants, have noted numbers of individuals of a species, or sizes of those individuals, fluctuating with "wet" and "dry" years. An early study (Purer, 1939) provided plant counts made during a very wet winter season and then some estimates of lower plant numbers in the succeeding year. Crampton (1959), who discussed abundance in *Orcuttia* species, noted "the amount of and consequent depth of standing water in the vernal pool is the most critical in their life cycle and reflects their yearly abundance." More recently, Zedler (1984) suggested that the distribution of standing water in time and space is the single most important factor affecting vernal pool plants. Recent studies such as that of Griggs (1980) on *Orcuttia*, have documented that populations change through time in relation to precipitation. Griggs sampled densities of several species of these grasses in pools in different parts of California over several years, and documented that these plants may not show up in pools at all in years of low rainfall. The present study, with counts of total individuals in one population of *Sidalcea hirusta* over four years, further documents wide fluctuation in population size, but indicates a lack of correspondence between population size and amount of rainfall (Table 1, Fig. 1).

Various components of rainfall other than total amount per season need to be identified as influencing responses by plants. Timing of the precipitation is most likely to be extremely important. Obviously other environmental factors, such as temperature, are also important. Responses to winter temperature and precipitation patterns are quite well studied for annuals of the California grasslands (e.g., Pitt and Heady, 1978), but there is very little information on this subject for the annuals of vernal pools.

Early rains during the fall, when temperatures are still high, can wet the soil but may not be heavy enough to produce standing water, or to maintain the moisture required by seedlings for continued growth into the winter months when rainfall is more certain. Rains in early September 1985 produced thousands of *Sidalcea* seedlings per square meter, but essentially none of these early seedlings survived through the dry, warm fall that year. On the other hand, early fall rains could be very advantageous to *Sidalcea*. Seed germination in September in response to early rains, which were followed by frequent rains throughout the winter, may have been responsible for the very large plants found in May-June of 1983. In 1983-84, rainfall was ample in November and December but was below average in both early (September-October) and later (January-June) parts of the growing season. In that year, population size was down, the number of fruits per plant was low, as was the number of seeds per fruit. Regular observations need to be made from the first rainfall that promotes seed germination, through the time that the soils are wetted to the water table and standing water is built up, to the time of final dry-down of the pool. It is now known that *Sidalcea hirsuta* starts growing with the first wetted soils and later grows part of the winter season in standing water.

Several studies emphasize the importance of standing water to vernal pool plants. A study analyzing species diversity (Holland and Jain, 1984) grouped pool species based on "depth of preference." To categorize individual species' responses to standing water, Zedler (1984) has described the distribution of pools plants along a moisture gradient which he termed the "water duration gradient." Along this gradient, plants ranged from those "upland" species rarely found within inundated areas, to those that occur within the longest inundation, mainly in bottoms of pools. *Sidalcea hirsuta*, which does not occur in the area he studied (San Diego County), appears to fit into the gradient where plants have physiological and morphological plasticity allowing them to withstand prolonged inundation (as in some species of *Downingia, Eryngium* and *Pogogyne* mentioned in this study). *Sidalcea hirsuta*, like several species mentioned by Zedler, produces submerged leaves that are different from the aerial leaves on the plants when they are in flower. *Sidalcea hirsuta* also fits Zedler's plant class (containing *Eryngium* and *Pogogyne*) that typically flower when the standing water has evaporated and the roots are drawing moisture from lower levels of soil. Zedler (1984) also emphasized that pool drying rates and soil moisture may help to determine the presence of a species. Greenwood (1984) noted "Overall it would seem that the length of the period in which the soil is saturated or nearly saturated is more critical than the length of time standing water occurs in the pond. Overall the duration of standing water may not be as crucial to the growth of individual plants as the availability of soil moisture in the root zone."

Peripheral Location in Pool

Sidalcea hirsuta in the Vina Plains area grows around the margins of large vernal pools that have standing water of long duration. These plants usually grow in standing water for part of the winter, but not in the deeper portions of the pool. Rather, they seem restricted to peripheral locations. Linhart (1972, 1988) emphasized that for vernal pool species, the periphery of the pool can be a more unpredictable environment than the center. Not only may peripheral plants have less water available than the plants of the same population closer to the center of the pool, but also the peripheral individuals may face abrupt or frequent changes in the soil water. Even daily fluctuations can occur at edges of smaller pools, with water levels falling after warm, sunny days and rising quickly after heavy rains. Linhart (1972) provided evidence for *Veronica peregrina* that there is a major genetic component to various differences that show up in peripheral versus central individuals in a population.

It might be instructive to study the progeny from the "outermost" and the "innermost" plants of this *Sidalcea* population, although the phenomena described by Linhart may not pertain to species growing peripherally in deep pools. In *S. hirsuta* the whole population is peripheral, occurring in this habitat that is more unpredictable. The most peripheral plants of the population are usually the smallest, while those plants growing deepest in the pool are usually larger and produce more fruits. The largest plants with the most fruits per plants occurred in 1983, when standing water lingered under nearly the entire population well into May (Table 1, Fig. 1). *Sidalcea hirsuta* has the ability to grow particularly large when standing water persists (the largest plant seen in 1983, not included in the plants sampled, bore 1421 fruits). Nevertheless, the plants do not extend more than a few meters into this large pool where there is some standing water nearer the center at least until May every year. The other major population of *S. hirsuta* at the Vina Plains Preserve is similarly peripheral, around a deep pool.

Some vernal pool species may be restricted to peripheral areas because of greater interspecific competition in other locations of the pool. Such competition is probably not occurring in *S. hirsuta*. Some plants of *Orcuttia pilosa* and *Eryngium vaseyi* var. *vallicola* do grow among the deepest-growing *Sidalcea* plants here, but these "innermost" plants of the *Sidalcea* population have considerable bare ground both among them and also extending from them inwards towards the abundant *Orcuttia* and *Eryngium* plants nearer the center of the pool. On the other hand, in the shallower areas of this *Sidalcea* population, grassland species (*e.g.*, *Bromus rubens*, *Hypochoeris glabra*) some times extend out among the sidalceas. These plants may well provide competition for water for the shallowest *Sidalcea* plants, especially during very dry springs like 1984 and 1985 (see Zedler, 1987, for discussion of shifts in non-pool and pool species in dry and wet years).

Another visible feature of the peripheral zone of the study pool is the coarse, cobbly substrate where most of the sidalceas grow. This cobbly area grades (toward the center) into a zone of more uniformly fine soils containing high clay content (Anita clay soil types according to Broyles, 1983) which crack deeply in the summer. Sidalceas are not found here. The nature of the peripheral substrate needs to be studied in relation to *Sidalcea* seed germination.

It may be that *Sidalcea hirsuta* is in part restricted to peripheral areas of this and similar deep pools with long-standing water because the seedlings can become established on wet mud, but not under water, and cannot tolerate standing water for long periods. Griggs (1981), Lin (1970), and Zedler (1987) have commented on differences in plant tolerance of the anaerobic environment that roots may encounter in mud under water. Seedlings of the outermost peripheral locations may be able to obtain enough oxygen for roots, as opposed to the seedlings in deeper water.

In addition to being peripherally located in this large pool, *Sidalcea hirsuta* occupied about the same length of pool over the four years. The only extension along the pool perimeter was seen in 1986, when the population was large and had a good mixture of small and large individuals. That spring, the total count was 16,695 plants, and 76 of these grew in a line extending north along the pool from the main area of the population. It is puzzling that the waves often occurring during winter in this large pool do not seem to cause movement and establishment of propagules around more of the pool (especially "downwind"). This lack of extension may be due either to seeds germinating and seedlings becoming rooted and anchored on the mud strictly before standing water is established, or to selective seeding in the sub-optimal areas.

Reproductive Ecology

Various features of the reproductive ecology and life cycle, especially those involving seeds and seedlings, need to be studied to more fully understand these results. For example, the lack of population expansion may relate to very limited dispersal fo the seeds, which have no apparent surface features associated with dispersal. Many of the single-seeded carpels remain within dried calyces. These may be flicked away if the plants are moved by animals or by the wind before the stems fall into the water in the winter months.

Information on seed germination may be especially important in understanding the fluctuations of population and seed crop sizes. Do the seeds germinate only on wet soil, or can they germinate under water, as is known for some vernal pool species? Early rains of about 54 mm from 7 to 11 September 1985 promoted germination of thousands of seeds below the dried upright stems, but standing water was not established until weeks afterward. Unfortunately these seedlings were not carefully followed to see how many died during the hot and dry period later in September. Similarly, seedlings that probably resulted after 70 mm of rain occurring from 1 to 24 September 1982 were not followed; it is probable that the large plants, with many fruits per plant, in the spring of 1983 (Table 1) were due in large part to the high total winter rainfall and also to this early rainfall, beginning in September (and remaining well-distributed through the winter).

It is not yet known if there is a long-term seed bank, or if most (or all) seedlings result from the last year's seed crop, as appears to be the case with the quickly germinating seeds of *Pogogyne abramsee* (Scheidlinger, 1984). Table 1 indicates that the seeds produced in any one crop could account for all the plants of the next season. Griggs (1980) has shown the importance of stored seeds in vernal pool soils for *Orcuttia* species and has also documented a necessity for symbiotic fungi to induce seed germination underwater in these plants. It may be that *Sidalcea hirsuta* seeds have neither long-term storage nor special germination requirements. Brief winter observations at the Vina Plains (*e.g.*, during the "dry" winter of 1983-84) showed seedlings of different sizes intermixed at highest levels within the peripheral zone occupied by *S. hirsuta*. This suggests that seeds can germinate in "waves," possibly in wet mud, but not under standing water. As pool waters evaporate in the spring, there are no newly-germinated *Sidalcea* seedlings, as is known for some vernal pool species. Griggs (1980) described waves of *Orcuttia* seeds germinating farther and farther out into the drying pool; this ultimately caused plants of very different stages of growth to occur within a distance of meters. Instead, in *S. hirsuta*, all plants flowered and set seed more or less synchronously, with the innermost plants in standing water or wet mud continuing to flower a few days longer than the outermost.

Young seedlings of *Sidalcea hirsuta* have rounded and unlobed cotyledons and lower leaves that contrast with the very deeply-dissected aerial leaves that appear later above water. This is not the striking polymorphy shown by three *Orcuttia* species (Griggs, 1981), where strap-shaped, floating aquatic leaves are succeeded by typical grass leaves as the pools dry down.

Mortality and survivorship of seedlings has not been studied for *S. hirsuta*. However, as has been noted widely for desert and Mediterranean annuals, most of the plants that grow to the flowering stage produce at least one fruit. Examination of both small and large plants indicates that almost all *Sidalcea* flowers that reach anthesis produce fruits. Flowers are protandrous. Moore (1971), who listed protandry for all five annual species of *Sidalcea*, also refered to the flowers as "heavily visited" by insects. At the Vina Plains, flowers were visited by a variety of insects, including small beetles, honeybees, small native bees, bumble bees, syrphid flies, and sphingid moths. The flowers were also thoroughly foraged in for pollen and nectar by large, black anthophorid bees that are probably the major pollinators here (Schlising, unpub. data).

Other features of reproductive ecology worthy of further study are variation among seed weights and seed numbers. Linhart (1972) has shown for *Veronica peregrina* that position of fruit on the plant can influence seed numbers (in that case, progressively fewer seeds from bottom to top). Proximal and distal fruits do have significantly different numbers of seeds in *S. hirsuta* (more in proximal fruits). Counts made for 1983-1986 may be biased because these counts were based on samples made in proximal portions of inflorescences, except in dry years when the randomly-selected plants sometimes had only one or two ("distal") fruits present. Perhaps more sophisticated methods of sampling are necessary to better estimate seeds per fruit or total seed production.

Scheidlinger (1984), among others, has noted that the ultimate fitness measure in a population of annual plants is seed production. The data presented for *Sidalcea hirsuta* are accompanied by many questions, but they clearly document fluctuation in total seed production over four years for this vernal pool annual. These data show that a hundred-fold difference in total seed crop is possible within a period of three growing seasons. Such seed crop fluctuation, in *S. hirsuta*, is based on plasticity in several responses of the plants. Seeds per fruit can vary significantly from year to year as can fruits per plant, in an even more dramatic fashion. Finally, as would be expected, number of fruiting individuals influences seed crop size. This study documents a 32-fold increase in fruiting individuals of this species in the same vernal pool over only two years.

ACKNLWLEDGEMENTS

The author thanks Diane Ikeda and Kent Clement for help with field sampling; Nancy Carter, Neil Schwertman, and Kay Schenk for help with statistics; Bob Groendyke for computer graphics; and Barbara Malloch Leitner of the Nature Conservancy for permission to conduct this study on the Conservancy's Vina Plains Preserve.

LITERATURE CITED

Alexander, D. G. 1976. Ecological aspects of the temporary annual pool fauna. *In* S. Jain [ed.], Vernal Pools, their Ecology and Conservation. Institute of Ecology Publication 9, University of California, Davis.

Broyles, P. F. 1983. A flora of the Nature Conservancy's Vina Plains Preserve, Tehama County, California. M. A. thesis, California State University, Chico.

Broyles, P. F. 1987. A flora of Vina Plains Preserve, Tehama County, California. Madroño 34: 213-231.

Crampton, B. 1959. The grass genera *Orcuttia* and *Neostapfia* : a study in habitat and morphological specialization. Madroño 15: 97-128.

Ebert, T. A., and M. L. Balko. 1984. Vernal pools as islands in space and time. In S. Jain and P. Moyle [eds.], Vernal Pools and Intermittent Streams. Institute of Ecology Publication 28, University of California, Davis.

Greenwood, N. 1984. The physical environment of series H, vernal pools in San Diego County, California. In S. Jain and P. Moyle [eds.], Vernal Pools and Intermittent Streams. Institute of Ecology Publication 28, University of California, Davis.

Griggs, F. T. 1980. Population studies in the genus *Orcuttia* (Poaceae). PhD thesis, University of California, Davis.

Griggs, F. T. 1981. Life histories of vernal pool annual grasses. Fremontia 9:14-17.

Holland, R. F., and S. K. Jain. 1977. Vernal pools. In M. Barbour and J. Major [eds.], Terrestrial Vegetation of California. Wiley-Interscience, New York.

Holland, R. F., and S. K. Jain. 1981. Insular biogeography of vernal pools in the Central Valley of California. Amer. Nat. 117:24-37.

Holland, R. F. and S. K. Jain. 1984. Spatial and temporal variation in plant species diversity of vernal pools. In S. Jain and P. Moyle [eds.], Vernal pools and intermittent streams. Institute of Ecology Publ. 28, University of California, Davis.

Hunter, J. E. 1986. Some responses of *Sidalcea calycosa* (Malvaceae) to fire. Madroño 33:305-307.

Jain, S. 1976. Some biogeographic aspects of plant communities in vernal pools. In S. Jain [ed.], Vernal Pools, their Ecology and Conservation. Institute of Ecology Publication 9, University of California, Davis.

Jain, S. 1984. Biosystematic and evolutionary studies in the genus *Limnanthes*: an update. In S. Jain and P. Moyle [eds.], Vernal Pools and Intermittent Streams. Institute of Ecology Publication 28, University of California, Davis.

Lathrop, E. W. , and R. F. Thorne. 1983. A flora of the vernal pools on the Santa Rosa Plateau, Riverside County, California. Aliso 10:449-469.

Lin, J. W. Y. 1970. Floristics and Plant Succession in Vernal Pools. M. A. thesis, San Francisco State College, California.

Linhart, Y. B. 1972. Differentiation within natural populations of California Annual Plants. PhD Thesis., University of California, Berkeley.

Linhart, Y. B. 1976. Evolutionary studies of plant populations in vernal pools. In S. Jain [ed.], Vernal Pools, their Ecology and Conservation. Institute of Ecology Publication 9, University of California, Davis.

Linhart, Y. B. 1988. Intrapopulation differentiation in annual plants. III. The contrasting effects of intra-and interspecific competition. Evolution 42: 1047-1064.

Martin, B. D., and E. W. Lathrop. 1986. Niche partitioning in *Downingia bella* and *D. cuspidata* (Campanulaceae) in the vernal pools of the Santa Rosa Plateau Preserve, California. Madroño. 33: 284-299.

McLaughlin, E. G. 1974. Autecological studies of three species of *Callitriche* native to California. Ecol. Monogr. 44: 1-16.

Moore, J. G. 1971. A systematic study of the annual species of *Sidalcea* (Malvaceae). M. A. Thesis, California State University, Chico.

Moran, R. 1984. Vernal pools in northwest Baja California, Mexico. In S. Jain and P. Moyle [eds.], Vernal pools and intermittent streams. Institute of Ecology Publ. 28, University of California, Davis.

Neder, J., W. Wasserman, and M. H. Kutner. 1985. Applied Linear Statistical Models, 2nd ed. Richard D. Irwin, Homewood, Illinois.

Ornduff, R. 1966. A biosystematic survey of the goldfield genus *Lasthenia* (Compositae: Heleniaeae). Univ. of Calif. Publ. Bot. 40:1-92.

Ornduff, R. 1976. Sympatry, allopatry, and interspecific competition in *Lasthenia*. In S. Jain [ed.], Vernal Pools, their Ecology and Conservation. Institute of Ecology Publ. 9, University of California, Davis.

Pitt, M. D., and H. F. Heady. 1978. Responses of annual vegetation to temperature and rainfall patterns in northern California. Ecology 59:336-350.

Purer, E. A. 1939. Ecological study of vernal pools, San Diego County. Ecology 20:217-229.

Rosario, J. A., and E. W. Lathrop. 1984. Distributional ecology of vegetation in the vernal pools of the Santa Rosa Plateau, Riverside County, California. In S. Jain and P. Moyle [eds.], Vernal Pools and Intermittent Streams. Institute of Ecology Publ. 28, University of California, Davis.

Scheidlinger, C. R. 1984. Population studies in *Pogogyne abramsii*. In S. Jain and P. Moyle [eds.], Vernal pools and Intermittent Streams. Institute of Ecology Publ. 28, University of California, Davis.

Schlising, R. A., and E. L. Sanders. 1982. Quantitative analysis of vegetation at the Richvale Vernal Pools, California. Amer. J. Bot. 69:734-742.

Schlising, R. A., and E. L. Sanders. 1983. Vascular plants of Richvale Vernal Pools, Butte County, California. Madroño 30 (Suppl.):19-30.

Stebbins, G. L. Ecological islands and vernal pools of California. In S. Jain [ed.], Vernal Pools, their Ecology and Conservation. Institute of Ecology Publ. 9, University of California, Davis.

Thorne, R. E. 1984. Are California's vernal pools unique? In S. Jain and P. Moyle [eds.], Vernal Pools and Intermittent Streams. Institute of Ecology Publ. 28, University of California, Davis.

U. S. Dept. Commerce. 1982-1986. Climatological data, annual summary. California. Vols. 86-90. Washington, D. C.

Weiler, J. H. 1962. A biosystematic study of the genus *Downingia*. PhD Thesis. University of California, Berkeley.

Zedler, P. H. 1984. Micro-distribution of vernal pool plants of Kearny Mesa, San Diego County. In S. Jain and P. Moyle [eds.], Vernal Pools and Intermittent Streams. Institute of Ecology Publ. 28, University of California, Davis.

Zedler, P. H. 1987. The ecology of Southern California vernal pools: a community profile. U. S. Fish Wildl. Serv. Biol. Rep. 85(7.11).

PLANT-ANIMAL INTERACTIONS

In the past there was a tendency among students of evolutionary biology to treat plants as completely self-sufficient, perhaps because they are autotrophic. More recently, plants often have been approached as if they were merely sessile animals. Fortunately, neither perspective holds at present. Most of the papers in this volume devoted to the ecology of plants touch upon the ecological and evolutionary impacts of animals upon plant biology. A brief overview of the articles illustrates this point. Animals are important pollinators of plants (Stebbins, Opler), and in the process affect seed set (Spira) and the genetic structure (Hamrick and Loveless) of populations. The absence of animal pollinators leads to the modification of breeding systems (Cruden and Lyon; Barrett; Cox). Conversely, dependence upon pollen and nectar for food affects the ecology of many animals.(Frankie et al.; Opler). Sometimes animal pollinators provide the selective impetus for very specific modifications in plant development and morphology (Stebbins). Seed dispersal is also an animal-mediated process where animals can have a detectable role in the organization of plant populations (Hamrick and Loveless; Linhart). Herbivory and parasitism are important challenges faced by plants, and they are evaluated from several perspectives in this section and elsewhere (e.g., Bock and Bock; Hendrix; Linhart; Spira; Turner and Pemberton; Uno). Animals are also discussed as anti-herbivore agents (Koptur; Turner and Pemberton).

Special attention is paid to insect herbivory in this section. There was a belief for some time that herbivory did not have a significant ecological or evolutionary effect upon plants, since nature was so green. This belief has been put to rest, thanks to a large body of contemporary literature that testifies to the ecological and evolutionary significance of herbivory upon plants. Recently, the argument has also been made that certain plants have mutualistic interactions with their herbivores, and that the plants may actually benefit from being eaten. Hendrix tests this in a series of careful experiments with several species of the Apiaceae (Umbelliferae). In this work he finds that these plants can respond to herbivore damage by shifting certain developmental patterns, but the overall effects of herbivory are unquestionably negative to the plants.

Chemical defenses against herbivory can be energetically expensive to the plant because they involve synthesis of compounds whose sole purpose is to counteract damage by herbivores, parasites, or disease-causing species. Certain chemical deterrents of herbivory are known to be either induced by the herbivore or produced in small quantities prior to injury. In her essay, Koptur asks if extra-floral nectaries may increase nectar secretion in order to attract predators and parasitoids of herbivores. Her results are ambiguous, but they lead Koptur to predict that herbivore-induced nectar production may be more common in annual than in perennial plant species.

Selective herbivory exerts a very powerful force upon plants. Because of their stationary nature, plants appear to be at the mercy of mobile hervibores. In fact, the very powerful selective forces exerted by herbivores and parasites upon plants have produced a wonderful variety of plant defense systems. In addition to chemical defenses, certain plants have been shown to use predatory animals for defense. The best documented group are Hymenoptera (Class Insecta), especially ants, which find food and shelter on a plant. They attack and deter other, potentially more destructive herbivores which approach their plants. Turner and Pemberton describe another class of predatory animals, mites (Class Arachnida), which act as defense agents. They document the various types of shelters or domatia in which the mites live while on the plants. Domatia of many shapes and arrangements occur in a number of unrelated plant genera and families.

FLORAL HERBIVORY AND THE APIACEAE: ANTAGONISTIC OR MUTUALISTIC INTERACTIONS?

Stephen D. Hendrix

The view that herbivory has a negative impact on plant fitness is supported by numerous studies (e.g., Morrow and LaMarche, 1978; Louda, 1984; Marquis, 1984; Kinsman and Platt, 1985). However, plants can react to herbivory during their lifespans via mechanisms hypothesized to compensate for decreases in fitness resulting from the loss of reproductive or vegetative tissue (Hendrix, 1979; McNaughton et al., 1983; Solomon, 1983; Paige and Whitham, 1987). The mechanisms involved in responses to herbivory can include shifts in the allocation of resources (Gifford and Marshall, 1973; Mooney and Gulmon, 1982), increases in chemical defenses (Rhoades, 1979; Carroll and Hoffman, 1980; Baldwin and Schultz, 1983), increases in extrafloral nectar production that attract ants (Koptur, this volume), increases in photosynthetic rates (Detling et al., 1979; Heichel and Turner, 1983), delayed senescence (Leopold, 1961), and changes in nutrient uptake (Chapin, 1980) and floral sex expression (Hendrix and Trapp, 1981). Both the variety of these responses and their occurrence in many plant families suggest that most, if not all, plants can respond to herbivory in one or more ways. Furthermore, these responses form the basis of a growing controversy concerning whether or not herbivory can increase or maximize plant fitness and thereby promote plant-herbivore mutualisms (Belsky, 1986; McNaughton, 1986).

Two important aspects of these responses which may shed light on the possibility that plant-herbivore interactions are mutualistic have yet to be adequately addressed. First, the degree to which responses to herbivory are a result of herbivore selection pressures is unknown. McNaughton (1984) hypothesized that the physiological responses resulting in compensatory growth or reproduction have evolved specifically as adaptations by plants facing predictable herbivore attack. However, it is also possible that some types of responses are a by-product of disturbances of normal patterns of resource allocation or hormonal control of growth and development. Thus, some plant species may respond to tissue loss regardless of the consistency of herbivore attack over evolutionary time. Second, the degree to which these responses offset losses in fitness, as measured by the number of offspring produced, is rarely determined (but see Paige and Whitham, 1987). Most studies on responses fo herbivory either do not address the question of fitness (Belsky, 1986) or the question is considerably complicated by the fact that the plants under study are perennial or reproduce vegetatively (McNaughton, 1979; Belsky, 1986).

Herein, I address these two aspects of plant responses to herbivory using the interactions of members of the Apiaceae and their flower-feeding herbivores as a model system. I use four species to address the first question: are the responses to floral herbivory in the Apiaceae the result of herbivore selection pressure? These species suffer from radically different levels of floral damage and differ in important characteristics unrelated to herbivore feeding pressures that may affect their ability to respond to damage. Thus, I can distinguish between herbivory and intrinsic factors as the causal agent of the responses. I use one species, *Pastinaca sativa* L., to address the second question: do compensatory responses offset the negative affects of herbivory?

MATERIALS AND METHODS
Natural History of Hosts and Herbivores

The four species of plants studied were *Pastinaca sativa*, *Heracleum lanatum* Michx., *Cicuta maculata* L., and *Zizia aurea* (Gray) Fern. (henceforth *Pastinaca*, *Heracleum*, *Cicuta*, and *Zizia*, respectively). All four species produce flowering stalks terminating in a primary (first) umbel. Secondary umbels terminate lateral branches. Large individuals of *Pastinaca*, *Cicuta*, and *Zizia* commonly produce tertiary umbels and occasionally *Pastinaca* and *Cicuta* produce quaternary umbels. Umbel orders develop in succession approximately 10-14 days apart. All fours species are andromonecious, and hermaphroditic flowers produce 2 one-seeded mericarps commonly referred to as seeds. Seeds of all species are dispersed by wind in late summer.

The four species differ in a number of characteristics, including life history and habitat (Table 1a). Also, the predominant sex of flowers in the primary umbel of *Pastinaca*, *Heracleum*, and *Cicuta* is hermaphroditic with the proportion of hermaphroditic to total flowers decreasing with successive umbel orders. In contrast, the proportion of hermaphroditic to total flowers in the primary umbel of *Zizia* is low, increases in secondary umbels, and then decreases in tertiary umbels, if they are produced.

Table 1. Summary of materials and methods. a) Natural history characteristics of host plants and floral herbivory and b) design of experiments testing responses to floral herbivory (see text).

Characteristic	*Pastinaca sativa*	*Heracleum lanatum*	*Cicuta maculata*	*Zizia aurea*
a) Natural history				
Life history	monocarpic perennial	polycarpic perennial	polycarpic perennial	polycarpic perennial
Habitat	disturbed roadsides and fields	open woodlands	wet prairies and meadows	wet prairies and meadows
Predominant sex of flowers in primary umbel	hermaphroditic	hermaphroditic	hermaphroditic	male
Floral herbivory	predictable	predictable	infrequent	infrequent
b) Experimental designs				
Location of experiments	greenhouse	field	field	field
Type of damage	natural	natural	artificial	artificial
Umbel order measured	tertiary	secondary	secondary	secondary

Floral herbivory varies considerably among the four species (Table 1a). Umbels of *Pastinaca* and *Heracleum* suffer predictable damage by *Depressaria pastinacella* (Duponchel) (Lepidoptera: Oecophoridae). Population levels of infestation range from 10-50% and whole umbels are often destroyed (S. Hendrix, personal observation). Umbels of *Cicuta* are attacked by *Depressaria cinereocostella* Clemens and *Papilio polyxenes*(Fab. (Lepidoptera: Papilionidae) (black swallowtail), but in Iowa both are rare (Hendrix and Marquis, in prep.). Furthermore, the amount of damage to *Cicuta* caused by *D. cinereocostella* is slight relative to damage to *Pastinaca* and *Heracleum* caused by *D. pastinacella*. *Zizia* is infrequently attacked by *P. polyxenes*. All these herbivores feed on the developing flowers and seeds, although black swallowtail larvae feed also on vegetative tissue.

Responses to Floral Herbivory

To make comparisons across species, only the primary umbel was removed. In *Pastinaca* and *Heracleum* this is the inflorescence most frequently attacked (Hendrix, 1979 and 1984). The effects of floral herbivory on *Pastinaca* (Hendrix, 1979; Hendrix and Trapp 1981), *Heracleum* (Hendrix, 1984), and *Cicuta* (Hendrix and Marquis, in prep.) were studied in 1977 and 1979, 1981, and 1984, respectively. The general methods (Table 1b) were similar to those described below for *Zizia*, except that in some species natural damage rather than artificial damage was used and some experiments were conducted under greenhouse conditions rather than in the field. Previous experiments with *D. pastinacella* indicated that herbivore and artificial damage do not differ in their effects (Hendrix and Trapp, 1981). In the experiments with *Cicuta* two temporal patterns of damage were used to simulate feeding by the two lepidopterans. Gradual damage simulated feeding by small larvae, such as *D. pastinacella*, and rapid damage simulated feeding by late instar black swallowtail larvae (Hendrix and Marquis, in prep.). The latter can consume an umbel in 24 hours.

For the study of *Zizia*, 30 individuals were randomly selected in a population located in Johnson Co., Iowa (T-80N, R-8W, SE 1/4 Section 5, Oxford Township) and then alternately assigned to either a control or experimental group in early spring, 1983. The experimental treatment consisted of removing the primary umbel by clipping off increasing numbers of umbellets from the primary umbel over a 11-13 day period beginning at the time of umbel expansion. This treatment simulates the typical temporal pattern and level of damage that occurs when insect herbivores such as *D. pasinacella* consume umbels. At anthesis, the numbers of staminate and hermaphroditic flowers per umbel were counted. Hermaphroditic flowers may be distinguished from staminate flowers by the presence of elongated styles. Three *Zizia* control plants and two experimental plants died before flowering was completed. Two additional experimental plants died after flowering but before seed set was completed. When the flowering stalk had completed elongation, the length of all above ground stem branches was measured as an indicator of plant size. Total stem length did not differ significantly between treatments (Student's t-test, $t = 1.15$, $df = 23$, $p > .20$. Seeds from all plants were collected at maturity in late July and counted.

Effects of Floral Herbivory on Fitness

In large individuals of *Pastinaca* (basal stem diameter of the flowering stalk > 12 mm), destruction of the primary umbel by *D. pastinacella* leads to complete compensation with respect to seed number (Hendrix, 1979; Hendrix and Trapp, 1981; see Fig. 4a, results). However, the mean biomass of tertiary seeds produced in response to herbivory is only 60% of the mean biomass of the primary seeds destroyed by feeding (Hendrix and Trapp 1981). Therefore, recruitment from primary and tertiary seeds may differ and the usual method of estimating fitness on the basis of seed number may be inaccurate.

To determine the effects of floral herbivory on fitness, I monitored the number of offspring recruited from primary, secondary and tertiary seeds. Duplicate experiments were begun in fall, 1982 and 1983. Different plants were used in each experiment because *Pastinaca* is monocarpic. Seeds were collected in late July from the primary, secondary, and tertiary umbels of ten undamaged plants and from the secondary and tertiary umbels of ten plants in which the primary umbel had been destroyed by *D. pastinacella*. One hundred and fifty 0.25 m^2 plots were established at the study site (Johnson Co., Iowa, T-80N, R-7W, SE 1/4 Section 25, Clear Creek Township). Plots were hand-tilled and leveled prior to sewing seeds. Nine hundred seeds from each umbel order of a parent plant were hand-sown in lots of 300 into each of three plots; assignment to plots was random. Ninety plots contained seeds from control plants (10 plants X 3 umbel orders X 3 replications) and 60 plots contained seeds from damaged plants (10 plants X 2 umbel orders X 3 replicates). Seeds were sown into plots in late August when natural seed dispersal is at its maximum. Plots were censused for seedling emergence and survival at two-week intervals the first year after seeds were sown (September - November; March - August), one month intervals in fall of the second year, and two week intervals in spring of the second year after seeds were sown. Less than 0.2% of seedling emergence occurred the third year after sowing (Hendrix and Trapp, in prep.).

RESULTS
Responses to Floral Herbivory

Table 2 summarizes the responses to floral herbivory of the four species with respect to 1) the percentage of hermaphroditic flowers produced per late-developing umbel, 2) percentage of hermaphroditic flowers per umbel setting seed, 3) seed set per late-developing umbel, and 4) seed set per plant. Three of the

310

Table 2. Summary of the responses of late-developing umbels and whole plants to destruction of the primary umbel in four species of Apiaceae. Each cell indicates response, type of statistical analysis, and significance of change.

Plant species

Response Characteristic	Pastinaca sativa	Heracleum lanatum	Cicuta maculata		Zizia aurea
	gradual damage	gradual damage	gradual damage	rapid damage	gradual damage
% hermaphroditic flowers/umbel	increased ANOVA $F = 1.18$ ***	increased ANOVA $F = 2.93$ *	no change Chi-square $\chi = 3.34$ ns	increased Chi-square $\chi = 10.27$ **	no change ANOVA $F = 0.15$ ns
% hermaphroditic flowers/umbel setting seed	no change ANOVA $F = 1.46$ ng	increased Mann-Whitney $U = 180$ ***	increased Chi-square $\chi = 10.98$ ***	increased Chi-square $\chi = 9.79$ **	no change ANOVA $F = 0.00$ ns
seed set/umbel	increased ANOVA $F - 26.22$ ***	increased t-test $t = 4.66$ **	increased Chi-square $\chi = 5.02$ *	increased Chi-square $\chi = 13.84$ ***	no change ANOVA $F = 0.35$ ns
total seed set/plant	no change t-test $t = 1.38$ ns	decreased t-test $t = 3.56$ **	decreased Chi-square $\chi = 16.76$ ***	no change Chi-square $\chi = 0.000$ ns	no change t-test $t = 0.07$ ns

*** $p < .001$
** $p < .01$
* $p < .05$

four umbellifers tested showed a significant increase in the percentage of hermaphroditic flowers in secondary or tertiary umbels following removal of the primary umbel (Table 2, Fig. 1). In tertiary umbels of *Pastinaca*, the percentage of hermaphroditic flower increased about 50% (Fig. 1a). and in secondary umbels of *Heracleum* the percentage of hermaphroditic flower number nearly doubled (Fig. 1b). Rapid damage to the primary umbel in *Cicuta* significantly increased the percentage of hermaphroditic flowers in secondary umbels but gradual damage did not (Fig. 1c). No tertiary umbels of *Cicuta* produced hermaphroditic flowers in any treatment. In *Zizia*, removal of the primary umbel did not significantly affect the percentage of hermaphroditic flowers in secondary umbels (Fig. 1d).

In two of the four species, the percentage of hermaphroditic flowers in late-developing umbels setting seed increased significantly following removal of the primary umbel (Table 2, Fig. 2). In control plants of *Heracleum*, secondary umbels set almost no seed while in experimental plants about 75% of the hermaphroditic flowers produced seeds (Fig. 2b). In *Cicuta* both gradual and rapid removal of the primary umbel resulted in about a doubling of the proportion of hermaphroditic flowers setting seed (Fig. 2c). However, fruit set in tertiary umbels of *Pastinaca* (Fig. 2a) and secondary umbels of *Zizia* (Fig. 2d) did not differ significantly between control and damaged plants.

The increase in the number of hermaphroditic flowers and/or proportion of these flowers setting seed in late-developing umbels resulted in significant increases in seed set per umbel in three of the four species (Table 2, Fig. 3). In *Pastinaca*, seed set by tertiary umbels doubled (Fig. 3a) and in *Heracleum*, seed set by secondary umbels increased 16-fold (Fig. 3b). Both gradual and rapid removal of the primary umbel in *Cicuta* resulted in increased seed set per secondary umbel (Fig. 3c). However, removal of the primary umbel of *Zizia* did not significantly alter seed set per secondary umbel (Fig. 3d.)

The total number of seeds produced by control plants and those with the primary umbel removed was statistically equivalent in three of the four species (Table 2, Fig. 4). In *Pastinaca*, control and damaged plants produced statistically equivalent number of seeds (Fig. 4a). Rapid damage to the primary of *Cicuta* did not significantly reduce seed number per plant but gradual damage did (Fig. 4c). In *Zizia* seed production by control plants was statistically equivalent to that of damaged plants (Fig. 4d). However, herbivore removal of the primary umbel in *Heracleum* significantly reduced total seed per plant by about 35% (Fig. 4b).

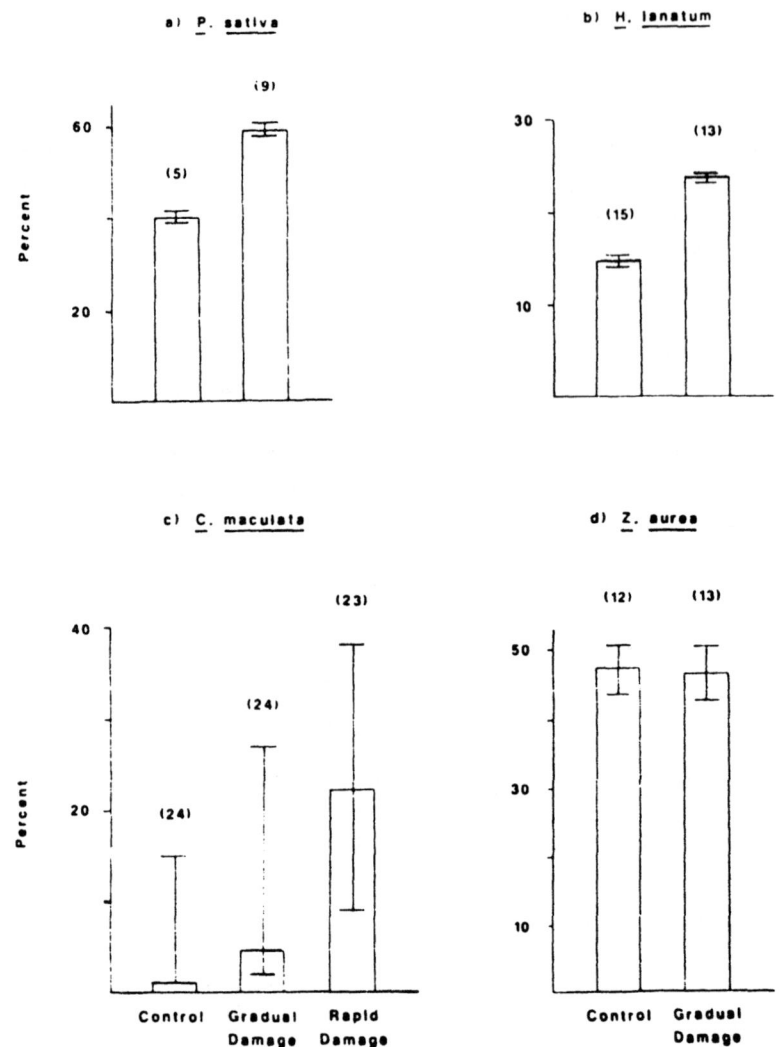

Figure 1. Effect of removal of the primary umbel on the percentage of hermaphroditic flowers per late-developing umbel. a) tertiary umbels of *Pastinaca sativa* (p < .001); b) secondary umbels of *Heracleum lanatum* (p < .05); c) secondary umbels of *Cicuta maculata* (control vs. gradual damage, n.s.; control vs. rapid damage, p < .01); d) secondary umbels of *Zizia aurea* (n.s.). Figures a, b, and d = mean of means per plant ± s.e.; c = median ± 25% and 75% quadriles of means per plant. Figure a redrawn from Hendrix and Trapp, 1981; Figure b redrawn from Hendrix, 1984; Figure c from Hendrix and Marquis, in prep. Values in parentheses represent number of plants sampled.

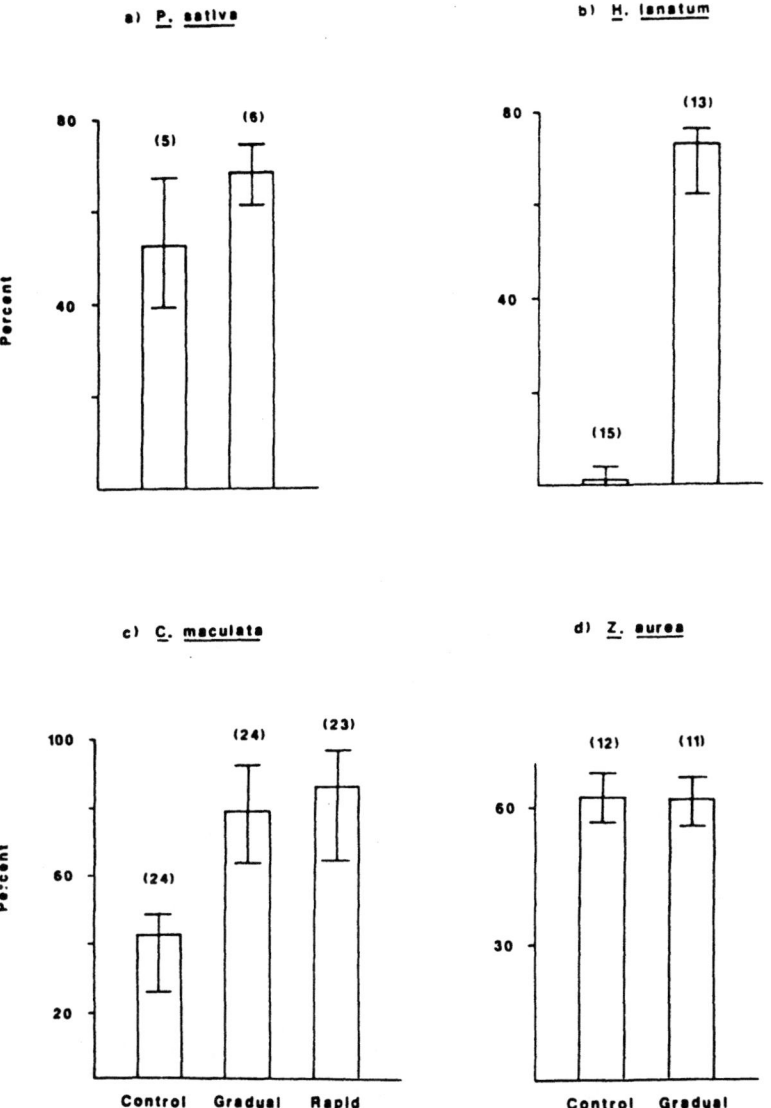

Figure 2. Effect of removal of the primary umbel on the percentage of hermaphroditic flowers per late-developing umbels setting seed. a) tertiary umbels of *Pastinaca sativa* (n.s.); b) secondary umbels of *Heracleum lnatum* (p < .001); c) secondary umbels of *Cicuta maculata* (control vs. gradual damage, p < .001; control vs. rapid damage, p < .01); d) secondary umbels of *Zizia aurea* (n.s.). Figure a redrawn from Hendrix and Trapp, 1981, figure c from Hendrix and Marquis, in prep; Figures a and d = mean of means per plant ± s.e.; figures b and c = median ± 25% and 75% quartiles of means per plant. See Figure 1 legend.

314

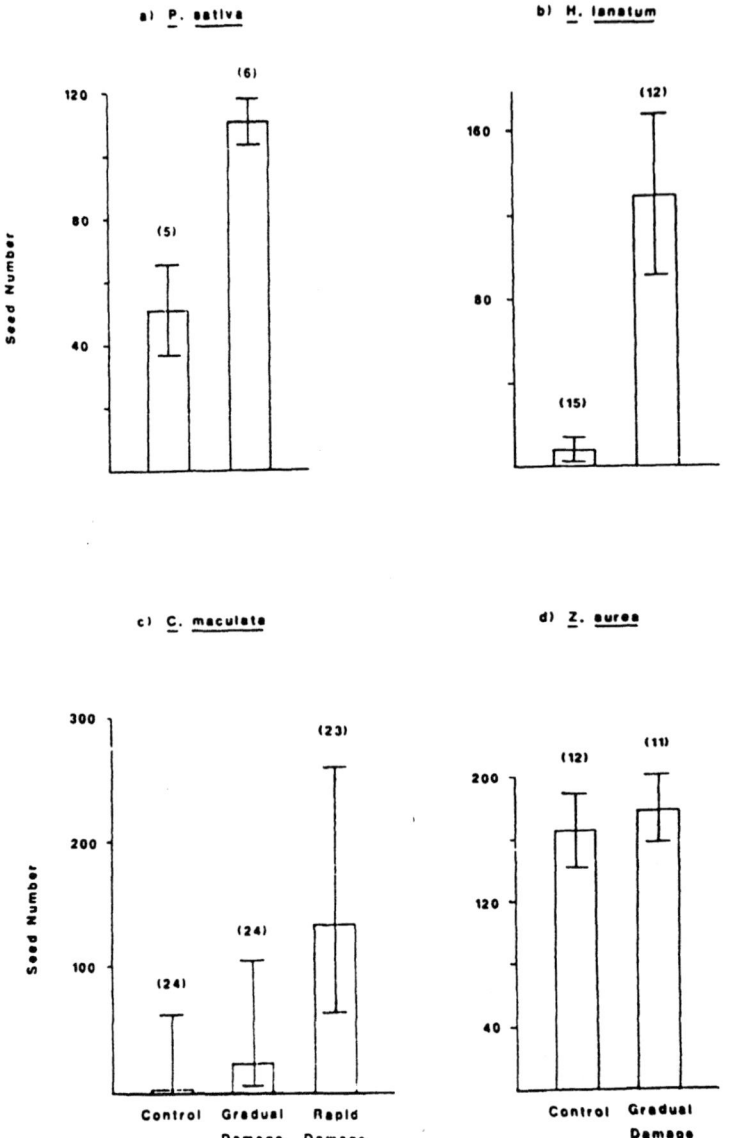

Fig. 3. Effects of removing the primary umbel on the number of seeds set per late-developing umbel. a) tertiary umbels of *Pastinaca sativa* (p < .001); b) secondary umbels of *Heracleum lanatum* (p < .01); c) secondary umbels of *Cicuta maculata* (control vs. gradual damage, p < .01; control vs. rapid damage, p < .001; d) secondary umbels of *Zizia aurea* (n.s.). See Figure 1 legend.

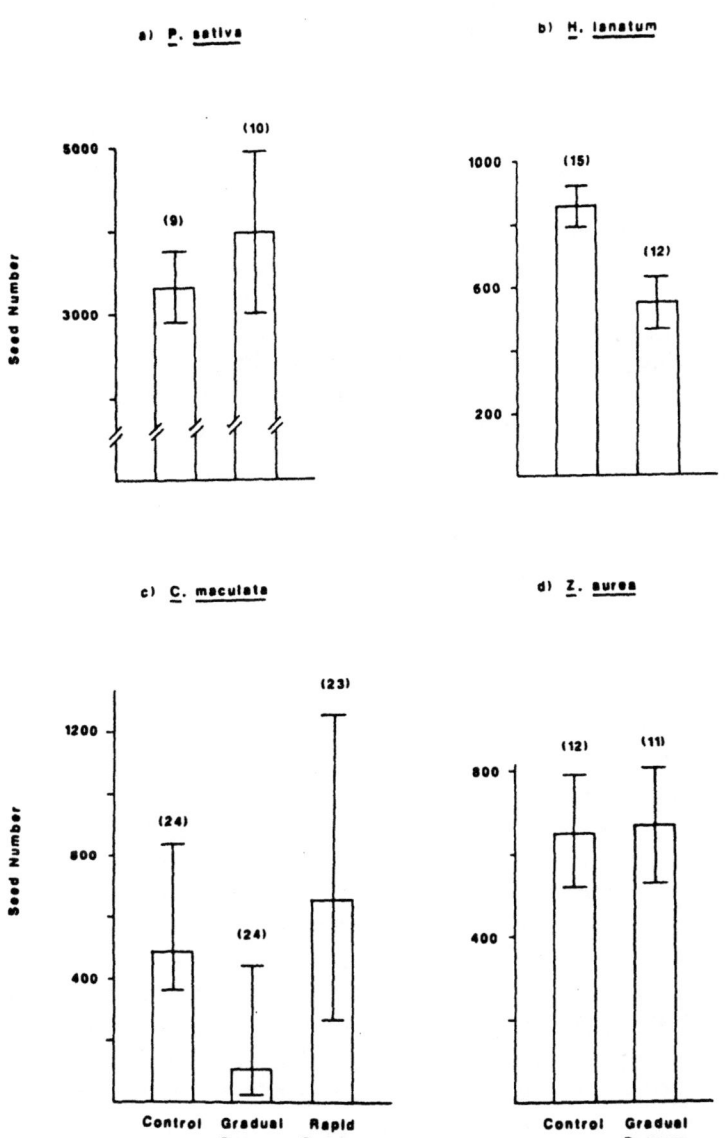

Figure 4. Effect of removal of the primary umbel on seed set per plant. a) *Pastinaca sativa* (n.s.); b) *Heracleum lanatum* (p < .01); c) *Cicuta maculata* (control vs. gradual damage, p < .001; control vs. rapid damage, n.s.); d) *Zizia aurea* (n.s.). Figure a redrawn from Hendrix, 1979. See Figure 1 legend.

Effects of Floral Herbivory on Fitness

Herbivore damage to the primary umbel did not significantly affect either the number of seeds produced by secondary umbels (Hendrix, 1979) or the recruitment characteristics of these seeds (Hendrix and Trapp, in prep.). Therefore, these results are not considered in this analysis. Also, the emergence and survivorship of seedlings from tertiary seeds (Hendrix and Trapp, in prep.) was unaffected by herbivory; and, in the statistical analysis reported here, recruitment from tertiary seeds of damaged and undamaged plants is combined. A more detailed analysis of population recruitment in *Pastinaca* will be reported elsewhere (Hendrix and Trapp, in prep.)

For seeds sown in the 1982 experiment, recruitment (number of plants alive/number of seeds sown) from primary seeds was significantly greater than that from tertiary seeds after two years (Student's t-test, $t = 4.71$ $P < 0.001$) (Table 3). However, in the 1983 experiment, recruitment from primary and tertiary seeds was statistically equivalent after two years (Student's t-test, $t = 0.78$, $P > 0.40$).

DISCUSSION
Responses to Floral Herbivory

For pairwise or diffuse co-evolution to occur, it is necessary that the interaction of the species or species groups result in the evolution of particular traits (Futuyma and Slatkin, 1983). It follows that if any given set of plant-herbivore interactions are a co-evolved mutalism, then the responses of host plants which offset the negative effects of herbivory should be the result of selection pressures by herbivores. The array of characteristics of the four species of umbellifers tested allows distinction between the hypotheses that the response(s) of these species to floral damage is due to either 1) herbivore selection pressure or 2) disruption of normal development patterns.

If herbivory itself is the causal agent behind the ability of these umbellifers to respond to herbivory, then the two species that suffer predictable attack (*Pastinaca* and *Heracleum*) should respond to removal of the primary umbel, while the two that suffer both infrequent and low levels of damage (*Cicuta* and *Zizia*) should not. Alternatively, if it is intrinsic developmental plasticity that determines the ability to respond to destruction of the primary umbel, then *Pastinaca, Heracleum,* and *Cicuta*, whose primary umbels contain mostly hermaphroditic flowers which ultimately produce seeds, should respond to floral herbivory. The species whose primary umbel contains mostly male flowers

(*Zizia*) should not because such umbels represent a relatively small resource sink compared to umbels that develop seed.

My results support the hypothesis that developmental plasticity following disruption of normal patterns of sex expression and resource allocation permits these umbellifers to respond to floral damage. Of the four species tested, only *Zizia* failed to respond to removal of the primary umbel (Table 2), indicating that in the Apiaceae the ability to offset floral damage is unrelated to the predictability and intensity of herbivory.

Alteration of resource allocation following artificial damage is common in many species (see Introduction) as are changes in sex expression in adromonoecious species (Hendrix and Trapp, 1981). Given that not all of these species suffer high, predictable levels of natural herbivory, it is likely that the ability to respond results from disruption of normal developmental patterns and physiological processes. In addition, other responses such as the ability of plants to increase photosynthetic rates, delay senescence, and alter nutrient uptake patterns, may be unrelated to herbivore selection pressure.

Although herbivore selection pressure does not affect a species' ability to react to removal of the primary umbel, the intensity of the responses may be increased by predictable herbivore attack. This is suggested by the fact that both *Pastinaca* and *Heracleum* responded to gradual destruction of the primary umbel by altering sex expression in late-developing umbels, but *Cicuta* did not. The failure to detect a shift in resource allocation in *Pastinaca* (Fig. 2a) may reflect small sample sizes and large variation within samples, rather than an inherent inability to respond to herbivory by shifting resources to tertiary umbels.

Effects of Herbivory on Fitness

If the interactions of the Apiaceae and their floral herbivores are mutualistic, it is also necessary that the fitness of damaged plants be greater than that of undamaged plants. Fitness, as measured by the number of seeds produced per plant is statistically equivalent in *Pastinaca, Zizia,* and *Cicuta* when the primary umbel is removed rapidly, but in none of these species is seed production by damaged plants significantly greater than that by undamaged plants (Fig. 4).

The possibility that the effects of floral herbivory on fitness may be neutral bears closer examination. In one of the *Pastinaca* recruitment experiments, recruitment from the smaller, tertiary seeds was statistically equivalent to that of primary seeds, indicating the effect of floral herbivory in some years may be neutral with respect to maternal fitness, through it is negative in others years (Table 3). However, when herbivores destroy the primary umbel, they consume large amounts of pollen, and in *Pastinaca* no compensation in pollen production occurs (Hendrix and Trapp, in prep.). Thus, paternal fitness is decreased.

Similarly, overall negative effects on fitness are likely to occur in *Cicuta* and *Zizia*. In *Cicuta*, the seeds produced in response to simulated herbivory were smaller than those destroyed, and in laboratory experiments, total germination of small seeds is less than that of large seeds (Hendrix and Sun, in prep.). Also, pollen is destroyed when the primary umbel is removed and it is likely that this pollen is not replaced. In *Zizia*, the primary umbel normally produces few, if any seeds, but as in *Pastinaca*, pollen production is probably reduced. Only one case of herbivory increasing plant fitness has been reported (Paige and Whitman, 1987), and it is likely that such cases are rare.

Despite the fact that fitness is reduced by floral damage in these species, the responses seen are important because they lessen the negative impact on fitness. In *Pastinaca*, *Heracleum*, and *Cicuta*, seed number per later-developing umbel increased significantly following removal of the primary umbel (Fig. 3), indicating that without any response to floral damage, the negative impact on plant fitness would have been even greater. Nevertheless, the interactions of these members of the Apiaecae and their flower feeding herbivores are antagonistic rather than mutualistic.

SUMMARY AND CONCLUSIONS

The ability of plants to respond to herbivory via a number of physiological mechanisms which may compensate for losses in fitness forms the basis of the hypothesis that some plant-herbivore interactions are mutualistic. If such interactions are mutualistic, then the responses to damage should be the result of herbivore selection pressure and fitness of damaged plants should be greater than that of undamaged plants. In the Apiaceae, the andromonoecious species *Pastinaca sativa*, *Heracleum lanatum*, and *Cicuta maculata* but not *Zizia aurea* respond to removal of their first inflorescence by increasing the percentage of hermaphroditic flowers in late-developing umbels and/or by increasing the percentage of hermaphroditic flowers in late-developing umbels that set seed. *Pastinaca*, *Heracleum*, and *Cicuta* have a common pattern of sex expression and subsequent resource allocation which differs from that of *Zizia*. However, only *Pastinaca* and *Heracleum* suffer significant and predictable floral damage in nature, suggesting that disruption of normal developmental patterns rather than herbivory determines a

Table 3. Recruitment, defined as numbers of flowering plants/number of seeds sown in *Pastinaca*. The numbers represent means of means per plant ± s.e. from primary and tertiary seeds two years after seeds were sown. Statistical comparisons made between seed origins within columns only. Values in parentheses represent number of plants sampled.

Seed origin	1982 Experiment	1983 Experiment
Primary Seeds (10)	1.37% ± 0.33	1.24% ± 0.34
	•••	n.s.
Tertiary Seeds (20)	0.22% ± 0.05	1.69% ± 0.39

•••$P < 0.001$

species' ability to respond to damage. In *Pastinaca*, population recruitment from undamaged and damaged plants did not differ significantly in one of two experiments, indicating that maternal fitness is not always reduced by herbivory. However, paternal fitness in *Pastinaca* is reduced because pollen destroyed by floral herbivores is not replaced. These results indicate that the interactions of the Apiaceae and their floral herbivores are antagonistic rather than mutualistic relationships.

ACKNOWLEDGMENTS

I wish to thank K. Brandt, C. Covert, K. Grove, R. Marquis, J. Metzer, T. Nielsen, P. Peller, R. Rhoades, I. Sun, and E.J. Trapp for assistance with field and laboratory work and/or with statistical analyses. The editors, J. Bock and Y. Linhart, as well as R. Cruden, T. Hegmann, and E. Nielsen provided many helpful comments on earlier versions of this chapter. This work was supported in part by NSF Grant BSR-8205543 and University of Iowa Summer Research Fellowships.

LITERATURE CITED

Baldwin, I. T., and J. C. Schultz. 1983. Rapid changes in tree leaf chemistry induced by damage: evidence for communication between plants. Science 221: 277-279.

Belsky, A. J. 1986. Does herbivory benefit plants? A review of the evidence. Amer. Nat. 127: 870-892.

Carroll, C. R., and C. A. Hoffman. 1980. Chemical feeding deterrent mobilized in response to insect herbivory and counteradaptation by *Epilancha tredecimnotata*. Science 209: 414-416.

Chapin, F. S. III. 1980. Nutrient allocation and response to defoliation in tundra plants. Arctic and Alpine Research 12: 553-563.

Detling, J. K., M. I. Dyer, and D. T. Winn. 1979. Net photosynthesis, root respiration and regrowth of *Bouteloua gracilis* following simulated grazing. Oecologia 41: 127-134.

Futuyma, D. J., and M. Slatkin. 1983. Introduction. In D. J. Futuyma and M. Slatkin [eds.]. Coevolution. Sinauer Associates, Inc., Sunderland, MA.

Gifford, R. M., and C. Marshall. 1973. Photosynthesis and assimilate distribution in *Lolium multiflorum* Lam. following differential tiller defoliation. Aust. J. Biol. Sci. 26: 517-526.

Heichel, G. H., and N. C. Turner. 1983. CO_2 assimilation of primary regrowth foliage of red maple (*Acer rubrum* L.) and red oak (*Quercus rubra* L.): response to defoliation. Oecologia 57: 14-19.

Hendrix, S. D. 1979. Compensatory reproduction in a biennial herb following insect defloration. Oecologia 42: 107-118.

Hendrix, S.D. 1984. Reactions of *Heracleum lanatum* to floral herbivory by *Depressaria pastinacella*. Ecology 65: 191-197.

Hendrix, S. D., and E. J. Trapp. 1981. Plant-herbivore interactions: insect induced changes in host plant sex expression and fecundity. Oecologia 49: 119-122.

Kinsman, S., and W. J. Platt. 1985. The impact of a herbivore upon Mirabilis hirsuta, a fugitive prairie plant. Oecologia 65: 2-6.

Koptur, S. 1989. Is extrafloral nectar production an inducible antiherbivore defense? In Y. Linhart and J. Bock [eds.]. The Evolutionary Ecology of Plants. Westview Pres, Boulder, CO.

Leopold, A. C. 1961. Senescence in plant development. Science 134: 1727-1732.

Louda, S. M. 1984. Herbivore effect on stature, fruiting, and leaf dynamics of a native crucifer. Ecology 65: 1379-1386.

Marquis, R. J. 1984. Leaf herbivores decrease fitness of a tropical plant. Science 226: 537-539.

McNaughton, S. J. 1979. Grazing as an optimization process: grass-ungulate relationships in the Serengeti. Amer. Nat. 113: 691-703.

McNaughton, S. J. 1984. Physiological and ecological implications of herbivory. In O. L. Lange, P. S. Nobel, C. B. Osmond, and H. Ziegler [eds.], Physiological Plant Ecology III, Encyclopedia of Plant Physiology, New Series, Vol. 12C: 657-677. Springer-Verlag, Berlin.

McNaughton, S. J. 1986. On plants and herbivores. Amer. Nat. 128: 765-770.

McNaughton, S. J., L. L. Wallace, and M. B. Coughenour. 1983. Plant adaptation in an ecosystem context: effects of defoliation, nitrogen, and water on growth of an African C4 sedge. Ecology 64: 307-318.

Mooney, H. A., and S. L. Gulmon. 1982. Constraints on leaf structure and function in reference to herbivory. Bioscience 32: 198-206.

Morrow, P. A., and V. C. LaMarche. 1978. Tree ring evidence for insect suppression of productivity in subalpine *Eucalyptus*. Science 201: 1244-1246.

Paige, K. N., and T. G. Whitham. 1987. Overcompensation in response to mammalian herbivory: the advantage of being eaten. Amer. Nat. 120: 407-416.

Rhoades, D. 1979. Evolution of plant chemical defenses against herbivores. In G. A. Rosenthal and D. H. Janzen [eds.]. Herbivores and their Interaction with Secondary Plant Metabolites. Academic press, NY.

Solomon, B. P. 1983. Compensatory production in *Solanum carolinense* following attack by a host-specific herbivore. J. Ecol. 71: 681-690.

IS EXTRAFLORAL NECTAR PRODUCTION AN INDUCIBLE DEFENSE?

Suzanne Koptur

INTRODUCTION

Extrafloral nectaries are plant glands located outside the flowers; they are widespread among the angiosperms, occurring in 68 of the 337 families, in 35 of 76 orders, 5 of 6 of the subclasses of dicotyledons, and 3 of the 4 subclasses of monocots (Elias, 1983). They occur on leaves, stipules, petioles, stems, bracts, sepals, and fruits (Bentley, 1977b), and their position is often related to their ecological role (see below). Nectaries are also found in some ferns, and may have similar ecological functions in ferns as in flowering plants (Koptur et al., 1982; Page, 1982; Lawton and Heads, 1984).

Evidence for ants visiting extrafloral nectaries and providing protection against herbivores is abundant and ever-increasing (reviewed by Bentley, 1977b; Buckley, 1982; Beattie, 1984; Jolivet, 1986). Ants visiting nectaries on vegetative parts may protect foliage (Bentley, 1976; Janzen, 1966, 1967; Koptur, 1979, 1984; Tilman, 1978; Kelly, 1986) which in turn can translate into greater seed set and increased fitness for the plant with nectaries (Koptur, 1979; Stephenson, 1982; Barton, 1986). Nectaries on or near reproductive structures can provide ant protection of developing ovules and seeds (Elias and Gelband, 1975; Bentley, 1977a; Deuth, 1977; Inouye and Taylor, 1979; Schemske, 1980, 1982; Keeler, 1981; Horvitz and Schemske, 1984).

Extrafloral nectaries can also attract predators or parasitoids of the herbivores (Leius, 1967; Price et al., 1980; Hespenheide, 1985; Weis and Abrahamson, 1985). Parasitoids can serve as important biological control agents in agricultural systems (Crepps, 1975; DeBach, 1964; Hassell, 1980, 1982). This sort of plant protection can be extremely important in natural systems as well, especially in areas where ants are not abundant (Keeler, 1985; Koptur, 1985).

Some plants can respond to defoliation by producing more defensive chemicals when attacked (Fowler and Lawton, 1985; Greig-Smith, 1986). The response of individual plants to defoliated areas with increased levels of feeding deterrents or toxins has been demonstrated in a number of species (Rottger and

Klingauf, 1976; Dixon and Barlow, 1979; Haukioja and Niemela, 1979; Wallner and Walton, 1979; Carroll and Hoffman, 1980; Barlow and Dixon, 1980; Edwards and Wratten, 1982, 1983, 1985; Haukioja 1982; Schultz and Baldwin 1982; Rhoades, 1983; Valentine et al. ,1983; Wratten et al.,. 1984; Karban and Carey, 1984; Raupp and Denno, 1984). The possibility that communication between plants can induce defenses in an undamaged neighbor in response to defoliation of another individual has been suggested, but not yet demonstrated with certainty (Rhoades, 1983; Baldwin and Schultz, 1983; Fowler and Lawton, 1985). If plant secondary chemicals can be mobilized to attacked organs, and their production increased soon after initial herbivore damage, then it seems likely that defense involving other metabolic products (water, sugars, amino acids, and other components of extrafloral nectar) could also respond to herbivore attack.

The "sap-valve" theory of extrafloral nectar secretion, which holds that nectaries are for eliminating "excess carbohydrates" and other compounds from plants, has been discounted in various investigations (D.A. Baker et al., 1978). Nectars contain a wider variety of sugars and amino acids, and in greater concentrations, than does simple phloem sap; therefore, the production of nectar involves some sort of active selection of constituents on the part of the plant. The constituents of floral nectars have been found to vary according to what pollinators frequent the plant species in question (Baker and Baker, 1983b), and are presumably the results of reciprocating evolutionary forces between plant and pollinator for the nectar composition that is most attractive to the visitor and most economical for the plant. Some plants have been found to have nectar of different compositions at different times of day; the nectars correspond to different guilds of visitors active at different times (Baker and Baker, 1983a). Some species of plants can respond to pollinator activity by secreting more nectar in flowers from which nectar has been removed (Cruden et al., 1983; Koptur, 1983).

The constituents of extrafloral nectars are similar, but not exactly the same as floral nectars of the same species (H.G. Baker et al., 1978). Few studies have examined the significance of the sugar ratios (Koptur, 1979; Smith, Lawton and Koptur, unpub. data), or amino acid complements (Koptur, 1979; Koptur, unpub. data), to ant visitors, but it has been demonstrated that the more concentrated the sugars in nectar, the more attractive they are to ants (Taylor, 1977, 1978), and that certain amino acids are preferred by certain ant species over others (Ricks and Vinson, 1970).

My goal is to test the hypothesis that plants with extrafloral nectaries will respond to damage by secreting more extrafloral nectar. Increased amounts of extrafloral nectar could lead to greater numbers of ants or parasitoids being recruited to the plant, which could lead to increased protection against herbivores. I have looked for extrafloral nectar induction in several systems, in which biotic protection by ants and/or parasitoids has been demonstrated experimentally.

METHODS

Vicia sativa L. (Fabaceae: Papilionoideae), the common vetch, is an annual herbaceous legume native to the Old World. The plants bear stipular nectaries, and in both exotic (Koptur 1979) and native (Koptur and Lawton, 1988) habitats, ants visiting these nectaries can provide protection against insect herbivores for foliage, flowers, and developing fruit. The flowers and fruits of most *Vicia* species are borne in the axils of the leaves, and in species with stipular nectaries (such as *V. sativa* and *V. sepium*) the flowers are sessile or on very short stalks, compared with species without nectaries (such as *V. cracca* and *V. hirsuta*) which have long peduncles. Nectar secretion in *V. sativa* begins when plants initiate anthesis, and continues through fruit maturation (Koptur and Lawton, 1988); therefore, the plants become established without the aid of biotic protection, but can benefit from ants and other nectary visitors throughout their entire reproductive life.

Fifty plants growing in a greenhouse at the University of York, England, were randomly divided into five groups of ten plants each. Initial nectar production was measured, using the technique developed by Irene and Herbert Baker for measuring small quantities: micropipettes pulled out to fine points (and therefore, uncalibrated) are used to draw up the nectar and spotted onto strips of filter paper; spot diameter can be correlated with volume (Baker, 1979). Using scissors to cut a fraction off of every leaflet, the plants were all defoliated to the level designated for the group (0%, 25%, 50%, 75% and 100% damage).

Nectar production was measured each afternoon for four days following the defoliations. Because this experiment was done in a controlled environment, it was not necessary to exclude nectary visitors or to bag plants. This experiment was initially attempted in the field, with much difficulty (rainy weather and difficulty in bagging small plants).

Ipomoea carnea (Convolvulaceae) is an emergent perennial morning-glory, growing in marshland in the lowland forests of

Costa Rica. Each leaf has 2 blotch nectaries at the base of the leaf blade on the abaxial surface near the petioles; the nectaries function on young and mature leaves on the upper part of both vegetative and reproductive shoots. Although standing in water to a depth of 0.5 m, most stems were occupied by black *Crematogaster* species of ants. Field studies of this (Keeler, 1977) and other *Ipomoea* species. (Keeler, 1980) have demonstrated benefit to these plants from ants visiting the foliar nectaries.

Ipomoea plants were studied with students from The Organization for Tropical Studies (O.T.S.) at Palo Verde, Guanacaste Province. We chose vegetative shoots from separate rhizomes and randomly designated plants to three groups of 20 plants each: 1) controls (no damage); 2) 25% of each leaf on the shoot removed with scissors; and 3) 50% of each leaf removed. Early in the morning, we excluded ants by shaking the plants and then coating the stem with tanglefoot (sticky resin). We also bagged the top ten or so leaves with pollen-tector heavy paper bags to exclude flying insects. We performed defoliations prior to bagging. In the afternoon (about 4 h. later) we measured nectar, using the spots on paper method.

After insignificant results were obtained in the first experiment, Mary Ann Lee and I repeated this experiment, using a design to control for variability. We chose 50 plants, excluded ants and bagged them early one morning. That afternoon, we collected accumulated nectar, tallied the amounts and ranked the plants in order of amount of nectar produced. Every second plant was designated experimental (to be defoliated 25%), the others were designated controls (no defoliation). The next morning we defoliated the experimentals (each leaf 25%) and rebagged the shoots. Each control was unbagged and rebagged. We returned to collect nectar in the afternoon.

Inga brenesii and *I. punctata* (Fabaceae: Mimosoideae) are tropical trees with foliar nectaries which function only on young and expanding leaves (Koptur, 1984), becoming dry when the leaves are fully hardened and mature. Both ants and parasitoids visiting these nectaries provide protection against insect herbivores in different situations (Koptur, 1984, 1985). Induction experiments were performed in May 1987 on these species in the field at Monteverde, Puntarenas Province, Costa Rica.

For each species, I chose 40 branches of new leaves (distributed on 8-10 trees) and designated 20 of these experimentals (25% defoliation) and 20 controls (no defoliation). On day 1, I excluded ants and bagged the branches. On day 2, I measured initial nectar accumulation, and then performed defoliations. On day 3, I measured nectar again.

RESULTS

Vicia. Variability within groups was high (Table 1); the results are summarized (means only) in Fig. 1. The only statistically significant difference was on day 1 after defoliation: analysis of variance showed that nectar production was higher with treatment ($p < .05$) and a posteriori tests between means showed the 25% and 50% treatments to be significantly different from the control group.

The effect disappeared the next day. With greater defoliation, there was no initial increase in nectar secretion over controls.

Ipomoea. The one-day experiment gave distributions of nectar secretion shown in Fig. 2; half the plants produced no nectar at all, making parametric statistics inappropriate. Because there were no differences between the groups (Kruskal-Wallis nonsignificant), we concluded it might be necessary to measure initial nectar secretion, use only actively secreting plants, and factor out pretreatment nectar production as a covariate.

The results of the two-day experiment (Table 2) showed that the two groups were virtually identical before the treatments (mean nectar volumes equal, due to our ranking of plants on the basis of nectar secretion). Although the mean nectar volume was higher in the experimental plants after defoliation, the differences were not significant by Student's t-test ($p < .05$).

Inga. *Inga brenesii* nectar volumes (Table 3) were very close in the two groups before the experiment, and just as close after the experiment (Kruskal-Wallis H not significant). Most leaves secreted nectar, but the data were not normally distributed, so this non-parametric test was used.

Inga punctata nectar volumes (Table 4) showed greater differences between groups, and changes in the direction of an inducible response, but variability was great, and no differences were significant. For both species of *Inga*, therefore, the null hypothesis is accepted.

DISCUSSION

Unpredictable weather and the difficulties of excluding ants and bagging vetch plants in the field led to the conducting of that experiment under controlled greenhouse conditions. This was the only experiment in which any significant results were obtained. Although it was much easier to exclude ants and bag the perennial

Table 1. *Vicia* nectar volumes from plants in greenhouse experiment, volumes are in microliters, $\bar{x} \pm$ s.d.

Days	Treatments				
	Control	25% damage	50%	75%	100%
0	.67 ± .37	.70 ± .62	.60 ± 57	.73 ± .56	.78 ± .68
1 *	.39 ± .41	.96 ± .40**	.58 ± .22	.52 ± .27	.40 ± .33
2	.20 ± .15	.37 ± .19	.37 ± .25	.33 ± .27	.25 ± .18
3	.17 ± .23	.35 ± .29	.19 ± .16	.15 ± .13	.20 ± .15
4	.20 ± .17	.24 ± .16	.23 ± .17	.18 ± .13	.17 ± .19

* Indicates only day in which there was any effect of treatment on nectar volume (by analysis of variance); ** - $p < .025$, * - $p < .05$.

• • • • •

Table 2. *Ipomea carnea* two day experiment. Nectar volumes produced in microliters ($\bar{x} +$ s.d.).

	Before	After
Experimental	.26 + .27	.56 + .59
Control	.26 + .28	.47 + .58
t-test difference	NS	t = .53, NS
	(n = 25 each)	

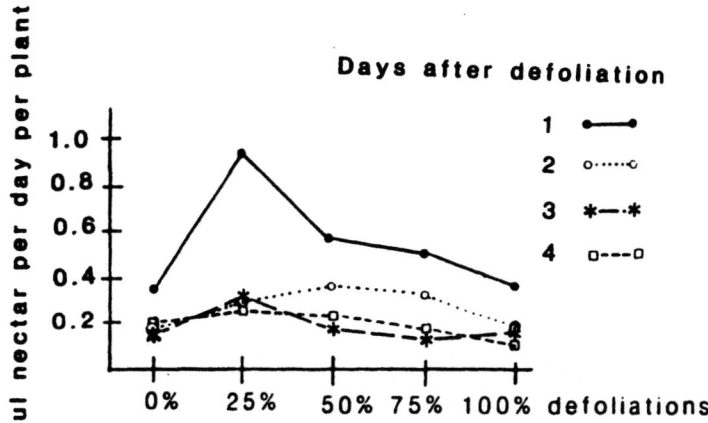

Figure 1. Nectar volumes from *Vicia* greenhouse experiment, to accompany Table 1. Only means were plotted. Note that each line depicts a day, with the treatment groups along the x-axis. The y-axis is mean nectar volume in microliters.

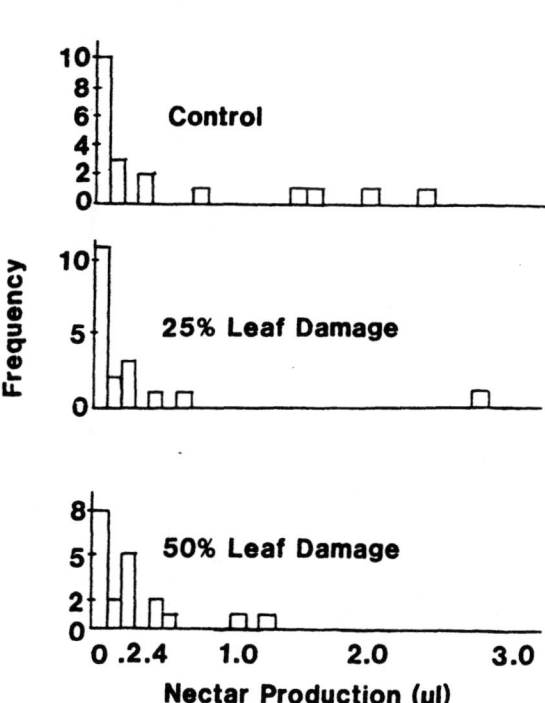

Figure 2. *Ipomoea carnea* one day experiment. Frequency distribution of nectar production for the different treatment groups.

Table 3. *Inga brenessi* nectar volumes in microliters.

	Before	After
Experimental	.16 + .09	.12 + .13
Control	.14 + .11	.10 + .12
Kruskal-Wallis H	.57	.25
probability	<.5 NS	<.7 NS

(n = 20 each)

• • • • •

Table 4. *Inga punctata* nectar volumes

	Before	After
Experimental	.04 + .08	.24 + .39
Control	.25 + .47	.18 + .48
Kruskal-Wallis H	.99	.30
probability	<.5 NS	<.3 NS

(n = 20 each)

woody plants in the field, any experiment conducted outside is subject to increased variability due to changes in ambient humidity, wind, and other factors. When small amounts of nectar were involved (the usual situation for extrafloral nectaries), environmental effects on variability were accentuated. This caveat must be borne in mind before the possibility of inducible nectar production is discarded for a given species. Under greenhouse conditions, an increase in extrafloral nectar production was seen only with moderate levels of defoliation. It may be that with greater leaf area loss, the reduction of photosynthate produced by a given leaf did not allow for increased nectar production. It has been suggested (Sarah Corbett, pers. comm.) that the response may simply be due to increased transpiration from cut leaf surfaces resulting in a sort of surge of liquid in the nectaries; I plan to test this by measuring floral nectar as well (since the flowers are situated in the axils of each leaf). If floral nectar increases when extrafloral nectar does, then perhaps it is only a physiological response associated with increased transpiration; if, however, floral nectar is not affected as is extrafloral nectar, the possible defensive role of this response will be supported.

I also plan to look for differences in nectar quality in response to defoliation. It is possible that, under stress, certain compounds change in concentration, and alter the attraction to biotic protective agents.

It is especially important to determine whether or not increased amounts of extrafloral nectar result in greater protection for the plant. This will be tested in field experiments in suitable systems. We will also look for this response in more systems where biotic protection has been found to be significant, in order to determine the general circumstances under which inducible extrafloral nectar production might occur.

At this point, I predict that herbaceous, annual plants will be more likely to show inducible nectar secretion than woody perennials. Annual species, which can reproduce only once, are known to expend more effort in reproduction than do perennials. They may also be able to economize on extrafloral nectar production, and produce more when herbivores are present. Plants which have extrafloral nectar as the sole reward for biotic protective agents will also be more likely to show this response than plants which offer a combination of rewards (*e.g.*, food bodies and domicile in addition to nectar). Only in situations where ants are not resident in plants can more ants be recruited to plants or to plant parts with increased nectar production. Less is known about the response of parasitoids to changes in nectar rewards, and so it is tempting to say that the inducible response will be more important in plants with ant protection than parasitoid protection resulting from extrafloral nectaries; but clearly, more information is needed.

Nectaries can play a role in agricultural systems in the biological control of crop pests (de Bach, 1964; Huffaker, 1971; Simmonds, 1971; Bentley, 1976). Ants and other insects attracted to extrafloral nectaries can reduce the effects of herbivores on crop plants (Bentley, 1977b, 1983; Risch and Carroll, 1982a, b), and although these interactions have not been exploited commercially, they may hold great potential. Recruiting natural protective agents can lessen the use of pesticides, reducing pollution and producing less-contaminated crops. A number of crop plants and their wild relatives (both temperate and tropical) have extrafloral nectaries: cotton (*Gossypium*) (Mound, 1962; Yokoyama, 1978; Adjei-maafo and Wilson, 1983), broad bean (*Vicia*) (Kupicha, 1976; Koptur, 1979), passionfruit (*Passiflora*) (Durkee, 1982; McLain, 1983; Smiley, 1986), peaches, plums, and cherries (*Prunus*) (Gregory, 1915; Dorsey and Weiss, 1920; Tilman, 1978), castor bean (*Ricinus*) (Reed, 1923; D.A. Baker et al., 1978), sweet potato (*Ipomoea*) (Beckmann and Stucky, 1981; Keeler, 1977, 1980), yam (*Dioscorea*) (Orr, 1923; Grout and Williams, 1980), rubber (*Hevea*) (Parkin, 1904, and buckwheat (*Polygonum*) (Salisbury, 1909). These traits could be utilized (or introduced from the genome of wild relatives into cultivars) for ecologically sound crop protection.

Stress on plants caused by insect herbivores has led to the evolution of many defenses, and the ability of some plants to regulate the amounts and locations of defensive substances is ecological fine-tuning that can permit adjustment of the individual to prevailing environmental conditions.

SUMMARY AND CONCLUSIONS

A number of chemical defenses of plants have been found to increase in response to presence of, or damage by herbivores, or both. Extrafloral nectaries attract ants and other beneficial insects that can provide biotic protection against a wide variety of herbivores. My goal was to test the hypothesis that plants predisposed to this type of defense (by virtue of having extrafloral nectaries) will respond to damage by secreting more extrafloral nectar. Increased amounts of nectar could serve to attract a greater number of ants or parasitoids and thereby increase biotic protection subsequent to incidents of damage.

I tested the hypothesis in three systems: 1) the annual, herbaceous legume *Vicia sativa*:; 2) the perennial woody morning glory, *Ipomoea carnea*; and 3) several *Inga* species, neotropical legume trees. Only in greenhouse experiments with *Vicia sativa* was the inducible response found: on the day following defoliation, plants defoliated to 25% and 50% levels secreted more extrafloral nectar on the average than control plants. Higher levels of defoliation (75% and 100% of the leaflets removed) did not induce any differences in nectar secretion. In field experiments with *Ipomoea* and two *Inga* species, there were no significant differences in nectar secretion between defoliated branches and controls.

More plants must be tested for this response before firm generalizations can be made. At this point, it appears that annual plants may be more likely to show inducible extrafloral nectar secretion than perennials, and herbaceous plants more than woody plants. A number of crop plants have extrafloral nectaries, and there is potential for use of this natural defense system in integrated pest management schemes. If plants are able to respond to varying levels of herbivory by adjusting their levels of extrafloral nectar secretion, this ability to economize when the defense is not needed can be an asset to increased growth and reproduction and, thereby, selective fitness.

ACKHOWLEDGMENTS

My first experiments with *Vicia sativa* were carried out as a project in Herbert Barker's Evolutionary Ecology class. Irene Baker helped with collections and measuring contents of extrafloral nectar. John Lawton encouraged me to do the first nectar induction experiments. Assistance and ideas also have been provided by Mary Ann Lee, Michael Blouin, Randy Snyder, Joe Slowinski, Kim Smart, Mary Rauchenberger (O.T.S. 85-3), and Jackie Pilliciotti. Fieldwork has been funded by the Latin American Caribbean Center of Florida International University, the Organization for Tropical Studies, and the National Science Foundation under a NATO postdoctoral fellowship awarded in 1983. Invaluable clerical support was provided by Nancy Koptur.

LITERATURE CITED

Adjei-maafo, I. K. K, and L. R. Wilson. 1983. Factors affecting the relative abundance of arthropods on nectaried and nectariless cotton. Environ. Entomol. 12: 349-352.

Baker, D. A., J. L. Hall, and J. R. Thorpe. 1978. A study of the extrafloral nectaries of *Ricinus communis*. New Phytol. 81: 129-137.

334

Baker, H. G., P. A. Opler, and I. Baker. 1978. A comparison of the amino acid complements of floral and extrafloral nectars. Botan. Gaz. 139: 322-332.

Baker, H. G., and I. Baker. 1983a. A brief historical review of the chemistry of floral nectar. In T.S. Elias and B.L. Bentley [eds.]. The Biology of Nectaries. Columbia University Press, New York.

Baker, H. G., and I. Baker. 1983b. Floral nectar sugar constituents in relation to pollinator type. In C. E. Jones and R. J. Little [eds.]. Handbook of Experimental Pollination Biology. Van Nostrand Reinhold, New York.

Baker, I. 1979. Methods for the determination of volumes and sugar concentration from nectar spots on paper. Phytochem. Bull. 12: 40-42.

Baldwin, I. T., and J. C. Schultz. 1983. Rapid changes in tree leaf chemistry induced by damage: evidence for communication between plants. Science 221: 277-279.

Barlow, N. D., and A. F. G. Dixon. 1980. Simulation of lime aphid population dynamics. Centre for Agricultural Publishing and Documentation. Wageningen, Netherlands.

Barton, A. M. 1986. Spatial variation in the effect of ants on an extrafloral nectary plant. Ecology 67: 495-504.

Beattie, A. J. 1984. The Evolutionary Ecology of Ant-Plant Mutualism. Cambridge University Press, London.

Beckmann, R. L., and J. M. Stucky. 1981. Extrafloral nectaries and plant guarding in Ipomoea pandurata (L.). G.F.W. May (Convolvulaceae). Amer. Jo. of Bot. 68(1): 72-79.

Bentley, B. L. 1976. Plants bearing extrafloral nectaries and the associated ant community: interhabitat differences in the reduction of herbivore damage. Ecology 57: 815-820.

Bentley, B. L. 1977a. The protective function of ants visiting the extrafloral nectaries in Bixa orellana (Bixaceae). J. Ecology 65: 27-38.

Bentley, B. L. 1977b. Extrafloral nectaries and protection by pugnacious bodyguards. Ann. Rev. of Ecol. and Syst. 8: 408-427.

Bentley, B. L. 1983. Nectaries in agriculture, with an emphasis on the tropics. In B. L. Bentley and T. S. Elias [eds.]. The Biology of Nectaries. Columbia University Press, New York.

Bueckley, R. 1982. Ant-Plant Interactions in Australia. Junk, The Hague.

Carroll, C. R., and C .A. Hoffman. 1980. Chemical feeding deterrent mobilised in response to insect herbivory and counteradaptation by *Epilachna tredecimnotata*. Science 209: 414-416.

Crepps, W. F. 1975. Influence of Specific Non-crop Vegetation on the Insect Fauns of Small-scale Agreoecosystems. M.S. thesis, , Department of Entomology, University of California, Davis.

Cruden, R. W., S. M. Hermann, and S. Peterson. 1983. Patterns of nectar production and plant-pollinator coevolution. In B. Bentley and T. S. Elias [eds.]. The Biology of Nectaries. Columbia University Press, New York.

DeBach, P. 1964. Biological Control of Insect Pests and Weeds. New York, Reinhold.

Deuth, D. 1977. The function of extrafloral nectaries in *Alphelandra deppeana* Schl. and Cham. (Acanthaceae). Brenesia 10/11: 135-145.

Dixon, A. F. G., and N. D. Barlow. 1979. Population regulation in the lime aphid. Zool. J. Linnean Soc. 67: 225-237.

Dorsey, M. J., and F. Weiss. 1920. Petiolar glands in the plum. Botan. Gaz. 69: 391-406.

Durkee, L. T. 1982. The floral and extrafloral nectaries of *Passiflora* II. The extra-floral nectary. Amer. J. Bot. 69: 1420-1428.

Edwards, P. J., and S. D. Wratten. 1982. Wound-induced changes in palatability in birch (*Betula pubescens* Ehrh. ssp. *pubescens*). Amer. Nat. 120: 816-818.

Edwards, P. J., and S. D. Wratten. 1983. Wound-induced defences in plants and their consequences for patterns of insect grazing. Oecologia (Berl.) 59: 88-93.

Edwards, P. J., and S. D. Wratten. 1985. Induced plant defences against insect grazing: fact or artifact? Oikos 44: 70-74.

Elias, T. S. 1983. Extrafloral nectaries: their structure and distribution. In B.L. Bentley and T.S. Elias [eds.]. The Biology of Nectaries. Columbia University Press, New York.

Elias, T. S., and H. Gelband. 1975. Nectar: its production and functions in the trumpet creeper. Science 189: 289-291.

Fowler, S. V., and J. H. Lawton. 1985. Rapidly induced defences and talking trees: the devil's advocate position. Amer. Nat. 126: 181-195.

Gregory, C.T. 1915. The taxonomic value and structure of peach leaf glands. Cornell Univ. Agri. Exper. Sta. Bull. 365: 183-322.

Greig-Smith, P. 1986. The trees bite back. New Sci., 1 May : 33-35.

Grout, B. W. W., and A. Williams. 1980. Extrafloral nectaries of *Dioscorea rotundata* Pior.: their structure and secretions. Ann. of Bot. 46: 255-258.

Hassell, M. P. 1980. Foraging strategies, population models and biological control: a case study. J. Animal Ecol. 49: 603-628.

Hassell, M. P. 1982. Patterns of parasitism by insect parasitoids in patchy environments. Ecolog. Entomol. 7: 365-377.

Haukioja, E. 1982. Inducible defences of white birch to a geometrid defoliator, *Epirrita autumnata*. In J. H. Visser and A. K. Minks [eds.]. Proc. 5th Internat. Symp. on Insect-Plant Relationships. Center for Agricultural Publishing and Documentation, Wageningen, Netherlands.

Haukioja, E., and P. Niemela. 1979. Birch leaves as a resource for herbivores: seasonal occurrence of increased resistance in foliage after mechanical damage of adjacent leaves. Oecologia 39: 151-159.

Hespenheide, H. A. 1985. Insect visitors to extrafloral nectaries of *Byttneria aculeata* (Sterculiaceae): relative importance of roles. Ecol. Entomol. 10: 191-204.

Horvitz, C. C., and D. W. Schemske. 1984. Effects of nectar-harvesting ants and an ant-tended herbivore on seed production of a neotropical herb. Ecology 65: 1369-1378.

Huffaker, C. B. 1971. Biological Control. Plenum Press, New York.

Inouye, D. W., and O. R. Taylor. 1979. A temperate region plant-and-seed predator system: consequences of extrafloral nectar secretion by *Helianthella quinquenervis*. Ecology 60: 1-8.

Janzen, D. H. 1966. Coevolution of mutualism between ants and acacias in Central America. Evol. 20: 249-275.

Janzen, D. H. 1967. The interaction of the bull's horn acacia (*A. cornigera* L.) with one of its ant inhabitants (*Pseudomyrmex ferruginea* F. Smith) in eastern Mexico. Kansas Univ. Sci. Bull. 47: 315-558.

Jolivet, P. 1986. Les Fourmis et les Plantes. Sociét´Nouvelle des Editions Boubée. Paris.

Karban, R., and J. R. Carey. 1984. Induced resistance of cotton seedlings to mites. Science 224: 53-54.

Keeler, K. H. 1977. the extrafloral nectaries of *Ipomoea carnea*. Amer. J. Bot. 64: 1182-1188.

Keeler, K. H. 1980. The extrafloral nectaries of *Ipomoea leptophylla* (Convolvulaceae). Amer. J. Bot. 67: 216-222.

Keeler, K. H. 1981. Function of *Mentzelia nuda* (Loasceae) postfloral nectaries in seed defense. Amer. J. Bot. 68: 295-299.

Keeler, K. H. 1985. Extrafloral nectaries on plants in communiteis without ants: Hawaii (USA). Oikos 44: 407-414.

Kelly, C. A. 1986. Extrafloral nectaries: ants, herbivores, and fecundity in *Cassia fasciculata*. Oecologia 69: 600-605.

Koptur, S. 1979. Facultative mutualism between weedy vetches bearing extrafloral nectaries and weedy ants in California. Amer. J. Bot. 66: 1016-1019.

Koptur, S. 1983. Flowering phenology and floral biology of *Inga* (Fabaceae: Mimosoideae). Systematic Bot. 8: 354-368.

Koptur, S. 1984. Experimental evidence for defense of *Inga* (Mimosoideae) saplings by ants. Ecology 65: 1787-1793.

Koptur, S. 1985. Alternative defenses against herbivores in *Inga* (Fabaceae: Mimosoideae) over an elevational gradient. Ecology 66: 1639-1650.

Koptur, S., A. R. Smith, and I. Baker. 1982. Nectaries in some neotropical species of *Polypodium* (Polypodiaceae): preliminary observations and analyses. Biotropia 14: 108-113.

Koptur, S., and J. H. Lawton. 1988. Interactions among betches bearing extrafloral nectaries, their biotic protection agents, and herbivores. Ecology 69: 278-283.

Kupicha, F. K. 1976. The infrageneric structure of *Vicia*. Notes from the Royal Botanic Garden, Edinburg, 34: 287-326.

Lawton, J. H., and P. Heads. 1984. Bracken, ants and extrafloral nectaries. I. The components of the system. J. Animal Ecol. 53(3): 995-1014.

Leius, K. 1967. Influence of wild flowers on parasitism of tent caterpillar and codling moth. Canad. Entomol. 99: 444-446.

McLain, D. K. 1983. Ants, extrafloral nectaries and herbivory on the passion vine, *Passiflora incarnata*. Amer. Midl. Nat. 110: 433-439.

Mound, L. A. 1962. Extrafloral nectaries of cotton and their secretions. Empire Cotton Growing Rev. 39: 254-261.

Orr, Y. 1923. The leaf glands of *Dioscorea maroura* Harms. Notes Royal Bot. Gard. Edinburgh 14: 57-72.

Page, C. N. 1982. Field observations on the nectaries of bracken, *Pteridium aquilinum*, in Britain. Fern Gaz. 12: 233-240.

Parkin, J. 1904. The extrafloral nectaries of *Hevea brasiliensis* Muell. Arg., an example of bud scales serving as nectaries. Ann. of Bot. 18: 217-226.

Price, P. W., C. E. Bouton, P. Gross, B. A. McPheron, J. N. Thompson, and A. E. Weis. 1980. Interactions among three trophic levels: influence of plants on interactions between insect herbivores and natural enemies. Ann. Rev. of Ecol. and Syst. 11: 41-65.

Raupp, M. J., and R. F. Denno. 1984. The suitability of damaged willow leaves as food for the leaf beetle, *Plagiodera versicolora*. Ecol. Entomol. 9: 443-448.

Reed, E. L. 1923. Extrafloral nectar glands of *Ricinus communis*. Botan. Gaz. 76: 102-106.

Rhoades, D. F. 1983. Responses of alder and willow to attack by tent caterpillars and webworms: evidence for pheronomal sensitivity of willows. In P. A. Hedin, [ed.]. Plant Resistance to Insects. American Chemical Society, Washington, D.C.

Ricks, B. L., and S. B. Vinson. 1970. Feeding acceptability of certain insects and various water soluble compounds to two varieties of the imported fire ant. J. Econ. Entomol. 63: 145-148.

Risch, S. J., and C. R. Carroll. 1982a. The ecological role of ants in two Mexican agroecosystems. Oecologia 55: 114-119.

Risch, S. J., and C. R. Carroll. 1982b. Effect of a keystone predaceous ant, *Solenopsis geminata*, on arthropods in a tropical agroecosystem. Ecology 63: 1979-1983.

Rottger, V. U., and F. Klingauf. 1976. Anderung im stoffwechsel von zuckerrubenblattern durch befall mit *Pegomya betae* Curt. (Muscidae: Anthomyidae). Zeitsch. Angewandt Entomol. 83: 220-227.

Salisbury, E. J. 1909. The extrafloral nectaries of the genus *Polygonum*. Ann. of Bot. 23: 229-241.

Schemske, D. W. 1980. The evolutionary significance of extrafloral nectar production by *Costus woodsonii* (Zingiberaceae): an experimental analysis of ant protection. J. Ecol. 68: 959-967.

Schemske, D. W. 1982. Ecological correlates of a neotropical mutalism: ant assemblages at *Costus* extrafloral nectaries. Ecology 63: 932-941.

Schultz, J. C., and I. T. Baldwin. 1982. Oak leaf quality declines in response to defoliation by gypsy moth larvae. Science 217: 149-151.

Simmonds, F. J. 1971. Biological control of pests. Tropical Science 12: 191-201.

Smiley, J. 1986. Ant constancy at *Passiflora* extrafloral nectaries: effects on caterpillar survival. Ecology 67: 516-521.

Stephenson, A. G. 1982. The role of extrafloral nectaries of *Catalpa speciosa* in limiting herbivory and increasing fruit production. Ecology 63: 663-669,.

Taylor, F. 1977. Foraging behavior of ants: experiments with two species of myrmecine ants. Behav. Ecol. and Sociobiol. 2: 147-167.

Taylor, F. 1978. Foraging behavior of ants: theoretical comsiderations. J. Theoret. Biol. 71: 541-565.

Tilman, D. 1978. Cherries, ants, and tent caterpillars: timing of nectar production in relation to susceptibility of caterpillars to ant predation. Ecology 59: 686-692.

Valentine, H. T., W. E. Wallner, and P. M. Wargo. 1983. Nutritional changes in host foliage during and after defoliation and their relation to the weight of gypsy moth pupae. Oecologia 57: 298-302.

Wallner, W. E., and G. S. Walton. 1979. Host defoliation: a possible determinant of gypsy moth population quality. Ann. of the Entomol. Soc. of Amer. 72: 62-67.

Weis, A. E. and W. G. Abrahamson. 1985. Potential selective pressures by parasitoids on a plant herbivore interaction. Ecology 66: 1261-1269.

Wratten, S. D., P. J. Edwards, and I. Dunn. 1984. Wound-induced changes in the palatability of *Betula pubescens* and *B. pendula*. Oecologia 61: 372-375.

Yokoyama, V. Y. 1978. Relation of seasonal changes in extrafloral nectar and foliar protein and arthropod populations in cotton. Environ. Entomol. 7: 799-802.

LEAF DOMATIA AND MITES: A PLANT PROTECTION-MUTUALISM HYPOTHESIS

Charles E. Turner and Robert W. Pemberton

INTRODUCTION

We first observed leaf domatia and associated predaceous mites while examining a then-unknown ornamental dogwood (*Cornus capitata* Wall.) from the Himalayas growing in Berkeley, California. We observed small pocket-like structures beneath the junctions of primary and secondary veins on the undersides of leaves from this tree, and what appeared to be predaceous mites within and running between these structures. Stimulated by this observation and curiosity about whether there might be some relationship between mites and these structures, we began to examine other plant species for their occurrence. Thus began our studies on a relatively unknown yet potentially significant relationship between mites and plants with leaf domatia.

Lundström (1887) introduced the term domatium (from a Greek word for small house) for a nonpathological plant structure that harbors other organisms beneficial to plants, and in contrast to structures such as galls that contain organisms harmful to plants. Lundström termed the domatia on plant leaves that he observed frequented by mites acarodomatia or mite domatia. The term domatium has also been applied to various plant cavities used by ants as nest sites (Lundström, 1887; Bequaert, 1922; Wheeler, 1942; Beattie, 1985). These myrmecodomatia or ant domatia occur on stems and roots as well as on leaves of the plants that bear them (Beattie, 1985). Though myrmecodomatia are typically larger than acarodomatia, the final arbiters of mite versus ant domatia are the animals themselves. Our use of the unspecified term domatium is limited to probable mite domatia on plant leaves.

Domatia are typically various small enclosing structures at the axils of veins on the undersides of leaves. Although different classification schemes have been devised (Lundström, 1887; Penzig and Chiabrera, 1903; Stace, 1965; Jacobs, 1966; Metcalfe and Chalk, 1979), most axillary domatia may be characterized as one or a combination of the following types: pits - cavities sunken into the laminar tissue (Figs. 1,4-5); pockets - pocketlike cavities beneath the sometimes extended vein tissue (Figs. 2,6); and tufts - dense tufts of hairs (Figs. 3,7). Hairs

342

Figure 1. Abaxial leaf surface of *Coprosma baueri* (Rubiaceae) with pit domatia. **Figure 2.** Abaxial leaf surface of *Distictis buccinatoria* (Bignoniaceae) with pocket domatia. **Figure 3**. Abaxial leaf surface of *Quercus agrifolia* (Fagaceae) with tuft domatia. Scale bars = 10 mm.

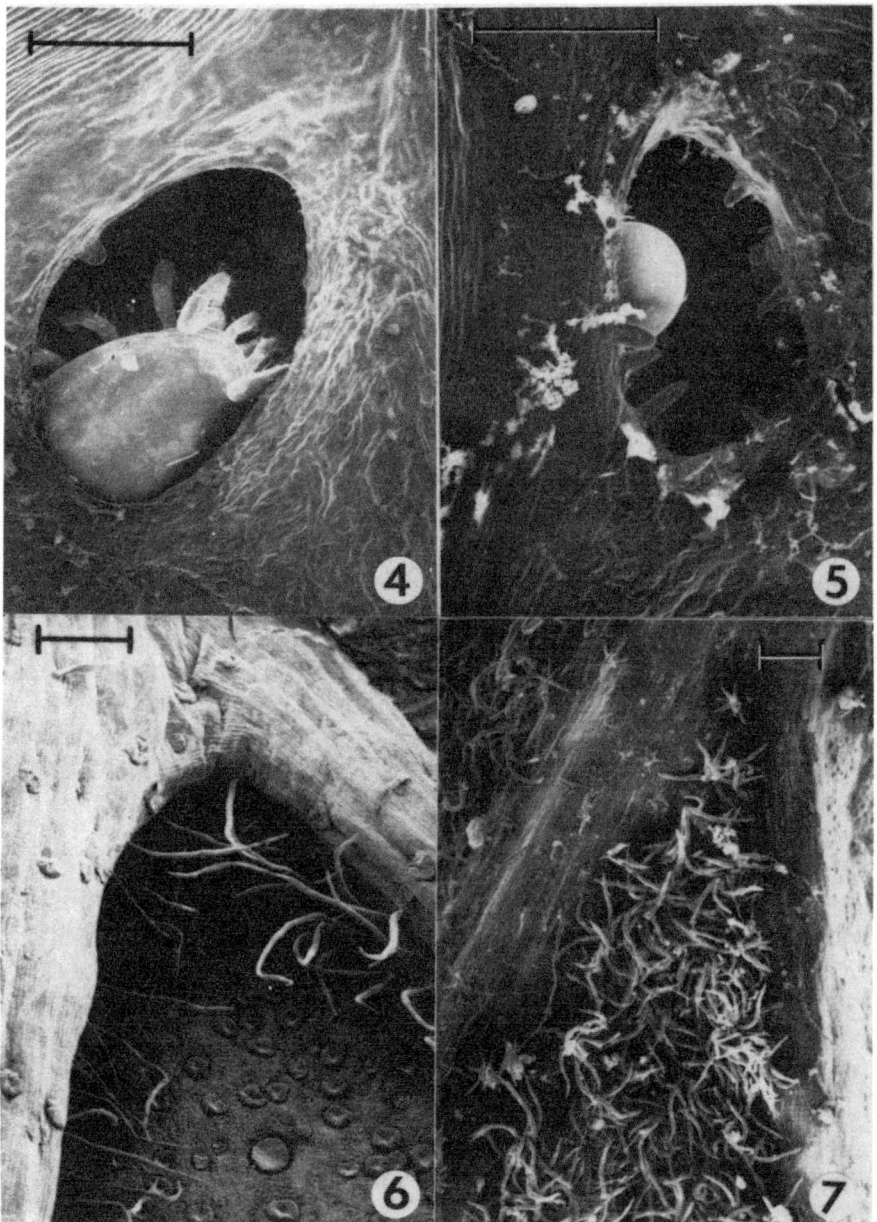

Figure 4. SEM of the predaceous mite *Amblyseius limonicus*
(Phytoseiidae) inside a pit domatium of *Coprosma baueri*
(Rubiaceae). **Figure 5.** SEM of an egg of the predaceous mite
Amblyseius limonicus inside a pit domatium of *Coprosma baueri*.
Figure 6. SEM of pocket domatium of *Distictis buccinatoria*
(Bignoniaceae). **Figure 7.** SEM of a tuft domatium of *Quercus
agrifolia* (Fagaceae). Scale bars = 250 μm.

are also often associated with pit and pocket type domatia. Some workers distinguish between pits that are invaginations into the mesophyll and pits that are formed by a raised dome-like structure (Jacobs, 1966). Nonaxillary folds - folded or rolledleaf margins, are also considered domatia by some workers (Lundström, 1887; Bequaert, 1922; Penzig and Chiabrera, 1903; Stace, 1965). Fold domatia are less common than the axillary domatia types. Axillary domatia most frequently occur at primary vein axis (between approximately equal primary veins or between primary and secondary veins), but also occur at the axils of secondary, tertiary and quaternary veins in some plant species.

As noted by Lundström (1887), domatia occur in a diverse array of woody dicotyledonous species. Domatia are heritable (Adamoli de Barros, 1960; Jacobs, 1966; To Ngoc Anh, 1966; Brouwer, 1979), and vary in number and form within and between individual plants of the same species, and with environmental conditions (Lundström, 1887; Hamilton, 1896; Admoli de Barros, 1963a,b; Jacobs, 1966). Lundström (1887) thought that mite presence was necessary for the development of domatia. Since that time other workers (Hamilton, 1896; Adamoli de Barros, 1960; Jacobs, 1966) have demonstrated that mites are not necessary for the development of domatia.

After briefly considering alternative hypothesis, Lundström (1887) proposed a mutualistic relationship between mites and plants with domatia, where mites that inhabit domatia feed on harmful organisms, particularly fungi on plant leaves. (Lundström also proposed that domatia might benefit plants through the absorption of excreted nitrogen or respired carbon dioxide from mites in domatia). Lundström did not offer much evidence to support his concept nor have others subsequently. Since Lundström's initial report (1887), mites and mite eggs have been observed in the domatia of many plants (Lagerheim, 1892; Hamilton, 1896; DeFonzo, 1897; Penzig and Chiabrera, 1903; Chevalier and Chesnais, 1941; Adamoli de Barros, 1966), but there have been few identifications of mites in domatia or attempts to elucidate their food habits. For various reasons many workers have not accepted Lundström's concept of a mite-domatia mutualism (Hamilton, 1896; Greensill, 1902; Shirley and Lambert, 1923; Chevalier and Chesnais, 1941; Jacobs, 1966; Schnell, 1970). The reasons for the lack of acceptance of Lundström's mutualistic hypothesis by these workers include a relative lack of supporting evidence, a lack of knowledge of the importance of predaceous mites, and the findings that not all leaves or individuals of domatia-bearing species possess domatia,

domatia do not always contain mites, and mites are not necessary for the development of domatia. A philosophical predisposition against a functional interpretation of domatia apparently also contributed to the lack of acceptance of Lundström's hypothesis by some of these workers (Jacobs, 1966).

To this day there has been little evidence provided on this question, and the ecological significance of domatia remains little studied or understood. One line of evidence necessary for an assessment of the significance of domatia is a thorough quantitative study of whether potentially beneficial mites are associated with domatia. In 1983 we initiated such a study (Pemberton and Turner, 1989) which is summarized next.

MITE-DOMATIA SURVEY STUDY
Methods

We sampled the domatia of 32 plant species selected to represent a diversity of taxonomic affinities (18 families: Anacardiaceae, Annonaceae, Aquifoliaceae, Bignoniaceae, Caprifoliaceae, Celastraceae, Combretaceae, Cornaceae, Elaeocarpaceae, Euphorbiaceae, Fagaceae, Hamamelidaceae, Lauraceae, Melastomataceae, Meliaceae, Oleaceae, Rubiaceae, Tiliaceae), places of origin (North, Central and South America, Europe, Africa, Asia and various Pacific islands), and domatia types (pits, pockets, tufts and folds) (Pemberton and Turner,1989). The plant species sampled were growing in natural and garden habitats in California, Costa Rica, and Hawaii. Some plant species were sampled in more than one area. For each sample, 10 leaves were removed for study. For each sample leaf, the domatia were described and counted, the mites and other arthropods within domatia were counted, collected and preserved for identification in alcohol, and mite eggs were counted.

Results

The results (Pemberton and Turner, 1989) of this survey are shown in Table 1. The domatia of 35 of 36 (97.2%) samples and 31 of 32 (96.8%) plant species sampled contained mites. Only *Liquidambar styraciflua* L. had no mites in its domatia, although we have since observed mites in the domatia of other nonsample plants of this species. Twenty six of 36 (72.2%) samples and 24 of 31 (77.4%) plant species had mite eggs in domatia. Mites were found in 28% of the approximately 2600 total domatia sampled, and almost half (48%) of the sampled leaves had some domatia occupied by mites. The mean combined mite and mite egg occupancy rate was 0.63 per domatium (34 samples counted). A mean of 2.78 mites (34 samples counted)

Table 1. Occurrence of mites within leaf domatia.

Plant Species (Family)	Native Region	Dom. Type[1]	x̄ No. D./lf[2]	Mite Species (Family) Within Domatia	% Leaves[3]	x̄ No. Mites Dom./Leaf[4]	% Dom. with Mites[5]	x̄ No. M & E per Dom.[6]	E or Y in Dom.[7]	Mite Food Hab.[8]
				CALIFORNIAN SAMPLES[11]						
Abelia X grandiflora Rehd. (Caprifoliaceae)	China	HCV	1.0	Tydeus sp. (Tydeidae)	100	6.4	100	≥7.0	B	F-P
Cinnamomum camphora Nees & Eberm. (Lauraceae)	China-Japan	P	5.4	unknown sp. (Euriophyidae)	30	1.0	≥6	≥0.26	B	H
Coprosoma baueri Endl. (Rubiaceae)	New Zealand	P	8.5[9]	Euseius hibisci (Chant) (Phytoseiidae)	70	1.2	≥8	0.16	2E	P
				Amblydromella rhenanoides (Athias-Henroit) (Phytoseiidae)						P
			6.3[9]	Euseius hibisci (Phytoseiidae)	60	1.4	≥10	≥0.22	E	P
Cornus capitata Wall.	Himal-alayas	HP	3.8[9]	Tydeus californicus (Banks) (Tyeidae)	-	7.3	74	≥1.97	E	F-P
			4.7[9]	Czenspinskia transversostriata (Oud.)(Saproglyphidae)	40	0.5	≥9	0.11	no	U
Fraxinus velutina Torr. (Oleaceae)	sw. USA	HCV	1.0[10]	Czenspinskia transversostriata (Saproglyphidae)	≥10	0.09	-	≥0.17	E	U
				Tydeus sp. (Tydeidae)						F-P
Grewia occidentalis L. (Tiliaceae)	South Africa	HP	6.8	Czenspinskia transversostriata (Saproglyphidae)	>33	2.1	28	0.33	2E	U
				Galendromus longipilis (Nesbitt) (Phytoseiidae)						P
				Pronematus sp. (Tydeidae)						P
Ilex sp. (Aquifoliaceae)		HP	7.7	Tydeus sp.(Tydeidae)	50	0.7	≥6	0.16	5E	F-P

Plant (Family)	Region			Mite species (Family)					
Laurus nobilis L. (Lauraceae)	Mediterranean	P	6.1	Aceria malpighianus (C&M) (Eriophyidae)	≥20	-	≥20	no	H
Distictis buccinatoria (Miers) Gentry (Bignoniaceae)	Mexico	Po	6.2	Euseius hibisci (Phytoseiidae)	50	0.9	≥8	no	P
Quercus agrifolia Nee. (Fagaceae)	California	T	5.3	Tydeus sp. (Tydeidae)	≥33	2.9	34	E	F-P
Schinus terebinthifolius Raddi. (Anacardiaceae)	Brazil	F	2.8[10]	Euseius hibisci (Phytoseiidae)	≥33	1.1	39	5E	P
Viburnum odoratissimum Ker. (Caprifoliaceae)	India-Japan	HP	8.1	Euseius hibisci (Phytoseiidae)	50	3.1	-	E	P
Viburnum sp. (Caprifoliaceae)	Japan	HP	5.7	Amblydromella rhenanoides (Phytoseiidae)	80	1.7	≥18	E	P
COSTA RICAN SAMPLES[12]									
Coffea arabica L. (Rubiaceae)	Africa-Arabia	P	-	Tydeus sp. (Tydeidae)	10	0.1	-	no	F-P
Rudgea cornifolia (Hum. & Bonpl.) Standley (Rubiaceae)	Central America	P	8.6	Agistemus sp. (Stigmaeidae)	50	1.4	10	4E	P
HAWAIIAN SAMPLES[13]									
Antidesma pulvinatum Hbd. (Euphorbiaceae)	Hawaii	HP	12.8	Brevipalpus sp. (Tenuipalpidae)	80	6.4	74	135E	H
Annona muricata L. (Annonaceae)	tropical America	HPI	11.2	Tydeus sp. (Tydeidae)	70	0.7	11	8E	F-P
Blakea sp. (Melastomaceae)	Panama	BP	2.2	Agistemus sp. (Stigmaeidae) Agistemus herbicolus (Chant)(Phytoseiidae) Phytoseius sp. (Phytoseiidae)	≥40	0.9	32	7E	P P
Bobea elatior Gaud.	Hawaii	P	-	Tydeus sp. (Tydeidae)	-	1.0	-	no	P F-P
Canthium odoratum (G. Forst.) Seem. (Rubiaceae)	Pacific Is.	P	4.6	Tydeus sp. (Tydeidae)	≥20	0.6	13	Y	F-P
Cedrela odorata L. (Meliaceae)	West Indies	P	4.0[10]	Brevipalpus sp. (Tenuipalpidae) Tydeus sp. (Tydeidae)	≥33	1.5	28	2E	H F-P

Plant species (Family)	Region			Mite species (Family)						
Coffea arabica L. (Rubiaceae)	Africa-Arabia	P	13.0	Agistemus sp. (Stigmaeidae)	100	2.6	27	0.54	44E	P
				Amblydromella haramoti (Prasad)(Phytoseiidae)						P
Coffea liberica Bull. (Rubiaceae)	tropical W. Africa	P	15.6	Bdella sp. (Bdellidae)	≥33	3.8	31	0.62	59E	P
				Ablyseius herbicolus (Phytoseiidae)						P
Coprosma longifolia Gray (Rubiaceae)	Hawaii	P	25.6	Tarsonemus sp. (Tarsonemidae)	40	0.8	3	0.03	No	U
Distylium lepidotum Nakai (Hamamelidaceae)	Ogasawara Is.	HP	5.9	Tydeus sp. (Tydeidae)	78	0.9	26	0.53	B	P
				unknown (Phytoseiidae)						P
Elaeocarpus japonicus Sieb. &Zucc. (Elaeocarpaceae)	China-Japan	Po	9.3	Agistemus sp. (Stigmaeidae)	≥10	1.1	10	0.12	No	F-P
				Tydeus sp. (Tydeidae)						
Gardenia taitensis DC. (Rubiaceae)	Society Is.	P	14.1	Brevipalpus phoenicis (Geyskes)(Tenuipalpidae)	≥40	9.6	45	0.89	29E	H
				Hemicheyletia bakeri (Ehara)(Cheyletidae)						P
				Parapronematus sp. (Tydeidae)						F-P
				Tydeus sp. (Tydeidae)						F-P
				Lorryia sp. (Tydeidae)						F-P
Morinda citrifolia L. (Rubiaceae)	Austral. Pacific	HP	7.1	Phytoseius sp. (Phytoseiidae)	≥20	1.2	20	0.27	7E	P
Perrottetia sandwicensis Gray (Celastraceae)	Hawaii	PPo[14]		Brevipalpus sp. (Tenuipalpidae)	≥20	0.9	14	0.20	21E	H
Randia cochinchinesis (Lour.) Merr. (Rubiaceae)	Guam SE Asia	P	9.5	Agistemus sp. (Stigmaeidae)	60	0.5	-	0.16	10E	P
Schinus terebinthifolius Raddi. (Anacardiaceae)	Brazil	F	3.1[10]	Tydeus californicus (Tydeidae)	100	14.3	97	5.13	11E	F-P
Tabebuia rosea-alba (Ridl.) Sandw. (Bignoniaceae)	tropical America	T	12.6	Mexecheles hawaiiensis (Baker)(Cheyletidae)	100	15.0	-	0.71	14E	P
				Phytoseius hawaiiensis Prasad (Phytoseiidae)						P
Terminalia catappa L. (Combreataceae)	East Indies	P	40	Phytoseius hawaiiensis (Phytoseiidae)	≥20	≥0.6	2	0.02	No	P
				Parapronematus sp. (Tydeidae)						F-P

1 Domatia Types HCV = Hairy Cross Vein. P = Pit. HP = Hairy Pocket. Po = Pocket. T = Tuft. F = Fold. HPi = Hairy Pit. BP = Basal Pockets. PPo = Pit and Pocket.

2 Mean number of domatia per leaf.

3 % Leaves with mites in domatia.: ≥ signifies minimum estimate.

4 Mean number of mites in domatia per leaf. When more than one mite sp. is listed per plant sp., data refer to all mite spp. combined.

5 Percent domatia with mites; ≥ signifies minimum estimate.

6 Mean number of mites and eggs per domatium. Numerical estimates as in footnotes 4 and 5.

7 Eggs (E) or young (Y) in domatium. B = Both; No = Neither. For some species, total numbers of eggs have been counted, and are indicated.

8 Mite food habits according to assumptions in text. P = predaceous. F = fungivorous. F-P = fungivorous and, to lesser degree, predaceous. H = herbivorous. U = unknown.

9 More than one sample site per plant species - C. baueri: site 1 = Berkeley, site 2 = Sacramento; C. capitata: site 1 - Berkeley hills, site 2 = Berkeley flatlands.

10 Species has compound leaves, sample unit is a single leaflet.

11 Samples from Berkeley and Oakland except Ilex sp., V. odoratissimum, and one C. baueri sample from Sacramento.

12 Samples from the Organization for Tropical Studies rainforest field station at La Selva (R. cornifolia) and a planting of coffee along the San Jose - La Selva Rd. (C. arabica).

13 Samples from Oahu except C. odoratum and M. citrifolia from Hilo, Hawaii, and T. catappa from Lahaina, Maui.

14 Pits and pockets both present, sometimes with associated hairs.

and 2.23 mite eggs (17 samples counted) were found in domatia per sampled leaf. The molted skins of mites also werefrequently observed within domatia.

A variety of mites were found in domatia. Specific information on the food habits of most mite species is not available, so we rely heavily in this analysis on the predominant food habits of groups of mites. The most frequently occurring mites were Phytoseiidae with 7+ species from 13 plant species, and Tydeidae with 5+ species from 16 plant species. Phytoseiidae are major predators of other mites including important groups of phytophagous mites such as tetranychid spider mites and eriophyid gall mites (McMurtry et al., 1970). In agricultural systems phytosiid mites are of recognized importance as biological control agents for phytophagous mites including spider mites on a wide variety of crops worldwide (McMurtry et al., 1970; Jeppson et al., 1975). Some phytoseiid mites also feed on thrips (Thysanoptera), the crawler stage of scale insects (Coccoidea), the nymphs of whiteflies (Aleurodidae) and the eggs of moths (Lepidoptera), whiteflies and thrips (McMurtry et al., 1970).

Tydeidae are primary fungivores, scavengers and pollen feeders, although there is increasing evidence that they are more predaceous than previously known on small insects, mites and their eggs, and some tydeid mites have been reported as damaging to plants (Baker and Wharton, 1952; Jeppson et al., 1975; Hessein and Perring, 1986). In our survey the following tydeid mite taxa were found in domatia: *Lorryia* sp., *Parapronematus* sp., *Pronematus* sp., *Tydeus* sp., *Tydeus californicus* (Banks) (Table 1). Hessein and Perring (1986), in summarizing the food habits of tydeid mites, reported that species of *Lorryia* and *Parapronematus* are fungivorous, a species of *Pronematus* is predaceous as well as fungivorous, and *Tydeus californicus* is predaceous on phytophagous eriophyid mites and possibly also is phytohagous itself at times.

Other primarily predaceous groups of mites found in domatia include 1 species of Bdellidae from 1 plant species, 2 species of Cheyletidae from 2 plant species, and 1+ species of Stigmaeidae from 5 plant species. Bdellidae, Cheyletidae and Stigmaeidae are predaceous on other mites (Jeppson et al., 1975), Bdellidae and Stigmaeidae are predaceous on small insects (Baker and Wharton, 1952; Krantz, 1978), and Bdellidae are known to feed on arthropod eggs (Krantz, 1978).

A small number of primary phytophagous mite taxa were found in the domatia of several plant species: 2 species of Eriophyidae occurred in the domatia of 2 plant species, and 1+ *Brevipalpus* species (Tenuipalpidae) were found in domatia from four plant species. A saproglyphid mite and a tarsonemid mite of unknown food habits were found in the domatia of 3 plant species and 1 plant species respectively. Small numbers of phytophagous insects (scale insects, mealybugs (Pseudococcidae), and thrips also were found in the domatia of some samples.

If we consider the predominantly predaceous and fungivorous groups of mites as potentially beneficial to plants because of their feeding on phytophagous arthropods and their eggs or on pathogenic fungi and their spores, then the Phytoseiidae, Tydeidae, Cheyletidae, Stigmaeidae and Bdellidae can be considered primarily potentially beneficial to the plants. We consider the Eriophyidae and *Brevipalpus* spp. (Tenuipalpidae) as potentially harmful to the plants, and make no assumptions about the saproglyphid or tarsonemid mites. Based on these assumptions, of the 23+ species of mites found within domatia, 17 species (73.9%) are considered beneficial, 4 species (17.3%) harmful, and 2 species (8.6%) of unknown consequence to plants. The domatia of 26 of 31 (83.3%) plant species containing mites had mite species potentially beneficial to the plants, and only 6 of 31 (19.3%) had mites potentially harmful to the plants. Of the sample plants with mites in domatia, 23 of 28 genera (82.1%) had potentially beneficial mites in their domatia. Mite eggs occurred in the domatia of 21 of the 26 plant species (80.7%) that had potentially beneficial mites. Samples with potentially beneficial mites had means of 2.9 mites (29 samples counted) and 1.6 mite eggs (14 samples counted) per sampled leaf. The results of this survey provide evidence of a strong association between domatia and potentially beneficial mites.

MITE OVIPOSITION STUDY

The frequent occurrence of mite eggs in domatia led to a limited experimental study of mite oviposition behavior in relation to domatia (Pemberton and Turner, 1989). Gravid females of the predaceous mite *Amblyseius limonicus* (Garman and McGregor (Phytoseiidae) were placed on excised leaves of *Coprosma baueri* Endl., an ornamental shrub native to New Zealand (Figs. 1,4-5). *A. limonicus* is a New World species known from California and Florida, and southward to Brazil; it has also been recorded from Hawaii and New Zealand (Collyer, 1964; McMurtry and Scriven, 1965; de Moraes et al., 1986). Two gravid females of the mite were placed on each of 10 leaves that were placed underside (abiaxial surface) up, thus exposing the pit domatia, on moistened cotton in petri dishes in the lab. Over the course of 5 days the mites produced a total of 106 eggs, with a total of 103 eggs (97.1%; $\bar{x}$ = 10.3 eggs per leaf, S.D. = 6.75) within domatia, and a total of only 3 eggs on the leaves outside domatia. The experiment showed a strong preference for oviposition within domatia for this predaceous mite.

ECOLOGICAL SIGNIFICANCE
Plant Protection-Mutualism Hypothesis
The mite-domatia interaction may function as a facultative mutualism in which domatia serves as shelters and nurseries for mites that reduce the number of phytophagous arthropods and pathogens on domatia-bearing plants. The results of our studies provide correlative support for the plant protection-mutualism hypothesis: Potentially beneficial mites are strongly associated with domatia, and predaceous mites preferentially oviposit within domatia. The hypothesized mutualism constitutes a structurally based mite-mediated plant protection system. This could be significant ecologically because mites are abundant and ubiquitous on plants, and domatia occur in more than 50 families of angiospermous plants throughout the world (unpubl. data). Another independent quantitative survey of domatia in northeastern Australia also found mites strongly associated with the domatia of 6 plant species and reported that domatia occur in at least 36 families of woody plants in Australian tropical rainforests (O'Dowd and Willson, 1986).

We consider the hypothesized mutualism to be facultative rather than obligate as indicated by the absence of domatia in some individual plants of domatia-bearing plant species (Lundström, 1887; Jacobs, 1966) or the absence of mites in domatia, and the presence of domatia-associated mites (such as the phytoseiid mite *Euseius hibisci* (Chant) from our study) also on the leaves of plant species without domatia (McMurtry et al., 1970). That is, plants and the leaves of typically domatia-bearing species can survive without domatia or mites in domatia, and domatia-frequenting mites can survive without domatia. The facultative and opportunistic nature of this interaction is also indicated by our findings in which (1) the domatia of plants growing in areas distant from their native areas are colonized by potentially beneficial mites. (2) Different mite species can occur in the domatia of the same plant species growing in different locations. For example, coffee (*Coffea arabica* L.) had phytoseiid and stigmaeid mites in Hawaii, but had tydeid mites in Costa Rica. (3) The same mite species occurred in the domatia of different unrelated plant species. For example, *Euseius hibisci* was found in the pit domatia of *Coprosma baueri*, in the pocket domatia of *Distictis buccinatoria* (Miers) Gentry and *Viburnmum odoratissimum* Ker., and in the fold domatia of *Schinus terebinthifolius* Raddi. (4) Domatia do not always harbor beneficial mites, and potential harmful mites and insects sometimes occur in domatia.

The proposed domatia-mite mutualism may be analogous to the extrafloral nectary-beneficial insect (especially ants) mutualism that has been shown to protect a diverse array of plants from herbivores (Bentley, 1977; Elias, 1983; Beattie, 1985) despite the occasional use of extrafloral nectaries by herbivores in some plants (Rogers, 1985). Both the extrafloral nectary and domatia-based mutualisms probably are facultative and the result of many independent cases of convergent evolution involving a diverse array of unrelated plants and arthropods throughout the world. The proposed facultative nature of the domatia-mite mutualism deflates some of the criticisms of the hypothesis (not all leaves or individual plants of domatia-bearing species possess domatia, domatia do not always contain mites, and mites are not necessary for the development of domatia), which are based on an assumption of an obligate mutualism. Lundström may have implicitly suggested an obligate mutualism to his critics by his belief that mites were necessary for the development of domatia.

Benefits to Plants and Mites

The evolution and maintenance of a mite-domatia mutualism requires that domatia-associated mites and domatia-bearing plants experience enhanced fitness through benefits from this interaction that result in increased reproduction. So far there have been no demonstrations of benefits to mites or plants from this system, and experimental tests of the plant protection-mutualism hypothesis are needed. We suggest that the most likely benefits to domatia-bearing plants are mite-mediated reductions in levels of foliar phytophagy, and the most likely benefits for domatia-associated mites are shelter from predation or desiccation. Mites that frequent domatia may benefit plants by feeding on phytophagous arthropods and fungi. Our mite-domatia associational data indicate that predaceous mites are an important component of this system. We found four primarily predaceous families (Bdellidae, Cheyletidae, Phytoseiddae, Stigmaeidae) of mites in domatia, including 12+ of 23+ mite species (52.1%) in 17 of 31 plant species (54.8%) with mites. Leaf herbivory has been shown to have a negative effect on reproduction in a variety of woody plants (Rockwood, 1973; Barnes and Moffitt, 1978; Stephenson, 1980; Myers, 1981; Reissig et al., 1982; Marquis, 1984), and protection from foliar herbivory is likely to result in increased plant reproduction and fitness.

The benefits to mites from domatia may be protection from predation or desiccation for adults, immature stages, or eggs. There is little information available on the importance of predation as a source of mortality for phytoseiid and other potentially beneficial mites, but spiders (Chant, 1956), staphylinid and coccinellid beetles (McMurtry et al., 1970), mirid and anthocorid bugs (Collyer, 1952; Krämer, 1961; Herbert, 1962), thrips (McMurtry et al., 1970), and chrysopid lacewings (Krämer, 1961) have been reported as preying on phytoseiid mites and eggs. Domatia may provide shelters that are inaccessible to these larger spider and insect predators. There also is limited information on the importance of desiccation to the adults or immature stages of potentially beneficial mites, but desiccation is known to be an important factor for phytoseiid mite eggs. Phytoseiid predatory mites have a higher and faster rate of egg hatch at higher relative humidities (McMurtry and Scriven, 1965; McMurtry et al., 1976; de Morase and McMurtry, 1981; Badii and McMurtry, 1984). We would expect that relative humidity is higher in the sheltered microenvironments of domatia. Many potentially beneficial mites including the Phytoseiidae appear to be behaviorally predisposed to use domatia. Phytoseiid mites generally spend most of their time and prefer to oviposit in sheltered sites of microrelief such as along veins and at vein junctions on the undersides of leaves in plants without domatia (McMurtry et al., 1970). Some phytoseiid mites seem to prefer leaves with more pubescence or with more pronounced veins (Collyer, 1958; Downing and Moilliet, 1967). The most important groups of phytophagous mites tend to make their own shelters - the webbing of tetranychids and the galls of eriophyids .

Economic Significance

In addition to their possible significance in natural systems, the proposed mite-domatia mutualism could be significant economically, since domatia occur in agricultural and other valuable plant species. Produce-yielding plants with domatia include coffee, varieties of grape (*Vitis vinifera* L.), walnut (*Jugland regia* L.), hazelnut (*Corylus avellana* L.), bay (*Laurus nobilis* L.) cashew (*Anacardium occidentale* L.), soursop (Annona muricata L.) West Indian cedar (*Cedrela odorata* L.), and rubber (*Hevea brasiliensis* Muell.) (Lundström, 1887; Penzig and Chiabrera, 1903; Chevalier and Chesnais, 1941; Adamoli de Barros, 1960, 1963b). Some major groups of ornamental trees with domatia include species of oaks (*Quercus* spp.), lindens (*Tilia* spp.), maples (*Acer* spp.), dogwoods (*Cornus* spp.), ashes (*Fraxinus* spp.) hawthorns (*Crataegus* spp.), cherries (*Prunus* spp.), hollies (*Ilex* spp.) and elms (*Ulmus* spp.) (Lundström, 1887; De Fonzo, 1897; Penzig and Chiabrera, 1903).

SUMMARY

The term domatium was introduced by Lundström in 1887 for plant structures that harbor organisms beneficial to plants. Lundström frequently observed mites associated with leaf domatia, which he termed acarodomatia. Acarodomatia are various small enclosing structures on the undersides of leaves in the form of hair tufts, pits, or pockets at vein axils, and nonaxillary folded or rolled leaf margins. Lundström proposed a mutualistic interaction between domatia-inhabiting mites and plants with domatia, but this hypothesis has not been accepted by most workers in part due to a lack of supporting evidence. We quantitatively sampled for mite presence in a variety of domatia types of 32 plant species from 18 plant families and a wide variety of places of origin growing in California, Costa Rica and Hawaii. About 74% of the mites from domatia of the sample plants were from groups (Bdellidae, Cheyletidae, Phytoseiidae, Stigmaeidae and Tydeidae) that could be considered potentially beneficial (probably predaceous or fungivorous) to plants; the domatia of ca. 84% of the sampled plant species containing mites (26 of 31 plant species) had potentially beneficial mites. Mite eggs occurred in the domatia of ca. 81% of the plant species that had potentially beneficial mites. This survey demonstrated that potentially beneficial mites are strongly associated with domatia. We carried out an experimental study of mite oviposition behavior in which gravid females of the predaceous phytoseiid mite *Amblyseius limonicus* posited ca. 97% of their eggs within the domatia of excised leaves of *Coprosma baueri*. These studies provide strong correlational evidence in support of a hypothesis of facultative mutualism between mites and domatia, in which domatia serve as shelters and nurseries for mites that reduce the number of phtophagous arthropods or pathogens on the leaves of domatia-bearing plants. This hypothesized mutualistic relationship could be important in natural and some agricultural systems, because mites are abundant and widespread on plants, and a diverse array of woody angiospermous plants in more than 50 families throughout the world have apparent acarodomatia. Domatia occur in a variety of economically significant plants, including coffee (*Coffea arabica*), walnut (*Juglans regia*), some varieties of grape (*Vitis vinifera*), and many woody ornamental timber plants. Experimental tests of the mutualism-plant protection hypothesis are needed to further our knowledge of mite-domatia systems.

ACKNOWLEDGMENTS

H.G. Baker first informed us that the structures containing mites that we observed in the leaves of *Cornus capitata* were probably what had been termed domatia. Identifications were kindly provided by H.A. Denmark (mites), S.H. Sohmer and D. Herbst (Hawaiian domatia plants), J.A. Beach (Costa Rican *Rudgea cornifolia*), and C.E. Kennett (the mite *Amblyseius linonicus*). The Foster Botanical Garden, Lyon Arboretum of the University of Hawaii, Waimea Arboretum on Oahu, Hawaii, the U.S. Forest Service Arboretum at Hilo, Hawaii, the La Selva Organization for Tropical Studies field station in Costa Rica, and the University of California Botanical Garden at Berkeley, allowed us to take domatia samples. L.A. Andres, J.A. McMurtry and R. Schmid critically reviewed the manuscript. The light photographs and the scanning electron micrographs of domatia and mites were taken by G.R. Johnson and L.A. Sunell, respectively.

LITERATURE CITED

Adamoli de Barros, M. A. 1960. Origem e formacao das domacias em *Coffea arabica* L. Anais da Escola Superior de Agricultura "Luiz de Queiroz" (Sao Paulo) 17: 131-138.

Adamoli de Barros, M. A. 1963a. Estudo comparativo das domacias de fôlhas normais e domacias de fôlhas cujas plantas foram cultivadas com deficiências e excessos de micronutrients (Fe, Mn, Mo, e Cu), em *Coffea arabica* L. variedade caturra K.M.C. Anais da Escola Superior de Agricultura "Luiz de Quieroz" (Sao Paulo) 20: 229-240.

Adamoli de Barros, M. A. 1963b. Ocorrencia de domacias em especies e hibridos da familia Vitaceae. Anais de Escola Superior de Agricultura "Luiz de Queiroz" (Sao Paulo) 20: 241-256.

Adamoli de Barros, M. A. 1966. Ocorrencia de domacias em especies e hibridos da familia Vitaceae II. Anais da Escola Superior de Agricultura "Luiz de Queiroz" (Sao Paulo) 23: 9-14.

Badii, M. H., and J. A. McMurtry. 1984. Life history of and life table parameters for *Phytoseiulus longipes* with comparative studies on *P. persimili* and *Typhlodromus occidentalis* (Acari: Phytoseiidae). Acarologia 25: 11-123.

Baker, E. W., and G. W. Wharton. 1952. An Introduction to Acarology. MacMilliam Co., New York.

Barnes, M. M. and H. R. Moffitt. 1978. A five-year study of the effects of the walnut aphid and the European red mite on Persian walnut productivity in coastal orchards. J. Econ. Entomol. 71: 71-74.

Beattie, A.J. 1985. The evolutionary ecology of ant-plant mutualisms. Cambridge Univ. Press, Cambridge.

Bentley, B.L. 1977. Extrafloral nectaries and protection by pugnacious bodygards. Ann. Rev. Ecol. Syst. 8: 407-427.

Bequaert, J. 1922. Ants on the American Museum Congo Expedition. A contribution fo the myrmecology of Africa. IV. Ants in their diverse relations to the plant world. Bull. Am. Mus. Nat. Histo. 45: 333-583.

Brouwer, Y.M. 1979. Domatia in seedlings. Flora Malesiana Bull. 32: 3239-3246.

Chant, D.A. 1956. Predacious spiders in orchards in southeastern England. J. Hort. Sci. 31: 35-45.

Chevalier, A., and F. Chesnais. 1941. Sur les domaties des feuilles de Juglandacees. Comptes Rendus, (Paris) 213: 389-392.

Collyer, E. 1952. The biology of some predatory insects and mites associated with the fruit tree red spider mite (*Metatetranychus ulmi* (Koch)) in southeastern England. I. The biology of *Blepharidopterus angulatus* (Fall.) (Hemiptera - Heteroptera, Miridae). J. Hort. Sci. 27: 117-129.

Collyer, E. 1958. Some insectary experiments with predacious mites to determine their effect on the development of *Metatetranychus ulmi* (Koch) populations. Entomol. Exp. Appl. 1: 138-146.

Collyer, E. 1964. The occurrence of some mites of the family Phytoseiidae in New Zealand, and descriptions of seven new species. Acarologia 6: 632-646.

DeFonzo, D. 1897. Contribuzioni alla concoscenza degli acarodomazii Il Naturalista Siciliano (Palermo), Anno Secondo: 85-92.

de Moraes, G. J., and J. A. McMurtry. 1981. Biology of *Amblyseius citrifolius* (Denmark and Muma) (Acarina - Phytoseiidae). Hilgardia 49: 1-29.

de Moraes, G. J., J. A. McMurtry, and H. A. Denmark. 1986. A catalogue of the mite family Phytoseiidae: references to taxonomy, synonomy, distribution and habitat. EMBRAPA-DOT, Brasilia.

Downing, R. S., and T. K. Moilliet. 1967. Relative densities of predacious and phytophagous mites on three varieties of apple trees. Canad. Entomol. 99: 738-741.

Elias, T. S. 1983. Extrafloral nectaries: their structure and distribution. In B.L. Bentley and T.S. Elias [eds.]. The Biology of Nectaries. Columbia Univ. Press, New York.

Greensill, N. A. R. 1902. Structure of leaf of certain species of *Coprosma*. Trans. Proc. New Zealand Inst. (Wellington) 35: 342-355.

Hamilton, A. G. 1896. On domatia in certain Australian and other plants. Proc. Linnean Soc. New South Wales (Sydney) 21: 758-792.

Herbert, H. J. 1962. Overwintering females and the number of generations of *Typhlodromus* (T.) *pyri* Scheuten (Acarina: Phytoseiidae) in Nova Scotia. Canad. Entomol. 94: 233-242.

Hessein, N. A., and T. M. Perring. 1986. Feeding habits of the Tydeidae with evidence of *Homeopronematus anconai* (Acari: Tydeidae) predation on *Aculopus lycopersici* (Acari: Eriphyidae). Int. J. Acarol. 12: 215-221.

Jacobs, M. 1966. On domatia - the viewpoints and some facts. Proc. K. Ned. Akad. Wet. (sec. C) 69: 275-316.

Jeppson, L. R., H. H. Keifer, and E. W. Baker. 1975. Mites Injurious to Economic Plants. Univ. California Press, Berkeley.

Krämer, P. 1961. Untersuchungen über den Einfluss einiger Arthropoden auf Raubmilben (Acari). Z. Angew. Zool. 48: 257-311.

Krantz, G. W. 1978. A Manual of Acarology.(2nd ed.). Oregon State Univ. Book Stores, Corvallis.

Lagerheim, G. 1892. Über neue Acarodomatien. Botan. Centralblatt. 49: 238-240.

Lundström, A. N. 1877. Pflanzenbiologische Studien II. Die Anpassungen der Pflanzen an Thiere. I. von Domatien Nova. Acta Regiae Soc. Sci. Upsaliensis (Ser. 3) 13(10): 1-88.

Marquis, R. J. 1984. Leaf herbivores decrease fitness of a tropical plant. Science 226: 537-539.

McMurtry, J. A., and G. T. Scriven. 1965. Life-history studies of *Amblyseius limonicus*, with comparative observations on *Amblyseius hibisci* (Acarina: Phytoseiidae). Ann. Entomol. Soc. Am. 58: 106-111.

McMurtry, J. A., C. B. Huffaker, and M. van de Vrie. 1970. Ecology of tetranychid mites and their natural enemies: a review. I. Tetranychid enemies: their biological characters and the impact of spray practices. Hilgardia 40: 331-390.

McMurtry, J. A., D. L. Mahr, and H. G. Johnson. 1976. Geographic races in the predaceous mite, *Amblyseius potentillae* (Acari: Phytoseiidae). Int. J. Acarol. 2: 23-28.

Metcalfe, C. R., and L. Chalk. 1979. Anatomy of the Dicotyledons. Vol. 1. Systematic Anatomy of Leaf and Stem, with a Brief History of the Subject (2nd ed.). Clarendon Press, Oxford.

Myers, J. H. 1981. Interactions between western tent caterpillars and wild rose: a test of some general plant herbivore hypotheses. J. Anim. Ecol. 50: 11-25.

O'Dowd, D.J., and M.F. Willson. 1986. Domatia on the leaves of Australian rainforest trees: a plant-mite mutualism? IV Int. Cong. Ecol., Syracuse, New York (abstract).

Pemberton, R. W., and C. W. Turner. 1989. Occurrence of predatory and fungivorous mites in leaf domatia: The basis of a possible facultative mutualism. Amer. J. Bot. 76: 105-112.

Penzig, O., and C. Chiabrera. 1903. Contributo alla conoscenza delle piante acarofile. Malpighia 17: 429-487.

Reissig, W. H., R. W. Weires, and C. G. Forshey. 1982. Effects of gracillarid leafminers on apple tree growth and production. Environ. Entomol. 11: 958-963.

Rockwood, L. L. 1973. The effect of defoliation on seed production of six Costa Rican tree species. Ecology 54: 1363-1369.

Rogers, C. E. 1985. Extrafloral nectar: entomological implications. Bull. Entomol. Soc. Am. 31: 15-20.

Schnell, R. 1970. Introduction a la phytogeographie des pays tropicaux. Les problemes generaux. Vol. 1. Les flores - les Structures. Gauthier-Villars Editeur, Paris.

Shirley, J., and C. A. Lambert. 1923. On *Comprosma baueri* Eudlicher. Proc. Roy. Soc. Victoria 35: 19-23.

Stace, C. A. 1965. Cuticular studies as an aid to plant taxonomy. Bull. Brit. Mus. (Nat. Hist.) Botany 4: 1-78.

Stephenson, A. G. 1980. Fruit set, herbivory, fruit reduction, and the fruiting strategy of *Catalpa speciosa* (Bignoniaceae). Ecology 61: 57-64.

To Ngoc Anh. 1966. Sur la structure anatomique et l'ontogenese des acarodomaties et les interpretations morphologiques qui paraissent s'en degager. Adansonia 6: 147-151.

Wheeler, W. M. 1942. Studies of neotropical ant-plants and their ants. Bull. Mus. Comp. Zool. Harvard 90: 1-262.

EVOLUTION AND ORGANIZATION IN COMMUNITIES

Ecologists commonly use two approaches to study biological communities. One of these captures a community at a point in time and carefully maps it. This allows comparison among such communities and their subsequent hierarchical classification into nested ecological units. This approach can provide a very useful description of community structure. But to understand how a community came to be organized in a particular fashion and how it functions, an evolutionary approach is needed. The evolutionary approach requires at least that communities be regarded as dynamic, changing entities. In addition, a recognition that species are variable is desirable. All the papers in this section fit into the second approach to the study of communities.

Each of the papers in this section is based upon the authors' several years of field studies of a particular community type. Evolutionary plant ecologists tend to start out studying a few populations of one or two species. Their efforts gradually are enhanced (some may say derailed) by biotic and abiotic "impingements" upon their original subjects. In these three cases, the authors have attempted to explain such impingements, thereby providing a broad interpretation of the communities they study. The section starts with a study conducted on the broad scale of whole mountain ranges, and concludes with an analysis of patterns of colonization and survival in the small, highly specialized and well-circumscribed communities of buffalo wallows. At an intermediate scale is a demonstration of the potential role of single species population dynamics upon multi-species assemblages, and the potential of genetic variability in shaping community-level interactions.

Stan Cook was Herbert Baker's first U.S. student. His thesis research focused on the reproductive ecology of the California poppy. His interests since have expanded to the community level. Here, Cook, Copsey, and Dickman report on their work in a more or less pristine area of old (450 yrs) montane coniferous forest of the northern Cascades. Their principal biological actors are two forest trees, Pacific silver fir (*Abies amabilis*) and mountain hemlock (*Tsuga mertensiana*), and a fungal pathogen (*Phellinus weirii*). They have documented by means of extensive field surveys and aerial photo interpretation, that alternate patterns of community succession can occur in these forests depending upon several factors, including time since last fire, presence or absence of the fungus, moisture levels within the specific habitat, and slope and aspect of the forested site. Some of the confounding factors include the fir's being more sensitive to fire, but less sensitive to the fungus than the hemlock. The results caution against simple, deterministic models of forest succession.

Linhart's paper also deals with forest communities. He uses his own extensive data and those of others to show that tree populations in temperate and tropical forests tend to be patchy. For all the insects, pathogenic fungi, and parasitic plants that use forest trees as hosts, the genetic patchiness of tree populations creates a phenotypic patchiness of resources that vary in their attractiveness or suitability. Variability among and within all these species that feed upon and reduce the fitness of forest trees creates highly diverse selection pressures. Such multidirectional diversifying selection is thought to be relevant to the high levels of genetic variability documented by Hamrick and Loveless in this volume. This paper presents a new way of making understandable the complexity of natural forest communities

Uno gives us a glimpse of the Great Plains when bison were the dominant ungulate of North American grasslands, and scattered bands of semi-nomadic people lived in a way that left almost no trace of their presence. From studying remnant buffalo wallows of these past times, he shows how they played a critical role in maintaining the species richness of the grassland flora. The wallows contain an array of native colonizing species often distinct from species of the undisturbed grassland. These colonists had considerable importance in maintenance of the vegetation cover. We know from modern experiences in the southwestern grasslands of the U.S. and elsewhere that grasslands denuded of vegetation are a prerequisite of desertification. Many of the wallow species are adapted for long distance dispersal, suggesting their importance in establishment not just in wallows but at other disturbed sites in grassland. The significance of this bison-initiated flora doubtless was of ecological and evolutionary significance to other animal taxa as well. Uno's analyses also indicate that the role played by various animals from bison to badgers and prairie dogs has been to create habitats with specific and predictable soil and moisture conditions within the great plains. These habitats have been the home of plants, some of which are highly dependent upon these periodic disturbances.

RESPONSE OF *ABIES* TO FIRE AND *PHELLINUS*

Stanton A. Cook, Alan D. Copsey, and Alan W. Dickman

INTRODUCTION

This is a study of the developmental mechanisms of a forest in which a fungal pathogen and fire are major mortality factors. We show how these mortality factors interact with the dominant tree species and physical factors to bring about different pathways of succession. We describe how diversity and dominance among the trees are governed by these interacting factors and speculate about adaptiveness of indirect effects among species.

The conceptual source for these objectives is within a framework that recently has been called "patch dynamics" (Pickett 1980; Pickett and White 1985). We review briefly the emergence of this paradigm to set the context for our study.

The Dynamic Mosaic Perspective

A.S. Watt explicitly represented a community as a mosaic or patchwork of dynamic local populations: within a patch of space, a temporal succession of organisms occurs as their reactive and interactive influences kill dominant individuals and alter probabilities for establishment; "...these patches form a mosaic and together constitute the community (Watt, 1947, p. 2)." There is a cycle in the stages of development of a patch: species within a patch replace on another but ultimately the replacement processes revert to the beginners. He used the term "phasic equilibrium" for that of "steady-state" (Watt, 1947 p 19) because he had referred to the stages of development of patches as "phases." The areas covered by these phases - and thus (by our extension) the diversity of species in the whole community - "...would be proportional to the duration of the phases, the total equalling the duration of the cycle of change." His example of beechwood (one of seven that he gave in 1947) clarifies his concepts and connects his idea of steady-state with that of succession.

There are four phases of development: gap (an open patch with unexploited resources); immature phases, sometimes dominated by other tree species (ash, oak, or birch); and the mature phase, that is occupied by beech with an understory of *Rubus*. In the eutropic beech forest of the Chiltern Hills, *Rubus* impedes regeneration of beech but not that of ash and oak, which "...have the power of growing through bramble..." and to suppress it. "This differentiation is very important for it ensures the constant presence of ash and oak in the beechwood ...And at the same time the maintenance of beech because under the canopy of oak and ash beech can grow up" (Watt, 1934, pp. 488-489). On oligotrophic soils of the Chiltern hills, "...there is a stable birch-beech wood, in which at any given time the proportion of birch to beech may vary, the birch will always be a constituent and one perhaps necessary to the continued existence of the beech. ...[T]he progressive component of change leading to succession [is] effectively offset by the inability of the beech to maintain continuity in space and in time over the whole area. The two phenomena, namely, plant succession and the internal cycle of change, are the same in principle" (Watt, 1944, p. 100). In this article of 1944, he generalized the connection between steady-state and succession with a diagram of patches occupied by phases and presented in dimensions of space and time (p. 101).

Successional change is initiated where there has been large-scale mortality and cessation of mortality. Crop-like or plantation-like structure, for example, results "...from a sudden release from grazing or from arable cultivation." Or burning of Scottish pinewoods imposes "...the structure of a crop or an even-aged pure plantation" (Watt, 1944, p. 102). He referred to agents of mass - i.e., contagiously distributed - mortality in various ways: "...exceptional factors of rare or sporadic occurrence, such as storms, fire, drought, epidemics which create a gap phase of exceptional dimensions..."(Watt, 1947, p. 13); or "...the disturbing factor..." (p. 19); or "fortuitous obstacles" (p. 2) or "checks" (p. 19). "At any given time, therefore, structure [of the mosaic of the community] is the resultant of causes which make for order and those that tend to upset it. Both sets of causes must be appreciated" (p. 2.).

Within a patch, productivity of plants rises and then declines in what Watt referred to as "upgrade" and "downgrade" series (cf. Watt, 1947, Fig. 11). "The patch becomes a microcosm of limited area with continuous but restricted food supply. Hence the population of that area behaves like a population of flies or human beings under similar restriction of space and food. It is unable to spread, but it may develop" (1947, p. 18).

Andrewartha and Birch (1954, ch. 14) built their comprehensive general theory of the numbers of animals in natural populations around the concept of a mosaic of local populations that range in density from zero (extinction) to carrying capacity: "A natural population occupying any considerable area will be made up of a number of such local populations or colonies. In different localities the trends may be going in different directions at the same time" (p. 657). Their Fig. 14.02 is equivalent to Watt's Fig. 11 (Watt, 1947), and, · with a somewhat different approach, they, as Watt (1944), diagrammed the spatio-temporal dynamics of populations as a mosaic of patches (Figs. 14.06 to 14.09).

By 1970, many students had become concerned with the community concepts of diversity and stability; and, at least in the United States, ecological thinking became dominated by several themes: niche theory based on competition and resource partitioning; hypothetico-deductive and mathematical theorizing; and "ahistorical" thinking (McIntosh, 1987), which appears to relate to the concept of a competitive equilibrium. But qualifying and corrective thoughts were also current.

MacArthur and Levins (1967, p. 377) wrote that "The diversity of coexisting species can be limited in at least three ways depending on the circumstances. First, there is a lower limit to the abundance of each species which sets an upper limit to the number of species (Schoener, 1965). Second, there may be an upper limit to the abundance of each species, set by the danger from predators and disease, which increases the possibilities for more species (Paine, 1966). Finally, environmental instability sets a limit to the degree of specialization and, for a given degree of specialization, competition may, but does not always, set a limit to the similarity of coexisting species." They were, clearly, well aware of the effects on diversity of mortality factors, not just competition.

Frederick E. Smith (1972) raised the important methodological criticism that "In most dynamic models of ecological systems the parameters do not vary in relation to spatial variations of the environment" (p. 310). For instance, "c", the parameter of efficiency of predation in the predator-prey equations of Lotka and Volterra, has but one value in space-time. "Mathematically, these are 'point' models, models of activity in one point in space. Their use, however, demands comparisons with populations that must occupy space, and it is more reasonable to view the models as 'homogenous,' in the sense that the space occupied by the system is homogenous at each moment with respect to all activities." He found that diversity and stability could not be achieved in point or homogeneous dynamic models, but they could in spatial or heterogeneous models in which parameters vary. He pointed out that his analysis was not particularly original. It agreed with Hutchinson, Andrewartha and Birch, Huffaker, et al. and Errington, and he pointed out that MacArthur and others were increasingly investigating the effects of environmental complexity and structure (Smith, 1972, p. 325). Nevertheless, he had explicitly demonstrated the necessity of interpreting community-scale phenomena of diversity and stability in a framework of a dynamic mosaic, (i.e., heterogeneity).

Consensus on the general utility of the concept of dynamic mosaic (patch dynamics) has followed growth of knowledge, from many systems, of the influences of "predators and disease" and "environmental instability" on species dominance and diversity and the course of succession. This growth has centered around several hypotheses. We review those that are pertinent to our study.

Intermediate Disturbance

According to the "intermediate disturbance hypothesis," density-independent mortality at intermediate frequencies creates a mosaic of patches of biotic assemblages at various stages of succession such that, overall, maximum diversity is achieved (Connell, 1978). Loucks (1970) had applied this concept (if not the name) in explaining how fires, at a suitable frequency, might maintain diversity of plants. Heinselman (1973) provided valuable empirical evidence of a mosaic of post-fire stands in the boreal forest of Minnesota. In the intertidal zone, logs, driven by surf, physically remove mussels and their associated community from rocks and so create a mosaic of successional patches (Paine and Levin, 1981). Connell (1978) attributed some of the present diversity of tropical forests to the mosaic of disturbance left in landscapes after abandonment of swidden agriculture. Primeval forests had been subjected to non-selective mortality and a patchwork of successional stands resulted. And, in coral reefs, he asserted that diversity is created by wave action and predation (presumably non-selective).

Predator-Mediated Diversity or Coexistence

Paine (1966) brought together evidence from literature, experiment, and observation of food chains in support of the hypothesis that local animal diversity is related to the efficiency with which predators prevent monopolization of the major requisites, i.e. prevent domination of resources (space on intertidal rocks in his study). Harper (1969) drew attention to Martin Jones's experiments with Welsh pastures, in which maximum diversity of plants resulted when sheep (catholic feeders) selectively ate the palatable competitively dominant *Trifolium repens* and *Lolium perenne*. Harper (1969, p. 60) also referred "...to the role of the pathogen in floristic diversity in natural rubber [tree] communities." Indeed, much consideration has been given the hypothesis that diversity of tropical trees is caused by frequency dependent preadaptation (more generally, mortality) on the most numerous or competitively dominant species - Connell (1978) has called this "compensatory mortality." He (1978) and his co-workers (Connell et al., 1984) found the results in the literature to be equivocal, and they could not assess the importance of evidence from their own investigation. Caswell (1978, p. 150) using a heterogeneous (sensu F.E. Smith), dynamic, spatially compartmented model demonstrated the possibility of "long-term, but non-equilibrial, coexistence among competitors under the impact of predation."

Predator-Mediated Dominance

In nutrient-poor Welsh pastures, the competitively dominant plants (*Agrostis tenuis* and *Festuca rubra*), that are best able to cope with low fertility, are coincidentally relatively unpalatable to sheep. Consequently, when poor pastures, or even poor pastures that are artificially fertilized, are intensively grazed, the competitive dominants persist and diversity is held low by the grazers (Harper, 1969). Even in tropical forests, dominance may be gained by one or a few species in the absence of non-selective (i.e., frequency independent) mortality. Ironwood, *Cyanometra alexandrei*, in Uganda is particularly interesting, because it is both a shade-tolerant dominant and avoided by browsing elephants. "They preferentially destroy young of the fast-growing early and middle succession trees, leaving the young of the late succession ironwood alone, thus hastening progression toward the low-diversity forest" (Connell, 1978 p 1307). Horse mussel, sea urchin, and kelp in the rocky subtidal zone in New Hampshire provide another example (Witman, 1985, 1987): kelp, fastening onto the mussels, cause them to be torn off the rocks by storm waves. But sea urchins that find refuge from predators by living among the mussels, feed on the kelp and so maintain dominance of mussels.

Alternate Successional Vectors

Predators and pathogens may direct succession along various pathways. This was quite obvious to Watt, who had investigated the roles of fungi and rabbits, for instance, in the dynamics of oak and beech forests (Watt, 1919, 1924) and who referred to soils of the Chiltern hills as "hardwood" soils because conifers were excluded from them by fungal pathogens (Watt, 1934). Humans have directly or indirectly altered the course of development of many communities. Estes and Palmisano (1974) discovered that presence or absence of sea otters was correlated with great differences in communities of algae and invertebrates in nearshore habitats of the Aleutian Archipelago. Dayton (1975) found corroboratory evidence. Simenstad, Estes, and Kenyon (1978) later concluded from archeological evidence that Aleut hunters, by greatly reducing the population of sea otters, had removed a key predator of herbivorous sea urchins, that, in turn, eliminated kelps from the community.

With this review, we have shown that by about the year 1978 there was widespread appreciation of the important role played by mortality factors in governing the properties of marine and terrestrial communities and a renewed awareness of the dynamic mosaic as a framework for accommodating the details of particular ecosystems. A shift occurred from point to spatial (homogeneous to heterogeneous) models and to including, along with competition, various other sources of mortality as determinants of the equilibrium community. Pickett (1980) enumerated the many mechanisms of plant communities that disrupt equilibria established among species solely through competition. Patches, created by various sources of mortality, allow coexistence; and "patch dynamics" provide a powerful and comprehensive paradigm for analysis of succession.

Context of Current Work

This study stems from observations by McCauley and Cook (1980) on the effects of a pathogen on diversity of a coniferous forest. The root-rotting fungus, *Phellinus (Poria) weirrii* (Murr.) Gilbertson, grows radially outward, vegetatively, from points of initial establishment, into old-growth stands dominated by mountain hemlock, *Tsuqa mertensiana* (Bong.) Carr. Some trees may survive the advancing front of fungi; others germinate and establish in the gaps that arise as the old-growth trees are killed by the fungus. Consequently, a forest growing within infestations tends to be older centripetally. McCauley and Cook sampled trees and their ages in transects across the boundaries of infestations and concluded (p. 23) that "...mortality caused by the fungus creates diversity" by reducing dominance of hemlock. Although none of the conifers that make up the forest was immune to the disease, some were more susceptible than others. The order from most to least susceptible is mountain hemlock, Pacific silver fir (*Abies amabilis* (Dougl.) Forbes), *Pinus contorta* var. *latifolia* Engelm., and *Pinus monticola* Dougl..

Hindsight shows a contradiction in these conclusions: *Abies amabilis* is also quite shade tolerant, more so than mountain hemlock (Fowells, 1965; Minore, 1979), and establishes beneath a canopy of mountain hemlock (Franklin and Dyrness, 1973). If it is both more shade tolerant and less susceptible to the fungus than mountain hemlock, then it should dominate the stands within infestations; and the fungus should diminish rather than increase diversity. We examine this contradiction here: there are indeed variable outcomes possible. Depending on site characteristics, the fungus directs development of vegetation along alternate successional paths - either it accelerates succession of hemlock to fir and creates a community of low diversity or it causes development of a diverse community of the sort described by McCauley and Cook (1980) and Cook (1982). Thus, both predator-mediated dominance and diversity exist.

MATERIALS AND METHODS
The Locality
The study area is at the headwaters of Salt Creek, tributary to the Willamette River and at the summit of the volcanic High Cascade Mountains in Lane County, Oregon. It is essentially pristine, not having been logged. Recent human-caused changes include a single road that was built through the basin in the early 1960's, suppression of forest fires for about 40 years, and introduction (Manion, 1981) of the white pine blister rust (*Cronartium ribicola* J.C. Fisch. ex. Rab.), which now primarily attacks young trees of *Pinus albicaulis* Engelm. and *P. monticola* Dougl.. These perturbations have not yet materially altered the forests, for they are of short duration relative to the ages of the stands and longevities of the trees. Also, the pines are a minor component of the stands. Elevations range from 1524 to 1829 m.

The study area is what Franklin and Dyrness (1973) call the "lower subzone of closed forest" of their Mountain Hemlock Zone, which is in contrast to an upper subzone of patchy mountain hemlock that is subalpine in character. The climate may be described as cool to cold temperate, as precipitation is high and much of it falls as snow, which accumulates to about 2.5 m and lasts from six to eight months. Glacial ice left the area about 12,000 years before present (Taylor, 1968). At 6700 years before present, Mount Mazama erupted (Mehringer, Blinman, and Peterson, 1977) 85 km (53 miles) to the south and deposited one meter of volcanic ash and pumice on the ground. The well-drained podsolic soil has been termed a sandy entic cryorthod in the Winopee Series (Boone, et al. 1988).

The Mortality Factors

Phellinus weirii **(Basidiomycetes).** This pathogen or saprophyte attacks most of the conifers of the Pacific Northwest and British Columbia. It may kill its host rapidly if it attacks its sap-wood, or slowly if it decomposes only the heart-wood. In either case it may persist in decaying wood as a saprophyte for a hundred or more years (McCauley and Cook, 1980). It establishes genets from sexual spores and grows from tree to tree through connections of roots; thus it radiates outward from foci of infection to form circular infestations (Fig. 1). This growth is at a rate of ca. 0.3 m per year within the study area (Cook, 1982). Mycelia persist within the centers of infection and attack successional trees that establish in the wake of the peripherally advancing mycelia. Crown fire evidently reduces that area of infestation by fungal genets but does not necessarily eliminate them. Genets in the research area range in age from 100 to 1300 years (Dickman, 1984). Ramets of single or diverse genets may coalesce. Presumably, in a forest where the fungus occurred throughout, discrete infestations like those in Figure 1 would not be evident.

Fire. Crown fires within the study area have been located and dated (Dickman, 1984; Fig. 2). Large fires occurred 120 and 175 years ago on the slopes of Maiden Peak and Twins Peak, respectively. And small fires occurred 60 and 115 years ago just north of Bobby Lake, which lies at the summit, in a saddle between the two peaks. Although *Phellinus* infestations exist within forests that are 175 years or less in age, they are difficult to discern because they are small, and the canopy on them superficially resembles that of the surrounding forests. Our study area (in the southern one-third of Fig. 2) is mainly covered by forests 350-460 years old.

The Species of Trees

Pinus contorta Dougl. (lodgepole pine) is the major post-fire pioneer. It persists for about two generations (260 yr), after which it is replaced by *Tsuga mertensia*, which begins to establish abundantly about 100 years after a fire - primarily on the north sides of lodgepole pines (Dickman, 1984). *Tsuga mertensiana* has intermediate shade tolerance (Minore, 1979). Both species range far to north, east, and south of the study area. *Abies amabilis* has a narrower range than they; it occurs only west of the Cascade mountains, from northwest California north to southeast-most Alaska (Fowells, 1965). Thus, in our locality, it

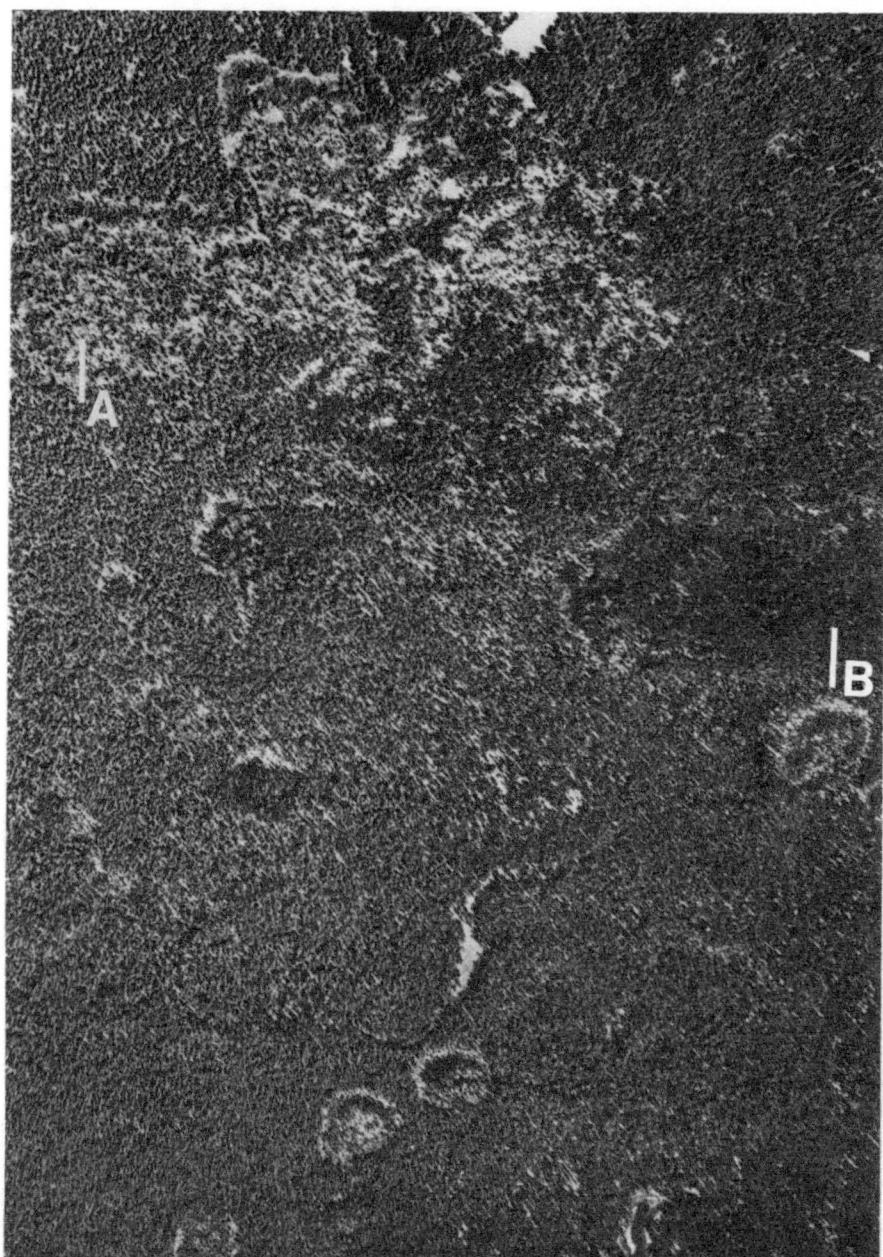

Figure 1. Aerial photograph of the south slope of Twins Peak. The *Abies*-poor infestation at "A" and the *Abies*-rich infestation at "B" are the subjects of intensive sampling. The white bars indicate the positions of the transects.

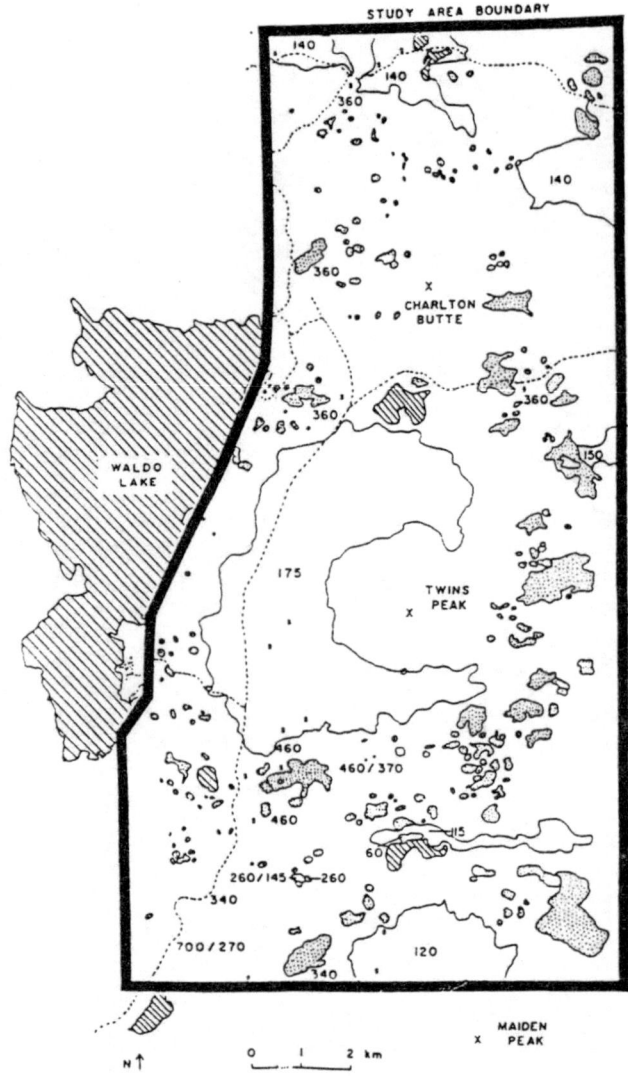

Figure 2. Fires and *Phellinus* infestations in the upper Salt Creek region. Lakes are symbolized by diagonal lines and roads by dashed lines. Post-fire stands with discrete boundaries are delineated with a solid line, and their ages are given in years. Some stands are of mixed age because of partial burns. The chronosequence of stands for study of post-fire succession was taken from within the entire area mapped; the letter "s" designates locations of sampling sites. *Phellinus* infestations that were detectable on aerial photographs are stippled. The area for our study of the response of *Abies* to *Phellinus* is in the lower one-third of the figure.

is near the southern limit of its range. It is the most shade-tolerant of the three species, requires ample moisture as a seedling, is easily killed by fire, and (because of its large seed) is dispersed for only short distances (Copsey, 1985; Fowells, 1965; Packee et al., 1982). As we shall show, it does not enter stands in significant numbers until about 350 years after stand-destroying fire.

Post-Fire Succession

A chronosequence of ten post-fire stands was selected (Fig. 2). Representative sub-areas within these stands were then chosen for intensive sampling so that 1) the maximum difference in elevation among them was little more than 200 meters and 2) their slopes and aspects were similar. In seven dense young stands up to 175 years old, 76 quadrats of dimensions 10 X 20 m were distributed along transects at 0-100 m intervals depending on the size of stand. In three low-density stands, that were 260 years and older, seven quadrats were sampled. These ranged from 1,800 to 10,000 m^2 in size (Dickman, 1984, Table 3). For each tree taller than 25 cm, species and exact diameter at 25 cm height were recorded. However, trees of less than 6 cm diameter were assigned to diameter classes of 0-1, 1-3, and 3-5 cm. We report here only results for *Abies amabilis*.

Post-*Phellinus* Succession

Fungal infestations fall into two classes according to densities of tree canopies inside them (Fig. 1). One class ("*Abies-*poor") has relatively low cover and appears light in an aerial photograph because light-colored soil is in view. Dominance is shared in this case by seven conifers. The other class ("*Abies-*rich") has high cover, appears dark in an aerial photograph, and is populated mainly by *Abies amabilis*. Sampling just one infestation from each class gave us time to describe in detail the vegetational dynamics of each.

Both infestations lie on a gentle south-facing slope and are level at their southern margins, where we laid out transects of contiguous quadrants 20 m wide extending north and south normal to the margins of the infestations. The transect at A, Fig. 1 (*Abies*-poor stand) extends from a point 30 meters inside of the surrounding 460-year-old hemlock-dominated forest northward across the advancing margin of fungus, 80 meters toward the center of the infestation. There on the north, it adjoins a square sample area of 0.5 hectares located at the center of the infestation. The transect at B of Fig. 1 (*Abies*-rich stand) lies 1.6 km south by east of that at A. It extends from 20 m within the old-growth hemlock forest northward across the advancing margin of fungus, 85 meters toward the center of the infestation. It adjoins at its north end a rectangular sample area of 0.4 hectares at the center of the infestation.

The following data were recorded within each square meter of these transects and the large central sample areas: live and dead trees taller than 25 cm, by species and diameter at a height of 25 cm. Locations of the bases of standing and fallen or decumbent trees were mapped according to the coordinates of the square meter in which they occurred. For small trees, diameter classes of 0-1, 1-3, and 3-5 cm were used. Note was taken of cause of morbidity or mortality. The most common causes of death were *Phellinus weirii*, *Echinodontium tinctorium* (Ell. and Ev.) Ell. and Ev., *Cronartium ribicola* J. C. Fisch. ex Rab., wind-throw, and felling from falling trees.

Fungi attack trees and their wood in characteristic ways. Therefore, it was usually possible to assign cause of death with certainty when trees had fallen and their wood was exposed (Bega, 1970; Foster and Wallis, 1969; Manion, 1981). We attempted to pull up, topple over or cut into the bases of dead yet standing trees to look for evidence of disease. *Phellinus* is not known to invade wood that has been attacked by other fungi, although the reverse is common (Hansen, 1979). Thus, if *Phellinus* was present at the base of a dead tree, it was likely to have been the cause of death.

Infestations were mapped in relation to topographical contours after they had been identified in aerial photographs and checked on the ground for presence of *Phellinus*.

RESULTS
Post-Fire Succession of *Abies amabilis*
Not until 350 years after a stand-destroying fire, does fir constitute as much as one percent of the number of individuals in a forest, and even then these are mostly small trees that live beneath the canopy of *Tsuga mertensiana* (Table 1). Since all of the area under consideration was burned within the past 460 years, only in certain circumstances had the fir reached the canopy: 1) within infestations of *Phellinus*, 2) on several steep north-facing slopes, and 3) beside the east shore of Waldo Lake.

Post-*Phellinus* Succession of *Abies*-Poor Infestation
Density of live hemlocks was low in the uninfested 460-yr-old forest (Fig. 3A), but the trees were large (Fig. 3B). All of these trees were killed by the advancing fungus. Regeneration of hemlocks occured in the opening at the margin; density increased

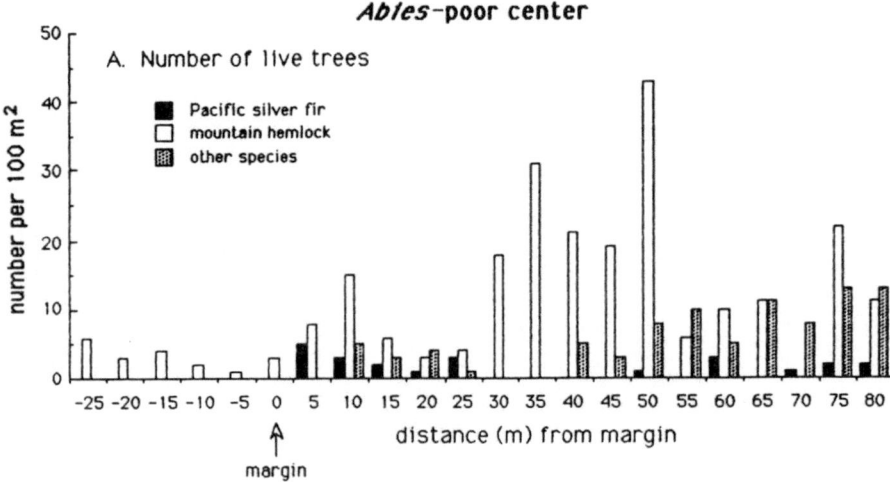

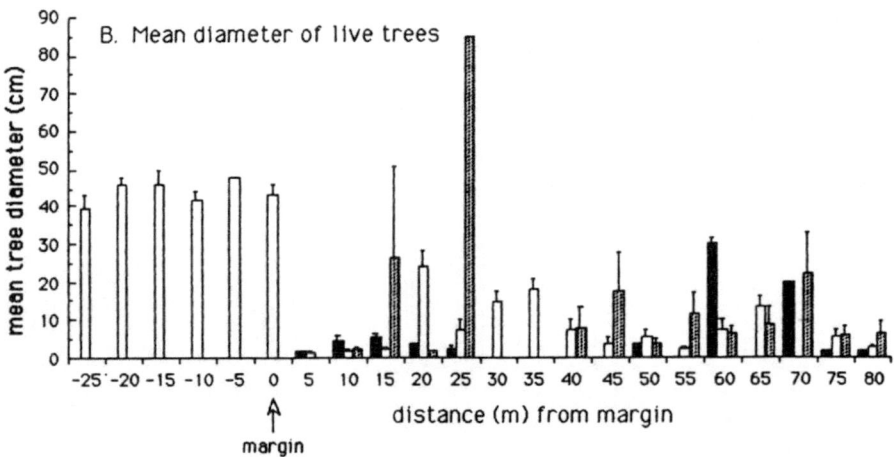

Figure 3. Live trees in a transect across the margin of a *Phellinus* infestation, that is poor in *A. amabilis.* The pathogen is moving toward the left in the figure, south into a 460-yr-old stand of mountain hemlock. A. Numbers per 5X20 m quadrat. B. Mean diameters of trees in quadrat. Error bars show standard error of the mean. Other species: *Pinus albicaulis* Englem., *P. contorta, P. monticola, Abies lasiocarpa* (Hook.) Nutt., and *A. procera* Rehdr.

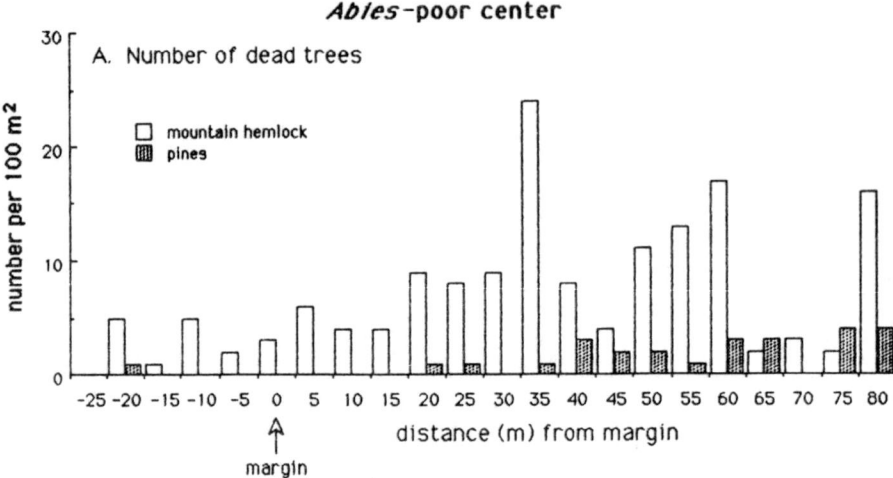

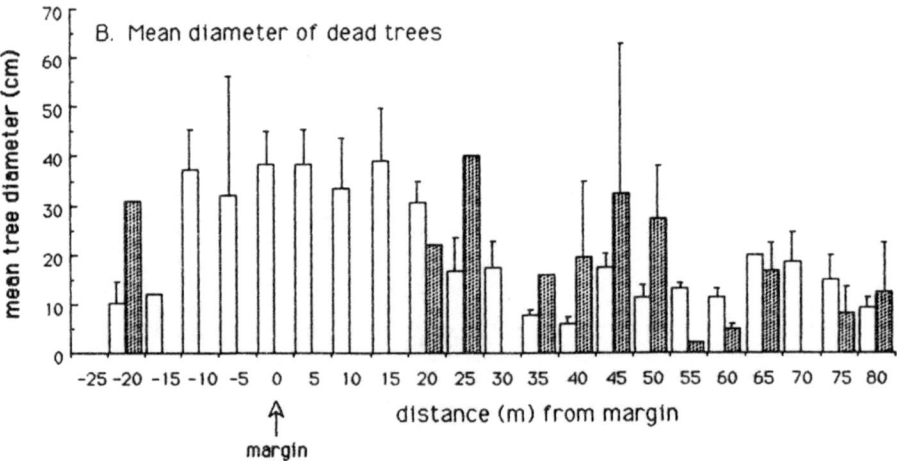

Figure 4. Dead trees in a transect across the martin of a *Phellinus* infestation, that is poor in *A. amabilis*. The pathogen is moving toward the left in the figure, south into a 460-yr-old stand of mountain hemlock. A. Numbers per 5X20 m quadrat. B. Mean diameters of trees in quadrat. Error bars show standard error of mean.

Table 1. Numbers of *Abies amabilis* individuals in ten stands of different ages and the percentages they are of all living trees in the stands. For details of sampling, see "Post-fire succession" in "Materials and Methods."

Age of Stand yr	Area sampled m²	Degrees of slope	Aspect	No. individuals per size class per hectare (diam in cm)					Abies as percent of all trees
				1	3	7	15	31	
60	1600	11-15	SSW						0
115	3400	6-15	S	10					0.2
120	1400	0-10	NNE	7					0.2
140	3200	0-15	SW			6			0.1
145/260	3400	0-10	S						0
150	1400	0-5	-						0
175	3600	0-10	SSW						0
260	2000	6-10	S		5	5			0.3
350	7200	0-5	-	11	6	3	1		2.0
460	14200	6-10	S	12	9	4		1	3.1

and decreased alternately in what may be damped oscillations toward the infestation's center. Mean diameter of living hemlocks decreased at the margin as old-growth trees were killed and rose and fell centripetally as its values gradually diminished to 5-10 cm. These patterns for numbers and diameters of living hemlock trees resulted from reinfestation at progressively earlier ages toward the infestation's center. Further evidence for this is found in data from dead trees (Fig. 4).

Abies amabilis occurred only within the infestation, patchily and in small numbers (Fig. 3A). At the margin, the trees were small; larger trees occurred centripetally (Fig. 3B). Individuals of other tree species (listed in figure caption) increased monotonically in number centripetally. They, like *A. amabilis*, reached mean growth sizes greater than those of the hemlocks toward the center. The great size and variance in diameter of other species was due primarily to presence here and there of exceptionally large individuals of *Pinus monticola*.

Number of dead hemlocks was relatively constant from within the old-growth forest northward to 20 m into the infestation (Fig. 4A). There it rose and fell in three progressively smaller peaks at 35, 60, and 80 m. These were at, or 5-10 m north of, peaks in numbers of living hemlocks (c.f. Fig. 3A) and, as we shall see below (Fig. 7), these were associated with mortality from either competition or *Phellinus*. Mean diameter of dead hemlock trees (Fig. 4B) was higher at the margin (ca. 40 cm) than within the old-growth forest; centripetally it fell and fluctuated around a value of about 15 cm. This reflects a decline in longevity.

No dead *Abies* were found on the transect. Number of dead pines increased slowly centripetally, and their diameters varied greatly, in large measure because of their small number.

Post-*Phellinus* Succession, *Abies*-Rich Infestation

Density of live hemlocks was low in the uninfested forests (Fig. 5A), but the trees were large (Fig. 5B). All of these trees were killed by the advancing fungus. Regeneration of hemlocks occurred in the opening at the margin; density increased centripetally but not to the extent that it did in the *Abies*-poor infestation. In contrast to *Abies*-poor stands, *Tsuga* did not exist occur near the center. Mean diameter of hemlocks increased to a location at 50 m on the transect and then declined. There was no evidence of oscillation as in the *Abies*-poor infestation.

Abies amabilis lived in the understory of the old-growth hemlock forest to the south of the margin (left in Fig. 5A), where there was a low density of saplings and small trees with diameters up to ca. 10 cm (Fig. 5B). Density rose to a maximum at 20 m, declined centripetally, and rose again in a gap at 75-80 m. Mean diameter was low at the margin because of establishment of a vast number of individuals. Mean diameter of *Abies* rose at the same rate as that of hemlock and eventually exceeded it as the hemlock was eliminated by *Phellinus*. Decline in mean diameter of firs coincided with the gap at about 80 m. No living trees of other species occurred on this transect.

Numbers of dead hemlocks appeared to be relatively high at 5-15 m (just north of the margin) and also at 30-40 m (Fig. 6A). The first peak was associated with death of large old-growth trees and the second with death of small successional trees (Fig. 6B). The population of dead trees was estimated from an assemblage of logs of various ages (as inferred from differing states of decay), and the mean diameters could reflect large boles, which decompose slowly. This effect persisted on this transect until about 25 m north of the margin. This was equivalent in time to 75 years. North of 40 m were medium to large sized hemlocks from the cohort of marginal regeneration that had died.

Number of dead firs rose to a maximum at 30 m, which was just 10 m northward of the site of maximum density of living firs. The diameters of these numerous dead firs were quite small. One tree of large diameter died at 15 m. Most likely it had begun growth in the old-growth hemlock forest and had survived the presence of *Phellinus* for some time. The mean diameter of dead firs rose along with that of hemlocks and firs at 75 and 85 m.

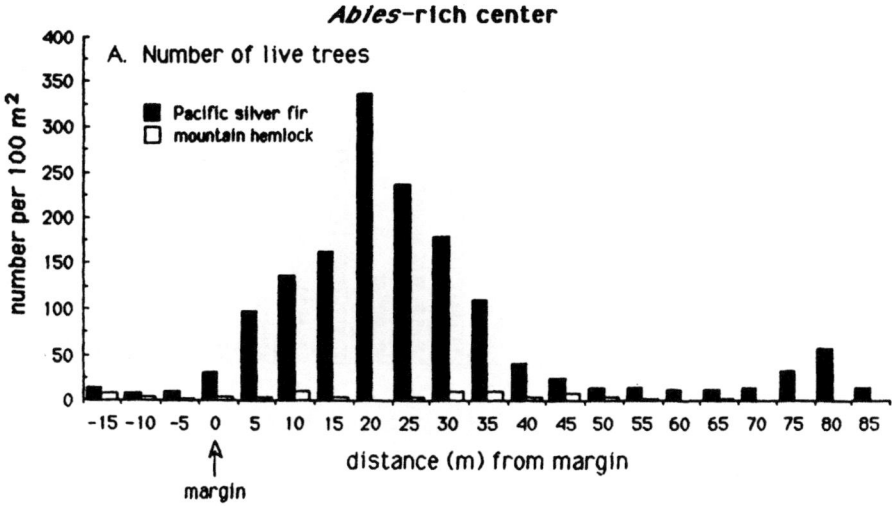

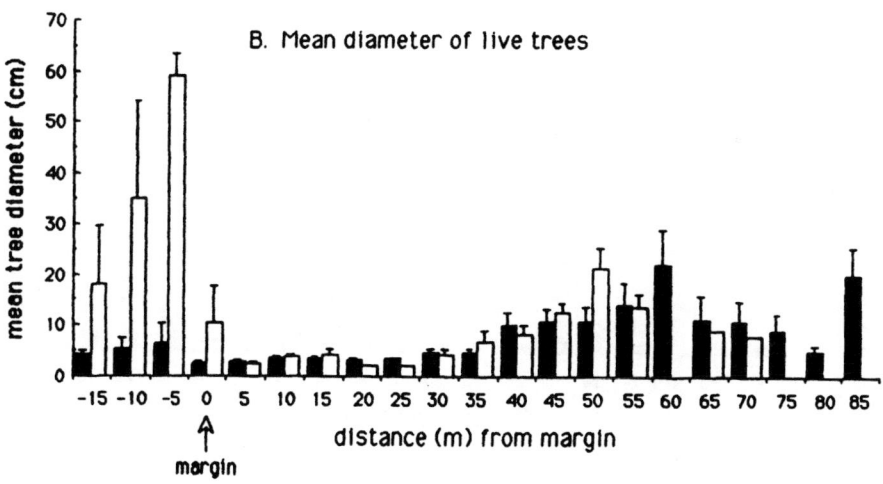

Figure 5. Live trees in a transect across the margin of a *Phellinus* infestation, that is rich in *A. amabilis*. The pathogen is moving toward the left in the figure, south into a 460-yr-old stand of mountain hemlock. A. Numbers per 5X20 m quadrat. B. Mean diameters of trees in quadrat. Error bars show standard error of mean.

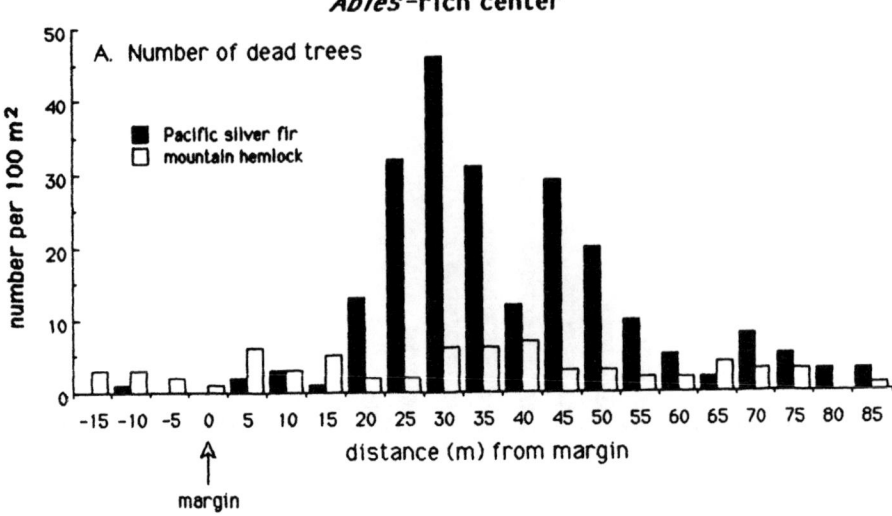

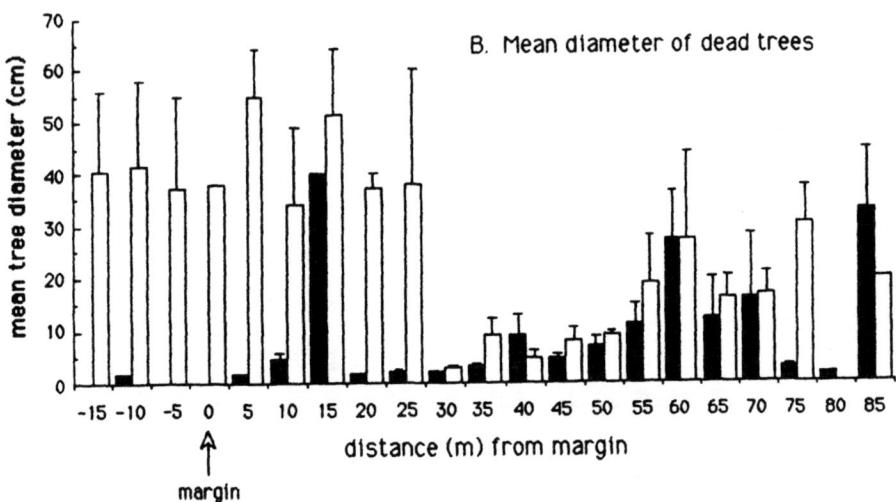

Figure 6. Dead trees in a transect across the margin of *Phellinus* infestation, that is rich in *A. amabilis*. The pathogen is moving toward the left in the figure, south into a 460-yr-old stand of mountain hemlock. A. Numbers per 5X20 m quadrat. B. Mean diameters of trees in quadrat. Error bars show standard error of mean.

Causes of Mortality in Infestations

Phellinus and "unknown" were the dominant classes of causes of mortality to trees on the transects (Figs. 7 A, B, and C). We believe most of the "unknowns" died from suppression, i.e. competition: if trees overtop one another in competition, then the diameters of living trees should exceed, on average, the diameters of those that have died. Where competition is the principal cause, the ratio of diameters of live trees to dead trees should exceed unity. At the margin of the infestation, however, living and dead hemlocks had the same diameters. And so, the ratio should be unity there. Finally, in the zone of primary regeneration (just centripetal of the margin) or in gaps, where samplings were numerous relative to large killed trees, the ratio should be less than unity. We used this ratio to summarize and compare the data of diameters of living and dead trees along the two transects (Figs. 7 D and E).

The ratio for hemlock on the *Abies*-poor infection center (7D) followed the expected pattern. In the old-growth forest (-25 m to -10 m), the ratio was about four. Even in this stand, self-thinning was taking place. It was associated with mortality of "unknown" cause to hemlock (Fig. 7A). Between -10 m and 0 m the ratio was about one. Here, large living hemlocks were surrounded by large *Phellinus*-killed hemlocks (cf. also figs. 3B and 4B). North of the margin, the ratio was much less than unity because of establishment of young hemlocks. From 20 to 30 m there was a large number of dead successional hemlocks of intermediate size (Fig. 4A and B); the ratio approached unity (Fig. 7D); for they had been killed by *Phellinus* (Fig. 7A). There was a high density of live hemlocks at 34 and 50 m; the ratio of diameters of living to those of dead trees rose above one (Fig. 7D). This is a region of intense competition and self-thinning that is associated with abundant death from the unknown cause (Fig. 7A). This zone is visible in Fig. 1 as a dark ring of dense trees inside the pale bare soil at the outer margin of *Abies*-poor infestations. North of meter 50, a sharp decline in number of living hemlocks (Fig. 3A) was associated with low values of the ratio (Fig. 7D) and frequent death from *Phellinus* (Fig. 7A). The regimen characteristic of the center of an *Abies*-poor infestation was present here: patchy mortality from the fungus followed by regeneration of hemlocks and other species in these patches. There was no mortality of *Abies* on the *Abies*-poor infection center.

The patterns of ratios and sources of mortality for hemlock along the transect of the *Abies*-rich center resembled those on the

382

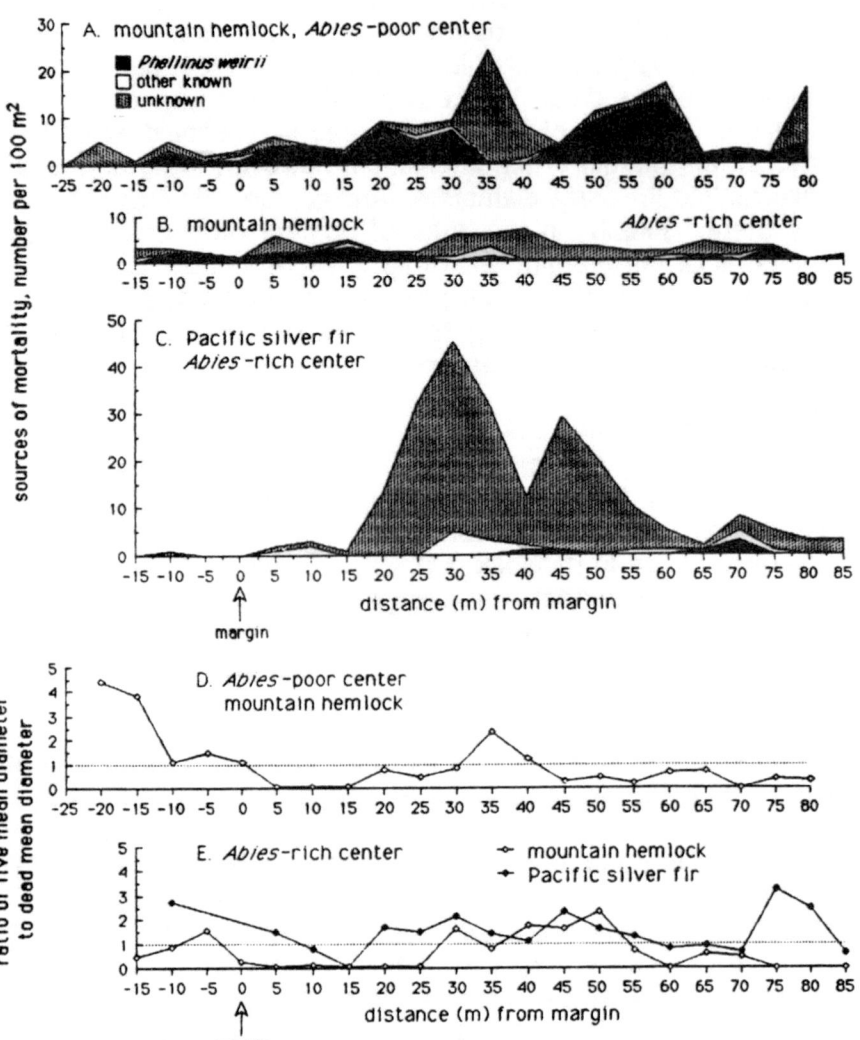

Figure 7. Causes of mortality (A-C) and ratios of live to dead mean diameters (D-E) for mountain hemlock and Pacific silver fir. See text for further details.

Abies-poor center (Fig. 7). Although more hemlock died of "unknown" causes in the *Abies*-rich infestation (0.48 as compared to 0.28), the difference was not statistically significant. On both transects, death from *Phellinus* extended some 13 m into the surrounding forest from the margin, where mortality was complete. The ratio of diameters of living to dead firs was less than one where large firs had died at 15, 60, 65, 70 and 85 m. Otherwise it was above one along the *Abies*-rich transect. It was high at 30 and 45 m, where there were many dead trees of small diameter (Fig. 6A) and where there were peaks in mortality from the "unknown" cause. This, for fir as for hemlock, was a zone of intense intra- and inter-specific competition. Also a peak was associated with the gap at 75-80 m, where a large number of small firs were competing for space (Fig. 5A). Firs were killed by *Phellinus* toward the center of the population, where they were large because the fungus only attacked the heart-wood of this tree. No ring-like pattern of dead trees resulted; the infestation was consistently covered by a dense canopy (Fig. 1 at B).

Comparison of Diversities

Shannon-Wiener diversity indices (H', log to base e) in the *Abies*-poor transect ranged from zero in the old-growth forest at the south where only hemlocks grow, to ca. 1.5 at 80 m to the north (Fig. 8A). The value for the 0.4 ha area at the center of the infestation and adjacent to the transect was 1.512. Clearly, the fungus increased diversity by directing stand development toward high species richness with equable distribution (cf. Fig. 3A). Between 5 and 25 m diversity was high in what has been termed the early, non-interactive stage of succession. Between 30 and 50 m there was complete (at 30-35 m) or near-complete numerical dominance by hemlock (Fig. 3)in the zone of intense competition that was discussed above.

The pattern of diversity at the *Abies*-rich center (Fig. 8B) differed greatly from that on the *Abies*-poor center. Diversity was higher in the old-growth forest because of presence there of firs beneath the overstory of mature hemlocks (Fig. 5A). As hemlocks were eliminated at the margin of the advancing fungus, hemlocks and firs became established, but firs so outnumbered hemlocks that diversity was minimal (cf. Fig. 5A at 20 m). Diversity increased to 45 m on the transect, evidently as proportionately more firs than hemlocks die from suppression. *Phellinus* killed the standing hemlocks, however. And they did not establish either in the understory or in gaps. Consequently, diversity was zero at 75-85 m. In the rectangular sampled area

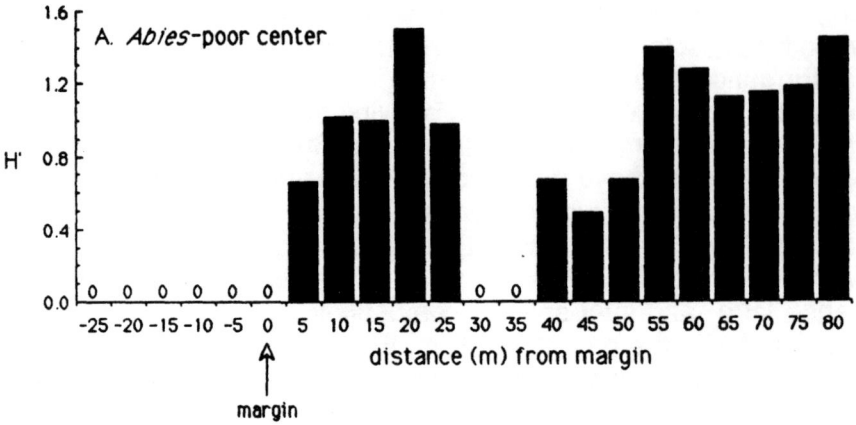

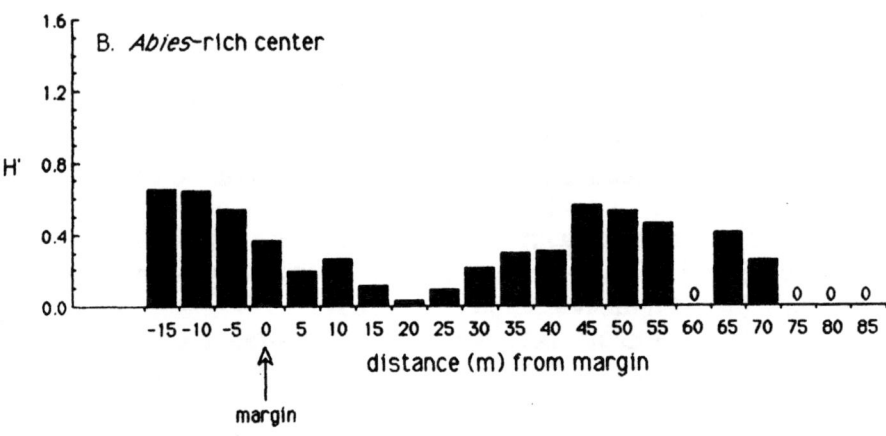

Figure 8. Shannon-Wiener diversity index (H') along the two transects. Values of H' are computed for each 5X20 m contiguous quadrant. Small zeros above the abscissas indicate that the observed diversity is zero, i.e., only one tree species is present. A. Diversity at the *Abies*-poor infestation. B. Diversity at the infestation, that is rich in *A. amabilis*.

of 0.4 ha, that is at the center of the infestation and adjacent to the transect at 85 m, h' was 0.141.

These measures of diversity probably approximate well those for the whole vascular plant community even although they are based solely on trees greater than 25 cm tall, because there are few angiospermous plants in forests of the mountain hemlock zone (Franklin and Dyrness, 1973).

Geographical Distribution of *Abies*-Poor and -Rich Infestations

Three major areas of *Abies*-rich infestations were on north-facing slopes. *Abies*-poor infestations occurred on south-facing slopes. This contrast was most striking on the east-west running ridge in the southwest corner of the study area (Fig. 9), where the *Phellinus* infestation straddled the ridge. *Abies*-dominated infestations also appeared to be limited at upper elevations. For instance, on the east slopes of the 2150 m peak at the west, only *Abies*-poor infestations occurred above 1800 m. Elsewhere, on gentle slopes at the center and northwest parts of the study area, (at our *Abies*-rich study area, for instance) the two sorts of *Phellinus*-infested communities occurred adjacent and interspersed.

DISCUSSION

The forest of the Upper Salt Creek watershed is changing in response to mortality from past fires and current spread of pathogens, especially *Phellinus weirii*. The spatial pattern of this change depends on properties of the tree species and their mortality factors and the environmental factors, as they are influenced by, for instance, elevation and aspect.

Stand-destroying crown fire is an agent of mass, non-selective mortality, a disturbing factor (Watt, 1947) that creates a large-scale mosaic of stands at various stages of succession. *Pinus contorta* was the major pioneer after fire. *Tsuga mertensiana* established abundantly about 100 years later - primarily on the north sides of lodgepole pines. After two generations (ca. 260 yr) of pines, mountain hemlock became dominant (Dickman, 1984; Dickman and Cook, 1989). After another one hundred years passed, *Abies amabilis* entered the stands as an understory tree. Even when the stand had developed 460 years since fire, the fir was only a minor, mainly understory, component. The long time for succession of Pacific silver fir here is similar to that at lower altitude in northwestern Oregon, where it establishes under *Pseudotsuga menziesii* and *Tsuga heterophylla* (Franklin and Dyrness, 1973 p 99).

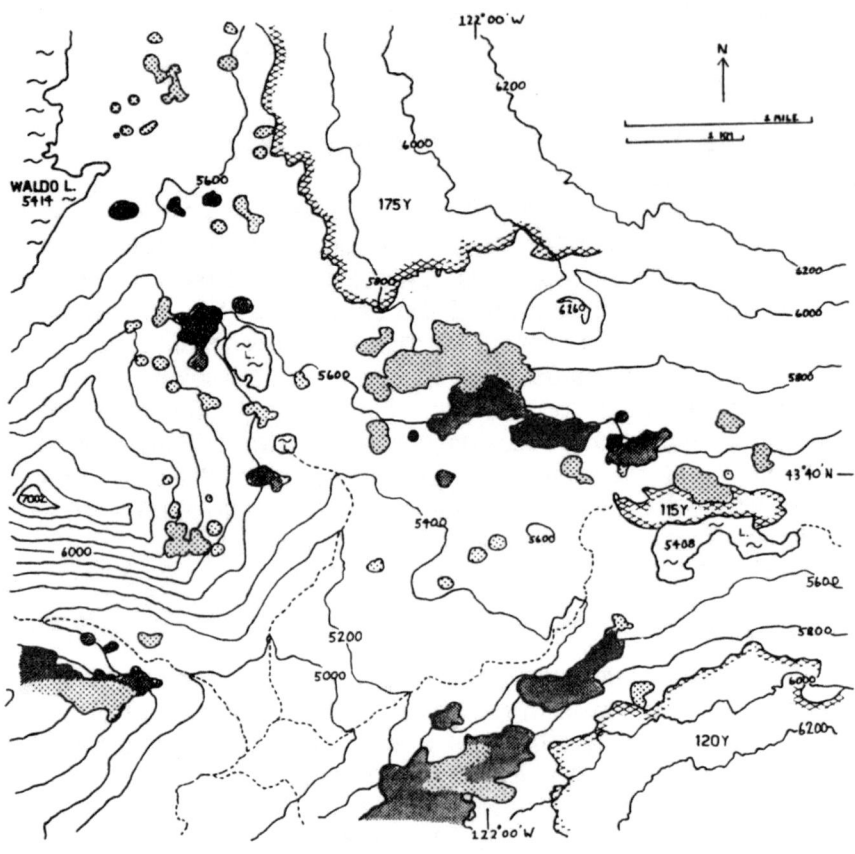

Figure 9. Map of *Abies*-poor (light stippled) and -rich (dark stippled) *Phellinus* infestations in the vicinity of upper Salt Creek in relation to topography. Lakes are indicated by "L." Elevations are in feet. Salt Creek and its tributaries are designated by dashed lines. Major recent fires (115, 120, and 175 yr) are outlined with crossed hachures.

However, on several steep north-facing slopes and at the eastern shore of Waldo Lake, *Abies amabilis* has became an overstory tree in absence of *Phellinus* and within 460 years of fire. Its rate of entry here was enhanced by proximity to refuges from fire, and dampness of site, which promoted establishment by seedlings.

Abies amabilis was restricted more in distribution by fire than other species because of its thin bark, low-growing canopy, shallow roots, heavy and weakly dispersed seeds, and its seedling's moisture requirement.

Two sorts of climax stands would likely develop in this area in absence of fire and *Phellinus* over a 600 to 700 year period, and both would have gap phase reproduction at equilibrium. Where the fir cannot exist (where conditions are too dry or otherwise adverse) hemlock would dominate at equilibrium (Table 2). *Abies amabilis* would dominate as equilibrial stands in suitable areas. *Abies procera* and *Pinus monticola* would be present at low density, the former on dry and the latter on moist sites. Succession would be faster under moist than dry conditions because of higher rates of establishment for all species. The rate of formation of gaps in climax stands would be set mainly by a combination of wind-throw and collapse of trees from infestation by the heart-rot fungus, *Echinodontium tinctorium,* and other pathogens.

Phellinus, in the absence of fire, brings about alternate successional vectors (Table 2). Where microsites for *Abies amabilis* are sparse (dry sites), *Phellinus* reduces dominance by *Tsuga*, allowing existence of the pines (*P. albicaulis, contorta*, and *monticola*), fir (*A. amabilis, procera,* and *lasiocarpa*), and rarely, *Picea engelmannii.* This increased diversity is achieved within infestations after several cycles of establishment and death of hemlock near the margins. Where infested, moist microsites are abundant for germination and establishment, and *A. amabilis* achieves dominance over hemlock because it is more tolerant of shade and *Phellinus.* This dominance exists in extensive *Abies -* rich infestations that lie on south- and north-facing slopes at the east side of the basin (Fig. 9). (There may be a belt of high moisture there because of prevailing wind patterns). Similar stands occur on north- and northeast-facing slopes to the west. Moisture may have brought about refugia from fire, for mature fir and *Phellinus.*

Predator-mediated diversity occurs where *Phellinus* kills hemlock in areas that are inhospitable to *A. amabilis* and where hemlock is the competitive dominant. Predator-mediated dominance comes about because the competitively dominant,

Table 2. Alternate vectors of post-fire succession when *Phellinus* inoculum is or is not available in the community, under contrasting conditions of moisture. Contorta = *Pinus contorta*. Tsuga = *Tsuga mertensiana*. Abies = *Abies amabilis*. Mixed conifers = preceding spp plus two spp of *Abies* and two spp of pines.

		Stage of Succession		
Moisture	Phellinus available	Initial	Transition	Equilibrium
Dry	No	Contorta	Tsuga	Tsuga
Wet	No	Contorta	Tsuga	Abies
Dry	Yes	Contorta	Tsuga	Mixed conifers
Wet	Yes	Contorta	Tsuga	Abies

shade-tolerant, fir is less susceptible to the fungus than is hemlock. The pathogen's effect is to hasten the dominance.

In these cases and those reviewed in the Introduction, certain forest species have been favored against their competitors by a third species. This has been called an "adaptive indirect effect" (Wilson, 1986) or "higher order interaction" (Futuyma, 1979). The relationship of benefitting party to benefitted was evidently unilateral in the cases of *Agrostis* or *Festuca* and sheep and *Cyanometra* and elephants. It was bilateral in the case of mussels and urchins, but, because self-sacrificing was not evident, Wilson (1986) called it "facultative mutualism". Between *Phellinus* and *Abies*, which is competing with *Tsuga*, the relationship may also be facultatively mutualistic. *Abies* may incur the cost of shortened life span by harboring the fungus in its heartwood, but it benefits by having its hemlock competitor attacked. The fungus may incur a cost by not exploiting sapwood, but it gains a refuge. This could benefit *Phellinus* selectively because it is reduced in range by fire and is only weakly vagile (Dickman, 1984; Dickman and Cook, 1989). Among the conditions that Roughgarden (1975) found to influence evolution of mutualism, there are benefits to host and guest, probability that a guest will find a host, probability that the host survives while the symbiont is associated with it, and extent to which the guest refrains from exploiting the host. These affect the probabilities of initial formation and subsequent evolutionary intensification of an association. Roughgarden applies his cost-benefit model to a gradient in symbioses between damselfish and sea anemones. The loosest association occurs in the intertidal habitat where the fish has to desert its anemone host at low water. The strongest mutualisms are in subtidal habitats. Vermeij (1983) provides other examples of hosts in marine environments that provide secure places to live to guests that are subject to intense predation. Fire in the subalpine forest is comparable to this predation. And *Abies*, surviving in refuges from fire, provide relatively secure places for the fungi to live. Refugia also provide a population configuration conducive to such evolution (Wilson, 1986). In periods between fires as the fir migrates outward, the fungus benefits from being spread, and the fir benefits from the fungus as an agent of interference in competition with hemlock.

This interpretation recalls Sander's (1968) contrast between stable, biotically determined (accommodated) communities with their many coevolved relationships, and unstable physically determined (disturbed) communities with their simpler interrelationships. Non-selective crown fires disrupt food webs and spatial structure of populations and inhibit coevolution. Refugia may provide places where coevolution can proceed. Comparison of tree-pathogen interactions in refugia with those in disturbed areas might well shed light on both community ecology and evolution.

CONCLUSION AND SUMMARY

This work supports the conclusion that the fungus *Phellinus*, plays a major role in population dynamics of forests of the Upper Salt Creek region. While it preserves diversity it also facilitates spread of a competitive dominant, *Abies*. *Phellinus* is guiding community development toward alternate climax communities, in absence of fire - climax, yes, because surely one must include parasites, pathogens, and predators as members proper to the ecosystem. *Phellinus* and *A. amabilis* experience migrational lag after fires. We see the extant landscape because fire is more deadly to fir than hemlock, and fungus is more deadly to hemlock than fir; but fungus is also less vagile than hemlock and more restricted in range by fire.

LITERATURE CITED

Andrewartha, H. G., and L. C. Birch. 1954. The Distribution and Abundance of Animals. Univ. of Chicago Press. Chicago, Illinois

Bega, R. V. 1979. Diseases of Pacific Coast conifers. Agri. Handbook No. 521. Forest Service, U.S. Dept. of Agriculture, Washington, D. C.

Boone, R. D., P. Sollins, and K. Cromack. 1988. Stand and soil changes along a mountain hemlock death and regrowth sequence. Ecology 69: 714-722.

Caswell, H. 1978. Predator-mediated coexistence: a nonequilibrium model. Amer. Nat. 112: 127-154.

Connell, J. H. 1978. Diversity in tropical rainforests and coral reefs. Science 199: 1302-1310.

Connell, J. H., J. G. Tracey, and L. J. Webb. 1984. Compensatory recruitment, growth, and mortality as factors maintaining rain forest tree diversity. Ecol. Monogr. 54: 141-164.

Cook, S. A. 1982. Stand development in the presence of a pathogen, *Phellinus weirii*. In J. Means [ed.]. Forest Succession and Stand Development Research in the Northwest. Forest Res. Laboratory, U.S. Forest Service, Corvallis, OR.

Copsey, A. D. 1985. Long-term effects of a native forest pathogen, *Phellinus weirii*: changes in species diversity, stand structure, and reproductive success in a *Tsuqa mertensiana* forest in the central Oregon High Cascades. Doctoral Dissertation. University of Oregon. Eugene.

Dayton, P. K. 1975. Experimental studies of algal canopy interactions in a sea otter-dominated kelp community at Amchitka Island, Alaska. Fishery Bull. 73: 230-237.

Dickman, A. W. 1984. Fire and *Phellinus weirii* in a mountain hemlock (*Tsuqa mertensiana*) forest: postfire succession and the persistence, distribution, and spread of a root-rotting fungus. Doctoral dissertation. University of Oregon. Eugene, OR.

Dickman, A. W., and S. A. Cook. 1989. Fire and fungus in a mountain hemlock forest. Canad. J. of Botany. In press.

Estes, J. A., and J. F. Palmisano. 1974. Sea otters: their role in structuring nearshore communities. Science 185: 1058-1060.

Foster, R. E., and G. W. Wallis. 1969. Common tree diseases of British Columbia. Forestry Branch Publ. No. 1245. Dept. of Fisheries and Forestry, Canada.

Fowells, H. A. 1965. Silvics of Forest Trees of the United States. Agricultural Handbook No. 271. U. S. Dept. Agriculture, Washington, D.C.

Franklin, J. F., and C. T. Dyrness. 1973. Natural Vegetation of Oregon and Washington. U.S.D.A. Forest Service Gen. Tech. Rep. PNW-8. Pacific Northwest Forest and Range Exper. Sta. Forest Service, U.S. Department of Agriculture, Portland, OR.

Futuyma, D. J. 1979. Evolutionary Biology. Sinauer Associates, Inc. Sunderland, MA.

Harper, J. L. 1969. The role of predation in vegetational diversity. Diversity and stability in ecological systems. Brookhaven Symp. in Biol. 22: 48-61. Brookhaven National Laboratory. Upton, NY.

Hansen, E. M. 1979. Survival of *Phellinus weirii* in Douglas-fir stumps after logging. Canadian J. of Forest Res. 9: 484-488.

Heinselman, M. L. 1973. Fire in the virgin forests of the Boundary Water Canoe Area, Minnesota. Quaternary Res. 3: 329-382.

Loucks, O. L. 1970. Evolution of diversity, efficiency, and community stability. Amer. Zool. 10: 17-25.

MacArthur, R., and R. Levins. 1967. The limiting similarity, convergence and divergence of coexisting species. Amer. Nat. 101: 377-385.

Manion, P.D. 1981. Tree Disease Concepts. Prentice-Hall, Inc. Englewood Cliffs, NJ.

McIntosh, R. P. 1987. Pluralism in ecology. Ann. Rev. Ecol. Syst. 18: 321-341.

McCauley, K. J., and S. A. Cook. 1980. *Phellinus weirii* infestation of two mountain hemlock forests in the Oregon Cascades. Forest Sci. 26: 23-29.

Mehringer, P. J., E. Blinman, and K. L. Peterson. 1977. Pollen influx and volcanic ash. Science 198: 257-261.

Minore, D. 1979. Comparative Autecological Characteristics of Northwestern Tree Species - a Literature Review. Gen. Tech. Rep., PNW-87. Pacific Northwest Forest and Range Exper. Sta., Forest Service, U.S. Dept Agriculture, Portland, OR.

Packee, E. C., C. D. Oliver, and P. D. Crawford. 1982. Ecology of pacific silver fir. In C.D. Oliver and R.M. Kenaday [eds.]. Symp. Proc.: The Biology and Manage. of True Fir in the Pacific Northwest. Forest and Range Exper. Sta.,Forest Service. U.S.Dept. .Agriculture. Portland, OR.

Pickett, S. T. A. 1980. Non-equilibrium coexistence of plants. Bull. Torrey Bot. Club 107: 238-248.

Pickett, S. T. A., and P. S. White [eds.]. 1985. The Ecology of Natural Disturbance and Patch Dynamics. Academic press, NY.

Paine, R. T. 1966. Food web complexity and species diversity. Amer. Nat. 100: 65-75.

Paine, R. T., and S. A. Levin. 1981. Intertidal landscapes: disturbance and the dynamics of pattern. Ecol. Monogr. 51: 145-178.

Roughgarden, J. 1975. Evolution of marine symbiosis - a simple cost-benefit model. Ecology 56: 1201-1208.

Sanders, H. L. 1968. Marine benthic diversity: a comparative study. Amer. Nat. 102: 243-282.

Schoener, T. 1965. Evolution of bill size differences among sympatric congeneric species of birds. Evol. 19: 189-213.

Simenstad, C. A., J. A. Estes, and K. W. Kenyon. 1978. Aleuts, Sea otters, and alternate stable-state communities. Science 200: 403-411.

Smith, F. E. 1972. Spatial heterogeneity, stability and diversity in ecosystems. In E. S. Deevey [ed.]. Growth by Intussesception, Ecological Essays in Honor of G. Evelyn Hutchinson. Trans. Connecticut Acad. of Arts and Sciences, New Haven, CT.

Taylor, E. M. 1968. Roadside geology, Santiam and McKenzie Pass highways, Oregon. In H.M. Dole. Andesite conference guidebook. Oregon Dept. of Geology and Mineral Industries Bull. 62.

Vermeij, G. J. 1983. Intimate associations and coevolution in the sea. In D. J. Futuyma and M. Slatkin. Coevolution. Sinauer Associates Inc. Sunderland, MA.

Watt, A. S. 1919. On the causes of failure of natural regeneration in British oakwoods. J. Ecol. 7: 173-203.

Watt, A. S. 1924. On the ecology of British beechwoods with special reference to their regeneration. Part II. The development and structure of beech communities on the Sussex Downs. J. Ecol. 12: 145-204.

Watt, A. S. 1934. The vegetation of the Chiltern Hills, with special reference to the beechwoods and their seral relationships. Part 2. The vegetation of the plateau. J. Ecol. 22: 445-507.

Watt, A. S. 1944. Ecological principles involved in the practice of forestry. J. Ecol. 32: 96-104.

Watt, A. S. 1947. Pattern and process in the plant community. J. Ecol. 35: 1-22.

Wilson, D. S. 1986. Adaptive indirect effects. In J. Diamond and T. J. Cased [eds.]. Community Ecology. Harper and Row, Publ., NY.

Witman, J. D. 1985. Refuges, biological disturbance, and rocky subtidal community structure in New England. Ecol. Monogr. 55: 421-445.

Witman, J. D. 1987. Subtidal coexistence: storms, grazing, mutualism, and the zonation of kelps and mussels. Ecol. Monogr. 57: 167-187.

INTERACTIONS BETWEEN GENETIC AND ECOLOGICAL PATCHINESS IN FOREST TREES AND THEIR DEPENDENT SPECIES

Yan B. Linhart

INTRODUCTION

The major premises of this essay are that (I) Forest tree populations often consist of family groups arrayed in mosaics of genetically and phenotypically distinct patches. (2) When a given species of parasite, fungus or herbivore (termed dependent species) attacks a tree population, the damage it produces is patchy because family groups of trees differ in phenotypic features such as attractiveness (e.g., nutrient levels in tissues) or repellency (e.g. levels of deterrent secondary compounds). Such phenotypic differences are determined, at least in part, by genetic differences. (3) Because of interspecific differences in the physiology of dependent species, interspecific differences exist in preference for (or avoidance of) different groups of phenotypes of a single host species. This is termed Species-Specific Host Selection. (4) A given dependent species is usually genetically variable. This can generate a situation where there is specific preference (or avoidance) of specific groups of phenotypes of one host species by specific groups of phenotypes of one dependent species. This is termed Phenotype-for-Phenotype Selection. (5) Both modes of selection operate in forest stands and generate complex evolutionary diversification acting upon tree populations in space and time.

The framework for the scenario is provided by the genetic patchiness of the host plants. Patterns of host use develop because of variability in behavior and physiology both among and within dependent species. Coevolution does not need to be invoked to explain the patterns observed.

Genetic analyses of forest tree populations indicate the presence of substantial spatial patchiness within these populations, even when they occupy relatively homogeneous habitats. This genetic heterogeneity is most likely the result of patterns of seed dispersal. In wind-dispersed species, seed dispersal tends to be relatively limited, and most seeds of a given parent tree tend to land and germinate nearby (Levin and Kerster, 1974; Levin, 1981). In animal-dispersed species, seed vectors usually consume several to many fruits from a single tree before moving on (Howe and Smallwood, 1982). Therefore, the seeds defecated or otherwise deposited in any one location often consist of genetically-related propagules, even when this location is some distance from the parent tree (see Hamrick and Loveless, 1986). These patterns of animal-generated seed deposition also can lead to the existence of family groups: i.e., groups of genetically-related individuals. From studies of seed dispersal and genetic structure, we can define a patch as consisting of a group of about 10 to 100 trees, usually occupying an area of approximately 25 to 5,000 m^2 (Levin and Kester, 1974; Levin 1981; Linhart et al. 1981b).

Genetic patchiness associated with family structure will occur even in the presence of very high outcrossing (e.g. Linhart et al. l98la,b; Mitton et al. 1981) and regardless of the number of pollen donors, because all progeny of a given seed tree are at least half-sibs. However, patchiness can be reinforced by localized gene flow via pollen, as noted in many species (e.g. Levin, 1981; Turner et al., 1982; Handel, 1983). When this familial mosaic is superimposed on the mosaic of environmental heterogeneity produced by factors such as variable soil chemistry and mycorrhizal activity, moisture and light conditions, competition, allelopathy, and gap-producing disturbances, the potential for spatial heterogeneity is further increased. The existence of this genetic heterogeneity has been documented in forests consisting primarily or solely of single species as well as multi-species ones (Table 1).

As a result of clustering of genetically-related individuals, forest stands consist of groups of individuals that resemble one another in those phenotypic features which have, at least in part, some genetic basis. These include morphological features such as leaf shape, thickness and pubescence, bark properties, branch architecture and biochemical properties such as relative amounts of sugars, amino acids, terpenes, tannins, etc.. The existence of significant genetic variability for these characters is documented in the plant breeding and forest genetics literatures (e.g. Allard 1960; Stern and Roche 1974; Dorman 1976; Wright 1976). As a result, the degree of phenotypic patchiness is likely to be proportional to the degree of genetic patchiness.

Forest trees serve as hosts to a large variety of disease organisms, parasites and herbivores. For the sake of simplicity, these species in the aggregate are referred to as *dependent species*. This term is considered useful because it includes both the species with negative impacts upon trees and other plants (the "pests" of forest managers) and those with beneficial roles (such as

Table I. Examples of evidence for the existence of genetic heterogeneity within forest tree populations. Relative species diversity of trees in forests occupied by species listed is noted at L (1-5 tree species comprise these forests); M (5 - 20 or so species); H (over 20, and usually over 50 species).

Species	Tree Species Diversity	Reference
Gymnosperms		
Picea abies (Pinaceae)	L	Tigerstedt, 1973
Pinus ponderosa (Pinaceae)	L	Linhart et al. 1981a, b
Pinus sylvestris (Pinaceae)	L	Tigerstedt et al. 1982
Pinus albicaulis (Pinaceae)	L	Furnier et al. 1987
Pseudotsuga menziesii (Pinaceae)	L	Shaw and Allard, 1981
Cupressus macrocarpa (Cupressaceae)	L	Kafton, 1977
Thujopsis dolobrata (Cupressaceae)	M	Sakai and Myiazaki, 1970
Angiosperms		
Shorea leprosula (Dipterocarpaceae)	H	Gan et al. 1977
Xerospermum intermedium (Sapindaceae)	H	Gan et al. 1977
Betula pendula (Betulaceae)	L	Scholtz 1960
Altingia excelsa (Hamamelidaceae)	H	Sakai, 1985
Fagus sylvatica (Fagaceae)	M	Cuguen et al. 1984
Coffea canephora (Rubiaceae)	H	Berthaud, 1984
Pithecellobium pedicellare (Mimosaceae)	H	O'Malley and Bawa, 1987
Piper amalago (Piperaceae)	H	Heywood and Fleming, 1986

mycorrhizae) and because it includes animals such as insect or mammalian herbivores, plants such as parasitic mistletoes, fungi such as disease-causing rusts and beneficial mycorrhizae, and bacteria. This term does not imply that a dependent species always depends on a single species of host but that its biological activities are focused upon one or more species of tree. Damage produced by dependent species within forest stands is often irregular or patchy. This irregularity is probably the result of a combination of patchy distribution of dependent species, known to be common among animals (Kareiva, 1986), and their apparent preference (or avoidance) of certain host phenotypes over others (e.g. Denno and McClure, 1983). This discrimination in turn is often due to either morphological or biochemical (i.e. nutritional or deterrent) features of individual phenotypes. If parasites and herbivores are to exert selective pressures on their tree hosts, it is important that the variable host characters associated with differential damage have a demonstrable genetic basis. In some cases, as in the so-called "gene-for-gene hypothesis" which usually involves resistance of given cereal varieties to specific races of fungal diseases, the genetic control of resistance is due to one or a few genes (Flor, 1956; Day, 1974). In other cases, the variability in resistance is multigenetic (Dimock et al. 1976; Hood and Libby, 1980). In other cases yet, the genetic basis has not been documented in the species studied, but can be inferred from studies of other species whose palatability varies either because of morphology (e.g. leaf thickness, trichome density), relative amounts of attractive compounds (e.g., sugars, amino acids), or deterrent compounds (e.g., terpenes, glycosides).

Any one species of host can be exploited by a variety of disease organisms, parasites and herbivores, and the resources exploited can be leaves, stem, wood, phloem, roots or other plant parts. In addition, many dependent species are likely to have different physiologies and behaviors, and therefore somewhat different *desiderata* with respect to a host plant. For example, Bernays (1982) points out that there are important differences among mammalian and among insect species in their sensitivities to hydrogen cyanide (HCN), daily energy requirements, and optimal ratios of protein to available carbohydrates plus fat; mammals also differ as a group from insects in these factors. In addition, insects may be relatively sensitive to terpenoids. Consequently, what makes a tree attractive to a fungus, a beetle and a deer probably involves different features of its phenotype. That is why I refer to these interspecific differences in host use as *Species-Specific Host Selection*. Examples are listed in Table 2, and suggest that, within a stand, phenotypes can be differentially

Table 2. Examples of species-specific host selection. Intra-specific variability in a host plant is associated with interspecific differences in host preference by species of parasites or herbivores (i.e. dependent species).

Host Species	Dependent Species	Preference Pattern	Reference
Pseudotsuga menziesii	Deer (*O. hemionus*) Hare (*L. americanus*)	No correlation between herbivore spp. in preferences for specific clones	Dimock et al. 1976
	Wooly aphid (*Gilletteella cooleyi*) Needle cast fungus (*Rhabdocline pseudotsugae*)	Populations resistant to aphid are susceptible to fungus	Stephan, 1987
Pinus radiata	*O. hemionus* Porcupine (*E. dorsatum*)	The herbivores prefer different clones	Hood and Libby, 1980
Dictyota dichotoma	Fish (*Lagodon rhomboides*) Polychaetae (*Platynereis dumerilii*)	*L.r.* deterred by diterpene alcohols, which stimulate feeding by *P.d.*	Hay et al. 1988
Camellia sinensis	80 spp. of Phytophagous Insects.	Guilds which vary in feeding patterns prefer hosts with different leaf characters	Banerjee, 1987
Pinus ponderosa	*O. hemionus* Rabbit (*S. nuttalii*) *E. dorsatum*	*O.h.* and *S.n.* prefer trees of same origin. *E.d.* preferences very different	Squillace and Silen, 1962
	O. hemionus Wooly aphid (*P. coloradensis*)	Feeding tree distributions show no significant correlation	Linhart, in prep.
	Porcupine (*E. dorsatum*) Squirrel (*S. aberti*)	*E.d.* and *S.a.* feed on trees with different resin characters.	Habeck, in prep. Snyder, in prep.
Trifolium repens	Molluscs, (primarily *Agriolimax reticulatus Arion ater Helix aspersa*) Weevils (Curculionidae) Fungi (*Cymadothea trifolii, Uromyces trifolii*)	Molluscs prefer acyanogenic morphs, *V.t.* damage mostly on cyanogenic morphs, *C.t.* show no differences, weevils are variable.	Dirzo and Harper, 1982b.
Cucurbita moschata	Beetles *Acalymma vittata* (Chrysomelidae) *Epilachna tredecimnotata* (Coccinellidae)	Cucurbitacins (probably) stimulate feeding by *A.v.* and inhibit feeding by *E.t.*	Carroll and Hoffman, 1980

susceptible to a variety of dependent species. Exceptions to this pattern of differential preferences do exist. For example, trees weakened by one parasite, such as a dwarf-mistletoe, may be especially prone to attack by *Dendroctonus* beetles (Mitton and Sturgeon, 1982). In a similar vein, certain host phenotypes may be vulnerable to attacks by many dependent species because they function poorly in all respects as a result of pleiotropic effects of genetic abnormalities, or chance establishment in unfavorable habitats. In addition, positive correlations in the spatial distributions of certain herbivore species have been noted (e.g., Whitham, 1983; Kareiva 1986). These exceptions are discussed below.

A parasite or herbivore species is not ecologically or genetically uniform. It is not an abstract "attack robot," but consists of populations of genetically-variable individuals. When this genetic variability within parasites or herbivores is superimposed upon a patchily variable host population, there develops the opportunity for very specific interactions. Host phenotypes or groups of phenotypes may be favored by specific phenotypes or groups of phenotypes of a single dependent species but avoided by other members of that species which favor hosts with different characteristics (Table 3). For this reason, I have termed such a pattern *Phenotype-for-Phenotype Selection*. A patchily variable pattern of selection upon a host can be especially pronounced if herbivore phenotypes are themselves in genetically-related groups. There is some evidence for such genetic clustering. Cases of aggregations of genetically-related individuals have been documented for a variety of herbivores including rodents (Anderson, 1970; Selander, 1970), butterflies (Rasmussen, 1978), slugs (Dirzo and Harper, 1982a), snails (Lamotte, 1959; Jones, 1973), and various other parasites (Price, 1980). Note also that in certain insects such as aphids, local groups can consist of parthenogenetically-produced, genetically-identical offspring of single mothers (Dixon, 1985). In other insects, genetic relatedness can be expected on the basis of egg-laying and/or dispersal patterns.

In the next sections, I present evidence for various components of this scenario, then note the exceptions, i.e. conditions under which it does not apply, and finally discuss the consequences of these multi-faceted interactions for genetic structure of forest tree populations and for the structure and organization of forest tree communities.

Table 3. Examples of phenotype-for-phenotype selection. Intra-specific variability in a host plant is associated with intra-specific variability of host preference within a single dependent species of herbivore, parasite or mycorrhizal fungus.

Host Species	Dependent Species	Preference Pattern	Reference
Flax, *Linum usitatissimum*	Rust, *Melampsora lini*	Specific genotypes of flax are susceptible to rust genotypes with specific pathogenicity alleles: "Gene-for-gene" pattern	Flor, 1956
Several monocot and dicot species	Several fungi	Gene-for-gene pattern	Reviews in Day, 1974; Barrett, 1983
Wheat, *Triticum aestivum*	Fly *Mayetiola destructor*	Gene-for-gene pattern	Gallun, 1977
Pinus ponderosa	Scale, *Nuculaspis californica*	Scale colonies survive better on original tree than on other trees to which transplanted	Edmunds and Alstad, 1978
Fagus sylvatica	Scale, *Cryptococcus fagisuga*	Scale colonies survive better on original tree than on other trees to which transplanted	Wainhouse and Howell, 1983
Plantago lanceolata	Pathogenic fungus, *Phomopsis subordinaria*	Significant host x fungus genotype interactions	de Nooij and van Damme, 1988
Pinus taeda	Mycorrhizal fungus *Pisolithus tinctorius*	Success of symbiosis and pine growth affected by host x fungus genotype interactions	Dixon et al. 1987
Rudbeckia laciniata	Aphid, *Uroleucon rudbeckiae*	Significant host clone x aphid clone interactions.	Service, 1984

A CASE STUDY IN STANDS OF *PINUS PONDEROSA*
Host Genetic Patchiness

Genetic analyses have been carried out in two ponderosa pine populations using electrophoretically-detectable variation at several protein loci. Both populations show significant spatial heterogeneity. For example, within a continuous ponderosa pine stand in Boulder Canyon, Colorado, all trees occupying an area of approximately two ha. were mapped, and their genetic constitution was determined for seven polymorphic loci. Most trees fall into one of six spatially-definable clusters, and these clusters differ significantly from one another at one or more loci (Table 4). Another arbitrarily-chosen area within a continuous ponderosa pine forest several km away shows a similar pattern of pronounced spatial heterogeneity (see Linhart et al., 1981a, 1981b for more complete descriptions of variability in these populations).

This spatial clustering and genetic patchiness are the result of limited seed dispersal and the associated localized seedling emergence (studied thanks to an albino allele, Linhart in prep.) and the patchy nature of forest disturbance and regeneration in this species (White, 1985).

Host Phenotypic Patchiness

Patchiness has been documented in patterns of cone production and in analyses of monoterpene composition of the xylem oleoresin. The results indicate that there are significant differences among tree clusters in cone production (Linhart et al., 1981a) and oleoresin amount and composition (Table 4, Linhart and Smith, unpubl. data). Rate of oleoresin flow is a good index of the volume of oleoresin output following tree wounding. In addition, individual trees differ in the proportions of various monoterpenes present in oleoresins (Smith, 1964, 1966). Consequently, intertree variation in oleoresin volume produces variation in amounts of monoterpenes released. Various monoterpenes have already been shown to play important roles in pine-insect interactions. For example, in ponderosa pine, myrcene is a precursor to the attraction pheromone of *Dendroctonus brevicomis* bark beetles, so that trees high in this compound suffer the consequences (Bedard et al., 1969). Conversely, high levels of limonene deter *Dendroctonus brevicomis* attacks (Smith, 1966; Sturgeon, 1979). Consequently interpatch difference in xylem oleoresin flow patterns can be expected to affect patterns of attack by certain dependent species.

Table 4. Differences between adjacent clusters of *Pinus ponderosa* in allele frequencies, oleoresin volume, and damage by two animals.

Feature	Cluster					
	A	B	C	D	E	F
Frequency of most common allele[1]						
PER (2)	0.79	0.57	0.75	0.71	0.74	0.72
F.E. (2)	0.63	0.67	0.75	0.74	0.46	0.55
C.E. (1)	0.72	0.77	0.56	0.77	0.70	0.77
GDH (1)	0.72	0.63	0.53	0.70	0.71	0.70
Oleoresin volume[2,3]	21.9 ± 2.6 (b)	19.4 ± 5.6 (b)	34.2 ± 5.2 (c)	23.9 ± 3.5 (b)	15.3 ± 2.7 (a)	28.5 ± 3.7 (c)
Relative damage (%)[2,4]						
Deer	49 (a)	49 (a)	30 (b)	33 (b)	34 (b)	48 (a)
Aphids	49 (a)	26 (b)	19 (b)	15 (b)	16 (b)	9 (b)

[1] Intercluster differences are significant for all loci shown (Linhart et al. 1981b).

[2] Intercluster differences significant. Cluster pairs compared with t-tests. Pairs that are significantly different from one another have different letters below them.

[3] Yield, in ml per vial inserted into trunk for 24 hr $\bar{x}$ ± S.E.

[4] Relative damage assessed as % trees within cluster with branches occupied by aphids (N=40 br./tree) or browsed by deer (N=10 Br./tree). Pairs that are significantly different from one another have different letters adjacent to them.

Patchy Damage by Dependent Species

In the Boulder Canyon population, the major vertebrate herbivore is the mule deer (*Odocoileus hemionus*); it browses on needles, and feeds more heavily on some trees than on others. These trees occur in groups. The most visible insect is the woolly aphid (*Pineus coloradensis*) which also shows a patchy distribution (Table 4). Neither of these animals has been shown to be attracted or repelled by specific monoterpenes; however, as discussed below, several mammal and insect species show strong feeding discrimination among individual ponderosa pines. In this population, deer and aphids attack trees with different genetic constitutions. These genetic differences also suggest the existence of complex diversifying selection, with certain genotypes being favored under some conditions and at a disadvantage at other times. The feeding preferences may act in a manner contrary to the potential selective effects of differential reproduction. Specifically, at one esterase locus (F.E.), allele 1 may be selected for because its carriers produce more cones but, at other times, it may be selected against because its carriers are browsed more heavily by deer. At another esterase locus (C.E.), carriers of one allele produce few cones, but are also parasitized less by the aphids.

More extensive documentation of patchy herbivore and disease attacks is available from a ponderosa pine stand in Arizona (Avery et al., 1976), where over 3300 trees were tagged and damage by several dependent species noted over a 50-year period. The species include squirrel (*Sciurus aberti*), porcupine (*Erethizon dorsatum*), a dwarf mistletoe (*Arceuthobium vaginatum*), a bark beetle (*Dendroctonus ponderosae*), and the fungus *Peridermium filamentosum*. Damage was recorded every five years from 1920 to 1955, then again in 1960 and 1970. The results show that different plots (each consisting of 200-300 trees in an area of slightly over 1 ha.) show evidence of different levels of damage by the various parasites and herbivores in the area (Table 5). A tree was recorded as damaged if it was attacked by a dependent species at any time during the 50-year period.

The distribution of damaged trees within a given plot at the Arizona site was patchy as well. Patterns of feeding by the squirrel and the porcupine are especially interesting because both are arboreal, and both feed on phloem to a significant degree. The squirrel strips phloem primarily from terminal twigs. The porcupine seems to prefer feeding on large branches and the main trunk. Distributions of attacked trees in a typical plot are shown in Fig. 1, where these distributions are clearly clustered. The two species usually fed on different trees. Of 655 trees attacked by

Table 5. Frequency of affected trees in 16 plots of *Pinus ponderosa* in N. Arizona. Each plot is approximately one ha. Frequency is expressed as % available trees showing attack by agent listed (data from Avery et al., 1976).

Plot	Total Trees Over 200 cm Diameter	Parasite or Herbivore[1]				
		Squirrel	Porcupine	Dwarf-Mistletoe	Bark Beetle	Rust
1	252	38	7	0.4	15	0.8
2	259	12	3	0.4	24	0
3	195	17	8	1.0	8	0
4	263	13	4	6.0	6	0.4
5	166	38	6	2.0	9	1.2
6	186	33	3	0	16	0.5
7	293	18	13	2.0	11	0
8	188	34	6	3.7	11	0
9	257	31	14	2.7	5	0
10	282	8	11	0	9	0
11	137	23	0	0	20	0
12	129	51	4	5.0	8	0
13	209	36	1	1.5	6	1.0
14	187	31	9	1.5	15	0
15	224	26	11	0	7	0.4
16	107	19	0	3.0	7	35.5
Results of G-test		P < 0.001	P < 0.001	N.T.[2]	P < 0.7	N.T.[2]

[1] Tassel-eared squirrel, *Sciurus aberti*; porcupine, *Erethizon dorsatum*; Dwarf-mistletoe, *Arceuthobiun vaginatum*; Bark beetle, *Dendroctonus ponderosae*; Rust, *Peridermium filamentosum*.

[2] Distributions not tested because of low numbers. However, note inter-plot differences.

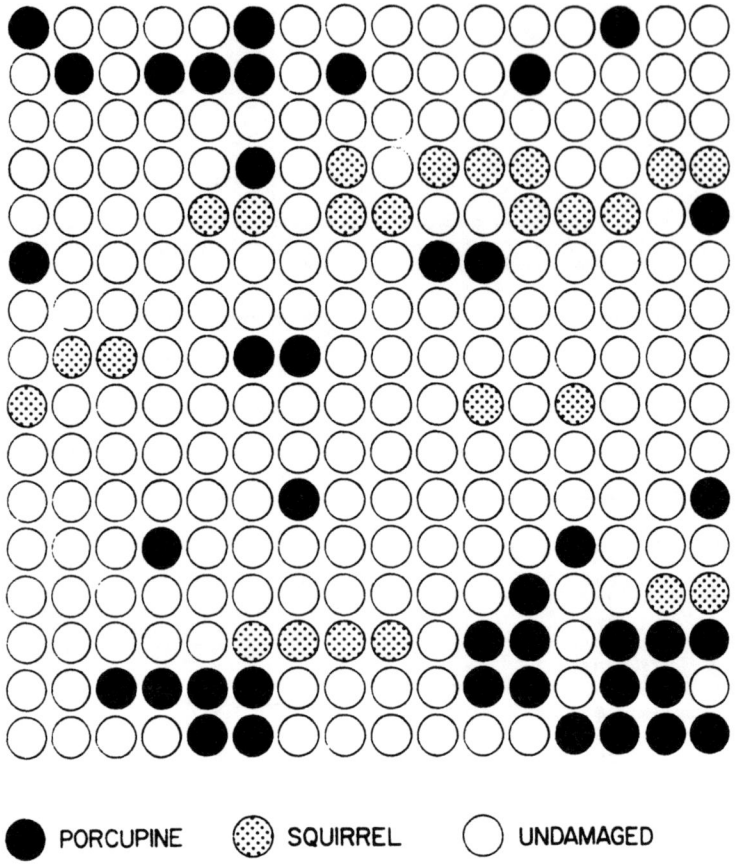

Figure 1. Approximate distribution of trees fed upon by squirrels (S) and porcupines (P) in a typical plot of ponderosa pine in N. Arizona (compiled from Avery et al., 1976). Each circle represents the approximate location of a tree, but trees have not been mapped precisely. Therefore this figure does not represent exact locations of trees. However, trees have been numbered; therefore, adjacent individuals are identifiable. Blank circles represent unaffected trees. It is typical of this and other plots that few or no trees show evidence of being utilized by both herbivores.

either phloem-feeder in the whole stand, only 18 were fed upon by both; if choices by the animals were completely random, 27 trees would be expected to show damage by both. The difference is not statistically significant. However the pattern observed suggests that the animals do not select the same feeding trees. Each tree was observed 10 times during the 50-year period of data collection, so there was plentiful opportunity for feeding by both animals to be observed on a given tree. That is, even if individuals of the two species avoided meeting one another, such avoidance could hardly focus around the same trees over a 50-year period, and is therefore inadequate to explain the results.

Feeding by the two species can be differentiated unambiguously. The squirrels feed within a selected tree where they clip off the terminal branches; the clipped-off foliage and stripped-off twigs litter the ground under these trees. In addition, favored feeding trees have a crown whose scraggly and worn appearance is markedly different from the full, dense crown and foliage of non-feeding trees. In contrast, porcupine feeding leaves distinctive large patches of exposed wood from which bark and phloem have been chewed off.

Herbivores are known to select (or actively avoid) individual feeding plants or sites using a wide variety of cues (e.g. Denno and McClure, 1983), and these patterns of squirrel and porcupine feeding provide no clues about the nature of these cues. Since there are no consistent differences in the sizes of the trees they feed on, these cues are most likely biochemical, and we are trying to determine what they are. In feeding experiments, *Sciurus aberti* have been shown to select trees with lower levels of cortical monoterpenes (Farentinos et al., 1981). In addition, preliminary results indicate that squirrels appear to use levels of xylem monoterpenes as cues (M. Snyder, in prep.); porcupines are not affected by these monoterpenes, but appear to use as cues phloem levels of certain metallic elements (S. Habeck, in prep.). The results suggest that even when two herbivores focus on the same resource, phloem in this case, they may use very different cues to determine which trees to feed on and which to leave alone. Both species can reduce host fitness considerably. Squirrel feeding trees have their pollen and seed outputs reduced by 90% compared to adjacent nonfeeding trees (M. Snyder, unpubl. data) and porcupines can girdle and eventually kill trees (Shubert, 1974).

Additional evidence for patchy distributions of damage in ponderosa pine forests include attacks by dwarf-mistletoe (Scharpf and Parmeter, 1978), *Dendroctonus* bark beetles

(Mitton and Sturgeon, 1982), *Nuculaspis* scale (D. Alstad, pers. comm.), and the pine white butterfly *Neophasia* (Weaver, 1961).

Genetic Variability in Susceptibility to Dependent Species

In addition to the monoterpene and protein variability noted above, ponderosa pine also varies in other traits relevant to attacks by herbivores and parasites. A differential susceptibility of individuals to attacks by dwarf-mistletoes (*Arceuthobium*) is said to have a genetic basis (Roth, 1974; Scharpf, 1984), though no specific characters associated with such resistance are identified. The resin midge (*Cecidomyia piniinopis*) lays its eggs into growing branch tips. Trees with shoots that have sticky or resinous surfaces are more prone to attack than trees with surfaces that are either smooth or covered with a waxy bloom (Austin et al., 1945). These shoot characters appear to be controlled by two genes (Duffield, 1985), and in a common-garden test of several families, heritabilities were estimated at 0.48 to 0.60 (Hoff, 1988).

Finally, ponderosa pine has been shown or suggested to vary genetically in its susceptibility to parasites and herbivores following studies of trees from many origins planted in common environments. These studies include susceptibility to browsing by deer, *Odocoileus hemionus* (Bates, 1927a) resistance to gall rust, *Peridermium harknessii,* (Bates, 1927b), susceptibility to tipmoth, *Rhyacionia* sp. (Higgins, 1927; Read 1980), and browsing impacts by deer (*O. hemionus*), porcupine (*Erethizon dorsatum*), and rabbits (*Sylvilagus nuttalii*) (Squillace and Silen, 1962).

Genetic Heterogeneity of Herbivores and Parasites

The scale insect *Nuculaspis* and the dwarf mistletoe *Arceuthobium* are both genetically variable. In *Nuculaspis*, colonies on individual trees appear to be genetically differentiated from one another, and transplant experiments have shown that colonies show higher survival on their tree of origin than when they are moved onto different ponderosa pines in other stands (Edmunds and Alstad, 1978), or within a stand (Alstad, pers. comm.). However, interpretation of some results is difficult, and more studies are needed (Unruh and Luck, 1987). In *Arceuthobium americanum* and *A. vaginatum*, groups of individuals parasitizing a given clump of trees are genetically significantly differentiated from other groups occupying nearby clumps (Linhart et al., in prep.). The extent to which *Arceuthobium* differentiation is the result of host selection (as it may be in

Nuculaspis) or the restricted dispersal known to occur in *Arceuthobium* (Scharpf and Parmeter, 1978) has yet to be determined. The genetic variability in these species provides the opportunity for specific interactions between host phenotypes and dependent species phenotypes.

EVIDENCE FROM OTHER WOODY SPECIES
Spatial Patchiness in Forests
Students of forest stands have noted the existence of gaps of different ages and compositions and the associated dynamic changes in forest structure on many occasions. These include the patterns noted in European forests by Jones (1945) and Watt (1947), Mayer and his co-workers' "Waldtextur" in Austria and Yugoslavia (e.g. Mayer and Neumann, 1981), Aubréville's observations in tropical forests (1938), and much recent work summarized in Pickett and White (1985). Whenever a gap opens, this provides the opportunity for regeneration, at least of certain species. Seeds from a nearby tree or trees are likely to be the primary beneficiaries, which provides the opportunity for establishment of genetically related groups of seedlings.

Genetic Patchiness of Forest Populations
Most species whose genetic constitution has been studied intensively at the local level in natural populations appear to show evidence of patchiness (Table 1). This includes species where the existence of such patchiness was thought to be absent, such as *P. abies* (Tigerstedt, 1973), but where, in fact, a re-analysis of the data indicates that patchiness exists (Linhart et al., 1981a). In addition, genetic analyses of pollination patterns in a number of species indicate the presence of significant inbreeding. In the species where such inbreeding has been detected, rates of outcrossing are usually very high. Consequently the observed inbreeding is most likely to have arisen from mating between nearby relatives, and is diagnostic of substantial family clustering and related types of genetic substructuring (e.g., Brown, 1979; Fins and Libby, 1982; Namkoong, 1984).

Genetic Patchiness in Dependent Species
Various herbivores and parasites are known to exhibit significant spatial patchiness of genetic constitutions. As expected, this patchiness appears to be especially common among sedentary species where it is often associated with family structure. These include scale insects (Edmunds and Alstad, 1978; Wainhouse and Howell, 1983), *Arceuthobium* dwarf-mistletoes (Linhart et al., unpubl. data), and molluscs including

the slugs *Agriolimax* spp. and *Arion* spp. (Dirzo and Harper, 1982a), and *Cepaea* snails (Lamotte, 1959; Jones 1973). Fungal pathogens also can be locally very variable (Burdon, 1987; Wolfe and Canten, 1987). Vagile species can be genetically substructured as well. Rodents, especially *Mus,* show this pattern (Anderson, 1970; Selander, 1970). Certain Lepidoptera of various species also are genetically structured into family groups either because of egg-laying or dispersal behaviors (Rasmussen, 1978; Fitzegerald and Peterson, 1988). In other Lepidoptera, reasons for structuring are not known, but the pattern is both very strong and very localized (Namkoong et al. 1982).

Genetic Variability in Host Susceptibility to Dependent Species

When dependent species attack plant populations, they often exhibit spatial variation in patterns of attack. The relative contributions of host genotype and environment to this variation cannot be ascertained without further analysis. When such analyses are done, a genetic component to insect or disease resistance usually is found. These analyses (which often include progeny tests and selection of resistant individuals, families or geographic provenances) have been especially common in commercially important forest trees (e.g., Fowells, 1965; Dorman, 1976; Wright, 1976; Prakash and Heather, 1986; Old et al., 1986; Stephan, 1987) although other species have been studied as well (e.g., McCrea and Abrahamson, 1987; Parker, 1988). Several detailed analyses have included clonal replication of genotypes as an important component of careful design. These studies provide important insights into the complexities of interactions between hosts and dependent species. For example, Fritz et al. (1987) and Fritz and Price (1988) have shown that there are marked differences between clones of the willow *Salix lasiolepis* in their susceptibilities to various gall-forming sawflies (Hymenoptera, Tentherenilidae). Maddox and Root (1986) carried out an exhaustive study of resistance to 16 diverse species of insects by *Solidago altissima.* The conclusion in both studies is that while there is a demonstrable genetic basis for resistance in certain situations, environmental factors can also be important. This is not surprising since the pathways from genotype to adult plant phenotype are lengthy, and both temporal and spatial environmental heterogeneity are often extensive. For example, seasonal and developmental variability in physiological activities produce within-plant spatial and temporal variation in nutrients, morphological characters, and secondary compounds (Denno and McClure, 1983). Host densities can affect temperature and moisture conditions (Pickett and White, 1985). Furthermore, the feature(s) of the host phenotype used by the dependent species to make its decision about whether or not to settle and feed on an individual host are also very poorly understood. Certainly food quality in the form of concentration of nutrients, and presence and amount of defensive compounds, is important. Finally, interactions between hosts and dependent species are often strongly affected by the activities of parasitoids and predators upon dependent species (Price et al., 1986).

One of my basic premises is that if susceptibility to attack by a dependent species has a genetic component, then genetically-related host individuals should resemble one another more closely in their levels of susceptibility than do unrelated ones. Similarities of clonal performance (e.g. the above studies by Maddox and Root, 1986; and Fritz et al., 1987) provide support for the premise. In addition, Rice (1983) demonstrated that, there is a positive association between genetic relatedness of *Pinus lambertiana* hosts and susceptibility to *Nuculaspis* scales.

Phenotypic Host Patchiness and Patterns of Attack by Dependent Species

Very few studies of intra-population patchiness have been done in woody species. A study of spatial variation in thyme (*Thymus*) indicates the existence of strongly patchy (and genetically-determined) biochemical phenotypes (Mazzoni and Gouyon, 1984). In addition, studies of flowering and seed production indicate that there is substantial patchiness in the spatial distribution of reproductive activity in several species (Shea, 1985; Linhart et al., 1987). In the herb *Trifolium*, patchy distribution of palatable and unpalatable morphs is associated with the patchy distribution of herbivores, although the direction of causality is unclear (Dirzo and Harper, 1982a).

When a fungus or insect species attacks a forest stand, a common pattern of attack is for one or a few individuals to land on a tree and attack it. Then, they or their progeny disperse to other trees. The end result is a patch of damaged trees. After some time has elapsed, several to many such patches can often be observed. This observation has been made for damaging agents as diverse as scolytid beetles (Schowalter et al., 1981; Mitton and Sturgeon, 1982), oak - insects (Futuyma and Wasserman, 1980), *Arceuthobium* dwarf-mistletoes (Hawksworth and Wiens, 1972; Scharpf and Parmeter, 1978), and various fungi (Wolfe and Caten, 1987). The usual assumption (discussed in the references cited above) is that the patchy distribution of damage is the result of one or a few individual propagules landing in a portion of a stand, followed by reproduction at that spot, then slow and localized dispersal from that spot. Both limited dispersal and sedentary habits undoubtedly contribute to the patchy nature of the damage observed. However the data presented here (Tables 2-5) strongly suggest that differential host susceptibility also plays a role.

The dependent species discussed above are all deleterious to trees. There is also a potential for dependent species whose effects are positive to respond to genetic patchiness. Specifically, mycorrhizal fungi are also known to have genetically-based differential abilities to form associations with individual hosts (Dixon et al., 1987; Thomas and Ghai, 1987). I expect that endophytic fungi (e.g., Carroll, 1988) will show similar patterns of variability and specificity.

Species-Specific Host Selection

One of the basic questions of this review is whether attractiveness of a host individual to one dependent species is associated in any way with its attractiveness to another species. Data are scarce (Table 2). In Douglas-fir (*Pseudotsuga menziesii*) certain clones are preferred by deer, others by hare, but there is no agreement between the two herbivores (Dimock et al., 1976). In this host species, populations that are resistant to wooly aphids are also susceptible to a needle cast fungus (Stephan, 1987). In *Pinus radiata*, deer and porcupine show different patterns of preference for certain clones. In none of the above examples can preference by dependent species or susceptibility by a host be related to specific features of host phenotype. However, in some situations, a class of compounds has a demonstrably deterrent effect upon one herbivore, but stimulates feeding by another herbivore species (Carroll and Hoffman, 1980; Hay et al., 1988).

In tea (*Camellia sinensis*), leaf geometry and surface characteristics vary. Leaves can be small, erect, and densely hairy, or large, horizontal and smooth, or semi-erect and intermediate in leaf area and pubescence. These leaf characters are under strict genetic control: an individual plant will retain a specific leaf type on all its branches, and there is no variation in leaf type with plant age, phenology or development (Banerjee, 1987). A survey of 80 species of phytophagous insects showed relatively strict preferences for specific leaf types by various guilds. Defoliators and leaf chewers fed primarily on plants with horizontal leaves. Leaf rollers and a leaf miner fed exclusively on semi-erect leaves. Sap feeders fed on all leaf types, but concentrated on erect and semi-erect leaves. Whether plants with these leaf types also differ in their chemistries is unclear. However, it may be that the choice of leaves is at least partly associated with the insects' mobility. Erect leaves attract primarily sessile species, while horizontal leaves attract mobile species, and semi-erect species harbor semi-mobile ones (Banerjee, 1987).

In the examples cited above, heterogeneity of host attractiveness is either known or presumed to have a genetic basis. However, such heterogeneity can also be somehow induced by one herbivore or parasite, thereby reducing the likelihood of attack by another species. For example, the fungus *Verticillium dahliae* and the spider mite *Tetranychus urticae* both attack cotton, *Gossypium hirsutum*. However, cotton plants exposed to either of these species are less likely to be infected and, if infected, suffer less severe damage by the other species (Karban et al., 1987). These authors report other cases of such interspecific interactions and resulting induced resistance in cotton. Hawksworth (1978) also notes that individual ponderosa pines attacked by the dwarf-mistletoe *Arceuthobium americanum* are less likely to support colonies of *A. vaginatum*, and vice versa; he suggests the existence of some interspecific physiological allelopathy by the two species. Recently, Crawley and Panttrasudhi (1988) have noted that interspecific competition between two insect herbivores on *Senecio jacobaea* produces a similar pattern.

Phenotype for Phenotype Selection

The existence of this type of selection presupposes genetic heterogeneity in dependent species, associated heterogeneity of host preference, and an interaction between the two. This pattern has been documented in several studies (Table 3, and Rauscher, 1983). The best-known examples of this association come from studies of cereal crops and fungi producing cereal rusts. Both hosts and parasites are genetically variable, and their interactions can be so specific that even on a small spatial scale, certain parasite races attack only certain host strains, leaving others alone (Day, 1974).

In studies of herbivory by several slug species (Dirzo, 1980; Whelan, 1982), locusts, *Locusta migratoria* (Bernays et al., 1976), and various other grasshoppers (Whelan, 1978), it has been noted that certain individuals repeatedly showed preference patterns at variance with those of their conspecifics. Dirzo (1980) noted that such "egregious slugs" suggest the possibility of genetically-based polymorphism in food preference. These observations, which have yet to be rigorously analyzed, do suggest the existence of intra-specific variability in herbivores for preference of specific phenotypes within given plant species. However, the existence of host X dependent species interactions is not an adequate demonstration of genetic specialization; differential fitness of specific dependent species genotypes on specific host plant genotypes is needed (see Rauscher, 1983, for details). In addition, there exists intra-specific variation by herbivores for preference of specific plant species over other species (e.g. Furniss and Carolyn, 1977; Denno and McClure, 1983; Futuyma and Peterson 1985; Berryman, 1988). In a number of insects, evidence for a strong association between specific insect phenotypes and specific host plant species has been well-documented. This has given rise to the idea that host races are common within insects. However, in some cases, the apparent host races may be sibling species. In other cases, a genetic basis for host preferences has not been established. A recent review of these complexities is provided by Futuyma and Peterson (1985).

In insects which are sessile and/or parthenogenetic at least during some portion of their life span, the possibility that there is phenotype-for phenotype selection is especially likely because of the reduced levels of genetic recombination associated with their life histories. As noted in an earlier section, the experiments of Edmunds and Alstad with *Nuculaspis* scale and ponderosa pine are suggestive, although the evidence needs to be firmed up (Unruh and Luck, 1987). The scale insect *Matsucoccus acalyptus* does not appear to form demes adapted to individual *Pinus monophylla* (Unruh and Luck, 1987). However, the scale *Cryptococcus fagisuga* does form demes adapted to individual beech *Fagus sylvatica* (Wainhouse and Howell, 1983). Cottonwood (*Populus*) trees can be either genetically susceptible or resistant to *Pemphigus* aphids. When aphids use susceptible trees, they have a complex life cycle which includes migration from the roots of an herbaceous host and sexual reproduction on the trees. Conversely, when aphids are faced with resistant trees, they stay away from trees and reproduce only asexually on their herbaceous host. Transfer experiments have demonstrated that both host and aphid traits have a genetic basis (Moran and Whitham, 1988).

In fungi, there also is evidence for specific host-fungus genotype interactions. The pathogenic *Phomopsis subordinaria* shows such specificity with *Plantago lanceolata* (de Nooij and van Damme, 1988). When mycorrhizal *Pisolithus tinctorius* infest loblolly pine (*P. taeda*), the success of the symbiosis and pine seedling growth are affected by interactions between specific fungal and host genotypes (Dixon et al., 1987).

EXCEPTIONS

There are documented exceptions to the patterns of differential host preferences by different species of parasites or herbivores. Most of these exceptions appear to fall into one of five categories. One category consists of situations where individual trees attacked by one species are weakened and therefore made prone to subsequent attack by the same or other species. For example, ponderosa pine weakened by dwarf-mistletoes, drought or physical damage can be especially prone to attack by *Dendroctonus* beetles or by the moth *Coloradia pandora pandora*; this appears to be the logical outcome of weakening of a tree's metabolism to the point that it is unable to produce enough resin to "pitch out" burrowing *Dendroctonus* as healthier trees can do (Mitton and Sturgeon, 1982), or tolerate the heavy defoliation of *Coloradia* (Wagner and Mathiassen 1985). Other situations in which weakening of a host predisposes it to attack by a variety of insects (Berryman 1988) or diseases (Boyce, 1961) also are known. A complicating factor is that the type of damage done to a host may affect its subsequent susceptibility. When the birches *Betula pendula* and *B. pubescens* are grazed or ring-girdled, usually by hares (*Lepus*) near ground level, the trees regrow relatively short shoots with increased levels of secondary compounds which apparently discourage subsequent feeding. Conversely, when twigs are browsed at some distance above ground, e.g., by moose (*Alces alces*), the trees respond by producing long, rapidly growing shoots. Such rapid growth may be useful to escape further browsing and keep up with competing neighbors. However these shoots may also contain fewer secondary compounds and are more attractive to moose and herbivorous insects, but not to hares (Danell and Huss-Danell, 1985).

The second class of exceptions consists of the modification of host physiology by one dependent species and increased attraction to another dependent species. The tortricid moth *Lobesia botrana* develops faster on grapes (*Vitis vinifera*) and apples (*Malus malus*) infected by the fungus *Botrytis cinerea* than on uninfected fruits. This may be due to a change in content of sugars, organic acids, or antibiotic substances (Savopoulou-Soultani and Tzanakakis, 1988). Individual ponderosa pines damaged by the fungus *Leptographium* are favored by porcupines (Spencer, 1964) apparently because of the higher sugar content in the sap of these individuals. Similarly, ponderosa pine attacked by *Arceuthobium* mistletoes appear to be prone to porcupine attack especially near mistletoe colonies (Linhart and Malville, unpubl. data) presumably because of high accumulation of carbohydrates in the area brought on by the presence of *Arceuthobium* (Tinnin, 1984).

The third class of exceptions consists of plants attacked by closely-related groups of species. For example, two species of *Phyllotetra* beetles show positive correlations in their distributions on their *Brassica* hosts (Kareiva, 1986). The same is true in two species of *Pemphigus* gall aphids on *Populus* (Whitham, 1983). The survey of insects on tea with different leaf morphologies indicates that whereas guilds with different feeding activities prefer different leaf morphologies, members of a given guild that consists of several species of one genus always show a preference for the same leaf type(s) (Banerjee, 1987). In these cases, it may be argued that certain groups of closely related species have similar physiologies, and therefore similar *desiderata*. Such similarities of physiologies are also evident in many *Dendroctonus* bark beetles (Mitton and Sturgeon, 1982). It is not clear at this point how common are such positive correlations between patterns of attack by different but related species upon specific individual hosts. Certainly, related species can also have very different patterns of establishment on individual hosts (e.g., Hawksworth, 1978; White, 1980).

The fourth category is similar to the third, but involves unrelated dependent species feeding on one specific plant part, and perhaps deterred or attracted by the same characters. The gall-forming flies on willow (Fritz et al. 1987) all use leaves and show, in some cases, significant positive correlations on certain clones. Rabbits and deer both feed on ponderosa pine foliage and appear to be positively correlated in their preferences (Squillace and Silen, 1962).

The fifth class of exception occurs when a parasite attack is of epidemic proportion, in which case most or all individuals in a stand are stricken. The reasons for this exceptional, seemingly non-discriminating pattern of attack are unknown but may be related to a combination of drought stress on hosts, with increased population sizes (and associated competition for food), along with changes in the genetic constitution of the parasite. When insect populations increase rapidly in numbers, the genetic variability observed in the expanding population can expand as well (Ford, 1962). This increased variability along with competition may enable some, and force, other phenotypes to feed on heretofore unpalatable individual trees. Another possible scenario involves herbivores or parasites which are endemic in a given area, and whose populations become increasingly well-adapted to their long-lived host patches. Eventually they overwhelm these patches, expand in numbers, and mate with populations on other patches. Genetic variability increases as a result, and an epidemic can start (Alstad, pers. comm.). Epidemics can also occur when a new species of parasite or a new species of host is introduced into new environments. Examples of the former include Dutch elm disease (a fungus, *Ceratocystis ulmi*, spread by the beetle *Scolytus multistriatus*), and chestnut blight (*Endothia parasitica*). Both were introduced from Eurasia, and have swept through North America showing little evidence of preferential patterns of attack. Examples of the latter include a variety of conifers planted outside their native range and attacked by fungi such as *Fomes annosus* (Boyce, 1961), and insects (Madden, 1988; Watt and Leather, 1988) which have a much less severe impact upon native trees.

    I believe that these exceptions do not disprove the existence of species-specific host selection or phenotype-for-phenotype selection, but help circumscribe the conditions under which such selection can operate.

DISCUSSION

    The patchy structure of forest stands is evident to any observant visitor. Heretofore, the emphasis has been on the fact that the patchiness is generated by disturbance, usually of a physical nature, producing gaps in the canopy, followed either by occupation of gaps by new individuals and new species (Pickett and White 1985; Sarukhán et al. 1985) or by intense flowering and fruiting activity within gaps by previously shaded individuals (e.g., Linhart et al., 1987).

    In this essay, I emphasize the point that patchiness in forest stands can also occur without disturbance as a consequence of patchy seed dispersal and the resulting family structure of groups of trees. In the past, considerations of what population structure ought to be in multispecies forests, especially tropical ones, led to the conclusion that conspecifics should be hyperdispersed as a result of the tendency by seed predators, herbivores and parasites to concentrate their damage on clustered seeds and seedlings (Janzen 1970). More recently, reviews involving detailed analyses of forest structure have documented clustered distributions of conspecifics in most (but not all) component species of temperate and tropical forests studied (e.g. Armesto et al. 1986; Hubbell and Foster, 1986). Even when conspecifics are not clustered, it is likely that many individuals in a local area are the offspring of one or a few seed trees. That is because within forest tree populations, a small minority of individuals often produce the majority of seeds (e.g., Linhart and Mitton 1985). This probably helps explain the existence of genetic patchiness (Table I) on a local scale. In some woody species, patchiness can be produced by clonal growth; these include poplars and aspen (*Populus* spp.), willows (*Salix* spp.), redwood (*Sequoia sempervirens*), and several alpine and boreal conifers (Shea and Grant, 1986).

Forest trees are potential hosts to a large variety of fungi, insects and other parasites and herbivores. Their long lives relative to these dependent species make them highly apparent and predictable as resources. Despite their long lives and apparency, forest stands are seldom overwhelmed by a parasite or herbivore, and this has been attributed to the high levels of biochemical diversity present in trees (Pimentel and Bellotti, 1976; Fox, 1981). This diversity is evident even within local stands, so that adjacent plants can be dramatically different in both amounts and compositions of defensive compounds (e.g., Smith, 1964; Mazzoni and Gouyon, 1984; Wisdom, 1982). The genetic patchiness within forest populations presents dependent species with patchily variable resources. The latter, by virtue of selecting different host phenotypes or groups of phenotypes, generate very complex diversifying selection in space and time. One species of tree can host dozens of fungal and many more insect species: almost 200 insects are cited in Furniss and Carolin (1977) for ponderosa pine, and over 230 for Douglas-fir. These insects include defoliators, miners, seed eaters, phloem feeders, and twig and bark borers, in addition to dwarf-mistletoe parasites and mammalian herbivores (Fowells, 1965). Not all these dependent species are active in one place or at one time. Neither is it likely that every dependent species has a unique suite of physiological and behavioral characteristics and can therefore generate a unique set of selection pressures. However, there is enough general evidence to suggest that at least certain classes of dependent species such as generalized versus specialized defoliators, or rodents versus ruminants, or insects versus mammals, can be expected to produce different selection pressures (e.g., Freeland and Janzen, 1974; Bernays, 1982; Denno and McClure, 1983; Hanski and Otronen, 1985). For these reasons, with the perspective of selection acting in many different directions, the very high level of genetic variability observed in both temperate and tropical tree species (Hamrick et al., 1979; Hamrick and Loveless, this volume) becomes comprehensible. The role of disease-inducing organisms as agents of diversifying selection was already recognized by Haldane (1949). He argued that the high frequency of polymorphisms in humans, another long-lived, apparent species, was explicable in this context.

The genetic mosaic of forest tree populations can, in turn, generate diversifying selection within the dependent species, or at least help maintain genetic variability in these species. If these interactions between host and a specific dependent species go on for several generations, then there may be a complex genetic feedback between the two, leading perhaps to coevolution *sensu* Janzen (1980). However, a co-evolutionary "arms race" is very seldom detected in plant-herbivore interactions (Futuyma and Slatkin, 1983; Denno and McClure, 1983). This may well be because such coevolution is likely to be very diffuse, particularly if there is a significant interaction between certain host genotypes and specific parasite genotypes (Table 3). In addition, because of the many dependent species that a given host may have to cope with during its evolutionary history, it is not surprising that such coevolution, if present, is extremely difficult to detect.

One major consequence of these interactions is that communities may be in a state of balanced evolutionary disequilibrium. It is a disequilibrium because following disturbance, conditions are unlikely to return to a pre-disturbance configuration. Factors preventing this return and conducive to disequilibrium include: a) following disturbance, the host phenotypes colonizing an area will be genetically different from pre-disturbance ones, b) at any one time, different herbivores and parasites may be at relatively high densities and therefore only certain phenotypes of the host are selected for or against, c) certain groups of hosts are simultaneously susceptible to one parasite species (or phenotype) and resistant to or at least tolerant of another parasite species (or phenotype). It is a balanced disequilibrium in the sense that usually no one element achieves dominance because a) the host clearly cannot evolve resistance to all its parasites and herbivores, and b) any one attacking species cannot attack all host plants with equal success because it is under a variety of ecological and genetic constraints. Finally, any one dependent species is unlikely to be at consistently high population densities in an area for any length of time; therefore it generates selective pressures on the host population on a periodic basis.

I do not wish to overstate the case for the importance of genetics in determining patterns of community organization in hosts and dependent species. Environmental factors, including predators, parasites and parasitoids of dependent species can play major roles in host suitability, and in population dynamics of dependent species (e.g. Denno and McClure, 1983; Loyn et al., 1983; Fritz and Price, 1988; Berryman, 1988). Environmental, herbivore-mediated induction of plant defences can occur (e.g. Haukioja, 1980). Soil moisture and fertility, and phenotypic plasticity can also contribute to variability in plant growth and development. This does not mean that genetic factors can be ignored: those who ignore these factors do so at their peril. Some time ago, Birch (1960) argued that ecologists and geneticists need to cooperate in order to understand the interplay of population dynamics and genetic variability. Yet various contributions to the current literature continue to treat both plants and their dependent species as simple Latin binomials, make predictably simplistic assumptions, and perpetrate the following sorts of fables: (1) insects are aggregated on certain plants solely because they all want to be in a cozy spot together; (2) the ecological roles of a given species are the same wherever the species lives, be it in maritime, continental or subtropical environments; (3) tree species A has 20 insect herbivores, whereas tree species B has 200, therefore A must have more efficient chemical defenses.

In addition, it must be understood that natural selection operates more often through differential reproduction than through survival versus death (Darwin, 1872, e.g., Ch. 4). Wallace's (1968) "hard" and "soft" selection arguments are also relevant. For this reason, reduction of reproductive output by some small percentage can be enough to generate significant selective pressures. Population geneticists have demonstrated that allele frequencies can change appreciably with selection coefficients in the 1% range, and coefficients of 10% or more can produce very rapid changes (Allard, 1960; Wallace, 1968; Endler, 1986). Consequently, levels of damage observed in natural populations are certainly adequate to produce evolutionary change. For example, as noted above, squirrels can reduce ponderosa pine reproduction in a given year by up to 90%. Herbivores can produce leaf loss up to 25% in *Piper* (Marquis, 1984). Leaf herbivores, even though they reduce leaf area by only 8-12%, can reduce acorn output in *Quercus* by 50-80% (Crawley, 1985).

The results and perspectives summarized in this essay also lead to a number of testable hypotheses. For example, 1) There are links between genetic variability among host plants, their phenotypic variability and their suitability as resources. 2) Comparisons of host choices among dependent species suggest that, when such species are systematically (and presumably physiologically) closely related, they may make similar host choices (e.g., Kareiva, 1986; Whitham, 1983; Banerjee, 1987). However, when they are physiologically dissimilar, they are more likely to choose different individual hosts (e.g., Dirzo and Harper, 1982b; Stephan, 1987; Banerjee, 1987). 3) The diversity of dependent species attracted to a given host generates multidirectional diversifying selection. Consequently, there are no "supertrees" that can withstand attacks by all their enemies. However, there can certainly be weaklings that, once debilitated, are especially vulnerable to other dependent species.

SUMMARY AND CONCLUSIONS

Seed dispersal patterns, sometimes in conjunction with pollen dispersal, and patchiness associated with physical conditions or disturbance, produce genetic mosaics in forest stands. The genetic mosaics produce, *ipso facto*, phenotypic mosaics, consisting of groups of trees with differential attractiveness to herbivores, parasites and diseases (dependent species).

Because different dependent species often have preferences for different host tree phenotypes, there is generated a species-specific host selection, which produces diversifying selection.

In addition, a given dependent species, because of its genetic (and phenotypic) variability, can show variable preference for specific host phenotypes. This is termed phenotype-for-phenotype selection, and it in turn, can generate further diversifying selection.

The very high levels of genetic and biochemical variability observed in many tree species are precisely what would be expected under these conditions of multiple and patchy selective pressures.

Spatial genetic and phenotypic patchiness in hosts may also select for patchiness in dependent species, which may in turn generate selection pressures increasing genetic patchiness in hosts. This potential feedback may be an important feature of forest community structure.

When a temporal component of variability is added to the phenotypic mosaics of hosts and parasites and the feedbacks between them, one is left with a vision of forest stands as constantly changing, multifaceted kaleidoscopes.

At least some components of this scenario are probably applicable to herbaceous species, since many of the examples cited come from studies of such species.

ACKNOWLEDGEMENTS

The ideas presented here have developed in the course of two decades of work in the forests of California, Costa Rica and Colorado. Companions during this work have included Irene and Herbert Baker, W. J. Libby, J. B. Mitton, P. Feinsinger, M. C. Grant, and J.L. Hamrick. I also thank D. Alstad, C. Bock, J. Bock, P. Feinsinger, M. C. Grant, J.L. Hamrick, J. Karron, P. Kareiva,

A.C. Lewis, J. B. Mitton, P. Morrow, P. Price, M. Snyder, V. Sork, C. Wisdom, and T. Whitham for references and critical reviews of previous drafts of this manuscript. Financial help was provided by N.S.F. (most recently, grant B.S.R. 8506077), the Council for Research and Creative Work of the University of Colorado, U.S. Forest Service, and the National Geographic Society.

LITERATURE CITED

Allard, R. W. 1960. Principles of Plant Breeding. New York, Wiley.

Anderson, P. K. 1970. Ecological structure and gene flow in small mammals. Symp. Zool. Soc. London 26: 299-325.

Armesto, J. J., J. B. Mitchell, and C. Villagran. 1986. A comparison of spatial patterns of trees in some tropical and temperate forests. Biotropica 18: 1-11.

Aubréville, A. M. A. 1938. La forêt coloniale: les forêts de l'Afrique occidentale francaise. Ann. Acad. Sciences Coloniales. Paris 9: 1-245.

Austin, L., J. S. Yuill, and K. G. Brecheen. 1945. Use of shoot characters in selecting ponderosa pines resistant to resin midge. Ecology 26: 288-296.

Avery, C. C., F. R. Larson, and G. H. Shubert. 1976. Fifty-year records of virgin stand development in southwestern ponderosa pine. U.S.D.A. Forest Service General Technical Report RM-22.

Banerjee, B. 1987. Can leaf aspect affect herbivory? A case study with tea. Ecology 68: 839-843.

Barrett, J. A. 1983. Plant-Fungus symbioses. In Futuyama and M. Slatkin [eds.]. Coevolution. Sinauer, Sunderland, MA.

Bates, C. G. 1927a. Varietal differences. J. For. 25: 610.

Bates, C. G. 1927b. A vision of the future Nebraska forest. J. For. 25: 1030-1040.

Bedard, W. D., P. D. Tilden, D. L. Wood, R. M. Silverstein, R. G. Brownlee and J. O. Rodin. 1969. Western pine beetle: field response to its sex pheromone and synergistic host terpene, myrcene. Science 164: 1284-1285.

Bernays, E. A. 1982. The insect on the plant - a closer look. In J. H. Visser and A. K. Minks [eds.]. Proc. 5th Int. Symp. Insect-Plant Relationships. Wageningen, Neth.

Bernays, E. A., R. F. Chapman, J. McDonald, and J.E.R. Salter. 1976. The degree of oligophagy in Locusta migratoria L. Ecol. Ent. 1: 223-230.

Berryman, A. A. [ed.] 1988. Dynamics of Forest Insect Populations. Plenum, New York.

Berthaud, J. 1984. Gene flow and population structure in *Coffea canephora* in Africa. In P. Jacquard, G. Heim and J. Antonovics [eds.]. Genetic Differentiation and Dispersal in Plants. Springer Verlag. Berlin.

Birch, L. C. 1960. The genetic factor in population ecology. Amer. Natur. 94: 5-24.

Boyce, J. S. 1961. Forest Pathology 3rd ed. McGraw Hill, New York.

Bradshaw, A. D. 1972. Some of the evolutionary consequences of being a plant. Evol. Biol. 5: 25-47.

Brown, A. H. D. 1979. Enzyme polymorphisms in plant populations. Theor. Pop. Bio. 15: 1-42.

Burdon, J. J. 1987. Diseases and plant population biology. Cambridge Univ. Press. Cambridge, U.K.

Carroll, C. R. and C. A. Hoffman. 1980. Chemical feeding deterrent mobilized in response to insect herbivory and counteradaption by *Epilachna tredecimnotata* Science 209: 414-416.

Carroll, G. 1988. Fungal endophytes in stems and leaves: from latent pathogen to mutualistic symbiont. Ecology 69: 2-9.

Crawley, M. J. 1985. Reduction of oak fecundity by low density herbivore populations. Nature 314: 163-164.

Crawley, M. J., and R. Pattrasudhi. 1988. Interspecific competition between insect herbivores: asymmetric competition between Cinnabar moth and the ragwort fly. Ecol. Ent. 13: 243-249.

Cuguen, J., B. Thiebault, F. Nitsiba, and G. Barriere. 1984. Enzymatic variability of beechstands (*Fagus sylvatica* L.) on three scales in Europe: evolutive mechanisms. In P. Jacquard, G. Heim and J. Antonovics [eds.]. Genetic Differentiation and Dispersal in Plants. Springer Verlag, Berlin.

Danell, K., and K. Huss-Danell. 1985. Feeding by insects and hares on birches earlier affected by moose browsing. Oikos 44: 75-81.

Darwin, C. 1872. The Origin of Species. 6th ed. Murray, London, U.K.

Day, P. R. 1974. Genetics of Host-Parasite Interactions. Freeman, San Fransisco.

Denno, R. F., and M. S. McClure, [eds.]. 1983. Variable Plants and Herbivores in Natural and Managed Systems. Academic Press, New York

De Nooij, M.P., and J.M.M. van Damme. 1988. Variation in pathogenicity among and within populations of the fungus *Phomopsis subordinaria* infecting *Plantago lanceolata*. Evolution 42: 1166-1171.

Dimock, E. J., R. R. Silen and V. E. Allen. 1976. Genetic resistance in Douglas-fir to damage by showshoe hare and black-tailed deer. For. Sci. 22: 106-121.

Dirzo, R. 1980. Experimental studies on slug-plant interactions I. The acceptability of thirty plant species to the slug *Agriolimax caruanae*. J. Ecol. 68: 981-998.

Dirzo, R., and J. L. Harper. 1982a. Experimental studies on slug-plant interactions III. Differences in the acceptability of individual plants of *Trifolium repens* to slugs and snails. J. Ecol. 70: 101-117.

Dirzo, R., and J. L. Harper. 1982b. Experimental studies on slug-plant interactions IV. The performance of cyanogenic and acyanogenic morphs of *Trifolium repens* in the field. J. Ecol. 70: 119-138.

Dixon, A. F. G. 1985. Structure of aphid populations. Ann. Rev. Entom. 30: 155-174.

Dixon, R. K., H. E. Garrett, and H. E. Stelzer. 1987. Growth and ectomycorrhizal development of lobolly pine progenies inoculated with three isolates of *Pisolithus tinctorius*. Silv. Genet. 36: 240-245.

Dorman, K. W. 1976. The genetics and breeding of southern pines. U.S. Department of Agriculture, Forest Service Handbook No. 471.

Duffield, J. W. 1985. Inheritance of shoot coatings and their relations to resin midge attack on ponderosa pine. For. Science. 31: 427-429.

Edmunds, G. F., Jr., and D. N. Alstad. 1978. Coevolution in insect herbivores and conifers. Science 199: 941-945.

Endler, J. A. 1986. Natural Selection in the Wild. Princeton Univ. Press. Princeton, NJ.

Farentinos, R. C., P. J. Capretta, R. E. Kepner, and V. M. Littlefield. 1981. Selective herbivory in tassel-eared squirrels: role of monoterpenes in ponderosa pines chosen as feeding trees. Science, 213: 1273-1275.

Fins, L, and W. J. Libby. 1982. Population variation in *Sequoiadendron*: seed and seedling studies, vegetative propagation, and isozyme variation. Silv. Genet. 31: 102-110.

Fitzgerald, T. D., and S. C. Peterson. 1988. Cooperative foraging and communication in caterpillars. Bioscience 38: 20-25.

Flor, H. H. 1956. The complementary gene systems in flax and flax rust. Adv. Genetics 8: 29-54.

Ford, E. B. 1962. Butterflies. Collins. London U.K.

Fowells, H. A. 1965. Silvics of forest trees of the United States. U.S.D.A. Forest Service Agricultural Handbook No. 271.

Fox, L. R. 1981. Defense and dynamics in plant-herbivore systems. Amer. Zool. 21: 853-864.

Freeland, W. J., and D. H. Janzen. 1974. Strategies of herbivory in mammals: the role of plant secondary compounds. Amer. Natur. 108: 268-289.

Fritz, R. S., W. S. Gaud, C. S. Sacchi, and P. W. Price. 1988. Variation in herbivore density among host plants and its consequences for community structure. Oecologia 72: 577-588.

Fritz, R. S., and P. W. Price. 1988. Genetic variation among plants and insect community structure: willows and sawflies. Ecology 69: 845-856.

Furnier, G. R., P. Knowles, M. A. Clyde, and B.P. Dancik. 1987. Effects of avian seed dispersal on the genetic structure of whitebark pine populations. Evolution 41: 607-612.

Furniss, R. L., and V. M. Carolin. 1977. Western forest insects. U.S.D.A. For. Serv. Misc. Pub. 1339

Futuyma, D. J. and S. C. Peterson. 1985. Genetic variation in the use of resources by insects. Ann. Rev. Entom. 30: 217-238.

Futuyma, D. J. and M. Slatkin, [eds.]. 1983. Coevolution. Sinauer. Sunderland, MA.

Futuyma, D. J. and S.S. Wasserman. 1980. Resource concentration and herbivory in oak forests. Science 210: 920-922.

Gallun, R. L. 1977. The genetic basis of Hessian fly epidemics. Ann. N.Y. Acad. Sci. 287: 223-229.

Gan, Y. Y., Robertson, F. W., and P. S. Ashton. 1977. Genetic variation in wild populations of rain-forest trees. Nature 269: 323-324.

Haldane, J. B. S. 1949. Disease and evolution. La Ricerca Scient. Suppl. 19: 68-76.

Hamrick, J. L., Y. B. Linhart, and J. B. Mitton. 1979. Relationships between life history characteristics and electrophoretically-detectable genetic variation in plants. Ann. Rev. Ecol. and Syst. 10: 173-200.

Hamrick, J. L., and M. O. Loveless. 1986. The influence of seed dispersal mechanisms on the genetic structure of plant populations. In A. Estrada and T.H. Fleming [eds.]. Frugivores and Seed Dispersal. Junk. The Hague, Neth.

Handel, S. N. 1983. Pollination ecology, population structure and gene flow. In L. Real [ed.], Pollination Biology. Academic Press New York

Hanski, I., and M. Otronen. 1985. Food quality induced variance in larval performance: comparison between rare and common pine-feeding sawflies (Diprionidae) Oikos 44: 165-174.

424

Haukioja, E. 1980. On the role of plant defenses in the fluctuations of herbivore populations. Oikos 35: 202-213.

Hawksworth, F. G. 1978. Biological factors of dwarf-mistletoe in relation to control. In Scharpf, R. F and J. R. Parameter [eds.]. Symposium on Dwarf Mistletoe Control through Forest Management. Gen. Tech. Rep. P.S.W. 31.

Hawksworth, F., and D. Wiens. 1972. Biology and classification of dwarf mistletoes (*Arceuthobium*) US Dept. Agric. For. Serv. Agric Handbook 401.

Hay, M. E., P. E. Renaud, and W. Fenical. 1988. Large mobile versus small sedentary herbivores and their resistance to seaweed chemical defences. Oecologia 75: 246-252.

Heywood, J. S., and T. H. Fleming. 1986. Patterns of allozyme variation in three Costa Rican species of *Piper*. Biotropica 18: 208-213.

Higgins, J. 1927. Facts and figures regarding the Nebraska planting project. J. For. 25: 1023-1030.

Hoff, R. J. 1988. Resistance of ponderosa pine to the Gouty Pitch Midge (*Cecidomyia piniinopis*) U.S.D.A. For. Ser. Res. Pap. INT. 387. Ogden, UT.

Hood, J. V. and W. J. Libby. 1980. A clonal study of intraspecific variability in radiata pine. I. Cold and animal damage. Aust. For. Res. 10: 9-20.

Howe, M. F., and J. Smallwood. 1982. Ecology of seed dispersal. Ann. Rev. Ecol. Syst. 13: 201-228.

Hubbell, S. P., and R. B. Foster. 1986. Canopy gaps and the dynamics of a neotropical forest. In M.J. Crawley [ed.]. Plant Ecology. Blackwell, Oxford, U.K.

Janzen, D. H. 1970. Herbivores and the number of tree species in tropical forests. Amer. Natur. 104: 501-528.

Janzen, D. H. 1980. When is it co-evolution? Evol. 34: 611-613.

Jones, E. W. 1945. The structure and reproduction of the virgin forest of the North Temperate zone. New Phytol. 44: 130-248.

Jones, J. S. 1973. Ecological genetics and natural selection in molluscs. Science 182: 546-553.

Kafton., D. 1977. Isozyme variability and reproductive phenology of Monterey cypress. Ph.D. Dissertation University of California, Berkeley, California.

Karban, R., R., Adamchak, and W. C. Schnathorst. 1987. Induced resistance and interspecific competition between spider mites and a vascular wilt fungus. Science 235: 678-680.

Kareiva, P. 1986. Patchiness, dispersal and species interactions: consequences for communities of herbivorous insects. In J. Diamond and T. Case, [eds.]. Community Ecology. Harper and Row.

Lamotte, M. 1959. Polymorphism of natural populations of *Cepaea nemoralis*. Cold Spring Harb. Symp. Quant. Biol. 24: 65-86.

Levin, D. A., and Kerster, H. W. 1974. Gene flow in seed plants. Evol. Biol. 7: 139-220.

Levin, D. A. 1981. Dispersal versus gene flow in plants. Ann. Missouri. Bot. Gard. 68: 233-253.

Linhart, Y. B., and J. B. Mitton. 1985. Relationships among reproduction, growth rates, and protein heterozygosity in ponderosa pine. Amer. J. Bot. 72: 181-184.

Linhart, Y. B., J. B. Mitton, K. B. Sturgeon, and M. L. Davis. 1981a. An analysis of genetic architecture in populations of ponderosa pine. In M.T. Conkle, [ed.]. Isozymes of North American Forest Trees and Forest Insects. U.S.D.A. Forest Service. Gen. Tech. Rept. PSW-48.

Linhart, Y. B., J. B. Mitton, K. B. Sturgeon, and M. L. Davis. 1981b. Genetic variation in space and time in a population of ponderosa pine. Heredity 48: 407-426.

Linhart, Y. B., P. Feinsinger, J. M. Beach, W. H. Busby, K. G. Murray, W. Z. Pounds, S. Kinsman, C. A. Guindon, and M. Kooiman. 1987. Disturbance and predictability of flowering patterns in bird-pollinated cloud forest plants. Ecology 1696-1710.

Loyn, R. M., R. G. Runalls, G. Y. Forward and J. Tyers. 1983. Territorial bell miners and other birds affecting populations of insect prey. Science 221: 1411-1412.

Madden, J. L. 1988. *Sirex* in Austrasia. *In* A.A. Berryman [ed.]. Dynamics of Forest Fnsect iInteractions. Plenum, New York.

Maddox, G. D., and R. B. Root. 1986. Resistance to 16 diverse species of herbivorous insects within a population of golden-rod *Solidage altissima*: genetic variation and heritability. Oecologia 72: 8-14.

Marquis, R. J. 1984. Leaf herbivores decrease fitness in a tropical plant. Science 226: 537-539.

Mayer, H. and M. Neumann. 1981. Struktureller und entwicklungsdynamischer Vergleich der Fichten-Tannen-Buchen-Urwälder Rothwald/Nieder Österreich und Cerkova Uvala/ Kroatien. Forstw. Cbl. 100: 111-132.

426

Mazzoni, C. and P. H. Gouyon. 1984. Horizontal structure of populations: Migration, adaptation and chance. An experimental study on *Thymus vulgaris* L. In P. Jacquard, G. Heim and J. Antonovics, [eds.]. Genetic differentiation and dispersal in plants. Springer Verlag. Berlin.

McCrea, K. D., and W. G. Abrahamson. 1987. Variation in herbivore infestations: historic vs. genetic factors. Ecology 68: 822-827.

Mitton, J. ., Y. B. Linhart, M. L. Davis, and K. B. Sturgeon. 1981. Estimation of outcrossing in ponderosa pine, *Pinus ponderosa* from patterns of segregation of protein polymorphisms and from frequencies of albino seedlings. *Silvae Genet.* 20: 117-121.

Mitton, J. B., and K. B. Sturgeon. [eds.]. 1982. Bark Beetles in North American Conifers. Univ. Texas Press. Austin, Texas.

Moran, N. A., and T. G. Whitham. 1988. Evolutionary reduction of complex life cycles: loss of host alternation in *Pemphigus* (Homoptera: Aphididae) Evolution 42: 717-728.

Namkoong, G. 1984. Genetic structure of forest tree populations In M. S. Swaminanthan [ed.]. Genetics: New Frontiers. Proc. 15th Int'l. Cong. Genetics. Oxford and I.B.H. pub. New Delhi, India.

Namkoong, G., J. A. Richmond, L. B. Nunally, B. C. McClain, and J. L. Tyson. 1982. Population genetic structure of Nantucket pine tip moth. Theor. Appl. Genet. 63: 1-7.

Old, K. M., W. J.Libby, J. H. Russell, and K. G. Eldridge. 1986. Genetic variability in susceptibility of *Pinus radiata* to western gall rust. Silv. Genet. 35: 145-149.

O'Malley, D. M., and K. S. Bawa. 1987. Mating system of a tropical rainforest tree. Amer. J. Bot. 74: 1143-1149.

Park, Y. S., D. P. Fowler, and J. F. Coles. 1984. Population studies of white spruce. II. Natural inbreeding and relatedness among neighboring trees. Can. J. For. Res. 14: 909-913.

Parker, M. A. 1988. Disequilibrium between disease-resistance variants and allozyme loci in an annual legume. Evolution 42: 239-247.

Pickett, S. T. A., and P. S. White. 1985. Natural Disturbance: the Patch Dynamics Perspective. Academic Press. New York.

Pimentel, D., and C. Bellotti. 1976. Parasite-host population systems and genetic stability. Amer. Nat. 110: 877-888.

Prakash, C. S., and W. A. Heather. 1986. Inheritance of resistance to races of *Melampsora medusae* in *Populus deltoides*. Sil. Genet. 35: 74-77.

Price, P. W. 1980. Evolutionary Biology of Parasites. Princeton Univ. Press. Princeton, N.J.

Price, P. W., M. Westoby, B. Rice, P.R. Atsatt, R. S. Fritz, J.N. Thompson, and K. Mobley. 1986. Parasite mediation in ecological interactions. Ann. Rev. Ecol. Syst. 17: 487-505.

Radwan, M. A. 1972. Differences between Douglas-fir genotypes in relation to browsing preference by black-tailed deer. Can. J. For. Res. 2: 250-255.

Rasmussen, D. I. 1978. Sibling clusters and genotypic frequencies. Amer. Natur. 113: 948-951.

Rauscher, M. D. 1983. Ecology of host-selection behavior in phytophagous insects. In R. F. Denno and M. S. McClure [eds.]. Variable Plants and Herbivores in Natural and Managed Systems. Academic Press, New York.

Read, R. A. 1980. Genetic variation in seedling progeny of ponderosa pine provenances. For. Sci. Monograph 23.

Rice, W. R. 1983. Sexual reproduction: an adaptation reducing parent-offspring contagion. Evolution 37: 1317-1320.

Roth, L. F. 1974. Resistance of ponderosa pine to dwarf-mistletoe. Silv. Genet. 23: 116-119.

Sakai, K-I. 1985. Studies on breeding structure in two tropical tree species In Population Genetics in Forestry. H.R. Gregorius, [ed.]. Springer Verlag. Berlin.

Sakai, K-I., and I. Miyazaki. 1970. Genetic studies in natural populations of forest trees. II. Family analysis: a new method for quantitative genetic studies. I.U.F.R.O. Quant. Genet. Meet. Brno, Czechoslovakia.

Sarukhan, J., D. Pinero, and M. Martinez-Ramos. 1985. Plant demography: a community-level interpretation. In J.White [ed.]. Studies in Plant Demography. Academic Press.

Savopoulou-Soultani, M. and M. E. Tzanakakis. 1988. Development of Lobesia botrana (Lepidoptera: Tortricidae) on grapes and apples infected with the fungus Botrytis cinerea. Env. Ent 17: 1-6.

Scharpf, R. F. 1984. Host resistance to dwarf-mistletoes. In F.G. Hawksworth and R.F. Scharpf. [eds.]. Biology of Dwarf-Mistletoes. U.S.D.A. For. Serv. Gen. Tech. Rept. RM-111.

Scharpf, R. F., and J. R. Parmeter. 1978. Symposium on dwarf-mistletoe control and management. U.S.D.A. General Tech. Rpt. P.S.W. 31.

Scholz, E. 1960. Die braun Maserbirke. Forst und Jagd Sonderheft/Samenplantagen 28-36.

Schowalter, T. D., D. N. Pope, R. N. Coulson, and W. S. Fargo. 1981. Patterns of southern pine beetle (Dendroctonus frontalis Zimm.) infestation enlargement. For. Sci. 27: 837-849.

Schubert, G. H. 1974. Silviculture of southwestern ponderosa pine: The status of our knowledge. U.S. For. Serv. Res. Pap. RM-123.

428

Selander, R. K. 1970. Behavior and genetic variation in natural populations. Amer. Zool. 10: 53-66.

Service, P. 1984. Genotypic interactions in an aphid-host plant relationship: *Uroleucon rudbeckiae* and *Rudbeckia laciniata*. Oecologia 61: 271-276.

Shaw, D. V., and R. W. Allard. 1981. Analysis of mating system parameters and population structure in Douglas-fir using simple-locus and multilocus methods. In M.T. Conkle [ed.]. Isozymes of North American Forest Trees and Forest Insects. U.S.D.A. Forest Service Gen. Tech. Rept. PSW 48.

Shea, K. L. 1985. Mating systems and population structure in Englemann spruce and subalpine fir. Ph.D. Dissertation, University of Colorado, Boulder, Colorado.

Shea, K. L. and M. C. Grant. 1986. Clonal growth in spire-shaped Engelmann spruce and subalpine fir trees. Can J. Bot. 64: 255-261.

Smith, R. H. 1964. Variation in the monoterpenes of *Pinus ponderosa*. Laws. Science 143: 1337-1338.

Smith, R. H. 1966. The monoterpene composition of *Pinus ponderosa* xylem resin and of *Dendroctonus brevicomis* pitch tubes. Forest Sci. 12: 63-68.

Spencer, D. A. 1964. Porcupine population fluctuations in past centuries revealed by dendrochronology. J. Appl. Ecol. 1: 127-149.

Squillace, A. E., and R. R. Silen. 1962. Racial variation in ponderosa pine. For. Sci. Monogr. No. 2.

Stephan, B. R. 1987. Differences in the resistance of Douglas-fir provenances to the woolly aphid *Gilletteella cooleyi*. Silvae Genet. 36: 76-79.

Stern, K., and L. Roche. 1974. Genetics of Forest Ecosystems. Springer Verlag. Berlin.

Sturgeon, K. B. 1979. Monoterpene variation in ponderosa pine xylem resin related to western pine beetle predation. Evolution 33: 803-814.

Thomas, G. V., and S. K. Ghai. 1987. Genotype-dependent variation in vesicular arbuscular mycorrhizal colonization of coconut seedlings. Proc. Indian Acad. Sci. (Plant Sci.). 97: 289-294.

Tigerstedt, P. M. A. 1973. Studies on isozyme variation in marginal and central populations of *Picea abies*. Hereditas 75: 47-60.

Tigerstedt, P. M. A., D. Rudin, T. Niemelä and J. Tammisola. 1982. Competition and neighboring effect in a naturally regenerating population of Scots pine. Silva Fennica 16: 122-129.

Tinnin, R. O. 1984. The effect of Dwarf-mistletoe on forest community ecology. In F.G. Hawksworth and R.F. Scharpf. [eds.]. Biology of Dwarf-Mistletoes. U.S.D.A. For. Serv. Gen. Tech. Rept. RM-111.

Turner, M. E., J. C. Stephens, and W. W. Anderson. 1982. Homozygosity and patch structure in plant populations as a result of nearest-neighbor pollination. Proc. Nat'l. Acad. Sci. U.S.A. 79: 203-207.

Unruh, T. R., and R.F. Luck. 1987. Deme formation in scale insects: a test with the pinyon needle scale and a review of other evidence. Ecol. Ent. 12: 439-449.

Wagner, M. R. and R. L. Mathiasen. 1985. Dwarf-mistletoe-pandora moth interaction and its contribution to ponderosa pine mortality in Arizona. Gt. Basin Natur. 45: 423-426.

Wainhouse, D., and R. S. Howell. 1983. Intra-specific variation in beech scale populations in susceptibility of their host *Fagus sylvatica*. Ecol. Ent. 8: 351-359.

Wallace, B. 1968. Topics in Population Genetics. Norton and Co., New York.

Watt, A. D., and S. L. Leather. 1988. The pine beauty in Scottish lodgepole pine plantations. In A.A. Berryman [ed.]. Dynamics of Forest Insect Populations. Plenum, New York.

Watt, A.S. 1947. Pattern and process in the plant community. J. Ecol. 35: 1-22.

Weaver, H. 1961. Ecological changes in the ponderosa pine forest of cedar valley in southern Washington. Ecology 42: 416-420.

Whelan, R. J. 1978. The influence of insect grazers on the establishment of post-fire plant populations. Ph.D. Thesis, University of Western Australia, Perth, Australia.

Whelan, R.J. 1982. Response of slugs to unacceptable food items. J. Appl. Ecol. 19: 79-87.

White, A. S. 1985. Presettlement regeneration patterns in a Southwestern ponderosa pine stand. Ecology 66: 589-594.

White, J. A. 1980. Resource partitioning by ovipositing cicadas. Amer. Nat. 115: 1-28.

Whitham, T. G. 1983. Host manipulation of parasites: within-plant variation as a defense against rapidly evolving pest. In Denno, R.F. and M.S. McClure. Variable Plants and Herbivores in Natural and Managed Systems. Academic Press. New York.

Wisdom, C. S. 1982. Effects of natural product variation on the interaction between a herbivore and its host plant. Ph.D. Dissertation, University of California, Los Angeles, California.

Wolfe, M. S., and C. E. Caten. 1987. Populations of Plant Pathogens. Blackwell. Oxford, U.K.

Wright, J. W. 1976. Introduction to Forest Genetics. Academic Press, New York.

DYNAMICS OF PLANTS IN
BUFFALO WALLOWS:
EPHEMERAL POOLS IN THE GREAT PLAINS

Gordon E. Uno

INTRODUCTION

Buffalo wallow is a colloquial expression used to designate a depression created by bison (*Bison bison*) as they roll in exposed prairie soil. These depressions are underlain by a subsurface hardpan that restricts the downward percolation of water into the lower soil profile. Thus, after heavy rainfalls, wallows fill with water and may remain full for days to weeks, depending on the ensuing rainfall pattern. Wallows, therefore, are ephemeral pools within perennial grasslands. Although the prairie has been studied extensively (England and Devos, 1969; Spedding, 1976; French, 1979; Risser et al., 1981), buffalo wallows as a community within the Great Plains have been mostly ignored. Authors of books about the American frontier have written anecdotally about the formation of wallows (e.g., Roe, 1951), and Barkley and Smith (1932) produced a preliminary report of wallow plans near Norman, Oklahoma. However, only recently has there been a renewed interest in these island-like communities within the prairie (Collins, and Uno, 983; Polley and Collins, 1984; Polley and Wallace, 1986). Current interest results from the distinctive nature of these water-filled depressions and the unusual assemblage of plants within them.

It has been estimated that 20,000,000 - 60,000,000 bison ranged over North American grasslands during prehistoric times (Larson, 1940). It follows that the number of wallows within the prairies also was tremendous. Little information exists about wallows because most have been plowed under or filled in as prairie grasses were replaced by more economically important species such as corn and wheat. Also, buffalo wallows lack the showy flora of similar habitats such as vernal pools of California, which have garnered so much attention (Jain, 1976; Jain and Moyle, 1984; Schlising, this volume).

The purposes of this study were to: (1) develop a model of wallow community development based on bison behavior and wallow location; (2) determine the species composition and number of individuals within wallows throughout the growing season; (3) estimate the size and determine the composition and development of the seed bank within buffalo wallows; and (4) test the effect of water on the composition and size of the plant community within wallows. This is part of a long-term study conducted at the Wichita Mountains Wildlife Refuge (WMWR), an approximately 24,000 ha mosaic of forest and mixed-grassland vegetation in Comanche County, southwestern Oklahoma (Buck, 1964; Crockett, 1964; Dooley and Collins, 1984). The refuge supports a herd of nearly 700 bison, which actively creates and reuses buffalo wallows.

MATERIALS AND METHODS

Viewed from above, most buffalo wallows are circular; some are slightly elliptical. The diameter of 8 circular wallows was measured, as well as the depth of the wallow in relation to the surrounding prairie.

To determine if seeds were dispersed into the wallow by wind, I lined six petri dishes with filter paper sprayed with aerosol adhesives. Traps were nailed to wooden stakes driven into the ground so that the traps rested 20 cm above one wallow bottom. The wallow was enclosed by four 1.5 X 3.5 m reinforced stock panels to prevent large mammals from entering. Filter paper with seeds attached was collected at 2-week intervals and replaced with new filter paper during the growing seasons of 1985 and 1986. Collected paper was returned to the laboratory where seeds were identified and counted. Martin and Barkley (1961) and Musil (1978) were used for seed identification.

The seed bank of wallows and the surrounding prairie was determined by taking soil cores (21 cm deep X 2 cm diameter) from inside and 1 m outside wallows. Each soil core was cut into equal sections: top, middle and bottom. These sections were split down the middle, and half of each section was placed in a petri dish, watered and kept in a growth chamber at 27°C and 12 hrs light/dark. Seedlings were counted, placed in a greenhouse, and grown until they were identified. The other half of each soil core section was washed through a fine-mesh wire screen and the remaining material sorted for seeds. Seeds were identified and counted.

Initial attempts to count seedlings in wallows proved impossible because of the large number of individual plants. The number of plants within wallows was estimated by removing randomly selected soil plugs (6 cm diam) from 10 different wallows at three different sites in the WMWR at 3-week intervals. These soil plugs were washed so the plants could be separated, identified, and counted.

To study the effect of standing water on the vegetation in wallows, in February 1987, I removed three adjacent soil plugs with plants from each of 10 locations within six wallows. One plug from each trio was placed into one of three large tubs, "underwater", "moist", or "alternating", and stabilized with sand. Plants and plugs in the "underwater" tub were covered with 15 - 20 cm of water for the entire duration of the experiment. Plants and plugs in the "alternating" tub were kept moist for five days of every week and covered with water for the remaining two days. Plants and plugs in the "moist" tub were watered as necessary but never covered with water. Plant species, size, number, and flowering time were compared among the three tubs.

RESULTS

The average dimensions for 8 circular wallows were 3.1 m diameter X 0.2 m deep. Most buffalo wallows in the WMWR are restricted to soil types with poor, shallow (less than 0.3 m) surface layers of soil.

Thirty-eight species of flowering plants and one pteridophyte were found in the wallows (Table 1). These species occur in two distinctly different groups: an early flowering group dominated primarily by weedy winter annuals and a later group, much less weedy, dominated by perennial grasses and wetland species.

A total of 6336 seeds was collected on the sticky traps over a two-year period, including 14 species during the first year and 11 species during the second year. In both years, during the time the traps were in place, only eight species of mature plants were found in the wallow. Some of the seeds collected (at least 22) had to be dispersed from outside of the wallow into the enclosure via the wind.

The seed bank within wallows had greater diversity and size than that of the surrounding prairie. Seeds of 25 species were found in wallow seed banks while only 15 species were found in the soil outside of the wallows (Table 2). Of these, 12 species were unique to the wallow seed bank, and two species unique to the seed bank outside the wallows. Seeds of only seven species germinated in the growth chamber. This indicates that, under conditions used, germination requirements were not met for most species of plants found in the soil cores.

The total number of seeds found in cores averaged 47.1 and 11.2 inside and outside wallows, respectively (Table 2). Most seeds were restricted to the upper third of the cores. Nevertheless, cores 21 cm deep were taken initially because bison easily penetrated to this depth as they crossed wet wallows. The

Table 1. Species found in buffalo wallows at the Wichita Mountains Wildlife Refuge in April and July 1983-1986.

Species	Life History	Habitat
APRIL		
DICOTS		
Chaerophyllum tainturieri Hook.	Annual	Weedy
Coreopsis tinctoria Nutt.	Annual	Weedy
Croton lindheimerianus Scheele.	Annual	Weedy
Euphorbia spathulata Lam.	Annual	Weedy
Geranium carolinianum L.	Annual	Weedy
Hedeoma hispidum Pursh.	Annual	Weedy
Helenium amarum (Raf.) Rock.	Annual	Weedy
Krigia oppositifolia Raf.	Annual	Weedy
Lepidium virginicim L.	Annual	Weedy
Myosurus minimus L.	Annual	Wetland
Plantago elongata Pursh.	Annual	Weedy
Plantago virginica L.	Annual	Weedy
Sibara virginica (L.) Roll	Annual	Weedy
Veronica peregrina L.	Annual	Weedy
MONOCOTS		
Agrostic elliottiana Schult.	Annual	Weedy
Alopecurus carolinianus Walt.	Annual	Wetland
Bromus japonicus L.	Annual	Weedy
Hordeum pusillum Nutt.	Annual	Weedy
Juncus torreyi Cov.	Perennial	Wetland
Vulpia octoflora (Walt.) Rydb.	Annual	Weedy
FERNS		
Marsilea mucronata A. Br.	Perennial	Wetland
JULY		
DICOTS		
Acacia hirta T. & G.	Perennial	Grassland
Ambrosia psilostachya DC.	Perennial	Weedy
Ammania coccinea Rottb.	Annual	Weedy
Chenopodium album L.	Annual	Weedy
Haplopappus ciliatus (Nutt.) DC.	Annual	Weedy
Iva annua L.	Annual	Wetland
Lythrum californicum T. & G.	Perennial	Wetland
Oxalis stricta L.	Perennial	Wetland
Rudbecki hirta L.	Annual	Weedy
Solanum elaeagnifolium Cav.	Perennial	Weedy
MONOCOTS		
Bothriochloa saccharoides (Sw.) Rydb.	Perennial	Grassland
Carex spp.	Perennial	Wetland
Cyperus acuminatus T. & H.	Annual	Wetland
Dicanthelium oligosanthes (Schult.) Gould	Perennial	Grassland
Eleocharis spp.	Annual	Weedy
Eragostis lugens Nees.	Perennial	Grassland
Juncus torreyi Cov.	Perennial	Wetland
Schizacharium scoparium (Michx.) Nash.	Perennial	Grassland
Tridens albescens (Vassey) Woot. & Strandl.	Perennial	Grassland

Table 2. Mean numbers of seeds found in soil cores taken from wallows and from the surrounding prairie 1m from wallow edge. The mean number of seeds per wallow core was significantly greater than per prairie core (T = 2.78, p < 0.02)

	Wallow Cores (n = 14)	Prairie Cores (n = 10)
Mean Seed Number Core (± 1 SD)	47.1 ± 58.1	11.2 ± 8.3
Top Section	34.0	8.8
Middle Section	12.1	1.7
Bottom Section	1.0	0.7
Estimated Number of Seeds/m^2	149,924	35,650
Number of Species in Core	25	15
Number of Unique Species	12	2
Number of Cores	14	10

Table 3. Comparison of total seedling emergence from cores under different water regimes. See text for further details.

	Water Regime		
	Underwater	Alternating	Moist
Species number	3	10	12
Number of plants	59	88	133
Percentage of plants Flowering on 5/21	20	61	79
Days until first plant flowered in tub	71	33	26

size of the seed bank was very large; estimating the seed bank to a depth of 21 cm, there are over 36,000 seeds/m² in the prairie surrounding the wallows and nearly 150,000 seeds/m² in each wallow.

Thousands of seeds germinated each year, producing wallows densely covered with seedlings and plants from late fall to late summer. An average of 69.9 plants per soil plug was found in the 10 wallows sampled in January (Fig. 1), which extrapolates to approximately 25,000 seedlings/m² in each wallow. Estimated cover of the 10 sampled wallows at this time was 46%, and because wallows average 7.5 m², each wallow is estimated to contain over 85,000 seedlings. The number of plants increased before the next sample date,and then decreased greatly. This drop in density coincides with the first large rainfall received in the area, which caused the wallows to fill with water (Fig. 1).

Seedlings were difficult to identify; therefore, they simply were grouped into monocots and dicots. Wallows from all sites contained both monocot and dicot seedlings, although there was a greater percentage of dicots (74%) than monocots (26%) in January. After the first rains in the WMWR, the percentage of monocots in the wallows increased as gasses and wetland species began to dominate. On the last sampling date, monocots made up 66% of the plants, although there were more dicot species represented.

Soil plugs for the experiment on the effects of standing water were taken from adjacent sites within wallows and thus were expected to contain similar plants. However, the underwater, alternating, and moist tubs ultimately contained different numbers and species of plants, and the plants in them began their flowering periods at different times (Table 3). In general, standing water for some or all of the time had a negative effect on the growth of most of the wallow plants. The most surprising species was *Coreopsis tinctoria*, Nutt. which dominated all three tubs and which, for almost 2 months, was the only species in the underwater tub. Perennial grasses were not found in the underwater tub, but were present in both other tubs.

DISCUSSION

Wallows frequently are initiated during bison rutting season (late June -early September) by a bison bull pawing the ground, urinating in the disturbed area, and then rolling repeatedly in the depression just formed (pers. obs.). This behavior might be represented one or two times before the bull wanders off. A wallow is formed only after recurrent wallowings in the same depression, but the creation and use of wallows is not restricted to

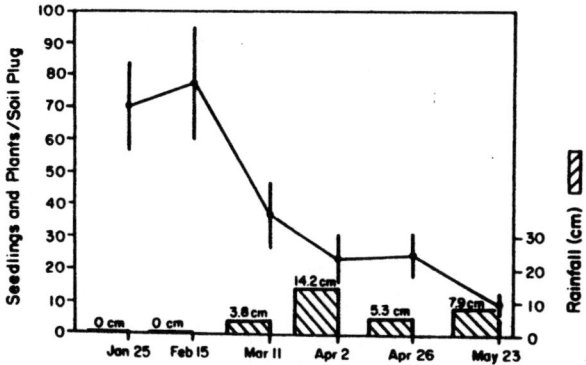

Fig. 1. Mean number of plants (± 1 SD) per soil plug (n = 10, 6 cm diam) taken from wallows, and total rainfall (cm) for two weeks prior to sample date. Relationship between number of plants and rainfall is nearly significant (F = 6.33, T = -2.52, p = .065).

bulls. Young and old bison of both sexes roll in dry wallows in warm weather. This dusting behavior may be related to protection from insect attack and to irritation caused by growth of a new coat (Roe, 1951). I have seen no bison rolling in mud or in wallows filled with water; however, bison do roll in wallows containing vegetation. The size of mature wallows has been reported to be quite large (e.g., 1400 m², Barkley and Smith, 1932), but in these cases I believe several adjacent wallows have merged into one large wallow through bison activity or soil erosion.

Mature wallows are inhabited by two distinctly different assemblages of plant species each year, and few of these are characteristic of mature prairies (Table 1). In the fall, seeds of winter annuals germinate, and the plants from these reproduce quickly in the spring. After spring rains fill the wallows, wetland species begin to dominate, flowering and fruiting throughout the summer. In spite of the unusual flora, species richness and diversity are lower inside wallows than in the adjacent prairie (Collins and Uno, 1983). Three hypotheses have been advanced to explain this phenomenon: (1)a lack of seeds dispersing into the wallows is responsible for a lower diversity; (2)germination requirements for seeds of many grassland species dispersed into the wallows are not met; and/or (3)high seedling mortality occurs due to fluctuating environmental conditions within the wallows. Results of my study tend to support the latter two hypotheses and refute the first.

Over 70% of the seeds in each soil core occurred in the upper 7 cm (Table 2). Soil cores from inside wallows contain more seeds and more species than do those from outside wallows. Also, seeds of species not found in wallows are found in the wallow seed bank and in the sticky traps placed in wallows. Thus, the lower species richness in the wallows does not result from lack of seeds dispersed into them. Seeds in wallows can come from three sources. They can be produced *in situ*, or enter wallows in at least two different ways: via buffalo chips that occasionally are dropped into wallows (Collins and Uno, 1985); and by the wind, as revealed from the sticky trap study.

The size of the seed bank within buffalo wallows ($150,000/m^2$ in the top 21 cm) was larger than that found in most other habitats studied, and is comparable to the seed populations in the litter and surface soil found in annual grassland communities in California (Young et al., 1981). This size was not unexpected because the weedy and wetland species occupying wallows produce large numbers of seeds. The soil cores used to estimate seed bank size, however, were much deeper than in other studies (e.g. Harper, 1977; van der Valk and Davis, 1978; Johnson and Anderson, 1986) and the estimate was based on actual counts of seeds and not on seedlings that germinated from the cores (e.g. Keddy and Reznicek, 1982; Parker and Leck, 1985). A large seed bank comparable to this study was reported from cultivated fields by Roberts (1981), where $90,000$ seeds/m^2 were found in the upper 15-25 cm of soil.

The distinctiveness of the wallow flora suggests the presence of environmental constraints on seed germination and seedling growth. In this study, fewer species germinated in the growth chamber tests than were found in soil samples screened from the same cores. This suggests, as expected, that not all the seeds within the seed bank will germinate at the same time, keeping diversity and richness levels lower than possible.

Low species richness also may be due to standing water or to water-level fluctuations within wallows. Water-level fluctuations have been hypothesized to affect species diversity within wetland communities (van der Valk, 1981), coastal plains (Keddy and Reznicek, 1982) and vernal pools in California (Jain, 1976). While zonation patterns are not as distinct within and around wallows as they are in vernal pools of California (Jain, 1976), a pattern can be distinguished (Polley and Wallace, 1986) whose basis has been correlated with moisture and nutrient levels (Polley and Collins, 1984). Buffalo wallows are ephemeral pools within the prairie. They collect and hold water for varying periods, from days to weeks at a time, depending on the rainfall pattern and the characteristics of individual wallows. The different water-holding capabilities of wallows may be due to their age, size and depth, compaction of soil, frequency of use by bison and other parameters. Of six wallows within 30 m of each other, on one sample date, one was full of water, one was dry, and four had a little standing water in the bottom. All wallows will hold water after a heavy rainstorm and may remain full if rainfall is frequent and heavy.

Standing water negatively affects certain plants within buffalo wallows (Table 3). The tubs in which plants were covered with water for some or all of the time had lower diversity, fewer plants, fewer plants that flowered, and later flowering plants than the tub with moist conditions. In the underwater tub, *Juncus torreyi* Coville was the dominant species (54% of all the plants), however, in the moist condition, only one *J. torreyi* individual was found out of 133 plants. This suggests that the standing water in the underwater tub, which mimics the conditions within wallows during the rainy season, inhibits the growth and development of many species of plants while favoring other water-tolerant species. It also helps to explain why adjacent wallows may contain different species and have different dominants. If each wallow has a distinctive ability to hold water, then some pools will be more ephemeral than others. This will select for different species within adjacent wallows even though they are served by the same seed pool source. *Coreopsis tinctoria* Nutt. was a dominant plant in all three tubs, indicating that it had the best "general-purpose-genotype" (Baker, 1965) of all species within wallows.

The plant counts made throughout the year (Fig. 1) indicate that the number of individuals within wallows increased in early spring, but showed a precipitous drop after a heavy rainstorm in which 3.8 cm of rain fell in a 24-hour period. One week later, all wallows sampled still had water in them, and many plants died before completing their reproductive period. Self-thinning in the densely populated wallows also might play a role in this decrease. However, in drier years winter annual species within these wallows continued to flower much later than March 11 (see Table 1) when they weren't covered by water. This evidence from the field also suggests that standing water negatively affects the number of wallow plants and their reproduction. Not only were fewer individuals found in the March 11 sample, but the average number of species per wallow was the lowest of all sample dates (3.3 versus a high of 5.0 on April 26). Thus, standing water also may affect diversity.

Based on observations within an area where wallowing is frequent, I propose a four-stage model of community development within wallows. In the WMWR, wallows are concentrated on unproductive soil (U.S.D.A. 1967). If vegetation is less dense on these soils, wallow formation may be easier. Stage I begins with the bison pawing clear an area of sparse vegetation exposing soil in which the bison may wallow. Wallowing compacts the ground forming a shallow bare area. At this stage, if no further wallowing occurs, weeds, such as *Circium arvense* (L.) Scop., will colonize the bare spot. During Stage II, the wallow continues to be hollowed out creating a depression 0.1 - 0.3 m deep. Winter annual weeds colonize these young wallows and reproduce the following spring. Stage III begins when the bottom of the wallow is compacted so that it holds water only long enough for wetland species to invade but not to survive and reproduce. Finally, mature wallows (Stage IV) contain winter annual weeds early in the growing season, wetland species after the spring rains, and perennial grasses that re-invade the wallow from the surrounding prairie. If no further wallowing occurs, the wallow may revert back to the original prairie, or it may never fill in, instead remaining distinguishable many years after the last bison has visited it (Polley and Collins, 1984). Several other factors may contribute to the continued existence of a wallow once it is formed. Rolling bison completely crush all plant life within a wallow, and their pawing removes new growth of perennial plants. Fires also can return a wallow to a bare condition, although it is quickly re-invaded by other plant species (Collins and Uno, 1983). In addition, standing water may kill perennial grasses. Thus, disturbance plays an important role in maintaining the distinct assemblage of wallow species. As Rabinowitz et al. (1984) have suggested, sparse, but better competitors, cannot overtake more common plants because disturbance may set back interactions. Within buffalo wallows, sparsely represented perennial grasses with perennating and vegetative organs cannot establish a strong foothold within the wallow unless the wallow is not used. Perennial grasses have invaded wallows at the WMWR protected by metal exclosures that prevent bison from entering (pers. obs.). In areas of active wallowing, annuals and wetland species should be strongly selected for because they can reproduce quickly, or during times when the wallow is not frequently used by bison, i.e., the cool, wet months.

CONCLUSIONS

The common concept of the Great Plains is a vast expanse of perennial grasses and forbs accounting for 95% of all prairie species (Risser, *et al.*, 1981). However, this characterization must be redefined because of the activity of animals such as badgers (*Taxidea taxus*) (Platt, 1975), pocket gophers (Geomyidae) (Mielke, 1977), prairie dogs (*Cynomys ludovicianus*) (Coppock, *et al.*, 1983), and bison. Because of their number and behavior, bison certainly have had a tremendous impact on the development of the prairie, especially in the evolution of communities dominated by species other than perennial grasses. The regional diversity of the mixed grassland is greatly increased by weedy and wetland species found within and around wallows (Collins and Uno, 1983). Buffalo wallows are pools of water, and thus are safe sites for wetland species within the prairie in addition to playa lakes, rivers, and ponds. Because of the ephemeral nature of the water, however, and because wallows are frequently disturbed, wetland perennials rarely persist in active wallows as long-lived plants. Thus, while bison help to create a suitable site for wetland species, they also regulate wetland and grass species' populations by destroying perennating parts. For perennial species, the seed bank represents an alternative means by which they can persist in buffalo wallows.

Within and around wallows, there is a great opportunity for the seed bank to play a significant role in vegetation dynamics because of frequent disturbance by bison. As England and DeVos (1969) stated, "Bison overgrazing, in association with trampling, rubbing and wallowing, contributed to the creation and maintenance of environmental conditions favorable to a variety of other wildlife by encouraging the growth of forbs." Because they are disturbed sites, buffalo wallows provide natural habitats for weeds in the prairie. There also is much disturbance by bison around wallows. Thus, this area serves as an additional habitat for some of the weedy species represented in the wallow flora. Unlike vernal pools in which non-dispersal is a strategy because a seed is likely to land in the inhospitable surrounding grassland (Holland and Jain, 1981), dispersal from a buffalo wallow into suitable germination sites of the adjacent prairie or other wallows is possible. The large size of wallow seed banks indicates their importance as seed sources for other wallow and prairie habitats which were frequently disturbed by bison and fire. The dispersal of seeds by wind and bison could distribute many species far from their source. This leads me to suggest that wallows have played an important role in the spread of native weedy and wetland species throughout the plains ever since bison roamed the prairie: 86% of the species found in wallows (Table 1) are native to the Great Plains. The influence of wallows would have been widespread and would have begun before humans helped to spread weeds in this area, such as along cattle drovers' trails (Baker, 1972). While they lack the showy flora, species richness, and endemics of other temporary systems such as vernal pools (Jain, 1976; Jain and Moyle, 1984), buffalo wallows do have a unique assemblage of species that differs distinctly from the surrounding prairie (Collins and Uno, 1983). Buffalo wallows have contributed greatly to the diversity, ecology, history, and interest in the grasslands of the Great Plains.

LITERATURE CITED

Baker, H. G. 1965. Characteristics and modes of origin of weeds. In H. G. Baker and G. L. Stebbins [eds.] The Genetics of Colonizing Species. Academic Press, New York.

Baker, H. G. 1972. Migrations of weeds. In D. H. Valentine [ed.] Taxonomy, Phytogeography and Evolution.

Barkley, F. A., and C. C. Smith. 1933. A preliminary study of the buffalo wallows in the vicinity of Norman Oklahoma. Proc. Oklahoma Acad. Sci.

Buck, P. 1964. Relationships of the woody vegetation of the Wichita Mountains Wildlife Refuge to geological formations and soil types. Ecology 45: 336-394.

Collins, S. L., and G. E. Uno. 1983. The effect of early spring burning on vegetation in buffalo wallows. Bull. Torrey Bot. Club 110: 474-481.

Collins, S. L., and G. E. Uno. 1985. Seed predation, seed dispersal, and disturbance in grasslands. Am. Nat. 125: 866-872.

Coppock, D. L., J.K. Detling, J. E. Ellis, and M. I. Dyer. 1983. Plant-herbivore interaction in a North American mixed-grass prairie. Oecologia 56: 1-9.

Crockett, J. J. 1964. Influences of soils and parent materials on grasslands of the Wichita Mountains Wildlife Refuge, Oklahoma. Ecology 45: 326-335.

Dooley, K. L., and S. L. Collins. 1984. Ordination and classification of western oak forests in Oklahoma. Am. J. Bot. 71: 1221-1227.

England, R. E., and A. Devos. 1969. Influence of animals on pristine conditions on the Canadian grasslands. J. Range Manage. 22: 87-94.

French, N. [ed.] 1979. Perspectives in Grassland Ecology. Springer-Verlag, New York.

Harper, J. L. 1977. Population Biology of Plants. Academic Press, London.

Holland, R. F., and S. K. Jain. 1981. Insular biogeography of vernal pools in th Central Valley of California. Amer. Nat. 117: 24-37.

Jain, S. [ed.] 1976. Vernal Pools: their Ecology and Conservation. Inst. of Ecology, Univ. of California, Davis. Publ. No. 9.

Jain, S., and P. Moyle [eds.] 1984. Vernal Pools and Intermittent Streams. Inst. of Ecology, Univ. of California, Davis. Publ. No. 28.

Johnson, R. G., and R. C. Anderson. 1986. The seed bank of a tallgrass prairie in Illinois. Amer. Midl. Nat. 115: 123-130.

Keddy, P. A., and A. A. Reznicek. 1982. The role of seed banks in the persistence of Ontario's coastal plain flora. Amer. J. Bot. 69: 13-22.

Larson, F. 1940. The role of the bison in maintaining the short grass plains. Ecology 21: 113-121.

Martin, A. C., and W. D. Barkley. 1961. Seed Identification Manual. Univ. of California Press, Berkeley.

Mielke, H. W. 1977. Mound building by pocket gophers (Geomyidae): their impact on soils and vegetation in North America. J. Biog. 4: 171-180.

Musil, A. F. 1978. Identification of Crop and Weed Seeds. U.S.Dept. Agric., Washington, D. C.

Parker, V. T., and M. A. Leck. 1985. Relationships of seed banks to plant distribution patterns in a freshwater tidal wetland. Amer. J. Bot. 72: 161-174.

Platt, W. J. 1975. The colonization and formation of equilibrium plant species associations on badger disturbances in a tall-grass prairie. Ecol. Monogr. 45: 285-305.

Polley, H. W., and S. L. Collins. 1984. Relationships of vegetation and environment in buffalo wallows. Amer. Midl. Nat. 112: 178-186.

Polley, H. W., and L. L. Wallace. 1986. The relationship of plant species heterogeneity to soil variation in buffalo wallows. Southw. Nat. 31: 493-501.

Rabinowitz, D., J. K. Rapp, and P. M. Dixon. 1984. Competitive abilities of sparse grass species: means of persistence or cause of abundance. Ecology 65: 1144-1154.

Risser, P. G., E. C. Birney, H. D. Blocker, S. W. May, W. J. Parton and J. A. Wiens. 1981. The True Prairie Ecosystem. Hutchinson Ross Publ. Co., Stroudsburg, PA

Roberts, H. A. 1981. Seed banks in soils. Adv. Appl. Biol. 6: 1-55.

Roe, F. G. 1951. The North American Buffalo. Univ. of Toronto Press, Toronto,Ont. Canada.

Spedding, C. R. W. 1971. Grassland Ecology. Oxford Univ. Press, London.

United States Department of Agriculture 1967. Soil Survey for Comanche County, Oklahoma. U. S. Gov't Printing Office, Washington, D. C.

Valk, A. G. Van der 1981. Succession in wetlands: a Gleasonian approach. Ecology 62: 688-696.

Valk, A. G. van der, and C. B. Davis. 1978. The role of seed banks in the vegetation dynamics of prairie glacial marshes. Ecology 59: 322-335.

Young, J. A., R. A. Evans, C. A. Raguse, and J. R. Larson. 1981. Germinable seeds and periodicity of germination in annual grasslands. Hilgardia 49: 1-37.

EVOLUTION IN CULTIVATED PLANTS

Agriculural research as carried out by ethnobotanists is closely linked to conservation. In this section we present three ethnobotanical studies. Robert Bird reports on his many years of work with the most widely studied New World food crop, maize. In this case, many of the most important wild ancestors are extinct, but abundant local races exist. These races are under threat from changes in traditional agricultural practices. The paper by Yun-Tsu Kiang and Y. C. Chiang deals with an extremely important Old World native, soybeans. Here, the cultivar coexists with its wild ancestors. The third crop, chayote, is reported on by Linda Newstrom. It has a short history of formal cultivation, and the distinctions between wild vs. cultivated forms are not clearcut. In all cases, the authors are interested in identifying and conserving genetic diversity, thereby insuring care of ethnobotanical resources along with the natural ones.

Robert Bird has spent many of the last 25 years studying ethnobotany in the New World tropics. His contribution contains heretofore unpublished data from his PhD thesis and from years of ongoing field work in the northeastern Peruvian highlands. His object of study, *Zea mays*, is perhaps the most important contribution of the New World to the Columbian Exchange. He shows how culture and habitat have served to keep varieties of maize distinct from one another through hundreds, perhaps thousands, of years. He worked with the Chupachu and Serrano people who farm side by side in the Huanuco region of the Andes. In addition to their distinctive races of cultivated maize, these people differ in their textile industries including choice of fiber, colors, and weaving styles, in dialect, in patterns of irrigation, and in other cultural attributes.

Bird used methods from numerical taxonomy to distinguish among Andean maize races. This task is difficult because over the centuries, people (and natural cross pollinations) have hybridized and back-crossed the varieties many times. Even so, the genetic traits of these races of maize give a picture of 2000 years of human history in this region that is not available elsewhere.

In the past 40 years these indigenous races of maize have been threatened by the introduction of "improved varieties" from outside the area. Many ethnobotanists and crop geneticists view these exotic forms with alarm because they have not undergone the intricate processes which have fit the ancient varieties to their habitats and to the agriculturalists who use them. It is appropriate to sound a cautionary note to "green revolutionists" who push exotic crops on people whose cultures have been stable through time.

Kiang and Chiang examined a close relative (*Glycine soya*: Fabaceae) of the cultivated soybean (*G. max*). Wild soybeans still exist in China, Japan, Korea, Taiwan, and the U.S.S.R.; the field work for this study was carried out in South Korea and Japan. The authors propose the Korean peninsula rather than China, the traditional choice, as the place of origin for soybeans because of the greater amount of genetic variation present in plants there.

Since the initiation of cultivation, wild soybeans may have had little impact on the cultivated form in contrast with the pattern seen for certain grains such as maize and wheat. The authors propose that two forces are keeping these species distinct from one another, despite their demonstrated ability to form fertile hybrids in nature: (1) Both wild and cultivated soybeans have a strong tendency to self-pollinate. (2) Wild soybeans show no preference for growing in disturbed places, i.e., along the edges of cultivated fields, as is often the case with wild relatives of other cultivars. The wild species offers a significant source of genetic variability for crop improvement.

Newstrom presents us with other aspects of contemporary ethnobotanical research. She has been studying chayote, a squash widely used for human food and fodder in Central and South America. She has studied this plant of Mexican and Guatemalan origin in Mexico and in Costa Rica. The fruits, tuberous roots, and leaves of chayote are eaten. The stem, the only part of the plant which people do not eat, is used for basket making. The fruits, and, to a lesser extent, the roots, are marketed in Latin America and in Europe. The plant is monoecious and facultatively viviparous, but seed germination is easily regulated by the regime employed for fruit storage. Enormous variation exists among wild and cultivated populations. This should allow crop breeding for such things as lowered bitterness in fruits in this useful plant.

Herbert Baker's interest in ethnobotany was fostered by his early years at the Univeristy of the Gold Coast, where he established the botanical garden, and developed a serious interest in tropical agriculture. This interest has persisted in his ongoing work in the New World tropics, and has been expressed through his many contributions to the literature of ethnobotany, tropical agriculture, and genetic conservation. These contributions here by his students are a further reflection of his interest.

MAIZE, MAN AND VEGETATION IN NORTH-CENTRAL PERU

Robert McK. Bird

INTRODUCTION

Variation in modern maize relates to both recent and ancient cultural events, and in turn these events have been channelled by the biological and physical environment. Large complexes of maize races are characteristic of certain regions; they are not randomly found across the Americas. New World cultures have been shaped by their natural environment, and have continually domesticated or improved animals and cultigens. This has linked them ever more closely to certain habitats. Thus the availability of these habitats becomes a major factor in a culture's extension into new regions. The department of Huánuco in north-central Peru provides some of the clearest evidence of this thesis.

The department of Huánuco is in a unique position in the Andes, encompassing not only numerous natural zones but also two major Sierran cultural zones and many cultures in the Montaña and Selva forest regions. Not far to the west and south of the capital city, Huánuco, are glaciers and frigid grazing land, to the north, cloud forest, to the east, tropical forest, and in the center, thorn forest. People near the city are related in many ways to people far to the north, but just to the west, higher in the mountains, the inhabitants are more like people in Bolivia and northern Argentina. Differences between these two ethnic regions can be measured in terms of agriculture, textiles, dialect, and other attributes. The maize of the department reflects this geography, often to an impressive degree.

Each aspect of a model which relates maize to the environment needs to be defined carefully from numerous field observations. Unfortunately, few workers have described the maize, peoples, vegetation or climate of north-central Peru, none in enough detail to allow construction of relational models. My research there, 24 months in 1964-1966 and one month in 1981, was multifaceted, but sometimes thinly spread. With the first study I travelled through much of highland Huánuco; with the second I was in northwest Huánuco, the Llata-Singa area. A theme of all the work has been the use of large sets of variables.

THE VARIABLES
Physical Geography

The highest peak in Huánuco, marking its western boundary, is Yerupaja (ca. 6634 m), the second highest in all the Tropics (American Geographical Society, 1938). Much of the terrain for 100 km to the northeast of Yerupaja lies between 4000 and 4400 m, cut deeply by numerous narrow valleys leading into the Marañón and Huallaga Rivers (Fig. 1). The eastern slopes drop 3500 m in a band only about 50 km wide. Air moving up from the Amazon region dumps up to 10 meters of rain per year on the Montaña forests of these slopes, leaving little for valleys just over the initial ridges (0.4 m/year at Huánuco).

There are enough meteorological stations in the Peruvian Andes (Tosi, 1960: Table 5; Robinson, 1963) to allow relationships between altitude, temperature, precipitation, distance from the Pacific Ocean, and latitude to be fairly reliably established. Available data show that temperature drops 5.7° C with each 1000 m rise in altitude, if the precipitation remains constant (Bird, 1970: 104; loc. cit.). Precipitation, and associated clouds, affect temperature according to the formula:

$$T_p = T + 5(\log P - 2)$$

where T_p is temperature ($^{\circ}$C) adjusted to that expected with 100 mm of annual precipitation, T is mean annual temperature, and P is mean annual precipitation (mm) (Bird, 1970: Fig. 27). Mean annual temperature has been adjusted for both elevation and precipitation and plotted against distance from the Pacific Ocean (Bird, 1970: Fig. 28). Rather surprisingly, temperatures less the effects of elevation and precipitation in the Selva and in that part of the Sierra east of the Continental Divide are very similar, without any latitudinal effect. This has led me to conclude that relationships of precipitation, temperature and elevation to vegetation and crops in Huánuco can be generalized anywhere in Peru east of the Continental Divide.

Recent topographic maps (1:100,000) have been produced by the Instituto Geográfico Militar (1973-1985). Unfortunately, these were not available to me during my field work. The altitudes of many of my study sites could now be measured much more exactly than by using altimeters, a few "base-point" towns, and understanding of tropical diurnal barometric shifts (Bird, 1970: 13).

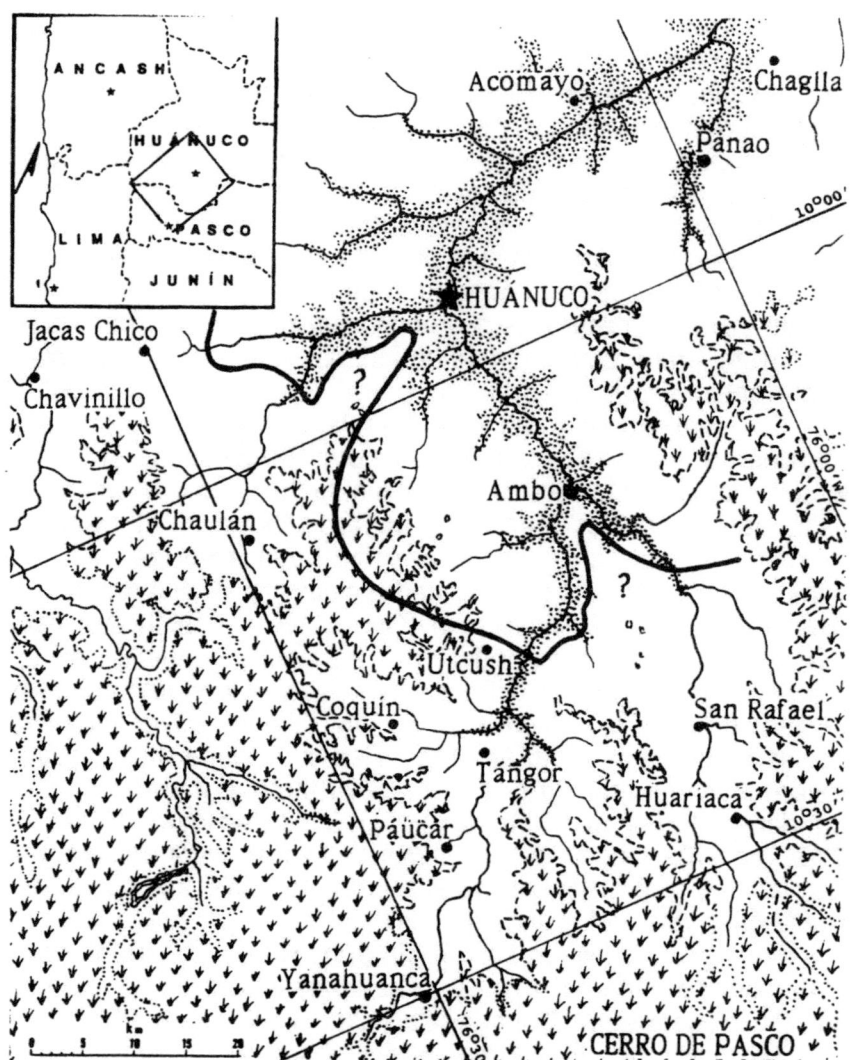

Figure 1. The upper Huallaga Basin. The warm vegetation belt and the zone of tropical crops lie within the 1500-2500 m band indicated by stippling. The area with "tufts" is mostly above 4000 m, including most of the Puna. The mild, cool and cold belts with almost all maize and potato fields lie between these two and are unmarked. The Chupach/Serrano boundary is estimated by the heavy line (Bird, 1984b). The Marañón River cuts through the upper left quadrant. Drawn from Bird, 1970, and from Instituto Geográfico Nacional maps, 1973-1985.

Vegetation Zones and Natural Boundaries

Linkages between the natural environment and crops can be quickly clarified by using shifts in vegetation. There are too few meteorological stations to establish patterns of rainfall and temperature in local detail, but vegetation is a clear indicator of ecological potential for cultigens. The habitats in which species are found and the frequencies and sizes of those species are very dependent on, and therefore reflect, the physical environment. More importantly, the vegetation of Huánuco does not present a continuum of change from zone to zone -- most of the indicator species initially selected as representative of a limited number of zones were later found to define boundaries where sets of species concurrently disappeared or appeared (Bird, 1970: 82-100).

Observations were made of the relative frequencies of the indicators in the available habitats at over 80 locations in Huánuco and Pasco, the department just to the south. In some cases, sets of species reach their upper limits where other sets are reaching their lower limits, defining boundaries as definitive as that between *Picea* and associated species and *Quercus* and associated species in Wisconsin (Curtis, 1959). Such thermally oriented boundaries cover 50-100 m elevation; the belts defined by them cover 450-600 m. Belts have been divided into zones by indicator boundaries which are pluvially related. Over forty species in the study area, all of them easily distinguished, have proven useful in defining one or more of the five thermal boundaries and/or five boundaries related to moisture (Bird, 1970: Table 4). Once the boundaries and zones were defined, belts were labeled with an appropriate thermal adjective and zones by an adjective describing moisture levels.

Boundaries between sets of cultigens were also established, with somewhat different results. They parallel but do not coincide with the vegetational boundaries (Bird, 1970: 100-103).

The warm belt in the region includes a narrow, very dry zone along the Huallaga valley bottom (Fig. 1). It is characterized by *Acacia* and *Espostoa* which, with *Anadenanthera* and *Jacaranda*, reach up to 2100-2500 m, where the mild belt starts. The mild belt in the Panao-Huariaca-Yanahuanca-Chavinillo quadrangle has two zones: dry *Caesalpinia-Schinus*, and moist *Furcraea-Salix* (Bird, in manu.). These zones reach ca. 2750-3120 m and 2400-2700 m respectively. Altogether 14 species define the mild belt's upper limit, which may coincide with the 13.5°C isotherm. Numerous New World cultigens including lima bean, cotton, avocado, manioc, gourd, and *Xanthosoma*, plus Old World

sugar cane, coffee, and banana, reach a common boundary 200 m
above the mild/warm vegetational boundary (Yamamoto, 1981).
    The cold/cool vegetational boundary is clearly established by
twelve species with common upper limits and by seven species
whose lower limits follow the same line. This lies close to the
10.0° C isotherm, at 3540 m in a dry area, at 3020 m where it is
very moist. Below the boundary are the fairly dry *Agave-
phryosporus,* moist *Duranta-Abatia,* and very moist *Pteridium-
Embothrium* zones. The moist *Polylepis-Barnadesia* and very
moist *Buddleia-Chusquea* zones lie above it (Bird, in manu.).
Maize and associated cultigens such as *Cucurbita ficifolia,* common
bean, cultivated lupine and two species each of *Cyclanthera* and
Amaranthus usually reach 10-50 m higher than the cold/cool
boundary.
    The Puna of the very cold and frigid belts is dominated by
bunch grasses interspersed with cushion plants, dwarf shrubs,
rosette plants and other odd life-forms (Weberbauer, 1945; Tosi,
1960). It is a complex of several zones in two belts lying between
3900 ± 100 m and 4800 ± 100 m in western Huánuco and Pasco.
In Pre-Columbian times large herds of llamas were typical of the
Puna; sheep dominate today (Yamamoto, 1985). Several potato
species and cultigens of *Oxalis, Ullucus, Tropaeolum* and *Lepidium*
dominate the crops.
    One can use this vegetational framework to search for close
linkage of maize races to temperature or precipitation, and to
improve models linking cultural trends to maize variation. Zonal
position of a maize field can usually be established easily because
indicator species can be seen in at least one of the several nearby
habitats and because boundaries are usually observable within a
half-hour's walk of a field.

Cultural Patterns
    In order to explain the great variation found within
cultigens, especially in Andean food staples, one needs to
understand the history and uniqueness of present cultures. The
people of the area have bred the present varieties and races for
their different uses and colors and have taken them into new
habitats. Moreover, the amount of difference between peoples is
reflected in the separation of their maize varieties. In the
highlands of Huánuco there are two principal indigenous ethnic
types: peoples flanking the warm, dry areas of the upper Huallaga
were historically called Chupachu, clearly distinguished from
several Serrano groups generally at higher elevations to the west
and south (Fig. 1; Bird, 1966, 1970: 62-81; 1984b). The

traits which characterize these peoples continue right to the boundary, making it very sharp.

Textiles provide one clear marker of ethnic differences, cotton being important in Chupachu weaving, colors and wool marking the Serrano (Bird and Mendizabal, 1986). Chupachu women weave ponchos; Serrano women do not. Settlement patterns differ: most Chupachu live scattered among their fields, while the Serrano crowd into towns. Architecture, costumes and dialects of Quechua add to the list (Bird, 1966, 1984b; Bird et al.1988). What most concerns us here are differences in ecological adaptation, agricultural practices and, especially, maize races.

Most Chupachu live on moist, moderate valley slopes at altitudes between 2200 and 3000 m in a very good agricultural zone, while Serrano towns are in a band above the headwater streams generally between 3300 and 3600 m. The steep fields of the Serrano are not as dry as the Chupachu valley fields nor as wet as the Chupachu upper corn fields. In the higher Serrano towns, potatoes are the major crop, while maize is the major crop of lower towns, for both Chupachu and Serrano (Fonseca M., 1966, 1972; Burchard, 1976; Mayer, 1985). Puna grasslands with their associated herds are very extensive in the Serrano region (Fig. 1), though much of the narrow very cold and frigid area east of the upper Huallaga and Marañón is too wet to be useful for herding or agriculture. The many tropical crops of the Huallaga valley bottom are almost exclusively cultivated by the Chupachu, who also have fields and gardens in the Montaña to which they travel. The Chupachu commonly irrigate maize and potatoes at intermediate elevations. The Serrano seldom irrigate maize and never irrigate potatoes.

Maize Patterns

The foundation for the study of Andean maize comes from the many publications on maize races in Latin America (listed and summarized in Hernandez X., 1973; Brown and Goodman, 1977. These reports are based on extensive collecting of landraces, growing samples in proximal plots and measuring vegetative and reproductive growth and form. Grobman et al. (1961) present further data for Peruvian races, including many measurements of cob structure and color patterns. An early, simplistic factor analysis of the Peruvian data (some from accompanying photographs) enabled me to reduce the set of variables through correlation analysis, while separating clusters of races. Specifically, the Ear Shape and Cuzco-Factors clearly set off three groupings, (1) Cuzco Large Flours and Flints having few rows of very wide kernels, (2) Central Andean Rotund Flours and

Altiplano Small Flours with many rows of elongate kernels, and (3) races with elongate ears having proportionately large ear bases, most commonly found in the North Andes. The formula for the Ear Shape Factor is:

$$EL \ (EL/KT) + (EL/ED) - EDMXPC$$

with EL = ear length, KT = kernel thickness, ED = ear diameter and EDMXPC = distance of maximum diameter from the ear base as percent of EL (all variables were roughly normalized and scored from 1 to 10). The Cuzco Factor is: 3/2 (KW + KW/KT) + 21- (RN + ED/KW), with KW = kernel width, and RN = row number, likewise standardized (Bird, 1970: 163).

In order to carry these analyses further, 65 ears representing the races and sub-races of Huánuco and most of the other races of Peru, many of them contributed by the Programa Cooperativo de Investigaciones en Maiz, were measured in 62 ways (Bird, 1970: 172-192). Fifty-eight ratios, indices, and other compoundings of the measurements were calculated. The measurements and combinations were split into two sets for separate factor analyses.

Eleven oblique (not entirely independent) factors were derived (Bird, 1970: 192-195). Half (31) of the measurements were closely associated (r ± 0.66 - 0.99) with at least one of these; half of the rest were moderately correlated (r ±= 0.33 - 0.65) with one or more factors. Because I used the ratios and other combinations of measurements in the factor analysis, the frequency that a measurement was used in combinations affected factor derivation. For instance, the Shank Factor is closely correlated with shank diameter (SD) and with the ratios SD/ED and SD/upper glume diameter. Such combinations are more useful in refining and clarifying patterns, adding nothing to the variation accounted for. The main value of this factor analysis has been to help me select measurements for later studies (Bird and Bird, 1980; Miksicek et al., 1981).

Further useful groupings of maize races were determined by means of a matrix of maize measurements assembled from the Latin American maize race reports (Goodman and Bird, 1977; Bird and Goodman, 1978). Variables which had not been measured on almost all races were dropped, as were races which had not been measured for all the remaining variables. Principal components (PCs) based on 219 races and sub-races were derived from a correlation matrix of eight ear variables (ear length, diameters of ear, cob and rachis, row number, kernel length, width and thickness). Factor scores of the races were calculated and were used to plot the races on PC x PC plots. Races were clustered using the factor scores of the five most significant PCs. The more unique, separable races were set aside -- at first the Amazonian interlocked Flours and small-kenneled races (kernel volume less than 235 mm^3; Fig. 2), and later eight large-kenneled races, mostly in the Cuzco Large Flours and Flints. New principal components were calculated, on the reduced set of races, and much better separation of the remaining race clusters were seen in new PC x PC plots (Bird and Goodman, 1978).

Numerical taxonomy based on raw measurements and on the factor scores of the ears has not provided a picture that is easily described (Bird, 1970: 196-204). No dendrogram could represent more than a part of the array of relationships within a complex taxon without breaking important links between races on different branches (Funk, 1985). The populations of maize are interrelated in a multidimensional net; the knots of the net are populations, the strands indicate the flow of genes. In some parts of the net, the knots and strands are numerous and crowded; some knots are very small; some strands are thick. Only a few populations of *Zea* contributed to the maize network in the early periods, but these were morphologically very different from one another (Wilkes, 1979; Bird, 1984a). After accidental hybridization and introgression, the resultant types were spread widely, evolving to fit local food preferences and to overcome habitat differences. As political influences of centralized states spread ever further, the connections in the phylogenetic net took on new dimensions (Bird, 1980)

Nineteen Latin American racial complexes containing 76 races were defined in this way (Fig. 2). Other races were considered too intermediate to assign to a complex, using only the eight available variables. These groupings of races plus some sub-complexes defined later have proven useful in describing maize geography and its relationships to cultural geography and history (Bird, 1980, 1984a; Bird et al., 1988). Although the clusters seem valid, clustering based on principal components and multiple discriminant function analyses of a greatly enlarged data base are needed, especially to substantiate the positions of intermediate races. Of the nineteen complexes, the Altiplano Small Flours, Central Andean Rotund Flours (CARF), Cuzco Large Flours and Flints, Andean Pointed White Popcorns, North Andean Flints and Flours, Rienda-Clavo and a set of Caribbean small dents are in the highlands of Huánuco and Pasco.

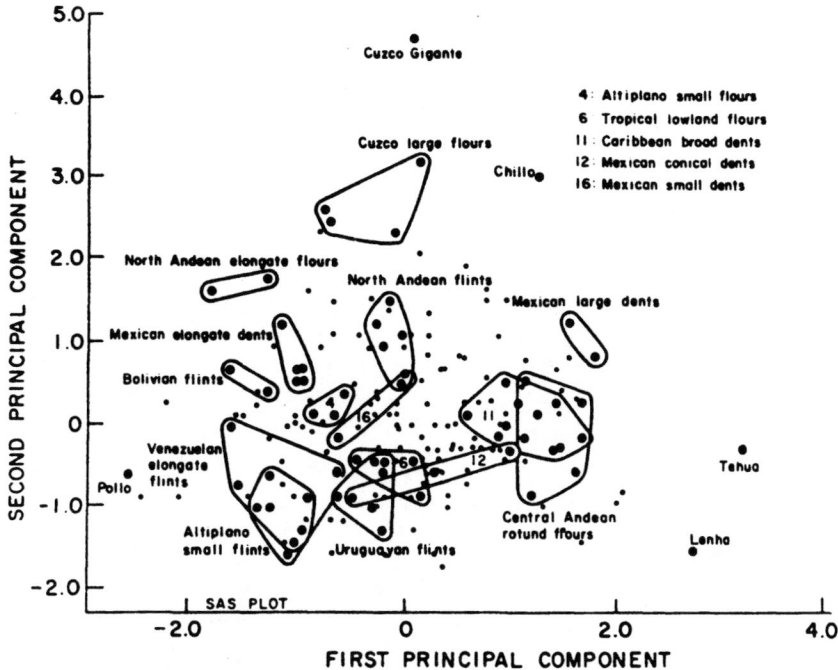

Figure 2. Plot on the first two principal components of Latin American maize ear measurements, from an analysis run after dropping small-kernelled races and the Amazonian Interlocked Flours. 190 races and sub-races are plotted, with less group overlap than on a plot made before dropping races. From Figure 3 of Bird and Goodman (1978).

Definition of the complexes in many areas is becoming more dependent on two large data sets, based respectively on chromosome knobs (McClintock, 1978; McClintock et al., 1981; Bretting et al., 1987) and isozymes (Doebley et al., 1983, 1985; Goodman and Stuber, 1983; Smith et al., 1985; Bretting et al., 1987). Unfortunately for intraregional comparison purposes, chromosome knobs of most Andean races seem to have been fixed in one pattern well before morphological evolution slowed, the Andean Two-Knob Complex with knobs only on the long arms of chromosomes 6 and 7 (McClintock, 1978). This pattern characterizes all races above 1800 meters elevation from southern Bolivia through Ecuador. The major exceptions are races in the Andean Pointed White Popcorns (APWP) (Fig. 3a) which almost always have many additional small knobs at up to six positions, plus an inversion typical of central Mexican highland maize and of Guatemalan Imbricado, also a pointed white popcorn (Grobman et al., 1961; Longley and Kato, 1965; McClintock,

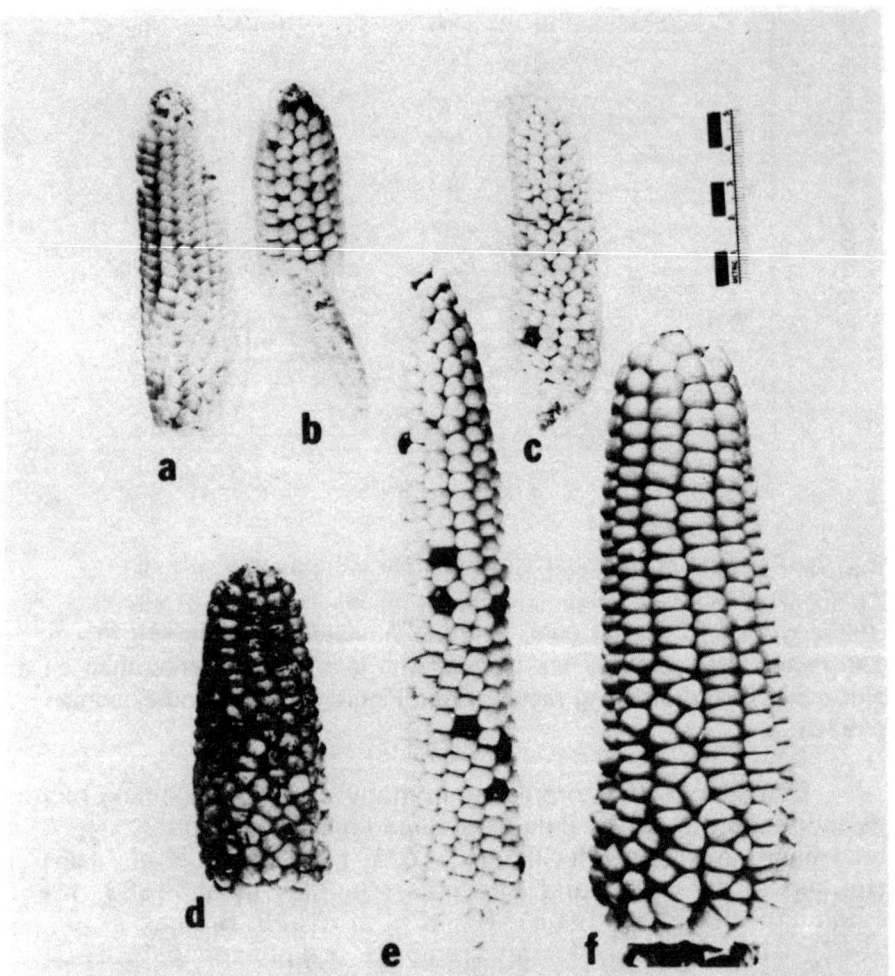

Figure 3. Flint and popcorn types typical of the highlands of Huánuco.
a) Confite Puntiagudo, an Andean Pointed White Popcorn (Bird 2395);
b) Serrano *kashpish* with kernels missing (Bird 2662); c) Chupachu
rosa hara or *pashta hara* (Bird 2740; = Perlilla); d) Serrano *wangsha*
(Bird 2664; e) Chupachu *kashpi* (Bird 2468); f) Chupachu *wangsa* (Bird
2485).

Figure 4. Flour types typical of the highlands of Huánuco plus Cuzco
Gigante. a) a large Serrano *gapia* in the CARF (Bird 2882); b) a small
Serrano *gapia* related to both Altiplano Small Flours and the Central
Andean Rotund Flours (CARF) (Bird 2554); c) Chupachu *yuraq* (white)
gapia (Bird 2496); d) Chupachu *chuspilyu* (Bird 6027); e) Cuzco Gigante
(Bird 2423). The scale is the same as in Figure 3.

1960, 1978). Data on isozymes of Andean races have been presented only for Bolivia yielding a pattern "notably consistent with the cytological study of chromosome knobs" (Goodman and Stuber, 1983: 169). The APWP have an equally unique isozyme pattern -- 12 alleles at 9 loci (out of 23 loci) are much more frequent. Andean races cluster together and differ greatly from lowland dents and flints.

In Huánuco much maize variation relates to an axis defined by Chupachu *kashpi* ("stick") flint and Serrano *gapia* ("soft") flour. The former has long narrow cobs, small round flint kernels and elongate cupules (Fig. 3e), probably a race of Rienda-Clavo, while the latter has high row number, long flour kernels and broad cupules (Fig. 4a,b), a variable race in the Ancashino-Paro sub-complex of the CARF. Position of maximum ear diameter and lower glume angle are also useful in this separation (Bird, 1970: Table 16).

There are three other types at the flint end of the axis: Serrano *kashpish*, a narrow-cobbed near-popcorn (Fig. 3b), Serrano *wangsha*, a small-kenneled, small-eared flint (Fig. 3c), both rare, and Chupachu *wangsha*, a large flint widest at the ear base related to the North Andean Flints and Flours (Fig. 3f). By many measures the last is about 1.5 times as large as Serrano *wangsha*, although row number and several ratios are similar (*ibid.*). Even though the Serrano and Chupachu share terms for flints, the four races appear very different, and they differ greatly in frequency. Half of a Chupachu family's harvest may be flint maize (Bird, 1970: 117-119, 215-219).

Flours of the region are just as variable. Serrano maize is dominated by flours which range widely in size and in pericarp color -- yellow-brown, tan, yellow, orange-brown variegated, and colorless (Fig. 4a,b). Kernel aleurone, the cell layer(s) just under the pericarp, is usually colorless in Pasco and southern Huánuco, but in northwest Huánuco purple is common. Among Chupachu flours white dominates (Fig. 4c), but *ushgulyu* , purple and white, and *chushpilyu* (Fig. 4d), very soft with purple, speckled aleurone are locally fairly common. Speckling is very rare in Serrano maiz. In kernel shape, texture and row number, and sometimes in ear form, *ushgulyu* and *chushpilyu* are similar to Serrano flours, but when grown in Berkeley, *chushpilyu* was the latest of the Huánuco varieties to shed pollen, whereas Serrano maize is relatively early. The Chulpi-Maiz Dulce sub-complex of the CARF, distinguished by a high frequency of the recessive sugary-1 allele, is represented in a small area near Chaglla. Cuzco Large Flours and Flints (Fig. 4e) are also rare -- the floury form is grown by a few innovative farmers who recently introduced it.

Popcorns are not common. There are two types of Chupachu popcorn -- the many introductions of Andean Pointed White Popcorn (Fig. 3a) and the apparently indigenous Perlilla (Fig. 3d; Grobman et al., 1961). The kernels of Serrano *kashpish* are very small and may pop occasionally.

Several gross factors besides climatic features determine the frequency of a race in an area -- especially preferences for certain foods and color arrays. Maize in highland Peru is most commonly eaten as *kancha*, parched, and *mote*, boiled, although maize beer, bread, fermented grain, corn-on-the-cob and forage are important products. The maize harvest may be processed and stored in many forms, and the recipes for food and drink preparation are numerous and often complex (Rick and Anderson, 1949; Martinez, 1960; Grobman et al., 1961; Stein, 1961; Bird, 1970: 113-119). The Chupachu prefer large-kenneled flints for *mote*; the Serrano prefer flours, especially white forms. Serrano beer is made from malted red-brown flinty maize.

Nearly black maize is used for an unfermented sweet drink prepared by boiling kernels and cob with sugar and fruits. The Chupachu, characterized by white and black or very dark blue clothing, have selected only yellow flints and white flours for their *mote* and *kancha*. The Serrano use many colors in their clothes, and their maize is exceptionally colorful.

DISCUSSION OF INTERRELATIONSHIPS
Dimensions, Boundaries and Barriers
Many dimensions have been defined both for maize morphology and for various aspects of the environment of maize in Huánuco. In the volumes defined by these dimensions, there are occasional discontinuities defining local races, vegetation zones or ethnic groups. Important deductions come from plotting their geographic distributions and looking for correlated trends and coincidental breaks.

There are major and abrupt shifts in maize on each of the three main branches of the upper Huallaga. In travelling up each tributary, long *kashpi* flint and very soft, speckled *chushpilyu* drop in frequency to nearly zero at the points where both multicolored *gapia* flour corn becomes common and flint maize with small kernels is found. The breaks coincide with the ethnic and historic boundary between the Chupachu and the Serrano, adding to the validity of patterns found in textiles, phonetics, and ethnic allegiances of 450 years ago (Fig. 1; Bird, 1984b; Bird and Mendizabal, 1986).

One might at first propose a correlation between size of maize plants across the upper Huallaga and altitude, therefore temperature. Plants at 2900 m on the east in the Chupachu area are much taller than those at 3300 meters 50 km to the west in a completely different ethnic area. This relative size of plant is genetic, confirmed when I grew many of these maizes at the University of California botanical gardens in Berkeley. By locating vegetational boundaries near the fields and thereby estimating temperature and precipitation, temperatures at these different altitudes are shown to be similar. So maize plant size is not correlated as much to temperature as to culture. However, cultural selection may be influenced by what is probably a longer frost-free growing season in the Chupachu area.

It is apparent that maize collections should be accompanied by information on temperature, precipitation (both probably estimated by vegetational boundaries), elevation and, especially, cultural position in order to get more precise statistical analyses. Each maize zone should contribute a significant number of collections representing the races available. In this way morphological trends in each race or sub-race could be compared to environmental clines. However, most of my collections are from houses and markets. It was difficult to sample soon before harvest in both high and low fields of even one village because the owners are usually hard to find, harvest times differ, and travel is difficult. Also, there is much more maize grown in the cool belt than at lower altitudes in a village's lands, in both the Serrano and Chupachu regions, partly because Spanish landholdings dominate much of the warm and mild belts, partly because more valuable crops compete for irrigation water in these generally dry to very dry belts.

Studies of gene flow would be very revealing in these contexts. Huánuco maize fields are very small, the number of races high, and interfamily variation is considerable. How can such different races of corn be maintained in close proximity? The large grains of maize pollen are not blown far (Craig, 1977: 677), but perhaps a new allele could travel 100 meters per year independent of human spread if strong afternoon, upvalley winds coincide with pollen shed. With movement of alleles possible, there must be barriers if there are separate races. Certainly some combinations of alleles are considered desirable, while others are selected against. Alleles less easily observed and genetically linked to those selected may be important. I suggest that alleles which somehow contribute to multigenic crossing barriers of the type demonstrated by Moll (1972) and Paterniani (1969) are important here, and that some of the color traits differentiating races may act as markers helping in the maintenance of boundaries and limits.

Modern Maize Related to Recent and Ancient Cultural Events

It is possible to link many of the complexes and sub-complexes of maize found in Huánuco to stages in human history of the past two millennia. For recent events, local interviews and recorded history may suffice, but more important, ancient, widespread events need careful study. The connections of Huánuco maize to maize complexes found throughout and beyond the Andes are measurable. Archaeological history and relationships have been studied in many regions in the Andes and nearby lowlands. By comparing the two sets of patterns, Huánuco and Andean maize geography has been tied to several stages in human development and to possible migration (Ramirez et al., 1960; Grobman et al., 1961; Bird, 1970, 1980, 1984a,b; Bird and Goodman, 1978; Bird et al., 1988). There has been no archaeobotany of Sierra maize which can confirm or negate such models.

According to local informants, pointed white popcorns in the Huánuco area have been introduced through commerce in this century, a pattern seen elsewhere. This reinforces McClintock's suggestion (1960) that these are introductions to the Andes derived from the Pyramidal complex of central Mexico (Bretting et al., 1987), a race complex preadapted to the high altitudes and, in the case of Pyramidal popcorns, cooking methods of the Andes. Mexican and Caribbean broad and small dents have been introduced, at least in part, by governmental programs of the last century, but only at low altitudes were they at all successful (Grobman et al., 1961: 302-318).

The Inca empire controlled north-central Peru for perhaps 60 years, and, although it effected great change in many parts of the Andes, it did little to disturb ethnic patterns in Huánuco (Morris and Thompson, 1985). This is reflected in the maize. It has been theorized that the Inca developed and spread the Cuzco Large Flours and Flints (Murra, 1960; Grobman et al., 1961; Bird, 1970: 139-210) and the Chulpi-Maiz Dulce sub-complex (Bird et al., 1988). In Huánuco these are rare. Cuzco flours are probably a modern introduction; Chulpi-Maiz Dulce may be ancient east of Panao (Fig. 1).

A complex set of linguistic and archaeological models leads to the hypothesis that a small empire, the Wari, expanded from south-central Peru to include Huánuco and Ancash on the north, perhaps 800 years before the Inca (Bird et al., 1988). It overran but did not dominate earlier Quechua-speaking people and apparently controlled small enclaves of people related to North Andean cultures. The purple aleurone *chushpilyu* of the Chupachu and the race Shajatu of northern Ancash may have been introduced by the Wari. Similar races are common in southern Peru and Bolivia, but not in the Serrano region of Huánuco.

The Chupachu/Serrano boundary may have been established almost 2000 years ago. It certainly was present 500 years ago, could not have been established by the Inca (A.D. 1425-1533), probably was present before the Wari (A.D. 550-850), and seems to relate to movements into the region of people making White-and-Red ceramics from the south (ca. 200 B.C.) and people making plainware from the north (A.D. 50-200) (Bird et al., 1988). This would explain the greater similarity of the Huánuco Serrano to peoples of southern Peru than to peoples of northern Peru, and the reverse relationships of the Chupachu, relationships measured by settlement pattern, ecological adaptation, present clothing and ancient ceramics (Fig. 5). It would also explain similarities of Chupachu *wangsa* flint to the North Andean Flints and Flours (Venezuelan, Colombian, Ecuadorian and northern Peruvian Sabanero), of Chupachu *kashpi* flint to Rienda-Clavo, and of Serrano *gapia* flours to Central Andean Rotund Flours found from Argentina through Peru to Ancash, a department northwest of Huánuco.

Cultural Events Channelled by Biological and Physical Environments

The terrain of Huánuco is dissected and abrupt. As air moves across the region, it drops great amounts of rain on slopes to the east, and then blows dry and hard up valleys in the rain shadow. Along Huánuco's thousand-meter slopes, the vegetation seems at first to shift in numerous continua, but on closer inspection sharp, predictable boundaries are seen. These define zones, each with a different economic potential. If one were to travel west from east of Panao to Llata along the 10.3°C isotherm, the cold/cool vegetational boundary would remain just below, and the upper limits of maize and associated cultigens would remain just above, even though elevation increased by 500 m. In this interval, precipitation would have decreased from ca. 2000 mm to 650 mm.

Peoples who have moved into the region have chosen where to live by the presence or absence of certain zones, in part because the domesticated plants and animals they brought with them were limited by temperature, moisture, and the biotic factors determined by these parameters.

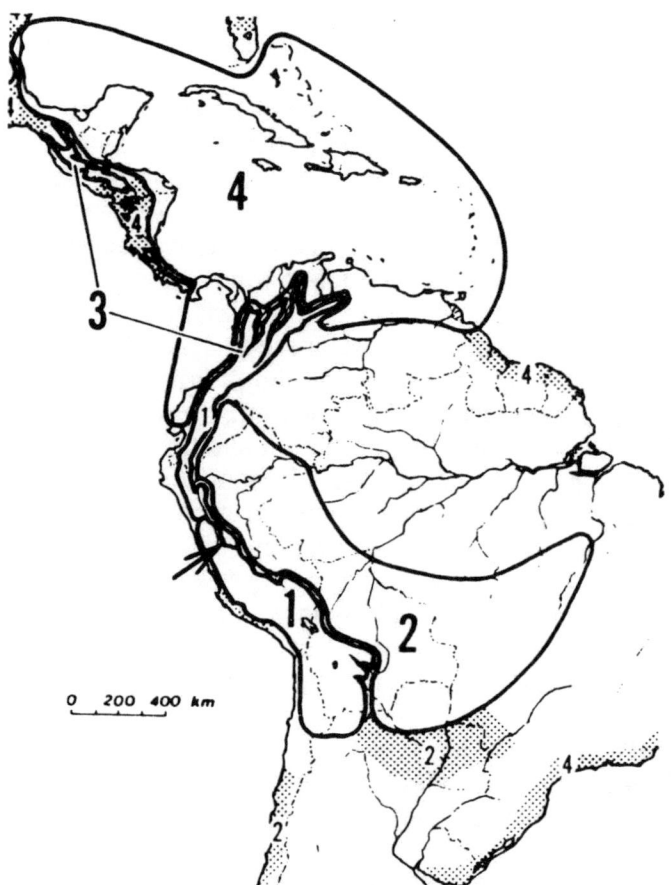

Figure 5. Map of maize-cultural regions and their extra-regional influences, as of 1870. The four South and Central American regions cover 1) the Central Andes, 2) the Southern Amazon basin, 3) the Northern Andes and highland Central America, and 4) the Caribbean and nearby lowlands. The Chupachu area is indicated by the arrow. Adapted from Figure 1 in Bird (1980).

People of the North Andes live more dispersed in their fields than do Central Andean people, at lower altitudes with more access to warm and hot zones where a wide range of special crops can be grown. On the other hand the extensive very cold to frigid pastures of the Central Andes have long been home to large llama herds, sources of meat, wood and pack animals. It can be argued 1) that much of the Serrano complex of cultigens, domesticated animals, and agricultural techniques and concepts came with the White-and-Red ceramic tradition people who at about 200 B.C.

terminated the Chavin tradition in the region (Browman, 1977;Burger, 1978) or 2) that much of the complex moved in from the south through close contacts over two millennia, or 3) both.

Likewise, the origins of Chupachu traits could have been sudden or gradual, but there seem to have been fewer opportunities for later contacts. The Chupachu territory is at the end of a narrow southward extension of North Andean culture, and no later contacts with northern Peru are noticeable in ceramics of the area. So, how could North Andean maize race complexes have been maintained in the Chupachu region without close contacts at least since the Spanish conquest and perhaps since the initial entry at ca. A.D. 150? Even though the prototypes of each maize complex now represented in Huánuco must have had much smaller ears 1800 years ago and though there have been many chances for foreign influence, both across the Chupachu/Serrano boundary and through very long-range introduction of races, there are many strong relationships of Chupachu maize varieties to maize of the North Andes. Either there was a strong flow of genes from far to the north, independent of archaeological ceramic styles, or old cultural concepts used to select seed corn each year have been very strong. Both factors may be involved.

SUMMARY

The patterns and relationships outlined here can be summarized as follows. Vegetation is determined by topography and atmosphere. Cultural immigration is partly determined by vegetation, and maize races reflect peculiarities in human cultural development. New definitions of vegetation-climate relationships depending on clear zonal boundaries can provide much better maps. Huánuco is home to many ethnic entities; the two major groups in the Sierra are separated by a sharp boundary. External relationships of these two groups can be determined, and a rough history over the past two millennia can be described. Several vegetation zones seem well correlated with the two ethnic areas. The presence or absence of many maize races in Huánuco is predicated on cultural geography, so much so that common points of origin and continuing interdependence seem inevitable.

ACKNOWLEDGMENTS

My dissertation, from which this is largely condensed, greatly benefitted from Herbert Baker's patient advice and revision. The 1964-1966 part of the fieldwork was with a project directed by John V. Murra, funded by the U.S. National

Science Foundation; the 1981 part was with a study directed by
César Fonseca Martel, funded by the National Geographic Society.

LITERATURE CITED

American Geographical Society. 1938. Map of Hispanic America.
1:1,000,000. Sheet C-18. (Cerro de Pasco).

Bird, R. McK. 1966. El Maíz y las divisiones étnicas en la sierra de
Huánuco. Cuad. Invest. Univ. Nac. H. Valdizán (Huánuco, Peru) 1:
34-44.

Bird, R. McK. 1970. Maize and its cultural and natural environment in
the Sierra of Huánuco, Peru. PhD dissertation, Dept. Botany,
Univ. California, Berkeley. 311 p. University Microfilms 71-
9767.

Bird, R. McK. 1980. Maize evolution from 500 B.C. to the present.
Biotropica 12: 30-41.

Bird, R. McK. 1984a. South American maize in Central America? In D.
Stone [ed.]. Pre-Columbian Plant Migration. Pap. Peabody Mus.
Archeol. and Ethnol., Vol. 76. Harvard Univ., Cambridge, MA.

Bird, R. McK. 1984b. The Chupachu/Serrano cultural boundary --
multifaceted and stable. In D. L. Browman, R. L. Burger and M.
A. Rivera [eds.]. Social and Economic Organization in the
Prehispanic Andes. Proc. 44th Int. Congr. Americanists, BAR
Internat. Ser. 194. Brit. Archaeol. Rpts., Oxford.

Bird, R. McK., and J. B. Bird. 1980. Gallinazo maize from the Chicama
valley, Peru. Amer. Antiquity 52: 285-303.

Bird, R. McK., D. L. Browman, and M. E. Durbin. 1988. Quechua and
maize: mirrors of Central Andean culture history. J. Steward
Anthrop. Soc. 15: 187-240.

Bird, R. McK., and M. M. Goodman. 1978. The races of maize V.
Grouping maize races on the basis of ear morphology. Econ. Bot.
31: 471-481.

Bird, R. McK., and E. Mendizábal Losack. 1986. Textiles, weaving, and
ethnic groups of highland Huánuco, Peru. In A. P. Rowe [ed.].
The Junius B. Bird Conference on Andean Textiles. The Textile
Museum, Washington, DC.

Bretting, P. K., M. M. Goodman, and C. W. Stuber. 1987. Karyological
and isozyme variation in West Indian and allied American mainland
races of maize. Amer. J. Bot. 74: 1601-1613.

Browman, D. L. 1977. External relationships of the Early Horizon
ceramic style from the Jauja-Huancayo basin, Junín. E. Dorado
2: 1-16.

Brown, W. L. and M. M. Goodman. 1977. Races of Corn. In G.F.
Sprague [ed.]. Corn and Corn Improvement. 2nd ed. Amer. Soc.
Agron., Madison, WI.

Burchard, R. E. 1976. Myth of the sacred leaf: ecological perspectives on coca and peasant biocultural adaptation in Peru. PhD dissertation, Dept. Anthropology, Indiana Univ., Bloomington, IN.

Burger, R. L. 1978. The occupation of Chavín, Ancash, in the initial Period and Early Horizon. PhD dissertation, Dept. Anthropology, Univ. California, Berkeley, University Microfilms 79-04390.

Craig, W. F. 1977. Production of hybrid corn seed. In G. F. Sprague [ed.]. Corn and Corn Improvement. 2nd ed. Amer. Soc. Agron. Madison, WI.

Curtis, J. T. 1959. The Vegetation of Wisconsin. An Ordination of Plant Communities. Univ. Wisconsin Press, Madison, WI.

Doebley, J. F., M. M. Goodman, and C. W. Stuber. 1983. Isozyme variation in maize from the southwestern United States: taxonomic and anthropological implications. Maydica 28: 97-120.

Doebley, J. F., M. M. Goodman, and C. W. Stuber. 1985. Isozyme variation in maize from Mexico. Amer. J. Bot. 72: 629-639.

Fonseca Martel, C. 1966. La comunidad de Cauri y la quebrada de Chaupiwaranga. Cuad. invest. Univ. Nac. H. Valdizán (Huánuco, Peru) 1: 22-33.

Fonseca Martel, C. 1972. La economía vertical y la economía de mercado en las comunidades alteñas del Perú. In J.V. Murra [ed.]. Visita de la Provincia de León de Huánuco en 1562. Univ. Nac. H. Valdizán, Huánauco, Peru.

Funk, V. A. 1985. Phylogenetic patterns and hybridization. Ann. Missouri Bot. Gard. 72: 681-715.

Goodman, M. M., and R. McK. Bird. 1977. The races of maize IV. Tentative grouping of 219 Latin American races. Econ. Bot. 31: 204-221.

Goodman, M. M., and C. W. Stuber. 1983. The races of maize VI. Isozyme variation among races of maize in Bolivia. Maydica 28: 169-187.

Grobman, A. W., Salhuana, and R. Sevilla, with P. C. Mangelsdorf. 1961. Races of Maize in Peru, Their Origins, Evolution and Classification. Nat. Acad. Sci. Nat. Res. Counc. Publ. 915. Washington, DC.

Hernández Xolocotzi, E. 1973. Genetic resources of primitive varieties of Mesoamerica *Zea* spp., *Phaseolus* spp., *Capsicum* spp. and *Cucurbita* spp. In O. H. Frankel [ed.]. Survey of Crop Genetic Resources in their Centres of Diversity. First Report. AGP: CGR/73/7. FAO-UN and IBP.

Instituto Geográfico Militar. 1973-1985. Carta Nacional del Perú 1: 100,000. IGM J631. Hojas 19-j (Singa), 20-j (La Unión), 20-k (Huánauco), and 21k (Ambo). Lima. [now the Instituto Geográfico Nacional].

Longley, A. E., and T. A. Kato Y. 1965. Chromosome morphology of certain races of maize in Latin America. Internat. Center for the Improvement of Maize and Wheat, Res. Bull. No. 1, Chapingo, Mexico.

Martínez, A. H. 1960. Vicos: los hábitos alimenticios. Rev. Mus. Nac. 29: 129-151.

Mayer, E. J. 1985. Production zones. In S. Masuda, I. Shimada, and C. Morris [eds.]. Andean Ecology and Civilization. Univ. of Tokyo Press, Tokyo.

McClintock, B. 1960. Chromosome constitutions of Mexican and Guatemalan races of maize. Carnegie Inst. Washington, Yearb. 59: 461-472.

McClintlock, B. 1978. Significance of chromosome constitutions in tracing the origin and migration of races of maize in the America. In D. B. Walden [ed.]. Maize Breeding and Genetics. J. Wiley & Sons, New York.

McClintlock, B., T. A. Kato, and A. Blumenschein. 1981. Chromosome Constitution of Races of Maize. Colegio de Postgraduados, Chapingo, Mexico.

Miksicek, C. H., R. McK. Bird, B. Pickersgill, S. Donaghey, J. Cartwright, and N. Hammond. 1981. Preclassic lowland maize from Cuello, Belize. Nature 289: 56-59.

Moll, R. H. 1972. Nonrandom fertilization by pollen in mixtures. Communicated in Maize Genet. Coop. News Letter 46: 152-158.

Morris, C. and D. E. Thompson. 1985. Huánauco Pampa: An Inca City and Its Hinterland. Thames and Hudson, New York.

Murra, J. V. 1960. Rite and crop in the Inca state. In S. Diamond [ed.]. Culture in History, Essays in Honor of Paul Radin, p. 785-821. Columbia Univ. Press, New York.

Paterniani, E. 1969. Selection for reproductive isolation between two populations of maize, Zea mays L. Evolution 23: 534-547.

Ramírez, E., D. H. Timothy, E. Díaz B., and U. J. Grant with G. E. Nicholson C., E. Anderson, and W. L. Brown. 1960. Races of Maize in Bolivia. Nat. Acad. Sci. -- Nat. Res. Counc. Publ. 747. Washington, DC.

Rick, C. M., and E. Anderson. 1949. On some uses of maize in the Sierra of Ancash. Ann. Missouri. Bot. Gard. 36: 405-412.

Robinson, D. A. 1963. Peru in Four Dimensions. Amer. Studies Press, Lima.

Smith J. S. C., M. M. Goodman, and C. W. Stuber. 1985. Relationships between maize and teosinte of Mexico and Guatemala: numerical analysis of allozyme data. Econ. Bot. 39: 12-24.

Stein, W. W. 1961. Hualcán: Life in the Highlands of Peru. Cornell Univ. Press, Ithaca.

Tosi, J. A. 1960. Zonas de Vida Natural en el Perú. Bol. Tecn. No. 5, Inst. Interamer. Ci. Agr. O. E. A., Zona Andina, Lima.

Weberbauer, A. 1945. El Mundo Vegetal de los Andes Peruanos. Estudio Fitogeográfico. Est. Exper. Agr. LaMolina, Lima.

Wilkes, H. G. 1979. Mexico and Central America as a centre for the origin of agriculture and the evolution of maize. Crop Improvement (India) 6: 1-18.

Yamamoto, N. 1981. Investigación preliminar sobre las actividades agro-pastorales en el Distrito de Marcapata, Departmento del Cuzo, Perú. In S. Masuda [ed.]. Estudios Etnográficos del Perú Meridional. Univ. of Tokyo Press, Tokyo.

Yamamoto, N. 1985. The ecological complementarity of agro-pastoralism some comments. In S. Masuda, I. Shimada, and C. Morris [eds.]. Andean Ecology and Civilization. Univ. Tokyo Press, Tokyo.

LATITUDINAL VARIATION AND EVOLUTION IN WILD SOYBEAN (*GLYCINE SOJA* SIEB. & ZUCC.) POPULATIONS

Y.T. Kiang and Y.C. Chiang

INTRODUCTION

The wild soybean (*Glycine soja* Sieb. & Zucc.) is a twining annual and is generally believed to be the progenitor of the cultivated soybean (*Glycine max* [L.] Merr.). Both the wild and cultivated soybeans have 2n = 40 chromosomes and are self-pollinating. Both have morphologically similar flowers that possess entomophilous characteristics. The natural outcrossing rate in wild soybean populations is about 1 to 2% (Kiang et al., 1987b). Self-pollination contributes to reproductive isolation between the two species, but the isolation is incomplete, and occasional natural hybrids are found in eastern Asia (Juvik et al., 1985; Kiang et al., 1987b). Hybridization studies on the inheritance of isozyme markers and linkage maps suggest that the two taxa would be considered as the same species if they were not so different morphologically (Fukuda, 1933; Karasawa, 1936; Tang and Tai, 1962; Kiang et al., 1985, 1987a; Kiang and Gorman, 1985; Chiang and Kiang, 1987b; Kiang, 1987).

The wild soybean occurs in China, Japan, the Korean peninsula, Taiwan, and the USSR. Although it has many characteristics of a weed, it does not depend on human disturbance for survival. It grows in a wide range of habitats including river banks and disturbed areas such as roadsides and wastelands. It often grows among other plant species. Having viny stems, it competes well with neighboring plants for sunlight by climbing on them. While collecting wild soybean seed on a river bottom in Taiwan in 1986, Kiang observed a clump of *Miscanthus floridulus* (Lobill.) Warb. consisting of 8 stems, each about 1.5 cm in diameter and 2 m long with all stems pressed down to the ground by the weight of the massive wild soybean vines. It is not uncommon to find wild soybeans growing next to cultivated soybean fields in Japan (Dr. N. Kaizume, personal communication) and South Korea (Kiang, unpublished observation). Since the wild and cultivated soybeans form hybrids, the two species together form the soybean gene pool.

Most cultivated crops have companion weeds and closely related wild species (Harlan and de Wet, 1963; Harlan, 1965). These weeds and wild species have played significant roles in the evolution of modern cultivated plants through genetic and ecological interactions with their cultivated relatives (Harlan, 1965, 1970). Because the wild soybean is closely related to the cultivated soybean, the effects of the wild soybean on the evolution of cultivated soybean deserve examination.

This paper reports on variation in morphology and reproductive patterns as a function of latitude in wild soybean populations and the possible role of the wild soybean in the evolution of the cultivated soybean.

MATERIALS AND METHODS

Twelve wild soybean seed sources originally collected from Japan and South Korea were included in this study. In mid-May of 1983, 20 plants from each seed source were grown in 15 cm diameter plastic pots, one plant per pot, with 45 cm spacing between plants in the glasshouse at the University of New Hampshire. A 120 cm long bamboo cane was inserted in each pot to support vines. The original geographic location of each seed source is shown in Fig. 1 and Table 1.

The following information on phenological, agronomic, and morphological characters was recorded:

I. Phenological characters
    1. number of days from sowing to germination
    2. number of days from germination to the first flower
    3. number of days between first flower and the first fresh pod set
    4. number of days between first fresh pod set and the first mature pod
    5. number of days from the first dry pod to the last dry pod
    6. life span in days

II. Agronomic characters
    1. proportions of 1-seed pods, 2-seed pods, 3-seed pods, and 4-seed pods
    2. total number of pods per plant
    3. total number of seeds per plant
    4. average number of seeds per pod
    5. average weight of 100 seeds
    6. harvest index (seed weight/total above-ground dry weight)

Table 1. The geographical location of the twelve seed accessions of South Korea and Japan.

Seed Accessions	Location	Latitude
K109 (PI487.430)	Hiratori, Hokkaido, Japan	40°32'N
E4 (PI487.428)	Morioka, Iwate Pref., Japan	39°42'N
K9 (PI407.192)	Chilcheon, Chunseong, South Korea	37°53'N
K7 (P107.181)	Maseogu, Yangju, South Korea	37°39'N
K102 (PI407.278)	Yongin, Yongin, South Korea	37°17'N
K28 (PI407.223)	Naecheon, Eumseong, South Korea	36°42'N
K52 (PI407.233)	Sinam, Yeongi, South Korea	36°37'N
K42 (PI407.262)	Changyeong, Chaneong, S. Korea	35°32'N
K101 (PI487.429)	Noborito, Kangawa Pref., Japan	35°30'N
K31 (PI407.252)	Milyang, Milyang, South Korea	35°30'N
M (PI486.220)	Mishima, Sizuoka Pref., Japan	35°6'N
K113 (PI487.431)	Ibusuki, Kagoshima Pref., Japan	31°12'N

472

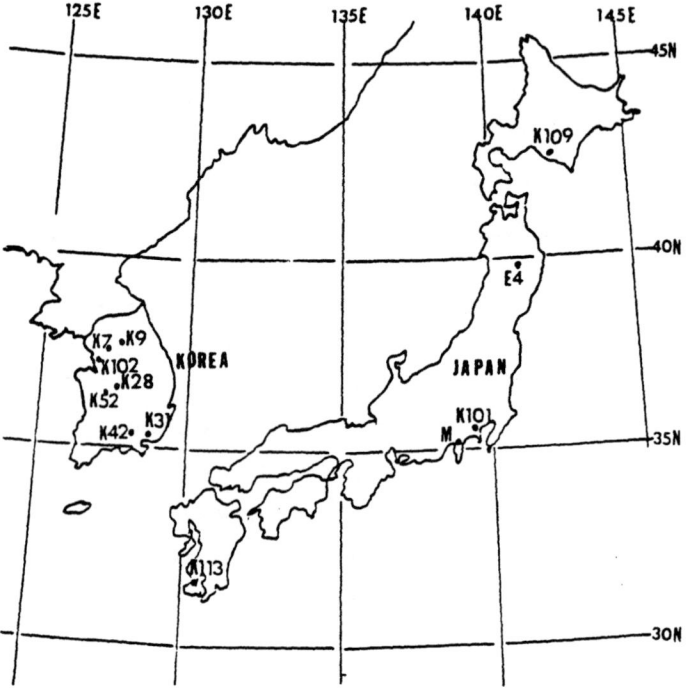

Fig. 1. The geographic locations of the 12 *G. soja* populations collected in Japan and South Korea.

7. plant height measured at 4 weeks after planting
8. plant dry biomass and number of nodules at 10 weeks
9. linear regression coefficient "b" (regression of cumulative number of mature
   pods in percentages on 2-day intervals)

III. Morphological traits
1. number of branches at one month
2. stem length from ground to first branched node
3. flower size: width of banner petal and length of flower tube
4. length and width of 3-seed pod
5. pubescence and density on mature leaf surface
6. average length and width of 30 leaflets
7. number of ovules per ovary

To obtain the average number of days between anthesis and seed maturity, at least 10 flowers per accession were tagged. Mature pods were harvested every 2 days, at which time total seeds per pod were recorded.

A random sample of 36 flower buds from 4 plants per seed source was examined under a dissecting microscope to count the number of ovules per flower.

The phenological, morphological, and agronomic data were examined with analyses of variance and principal component analysis. Linear regression and correlation coefficients also were calculated. The means of all traits examined were compared between Japanese populations and Korean populations.

RESULTS AND DISCUSSION
Phenological Data

All phenological data examined were significantly different among the 12 populations (Table 2). A significant inverse correlation was found between the number of days from germination to the first flower and latitude ($r = -0.955$). Flowering times are often correlated with abiotic factors such as photoperiod and temperature (Rathcke and Lacey 1985). Plant life span and latitude were also highly correlated ($r = -0.842$). The more northern the origin of the seed source, the earlier the first flower appeared, and the shorter the life span; the significant negative correlation between life span and latitudinal habitat has also been reported in wild soybean strains from Siberia, Northeastern China, Japan, and South Korea. The length of the frost-free period of the habitat was suggested as the decisive

Table 2. The means (number of days) (and ± S.E.) of phenological data of 12 *G. soja* populations.

	Days to germination	Germin. to 1st flower	1st flower to 1st pod set	1st pod to 1st dry pod	1st dry pod to last dry pod	Life span	Anthesis to seed mat.
K109	3.88 (0.08)	53.1 (0.59)	19.0 (0.58)	38.9 (0.43)	33.0 (2.73)	144.0 (1.95)	41.4 (1.32)
E4	4.53 (0.21)	61.8 (1.60)	11.8 (1.01)	40.8 (0.93)	38.1 (1.01)	152.5 (0.91)	44.2 (1.18)
K9	4.26 (0.11)	83.6 (0.98)	10.6 (0.70)	33.2 (0.50)	26.9 (0.48)	154.3 (0.22)	41.2 (0.90)
K7	4.67 (0.21)	86.9 (0.92)	8.9 (0.54)	34.7 (0.85)	23.7 (0.55)	154.2 (0.28)	43.5 (0.33)
K102	4.32 (0.18)	76.8 (0.91)	16.3 (0.91)	35.8 (0.69)	25.2 (0.57)	154.1 (0.33)	40.1 (1.07)
K28	4.63 (0.12)	85.0 (0.63)	10.4 (0.24)	33.4 (0.53)	25.4 (0.52)	154.2 (0.62)	42.0 (0.90)
K52	4.64 (0.20)	78.0 (1.31)	14.9 (1.43)	34.3 (1.19)	25.6 (0.41)	152.8 (0.73)	40.6 (0.28)
K42	4.28 (0.12)	83.8 (0.64)	9.8 (0.50)	33.0 (0.50)	26.4 (0.80)	153.0 (0.28)	39.5 (0.78)
K101	5.00 (0.19)	100.9 (0.23)	9.2 (0.40)	41.4 (0.40)	12.0 (0.48)	163.5 (0.75)	40.4 (0.58)
K31	4.91 (0.21)	89.5 (0.71)	10.6 (0.57)	39.6 (0.84)	22.1 (0.73)	161.8 (0.40)	41.5 (0.48)
M	5.17 (0.19)	109.4 (0.62)	6.4 (0.56)	40.6 (0.46)	18.8 (0.58)	175.2 (0.53)	42.6 (0.25)
K113	4.93 (0.21)	116.8 (0.48)	7.3 (0.56)	35.4 (0.43)	19.8 (0.55)	179.3 (0.58)	41.0 (0.23)
F_1test	8.31**	454.3**	25.4**	22.3**	38.1**	24.5**	2.7**
r	-0.72**	-0.92**	0.72	0.15	0.71**	0.84**	0.84**

* * Statistically significant at 1% level
r = correlation coefficient of the means with latitude

factor (Fukui et al., 1978). These patterns presumably reflect adaptations of wild soybean populations to the short growing season of high latitudinal habitats.

An additional significant correlation was observed between latitude and the number of days between first flower appearance and first pod set (r = 0.724). A large number of flowers aborted before pod set was observed in northern populations. In K109 (42°36'N), the duration between flowering and the first pod set was 45 days when seeds were sowed on 4/24/83; 19 days when sowed on 5/24/83; and 8 days when sowed on 6/20/83. That is, K109 started to grow pods no earlier than late July no matter how early it started to flower. The cause of long duration of flower abortion is unknown. Environmental factors such as photoperiod could play a major role. Long days seem to inhibit pod development in soybean and *Cassia* species (T. Lee, personal communication). Although the duration from anthesis to seed maturity was different statistically among these 12 populations, in general the difference in the seed filling period was relatively small with no clearcut north-south pattern. The difference between E4 (39°42'N) and K42 (35°32'N) was about 4.7 days, the largest among the 12 populations (Table 2). Thus, the significant difference in life span is mostly due to the different duration from germination to the first flower.

The correlation coefficient between latitude and the first principal component of phenological data, which accounts for 59.6% of the total phenological variation among the 12 populations is r = -0.912 which is significantly different from zero at the 1% level (Fig. 2). Therefore, around 60% of the phenological variation among these 12 *G. soja* populations is highly associated with their latitudinal locations. No significant correlation was found between variability of phenological traits as expressed by coefficient variation (C.V.) and latitude, except for the number of days from germination to the first flower (r = 0.77).

Agronomic Data

There was significant variation among the 12 populations for each of the nine agronomic characters (Table 3). Among the three seed yield components (# seed/pod, weight of 100 seed, and number of pods/plant) only the number of pods per plant was significantly positively correlated with the total seed yield per plant. The other two components were weakly and negatively correlated with the total seed.

Table 3. Agronomic data (means) of 12 G. soja populations collected from the greenhouse experiments.

	No. Seed /Pod	100 Seed Wt. (g)	No. Pods/ Plant	Seed Yield/ Plant (g)	Harvest Index	b^a	No. nodules /plant	Total dry Wt(g)/ plant	Ht (cm) 3 wks. old plant
K109	2.72	3.03	214.8	17.68	0.394	0.048	33.0	11.5	41.85
E4	2.55	2.12	326.3	17.65	0.359	0.041	34.5	12.8	47.63
K9	2.45	2.09	457.8	23.35	0.401	0.055	15.8	14.0	26.18
K7	2.45	2.22	397.1	21.61	0.372	0.058	13.5	15.9	17.13
K102	2.46	2.26	244.9	13.66	0.318	0.050	0	12.4	17.85
K28	2.44	2.27	436.6	24.09	0.386	0.069	26.0	11.9	9.90
K52	2.10	2.34	442.7	20.14	0.356	0.053	4.8	10.3	7.78
K42	2.47	2.29	386.4	21.85	0.382	0.065	40.5	12.2	29.3
K101	2.07	1.64	569.4	19.14	0.285	0.117	14.5	8.2	17.85
K31	2.00	2.29	503.9	23.12	0.374	0.053	21.0	9.3	7.08
M	2.22	2.34	436.1	22.64	0.285	0.071	16.3	9.7	27.18
K113	2.02	2.50	365.3	18.36	0.234	0.071	16.5	13.0	22.48
F test	3.6**	47.1**	43.6**	12.7**	81.2**	9.14**	3.6**	2.28**	5.76**
r	0.86**	1 - 0.34	-0.15	0.21	0.68**	-0.14	0.29	0.18	0.52

a regression coeff. b of % accumulated number of dry pod vs. two days harvesting interval

*statistically significant at 5% level **Statistically significant at 1% level

r = correlation coefficient of the means with latitude

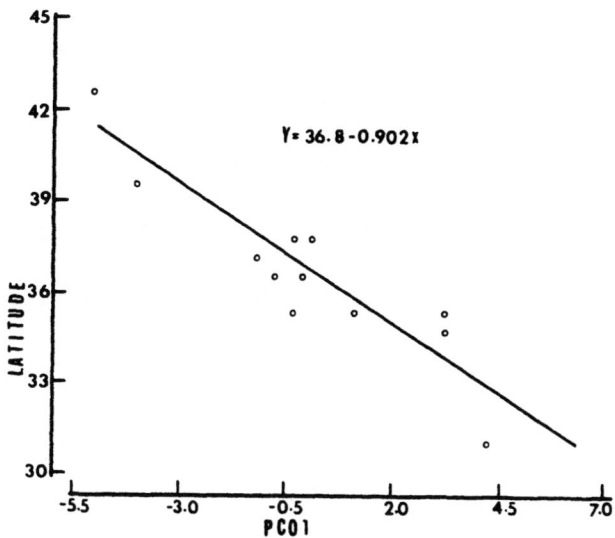

Fig. 2. Plot of latitudinal locations on the first principal component (PC01) axis, which accounts for 59.6% of phenological variation among the 12 *G. soja* seed populations.

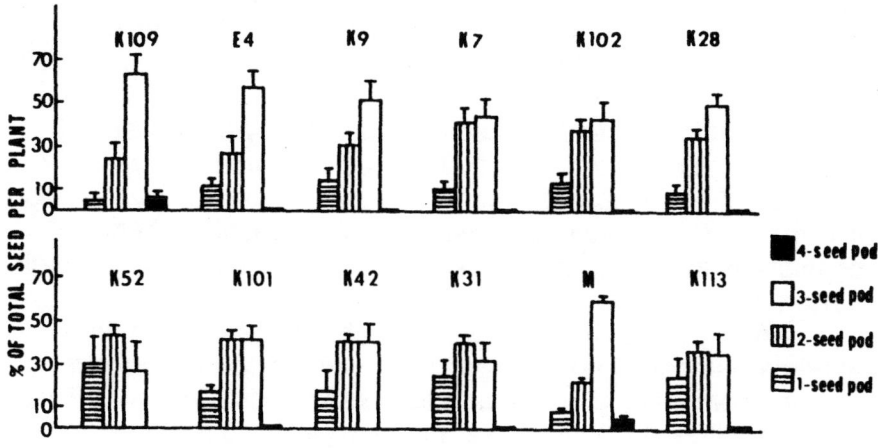

Fig. 3. Patterns of seed packaging among the 12 populations of *G. soja* (Bars indicate ± S. E.).

The linear regression coefficient b was obtained by calculating linear regression of the cumulative proportion of mature pods at two-day intervals (Table 3). The larger the b value, the more synchronous is seed set within a plant. The values of b among the 12 populations were significantly different, with K101 (35°30'N) having a very steep b and K109 (42°36'N) and E4 (39°42'N) having average to low b values. If the number of days from the fist flower to the last dry pod represents plants engaging in reproductive function, then the northern populations spend a significantly larger proportion of their life span engaging in reproductive function than the southern populations (Table 2). For example, population K109 employed 63% of its lifetime engaging in reproductive function while the reproductive period of K113 population was only 35% of its life span.

The 12 populations were very different in their pattens of seed packaging. For example, K52 had 30.2% of total pods with 1 seed, 42.9% with 2 seeds, and 26.9% with 3 seeds; while K109 had 5.3% of total pods with 1 seed, 24.5% with 2 seeds, 64.2% with 3 seeds, and 5.9% with 4 seeds (Fig. 3). The seed-packaging patterns of K109 and M were similar, as were patterns in K7 and K102, K52 and K31, and K101 and K42. Most of the flower buds examined consisted of 3 ovules, with a small proportion having 4 ovules. An average of 3 ovules per ovary was observed in all populations. Therefore, differences in the average number of seeds per pod was due to ovule and seed abortion. The ovule abortion could be due to failure of pollination and fertilization, and seed abortion could result from resource competition or genetic incompatibility (Van Shaik and Probst, 1958; Stephenson, 1981; Lee and Bazzaz, 1986).

The different seed packaging patterns may have resulted from ecological differentiation, such as resource limitation, unpredictable seasonal changes, seed and pod predators, etc. More information on the ecological conditions important to these populations may help us to understand the observed seed packaging patterns.

Among many of the quantitative characters studied, the average number of nodules per plant was positively correlated with the percentages of 3-seed and 4-seed pods (Table 4). That is, among these 12 populations, the more nodules per plant a population had, the higher the percentage of 3-seed and 4-seed pods.

There also was a positive correlation between flower size and the percentage of 3-seed and 4-seed pods. If the larger flowers have larger ovaries, the larger ovaries may in turn have more

Table 4. Correlation coefficients between several quantative characters studied in 12 *G. soja* populations.

	Percentages of			
	1-seed pod	2-seed pod	3-seed pod	4-seed pod
Av. # nodules/plant	-0.560	-0.745**	0.666*	0.827**
Width banner petal	-0.565	-0.776**	0.690*	0.825**
Flower length	-0.648*	-6.20*	0.700*	0.557
Flower tube length	-0.510	-0.594	0.630*	0.362

* significant at 5% level
** significant at 1% level

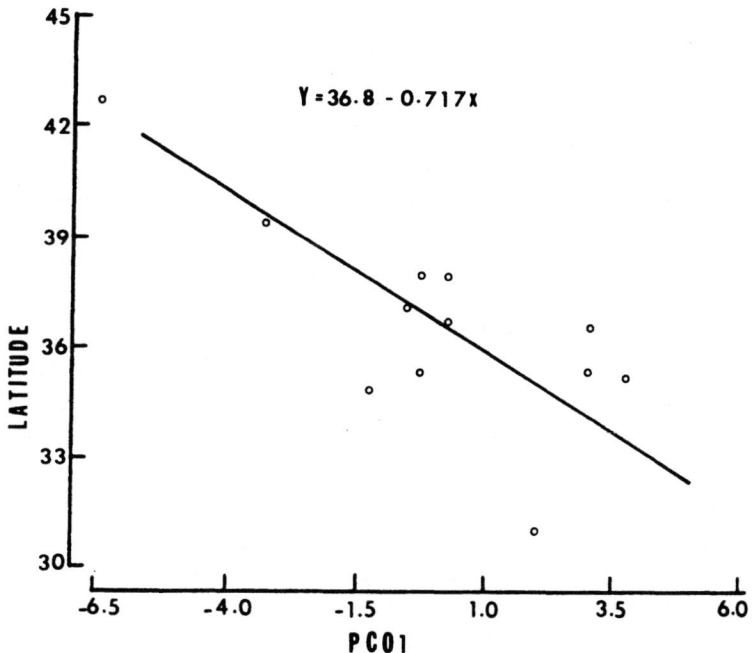

Fig. 4. A plot of 12 *G. soja* populations with their latitudinal locations against the first principal component (PC01) of the agronomic variation. The PC01 accounts for 43.1% of the total agronomic variation observed.

room or nutrition for seed development, thereby reducing seed abortion. This argument seems to be supported by the observation that average seed weight and the mean length of flower tubes were significantly correlated ($r = 0.57$).

Many patterns of variation of agronomic characters were statistically associated with latitude. For instance, more northerly population had lower seed yields per plant ($r = -0.74$) and the number of pods per plant ($r = -0.74$), but more seed per pod ($r = 0.80$) and higher harvest indices ($r = 0.70$). The plot of 12 populations with their respective latitudinal locations and the first principal component of agronomic data, which account for 43.1% of the total variation of agronomic traits, is presented in Fig. 4. The correlation between the two axes is significant ($r = -0.75$). There was no significant correlation between variability of argonomic traits (as expressed by the coefficient of variation) and latitude.

Morphological Data

All morphological traits examined exhibited statistically significant variation among the 12 populations (Table 5). No consistent trend of morphological variation in relation to latitude was observed, nor any relationships between variability (C.V.) and latitude. The seven South Korean populations were collected from the mid-latitude of a relatively small area with a very different longitude from Japanese populations. When South Korean and Japanese populations as groups were compared, the number of branches was significantly higher in South Korean populations (at 1% level), whereas Japanese populations had higher densities of leaf pubescence (at 1% level).

The first principal component, accounting for 33.1% of the total morphological variation, was not significantly correlated with latitude. In contrast, 59.6% of the phenological variation and 43.1% of the agronomic variation were highly associated with latitude. Most phenological and agronomic characters measured are related to plant reproduction. Whether a plant can exist and reproduce in an area is mainly determined by environmental factors such as day length, temperature, water, growing season, competition, disease and predation. Most of the variation in physical environmental factors are associated with latitude. In contrast, many morphological characters, such as leaf size, shape, number of branches and stem length, generally believed to be more plastic than reproductive organs, are the result of adaptation of plants for growth and competitive survival under local conditions, which may vary independently of latitude. For instance, leaf size and shape as well as the density of pubescence

Table 5. The means of morphological data from 12 G. soja populations

	No. of branches	Stem 1st node ht. (cm)	Width banner petal (mm)	Flower length (mm)	Flower tube L. (mm)	3-seed pod length X width (cm)	Leaf pubescence density /(mm)^2	Length/width of leaflet central	lateral
K109	0.25	6.81	5.63	6.78	2.94	1.75	19.5	3.27	2.28
E4	0.75	10.25	4.97	6.78	3.03	1.20	26.2	3.26	2.43
K9	2.00	2.36	5.06	6.31	2.50	1.15	17.0	2.29	1.75
K7	2.00	1.79	4.97	6.75	2.38	1.51	16.0	2.35	1.85
K102	1.90	2.58	5.03	6.22	2.94	1.37	7.4	2.05	1.82
K28	1.95	1.71	5.72	6.94	2.50	1.64	14.7	1.70	1.45
K52	1.85	1.24	4.34	5.56	2.34	1.55	11.0	3.45	2.54
K42	1.60	2.65	4.81	6.56	2.47	1.28	13.3	2.07	1.71
K101	1.85	1.57	4.47	5.69	2.06	1.11	21.5	2.39	1.87
K31	2.65	0.99	4.47	6.00	2.25	1.18	18.2	2.81	2.08
M	1.80	1.78	5.84	6.97	2.56	1.44	17.7	1.81	1.64
K113	0.80	3.33	5.53	6.72	2.50	1.75	20.6	1.80	1.64
F test	14.4**	17.7**	26.9**	25.5**	16.8**	151.9**	68.7**	40.3**	12.73**
r	0.36	0.57**	0.02	0.15	0.58*	0.01	0.09	0.60*	0.18

*statistically significant at 5% level **statistically significant at 1% level
r = correlation coefficient of the means with latitude

affect the heat exchange and photosynthesis (Givnish, 1979; Barbour et al., 1980) or patterns of herbivory (Denno and McClure 1983).

Field observations made in South Korea, Japan and Taiwan indicate that morphological traits were rather uniform within populations but variable among populations. Based on electrophoretic studies (46 loci) these 12 populations showed rather low genetic variation (H = 0.03) (Chiang 1985). This may be due to these populations being descended from a small number of seeds originally collected in the wild.

The population means of Japanese and South Korean sources were compared by an F test. Among 23 traits examined, six showed significant differences between Japanese and South Korean population means. They are number of days from first pod set to first dry pod, harvest index, height of 3 week plant, number of nodules, number of branches at 1 month, and leaf pubescence density (Table 6).

DISCUSSION

The reproductive biology of wild soybeans is significantly influenced by latitude. As a consequence of adaptation to different latitudes, wild soybean populations seem to adopt different patterns of resource allocation for seed production. Plants from northern latitudes are adapted to a short growing season, and they initiate flowers early when the plants are relatively small. Because of limited vegetative development, these plants can support relatively fewer flowers and pods at a given time. In addition to small vegetative parts, which capture less solar energy, the assimilates need to be allocated simultaneously to continuing vegetative growth and seed filling. In this manner, plants ensure seed production in years of early frost, and they can produce more seed in years with longer growing seasons. This appears to lead to less synchronous flowering. In contrast, plants from habitats with longer growing seasons have a longer period of vegetative growth, and accumulate more resources and photosynthates before flowering. With a large vegetative structure and an increased capacity for stored energy, these plants can afford to allocate a larger proportion of resources to flower, seed and pod development. Therefore, they can support many pods at a given time. This appears to permit synchronous flowering. Synchronous seed setting is a desirable agronomic character for mechanized culture and harvest. In addition, Augspurger (1981) found that synchronous flowering in *Hybanthus prunifolius* increased the potential of outcrossing and seed production and it also reduced seed predation.

Table 6. Comparison of means of traits between Japanese and South Korean populations of G. soja.

Traits	Japanese population means	S.Korean population means	$F_{1,10}$ values	Significant levels
Sowing to germination (days)	7.20	6.85	2.76	n.s.
Germination to first flowering (days)	89.12	83.40	0.34	n.s.
First flower to first pod set (days)	10.74	11.60	0.16	n.s.
First pod set to first dry pod (days)	39.42	34.85	55.11	1%
First dry pod to last dry pod (days)	24.32	25.00	0.05	n.s.
Life span (days)	162.90	154.9	1.98	n.s.
No. seed/pod	2.44	2.34	0.48	n.s.
Average 100 seed weight (g)	2.33	2.25	2.65	n.s.
No. pods/plant	382.38	409.91	0.20	n.s.
Harvest index	0.331	0.37	20.35	1%
Total dry weight/plant (g)	11.04	12.29	0.98	n.s.
Root dry weight/plant (g)	2.15	2.78	4.37	n.s.
3 wks height (cm)	11.04	12.29	5.79	5%
No. nodules/plant	22.96	17.37	6.11	5%
No. branches/plant	1.57	3.79	18.43	1%
Stem length from ground to first branched node (cm)	4.75	4.61	n.s.	
Flower tube length (mm)	2.62	2.48	0.09	n.s.
3 seed-pod length (cm)	2.74	2.64	0.50	n.s.
Pubescence length (mm)	1.50	1.42	0.53	n.s.
Pubescence density (mm^2)	21.10	13.94	11.98	1%
10th central leaflet length (cm)	5.85	5.21	2.17	n.s.
10th central leaflet width (cm)	2.75	2.62	0.21	n.s.
Central leaflet length (random sample, cm)	2.51	2.39	0.10	n.s.

Wild soybean seed contains high protein levels (40-50% of dry weight), so that nitrogen nutrition may be a limiting factor in seed production. Plants with a large number of nodules presumably can supply more nitrogen resulting in lower seed abortion. However, no significant positive correlation was found between the mean number of nodules per plant and either average seed weight or average seed yield per plant.

The average number of seeds per pod and the harvest index (seed weight/above ground biomass) in the populations of northern origin are higher than those of the southern populations. The smaller numbers of seed per pod in the southern populations is an outcome of ovule and seed abortion. Ovule or seed abortion tends to increase the cost of reproduction in terms of pod wall material required per seed (Lee and Bazzaz, 1986). Thus, populations from northern habitats seem to have evolved to economize energy expenditure in seed production by reducing ovule and seed abortion. The higher harvest index indicates that a higher proportion of the total resource is allocated to seed production. Based on the data, northern populations showed higher reproductive efficiency, although the total number of seeds per plant was smaller than that of southern populations.

Vavilov (1926, 1951), from field observations and detailed morphological studies, found great diversity in Chinese soybeans, and concluded that the soybean had a Chinese center of origin. In contrast, our extensive survey of soybean germplasm from China, Japan, South Korea, Taiwan and USSR showed that the germplasm of both cultivars and wild soybeans from South Korea had the highest genetic variation and that from USSR lowest. For these reasons, we suggested that the Korean Peninsula is one of the centers, if not the center, of soybean gene diversity (Kiang, et al., 1987b). Natural populations of wild soybean consist of different genotypes which often grow in proximity. Although soybeans are predominantly self-pollinating, occasional cross pollination by insects can produce outcrossed progeny. In contrast, the cultivated soybean is grown as a single genotype in a large area and occasional cross pollination will not result in hybrid seed. The observed frequency of heterozygous seed in *G. soja* is 2.4% while in cultivars, it is 0.06% (Kiang, et al., 1987b). In experimental fields the average frequency of outcrossed progeny in cultivated soybean was significantly higher in a mixed genotype stand than in a pure stand, and the proportion of hybrid seed increased significantly when plants were surrounded by different genotypes in close proximity (Chiang and Kiang, 1987a).

Cruden (1977, and Cruden and Lyon, this volume) showed that pollen/ovule ratios are correlated with breeding system, and divided flowering plants into five classes: xenogamy, facultative xenogamy, facultative autogamy, obligate autogamy, and cleistogamy. He concluded that the evolutionary shift in breeding system from one class to another was accompanied by change in the mean number of pollen grains per ovule (P/O). Based on Cruden's classification, the P/O ratio of soybeans (1.215 $\pm$ 78 - 1,410 $\pm$ 88) falls between facultative xenogamy and xenogamy (Chiang and Kiang, 1987a). Yet, based on the breeding system, soybeans should belong to the facultative autogamy class. We suggested that soybeans probably shifted their breeding habit from outcrossing to highly selfing in recent evolutionary history, and thus they still retain entomophilous flowers and P/O ratios which are characteristic of outcrossing species (Chiang and Kiang, 1987a).

The evolutionary dynamics of cultivated plants such as maize, wheat, sugar cane and cotton have been affected significantly by introgression with their weedy and wild relatives (Harlan and de Wet, 1963; Zohary, 1965; Harlan, 1965, 1970; Baker, 1978). Although natural hybrids between cultivated and wild soybean have been reported (Juvik, et al., 1985; Kiang et al., 1987b), the effects of such hybridization on the evolution of cultivated soybean have not been studied in detail. However, this hybridization seems to have given rise to some novel variants. One weedy form, which has characters intermediate between the cultivated and wild soybeans, was found in Manchuria, and named *Glycine gracilis* (Skvortzow, 1927). The diploid chromosome number and size of *G. gracilis* were the same as *G. max* and *G. soja* (Fukuda, 1933). *G. gracilis* is found wherever *G. max* and *G. soja* overlap in distribution. *G. gracilis* most likely has evolved as the result of hybridization between *G. max* and *G. soja* (Hymowitz, 1970).

The rarity of natural hybrids between wild and cultivated soybeans presumably is due to self-pollination in both species. In addition, wild soybean does not depend on human disturbance for survival; thus it is not closely associated with cultivars ecologically. Hybrid swarms involving the two soybean species have never been reported. Unlike outcrossing crop species and their relatives such as maize (Harlan, 1970), introgression following hybridization is rare in cultivated and wild soybeans. Furthermore, the hybrids and their offspring probably cannot successfully establish in non-disturbed areas. Oka (1983) sowed 36 lines of F_3 hybrid seeds of *G. max* and *G. soja* in abandoned fields in Japan and Taiwan. In the second year plant numbers ranged from 1 to 587/m² among the 36 lines. In contrast, the wild parent had 635/m² and the cultivated parent none. In the third year, in one abandoned field in Japan the plant number ranged from 0 to 146/m² (wild parent 765/m²); in the grassy field margin in Japan and the abandoned field in Taiwan, the numbers of hybrids were much smaller. The average number of hybrid plants per m² was significantly smaller than that of the wild parent; the number was also significantly smaller in the third year. The numbers of plants of any given line that became established in the second and third years were correlated, indicating that the number of established seedlings was genetically controlled. Several traits were correlated with the mean number of established seedlings: germinating activity ($r = -0.36$) (wild soybeans have seed dormancy while cultivars do not), pod dehiscence ($r = 0.67$) (wild soybean dry pods dehisce, while most cultivars do not), reproductive allocation ($r = 0.69$), and number of dispersed seed/m² ($r = 0.52$). Oka (1983) concluded that regenerative success was genetically controlled, and more than half of its variance could be explained by the values of pod dehiscence. Having characteristics of cultivated soybeans, the hybrids may not be able to establish in undisturbed natural conditions. Hybrids can be easily identified in cultivated fields and are readily eliminated by farmers because of their undesirable traits such as viny stems, small seeds, and pod shattering.

CONCLUSIONS

Human cultivation, dispersal, breeding and selection have no doubt dominated the direction of evolution in cultivated soybeans. The impact of genetic and ecological interaction of the cultivated soybean and wild soybean on the evolutionary dynamics of these two species may not be as significant as in other cultivated plants, such as maize, sugar cane and wheat. Nevertheless, the wild soybean population is a valuable reservoir of genetic diversity for the improvement of cultivated soybeans.

SUMMARY

Twelve wild soybean (*Glycine soja* Sieb. & Zucc.) seed sources collected from populations in Japan and South Korea were examined for morphological variation and patterns of reproduction. Except for angles and length of pubescence, all morphological traits studied were significantly different among the 12 populations; but no clear north-south pattern of variation was observed. All phenological and agronomical traits examined were significantly different among the 12 populations. A significant correlation was found between the number of days to the first flower and latitude ($r = -0.955$). Plant life-span and latitude were also significantly correlated (r = -0.842). The first principal component, which accounted for 59.6% of the total phenological variation, was significantly correlated with latitude (r = -0.912). The pattern of seed packaging among the 12 populations varied significantly. The northern populations had significantly higher numbers of seed per pod, as well as lower seed and ovule abortion, than the southern populations. The northern populations made higher reproductive allocation (proportion of life span used in reproductive function, and harvest index) than the southern populations. The average numbers of nodules were positively correlated with the percents of 3-seed and 4-seed pods. Among the seed yield components (number of seed per pod, number of pods per plant, seed weight) only the number of pods per plant was positively correlated with the seed yield. There was no significant correlation between variability (as expressed by CV) and latitude except for the number of days from germination to the first flower (r = 0.77). Among 25 traits examined, means of only 5 traits showed significant differences between Japanese and Korean populations. Available evidence suggests that interactions between the cultivated and wild soybean on the evolutionary dynamics of these two species may not be as significant as in other cultivated plants, such as maize and wheat.

ACKNLOWLEDGMENTS

We thank Professor N. Kaizuma of Iwate University, Japan, and Dr. Young Soo Ham of Institute of Forestry Genetics, Korea, for generously providing seeds for this research. We especially thank Professor Yan B. Linhart for carefully reading the earlier version of the manuscript and making comments and suggestions, and for asking questions that were most helpful in improving this paper.

This work was supported in part by USDA Grant 82-CRSR-21045.

LITERATURE CITED

Augspurger, C. K. 1981. Reproductive synchrony of a tropical shrub: Experimental studies on effects of pollinators and seed predators on *Hybanthus prunifolins* (Violaceae). Ecology 62: 775-788.

Baker, H. G. 1978. Plant and Civilization. Wadsworth Publishing Co., Inc. Belmont, CA.

Barbour, M. G., J. H. Burk, and W. D. Pitts. 1980. Terrestrial Plant Ecology. Benjamin/Cummings Publishing Co. Inc., Boston, MA.

488

Chiang, Y. C. 1985. Genetic and Quantitative Variation in Wild Soybean (*Glycine soja*) Populations. PhD thesis. University of New Hampshire, Durham.

Chiang, Y. C., and Y. T. Kiang. 1987a. Geometric position of genotypes, honeybee foraging patterns and outcrossing rate in soybean. Bot. Bull. Academia Sinica 28: 1-11.

Chiang, Y. C., and Y. T. Kiang. 1987b. Inheritance and linkage relationships of 6-phosphogluconate dehydrogenase isozymes in soybean. Genome 29: 786-792.

Cruden, R. W. 1977. Pollen-ovule ratio: a conservative indicator of breeding systems in flowering plants. Evolution 31: 32-56.

Denno, R. F., and M. S. McClure. 1983. Variable Plants and Herbivores in Natural and Managed Systems. Academic Press, New York, NY.

Fukuda, Y. 1933. Cytological studies on the wild and cultivated Manchurian soybeans. Japan. J. Bot. 6: 489-506.

Fukui, J. S., S. Kaizuma, and N. Kaizuma. 1978. Comparative investigation on interstrain variation in the growing periods of Siberian (USSR) Northeastern Chinese, South Korean and Japanese strains of wild soybean, Glycine soja Sieb. and Zucc. J. Faculty Agriculture (Iwate Univ.) 14: 71-79).

Givnish, T. 1979. On the adaptive significance of leaf forms. In O. T. Solbrig [ed.]. Topics in Plant Population Biology. Columbia University Press, New York.

Harlan, J. R. 1965. The possible role of weed races in the evolution of cultivated plants. Euphytica 14: 173-176.

Harlan, J. R. 1970. Evolution of cultivated plants. In O. H. Frankel and E. Bennett [eds.]. Genetic resources in plants - their exploration and conservation. FA. Davis Co. Philadelphia.

Harlan, J. R., and J. M. J. de Wet. 1963. The compilospecies concept. Evolution 17: 497-501.

Hymowitz, T. 1970. On domestication of the soybean. Economic Botany 24: 408-421.

Juvik, G.A., R.L. Bernard, and H.E. Kaufman. 1985. Directory of germplasm collections. 1. II. Food Legumes (Soybean). International Board for Plant Genetic Resources.

Karasawa, K. 1936. Crossing experiments with *G. max* and *G. ussuriensis*. Japan. J. Bot. 8: 113-118.

Kiang, Y. T. 1987. Mapping three protein loci on a soybean chromosome. Crop Sci. 27: 44-46.

Kiang, Y. T., and M. . Gorman. 1985. Inheritance of NADP-active isocitrate dehydrogenase isozymes in soybeans. J. Heredity 76: 279-284.

Kiang, Y. T., M. B. Gorman, and Y. C. Chiang. 1985. Genetic and linkage analysis of a leucine aminopeptidase in wild and cultivated soybean. Crop Sci. 25: 319-321.

Kiang, Y. T., Y. C. Chiang, and C.J. Bult. 1987a. Genetic study of glutamate oxaloaccetic transaminase in soybean. Genome 29: 370-373.

Kiang, Y.T., Y.C. Chiang, J.Y.H. Doong, and M.B. Gorman. 1987b. Genetic variation of soybean germplasm. In S.C. Hsieh [ed.]. Crop Exploration and Utilization of Genetic Resources. Published by Taichung District Agricultural Improvement Station, Taiwan.

Lee, T. D., and F. A. Bazzaz. 1986. Maternal regulation of fecundity: non-random ovule abortion in *Cassia fasciculata* Michx. Oecologia 68: 459-465.

Rathcke, B., and E. P. Lacey. 1985. Phenological patterns of terrestrial plants. Ann. Rev. Ecol. Syst. 16: 179-214.

Oka, H. I. 1983. Life-history characteristics and colonizing success in plants. Amer. Zool. 23: 99-109.

Skvortzow, B. V. 1927. The soybean - wild and cultivated in Eastern Asia. Manchurian Res. Soc. Publ. Serv. A. Nat. Hist. Sec. No. 22: 1-8.

Stephenson, A. G. 1981. Flower and fruit abortion: Proximate causes and ultimate functions. Ann. Rev. Ecol. Syst. 12: 253-279.

Van Shik, P. H., and A. H. Probst. 1958. Effects of some environmental factors on flower production and reproductive efficiency in soybeans. Agronomy J. 50: 192-197.

Tang, W. T., and G. Tai. 1962. Studies on the qualitative and quantative inheritance of interspecific cross of soybean. Glycine max and G. formosana. Bot. Bull. Acad. Sin. 3: 39-60.

Vavilov, N. 1926. Studies on the Origin of Cultivated Plants. Leningrad.

Vavilov, N. 1951. The Origin, Variation, Immunity and Breeding of Cultivated Plants. Chronica Botanica Vol. 13. The Ronald Press Company, NY.

Zohary, D. 1965. Colonizer species in the wheat group. In H.G. Baker and G.L. Stebbins [eds.]. The Genetics of Colonizing Species. Academic Press, NY.

REPRODUCTIVE BIOLOGY AND EVOLUTION OF THE CULTIVATED CHAYOTE
SECHIUM EDULE: CUCURBITACEAE

L. E. Newstrom

INTRODUCTION

Chayote, *Sechium edule* (Jacq.) Sw., is in one of the more important genera species in the Cucurbitaceae (Whitaker and Davis, 1962). The commercial importance of squashes (*Cucurbita*), cucumbers and melons (*Cucumis*), watermelons (*Citrullis*), bottle gourds (*Lagenaria*), and sponge gourds (*Luffa*), has resulted in much attention directed to these crops. In comparison, chayote is relatively little known. Used by the Aztecs in pre-Hispanic Mexico, chayote most likely originated in Mexico and Guatemala (Newstrom, 1986). It is an inexpensive staple food used mostly in Latin America (Leon, 1968), but it is cultivated widely throughout tropical and subtropical regions of the world. Commercial production occurs in several countries including Mexico, Costa Rica, Brazil, Puerto Rico, and Italy (Purseglove, 1968; Lopes, 1979; Casseres, 1980). Commercial export of chayote fruits has increased dramatically in the last 15 years, particularly in Costa Rica where vegetative propagation has produced standardized fruits for the market.

Chayote is a multi-use plant (Baker, 1970; Table 1). This species grows best between 300 and 2000 m (Messiaen, 1979; Casseres, 1980). It requires high humidity (80-85%), rainfall between 1500 and 2000 mm per year (or irrigation), and mild temperatures of 20° - 25° (with limits of 12° and 28° C) (Messiaen, 1979; Casseres, 1980; Bukasov, 1981). Chayote is frost sensitive and susceptible to nematodes, fungal and viral diseases, and spider mites (Rao, 1964; Lordello, 1972; Pereira *et al.*, 1973; da Ponte, 1973; Sullia and Khan, 1980). Although perennials, chayote plants are kept for less than 3 years in commercial production because of these disease problems. Starch-storing tubers, which develop in the second year, can be removed without damaging the plant. These tubers are most used in Mexico and Guatemala but are not exported. The plant itself repeatedly resprouts from a rootstock after it has died back due to drought or cold.

Table 1. The uses of chayote, *Sechium edule*, a multi-use plant. (From Cook 1901; Somay 1921; Hoover 1923; Martinez 1959a, 1959b; Bueno et al. 1970; Medina, 1974; Ruberté and Martin, 1975; and Morton 1981).

Plant Part	Human Food	Medicine	Animal Food	Ornament
Whole Plant				For Shade (Italy, Java)
Tubers	Starch for vegetable, candy			
Stems				Silver straw hats, baskets (Ile de Réunion)
Leaves		Infusion for hypertension, Arteriosclerosis, Urinary calcifications, Intestinal and skin inflammations.	Forage for pigs, cattle, poultry	
Shoots	Pot herb high in Vits. A,B,C & in Ca, Fe			
Flowers			Nectar for insects	
Fruits	Table vegetable fiber & minerals, Industrial food filler, Substitute for apples, artichokes	Fruit flesh cauterizes wounds	Forage for pigs, cattle, poultry	
Seeds	Delicacy, protein			

In this chapter, aspects of the evolution and domestication of chayote will be discussed. The reproductive biology of chayote has been studied recently (Newstrom, 1986): a summary of pollination and seed germination is presented. The newly discovered populations of chayote in the wild (Cruz Leon and Querol Lipcovich, 1985; Newstrom, 1986) are compared to the cultivated crop to provide information on the evolutionary history of the crop. The ethnobotany and folklore of chayote are included because they may provide insights into the biology and domestication of the crop.

REPRODUCTIVE BIOLOGY
Pollination

Chayote is a large perennial, monoecious vine with a single seeded viviparous fruit; it is self-compatible (Newstrom, 1986) and cross-pollinated with a generalist pollinator syndrome (Fidel and Tristan, 1931; MacGregor, 1976; Wille and Orozco, 1983; Newstrom, 1986).

The male and female flowers are almost identical, with rotate corollas and ten pouch-like nectaries on the floor of the saucer-shaped hypanthia. In flowers bagged before dawn as buds, the average cumulative volume of nectar at 0900 hrs taken from male flowers was 7.7 µl of nectar with 32% w/w sugar (n = 11); from female flowers, 4.8 µl of nectar with 37% w/w sugar (n = 8) (Newstrom, 1986). The small amount of nectar available in the afternoon in bagged male and female flowers was over 50% w/w sugar, possibly due to evaporation.

The major pollinators of the cultivated chayote belong to the Hymenoptera, particularly Apidae. The primary species were *Trigona* bees in Costa Rica and Mexico (Wille and Orozco, 1983; Newstrom, 1986). Introduced *Apis mellifera* was the most abundant pollinator in California and in commercial farms in MesoAmerica wherever pesticide applications were frequent (Newstrom, 1986). Other large bees such as *Bombus medicus* also were pollinators. Secondary pollinators, carrying smaller pollen loads and/or traveling shorter distances, were small bees, small wasps such as *Polybia* and *Stelopolybia*, and various Diptera and Coleoptera. Many ant species collected nectar; some also carried small amounts of pollen. Some of the large hairy ants are capable of achieving pollination since they were often the first to arrive in the morning. They would not be important as outcrossing agents since they did not tend to travel from plant to plant.

A crop in need of insect pollination can present difficulties to farmers because pesticides often kill pollinators. On those farms using pesticides, few wild bees and wasps forage, and abundant honey bees occurred only on farms with beehives interspersed in the plantation. In Costa Rica and Mexico, we saw farms containing so few pollinators due to over-use or improper timing of chemical pesticide applications that the yield probably was reduced.

A common folklore about chayote is that both a "male" and a "female" fruit must be planted together to get good fruit set. The "male" fruits are reported to have only one shoot sprouting from the seed but the "female" fruits have more than one, though the basis for this observation is unknown. I have not observed a fruit with more than one shoot, and polyembryony is not known to occur. Chayote is able to self-pollinate but it is not known if cross pollen is more effective than self pollen (Newstrom, 1986).

Germination

It is typical for unharvested chayote seeds to germinate in the fruit while on the parent plant. This lack of dormancy is important. From the practical aspect of seed storage, chayote is viewed as a "recalcitrant" seed, meaning that it does not tolerate desiccation, and hence a rest period for long-term storage cannot be induced (Harrington, 1972; King and Roberts, 1979). In a fruit harvested at 4 to 6 weeks of age, the shoot will emerge in one week at the distal end unless kept in cold storage (Newstrom, 1986). After harvesting, a germinated fruit can stay alive for one or more months, depending on its size. I have kept fruits at room temperature on a dry shelf for 2 to 3 months while the shoots continued to grow (up to 1 m long) and the fruit shrank and shriveled. Fruits kept at room temperature in a dark moist environment sprouted roots with no shoot; those in a light, dry environment sprouted a shoot with no roots (Fig. 1). Chayote seeds cannot be excised from the fruit unless it has already sprouted; therefore, a seed cannot be stored without the fruit. When planting chayote, one usually plants the entire fruit, but in Michoacán, Mexico, the mature seeds are removed from the fruit, along with over half of the cotyledons, and are planted in moist soil. The farmers say that if they plant this portion of the seed without the fruit, more fruits than tubers will be produced, but if they plant the entire fruit more tubers will be produced. Besides being intolerant to desiccation, chayote fruits and seeds also suffer from chilling injury, another practical impediment to long term storage (King and Roberts, 1979). There is a limit to how long and how cold a fruit can be kept with the seeds remaining viable. Fruits kept at or below 2.5° C are unable to sprout when returned

Figure 1. Mature fruits of chayote (*Sechium edule*) showing the vivipary with with only roots sprouting (right, dark conditions), and with only shoots sprouting (left, light conditions).

to 20° C (Littman et al., 1981). Storage of chayote seeds in their fruits is thus a matter of months, not years.

From a developmental standpoint, this lack of dormancy implies that chayote is viviparous (Gnagnarini and Lorenzi, 1985). Guppy (1912) thought that the inherent capability of seeds to dispense with their rest period (as in *Hedera* and *Quercus*), was so widespread that it was the primitive condition in all plants. The term 'true vivipary or 'reproductive vivipary' is today limited to seeds that normally germinate with shoots emerging from the fruit while it is still attached to the parent plant (Henckel, 1979; Juncosa, 1982, 1984). The only truly and consistently viviparous dicotyledonous species are thought to be four genera from the mangrove family, Rhizophoraceae, the sea grass, *Amphibolis*, and some palm species (Juncosa 1982, 1984). Cryptovivipary, the emergence of the embryo from the seed coat but not the fruit before separation from the parent plant, is found in several other mangrove genera: *Aegiceras, Avicennia* and *Pelliciera* (Juncosa, 1982). These two types of vivipary, and a third type, reproductive-vegetative vivipary, are considered to be euvivipary (Henckel, 1979). In reproductive-vegetative vivipary, also known as clonal vivipary (Juncosa, 1982), brood buds appear in place of flowers in plants of the high arctic, high altitudes, and arid climates. There is also a phenomenon called pseudo-vivipary that occurs in two forms: vegetative pseudovivipary with broodbuds on stems, leaves, and roots, and pseudo-reproductive vivipary observed on dead maternal plants, such as grain germination in harvested bundles in humid weather (Henckel, 1979). Finally, Juncosa (1982) describes a teratological vivipary that involves premature germination known mainly in cultivated fruits. Vivipary in this case is not a normal part of the life cycle of the plant and occurs only in a portion of the fruits. Vivipary in chayote has been considered a teratology in the past (Penzig, 1922; Crovetto, 1948).

Vivipary in mangrove swamps has evolutionary significance because the saline water and movement of the tides would result in selection for long fruit retention and rapid germination (Joshi, 1933; Rabinowitz, 1978; Juncosa, 1982, 1984). The adaptive value of dormancy for wild plants outside of the mangrove swamps is to maintain a seed supply in the soil, but it is non-adaptive for cultivated plants in which there has been selection for uniformity of germination with a relatively short time to first appearance of the seedlings. As a result, dormancy is much reduced under cultivation (Harlan, 1975; Murray, 1984). One of the evolutionary trends in the process of domestication, then, is the loss of dormancy (Harlan, 1975). This loss of dormancy can result in vivipary on an occasional basis such as in wheat, citrus, tomato, etc., or on a regular basis such as in chayote.

Changes in dormancy can result from altered seed coat structure and permeability and from the elimination of hormonal inhibition that would normally ensure a period of after-ripening (Murray, 1984). Seed coat structure is variable in chayote. Some cultivars of chayote have lignification in the testa typical of the Cucurbitaceae, and other cultivars do not (Reiche, 1921; Singh, 1965; Coria and Engleman, 1983).

Merola (1949) suggested that the seed takes up water and nutrients from the pericarp. Giusti et al. (1978) showed that radioactive phosphorus from the pericarp was translocated to the cotyledons through haustoria. They proposed the new term 'spermocarp' for a fruit that behaved like a seed.

The role of abscissic acid (ABA) in preventing precocious germination has been studied in chayote by Gnagnarini and Lorenzi (1985). They found that rather than the embryo being insensitive to ABA as in some *Rhizophora* species, there was instead an absence of accumulation of ABA in maturing cotyledons, which may explain the lack of dormancy in chayote. Investigations of the physiology of other hormones, such as gibberellins, involved in seed development in chayote have been conducted in Italy (Lorenzi and Ceccarelli, 1983; Ceccarelli and Lorenzi, 1983; Albone et al., 1984; Lorenzi and Ceccarelli, 1986). Aung and Flick (1976) have produced a seedless chayote fruit by application of gibberellin on the stigma.

Thus, in chayote, several mechanisms exist to prevent dormancy: 1) lack of lignified testa, 2) translocation of nutrients and water form the pericarp to the seed, and 3) lack of accumulation of ABA. Two questions that need investigation are 1) what proportion of the fruits of any given cultivar drop before sprouting, and 2) how is abscission layer formation coordinated with emergence of the shoot from the fruits? There probably are differences in the mechanisms preventing dormancy and the extent of vivipary among different cultivars of chayote. Environmental conditions that encourage vivipary in the crop have not been studied.

Evolution in cultivation is very different from evolution in the wild (Harlan, 1975). Explanation of the ecological or evolutionary significance of vivipary in cultivated chayote must be considered separately from its significance in populations and related species found in the wild. Reiche (1921), Merola (1949), Brucher (1977) and others have suggested various explanations, including the idea that water and nutrient reserves in the fruit enable the seed to survive dry periods. However, the role and extent of vivipary (teratological or true reproductive vivipary) in chayote, and its extent and evolutionary or ecological significance in the wild, are not well understood.

EVOLUTION OF THE CROP
Variation in Cultivation

The diversity of fruit morphology in chayote is high, particularly in the land races in regions of Mexico where traditional agriculture is prevalent (Newstrom, 1986). There is variation in length (95 cm to 20 cm), color (continuous range from white to dark green), shape (9 distinct forms), surface features (including spines, lenticels, furrows and ridges) (Maffioli, 1981; Engels, 1983; Cruz Leon and Querol Lipcovich, 1985; Newstrom, 1986), flavor, texture, and amount of fibers surrounding the seeds. Until recently, variation in flavor has been neglected (Cruz Leon and Querol Lipcovich, 1985). The external appearance of the fruit does not predict the flavor, which ranges from insipid, through bland-sweet, to a delicate starchy taste similar to that of a new potato (Newstrom, 1986). Commercial cultivars are bland tasting because they are sold for industrial use as well as for a table vegetable. More flavorful lines could be developed to increase the popularity of chayote as a table vegetable.

There also is variation in flower characters, but this has not been well documented. Sex ratio varies among cultivars and can be altered in several different ways (Newstrom, 1986). Differences in production characteristics such as time of flowering and yield per crop have been studied by Cruz Leon and Querol Lipcovich (1985). The folk taxonomy of chayote indicates that the crop is cross pollinated and does not breed true. There are many names, but they tend to be regional and it is uncertain to which land race of cultivar a name refers. In Mexico some names are *upopo, chayotito gachopin, chayote pelon,* and *chayotillo.* *Chayotillo* apparently refers to a small round white cultivar with excellent flavor that is called *perulero* in Guatemala and *coscoros* in Costa Rica. Chayote cultivars and land races have not been scientifically described, and their geographical distribution is not documented, so no comparisons can be made to relate folk taxonomy to scientific names.

Variation Outside Cultivation

Several chayote populations outside cultivation have recently been discovered in Mexico (Cruz Leon and Querol Lipcovich, 1985; Newstrom, 1986). These populations are similar to the cultivated chayote except that the wild fruits are bitter. The wild populations are in disturbed habitats at edges of fields, on steep banks along roadsides, or steep slopes in canyons where cultivation of maize is practiced by the indigenous farmers.

In countries such as Venezuela, where chayote is known to have been introduced, the populations growing in the wild are certainly feral (Brucher, 1977; Newstrom, 1986). Bitter fruits have not been reported in these populations. However, the wild populations are not necessarily feral in Mexico and Guatemala, where the closest wild related species, *Sechium compositum* (Stand. & Steyerm.) C. Jeffrey, is native. However, neither can it be concluded that they are truly wild, a persistent problem in studying the evolutionary history of many crops. The Mexican populations may be ancestral to the crop or they may be derived from the crop as escapes with bitter fruits due to mutation or hybridization with a bitter wild species such as *S. compositum*. One of the difficulties in studying populations of chayote in the wild is that it is impossible to determine the number of individual genotypes in any one population without destructive sampling. The scrambling vines intertwine and several shoots resprout from the same plant each year. Fruits that drop under the parent plant may become established and be confused with a new sprout from the parental plant.

Data have been collected on morphological and chemical characters of fruits and flowers in five populations found in the wild. Placing *S. edule* and *S. compositum* at two extremes of a continuum, the five populations were grouped according to their closeness in morphology and chemistry to either species (Table 2). The fruits and tubers of cultivated chayote are edible, but *S. compositum* has bitter, inedible fruits and tubers that are used for soap. The morphological characters were scored to produce a hybrid index (Anderson, 1953; Anderson and Stebbins, 1954) using a score of 0 for *edule* characters and 5 for *compositum* characters.

The flowers of the five populations were similar to chayote flowers, but stamen characters varied (Table 2). The flowers of La Esperanza, Oaxaca, and the Tetla, Veracruz, populations were the closest to *S. edule*, with thick curled anther branches showing pollen dehiscence on top. The Manzanillo, Veracruz, Chiltepec, Oaxaca, and Valle National, Oaxaca, populations had thin straight anther branches similar to *S. compositum*. In addition, the Chiltepec flowers were unique because they had distinctly notched petals.

Fruits of all five populations were bitter but their external morphology was most similar to several chayote land races. In Table 2, the morphology of the fruit was scored as 0 for all five

Table 2. Comparison of the flower and fruit morphology, and biochemical composition of fruits in cultivated *Sechium edule*, related populations from the wild, and the wild *S. compositum* in Mexico. Measured on dry weight of fruits. Numbers in brackets indicate hybrid index scores. Flower morphology: 0 = thick curled anther branches, 5 = thin, straight anther branches. Fruit morphology: 0 = external characters within variation found in cultivars, 5 = characters not found in cultivation. Bitterness: 0 = nonbitter, 5 = extremely bitter to taste. The percent sugar is by weight. The sugar ratios are sucrose/glucose + fructose (S/G + F) and glucose/fructose (G/F).

Population Location	Flow Morph	Fruit Morph	Bitter	% Sugar W/W	Ratio S/G+F	Ratio G/F	Amino Acid μmoles/100 gms	Total Hybrid Index
S. edule Cultivated	(0)	(0)	(0)	10.0 (0)	.001 (0)	.34 (0)	6.07 (0)	(0)
S. edule La Esperanza	(0)	(0)	(5)	16.4 (0)	.001 (0)	.53 (0)	19.43 (2)	(7)
S. edule Tetla	(0)	(0)	(5)	6.4 (3)	.018 (3)	.53 (0)	43.59 (4)	(15)
S. edule X? Manzanillo	(5)	(0)	(5)	15.5 (0)	.001 (0)	.81 (1)	35.65 (4)	(15)
S. edule X? Valle Nacional	(5)	(0)	(5)	4.9 (3)	.001 (0)	1.17 (2)	41.43 (4)	(19)
S. edule X? Chiltepec	(5)	(0)	(5)	5.6 (3)	.020 (3)	1.17 (2)	42.72 (4)	(22)
S. compositum Motozintla	(5)	(5)	(5)	0.05 (5)	.039 (5)	4.67 (5)	54.45 (5)	(35)

populations because their overall fruit morphology was identical to various land races of cultivated chayote. The characters of *S. compositum* , scored as 5's, (specifically, lack of a distal cleft and spines confined to ridges), were not found in cultivation. The small spheroid fruits of both the La Esperanza and Tetla populations were covered by dense hard spines. The fruits of the Manzanillo population also were similar to chayote cultivars, but they were variable within the population in their size, shape, and density of spines. The fruits of the Chiltepec and Valle Nacional populations from Oaxaca, which were uniform within and between populations, were smooth, light green, flattened obovoid with a distal cleft identical to various chayote land races.

The composition of the fruits sort out in a similar way to the morphological characters. The La Esperanza and Tetla populations were generally closest to chayote. The Manzanillo was intermediate, and the Chiltepec and Valle Nacional were closest to *S. compositum* (Table 2). The ratio of glucose to fructose follows the morphological similarities more closely than the percent of sugar (w/w). The La Esperanza and Tetla fruits had a ratio of 0.53, the closest to *S. edule.*

The amino acid concentration was relatively high in fruits of all five wild populations: 19 to 43 µmoles/100 gms compared to the low 6.07 in *S. edule* and the very high 54 in *S. compositum* (Table 2). The qualitative test for proteins and phenolic compounds showed that the Manzanillo population had a large amount compared to the other populations, and to *S. edule* or *S. compositum.* Lipids, cardiac glycosides and organic acids were not detected in any of the populations, nor in *S. edule* or *S. compositum.*

Cucurbitacins

The bitterness in *S. compositum* and the five populations considered here is most likely due to the presence of cucurbitacins, because tasting the bitter fruit flesh paralyzes the tongue (pers. obs.), a known effect of cucurbitacins (Metcalf, 1986). The distasteful and highly toxic cucurbitacins are a group of about 20 oxygenated tetracyclic tri-terpenes, the most bitter substances known (detectable in 1 ppb by humans) (Guha and Sen, 1975; Metcalf, 1986). They evolved in the Cucurbitaceae as protectors against herbivore feeding (Rhodes et al., 1980; da Costa and Jones, 1971; Metcalf, 1986). However, to diabroticite beetles of the tribe Luperini, family Chrysomelidae, the cucurbitacins act as kairomones promoting host selection and compulsive feeding (Metcalf et al., 1980., 1982). The association between these beetles and cucurbit plants benefits the beetles because they sequester cucurbitacins, thus providing themselves with chemical protection form predatory attacks by certain insects (Howe et al., 1976; Ferguson and Metcalf, 1985).

It is thought that the edible cultivars of the Cucurbitaceae originated from wild bitter species following one or more recessive mutations (Enslin, 1954). The reverse mutation could also occur and may be responsible for the bitter plants found in cultivars of zucchini (Rymal et al., 1984). The inheritance of cucurbitacins was originally thought to be regulated by a single dominant gene Bi (Rhodes et al., 1980) as demonstrated in crosses using cucumber (Cucumis), watermelon (*Citrullis*), bottle gourd (*Lagenaria*), and squash (*Cucurbita*) (see references in Lee and Janick, 1978; Borchers and Taylor, 1988). However, it was proposed that an additional suppressor gene, su[Bi], prevents the synthesis of the cucurbitacins in fruits derived from bitter seedlings (Rhem and Wessels, 1957; Lee and Janick, 1978). Borchers and Taylor (1988) recently showed that, in a *Cucurbita mixta* and *C. pepo* cross, there are three dominant complementary genes which interact to produce an F_1 having bitter fruits from two nonbitter parents. The bitterness depends on dominant alleles at all three loci. Other genes controlling the quantity and chemical nature of the cucurbitacins have been proposed (see references in Rhodes *et al.*, 1980). If a series of genes are responsible for the sequential biosynthesis of cucurbitacins, then nonbitter species or cultivars can arise from a number of different recessive mutations (Borchers and Taylor, 1988).

Status of Populations

Present evidence (Table 2) suggests that the status of the five populations discovered is as follows: two (La Esperanza and Tetla) are most likely ancestral or escapes, one (Manzanillo) is most likely of hybrid origin and two (Valle Nacional and Chiltepec) are either hybrids or are morphs of *S. compositum.*

The two closest populations both morphologically and chemically (Table 2) are from La Esperanza and Tetla. Their fruit types are typical of the expected primitive type: small, round, and green, with dense hard spines. They are the most difficult chayote fruits to handle because the spines are very sharp and firm. Smaller, well-defended fruits are generally thought to be the primitive type for crops (Hawkes,1983; Heiser, 1988), supporting the idea that these two populations are candidates for the wild progenitors of chayote.

A mutation allowing the expression of bitterness may occur readily in escaped chayote populations. Evidence for this could be found by looking for bitterness in the feral populations in countries such as Java (Backer and Bakhuizen van der Brink, 1963) and Ile de la Réunion (de Cordenoy, 1895), where we know that chayote was introduced. If bitterness is found there, then the mutation back to bitterness may occur readily and the populations in the wild in Mexico may be feral.

On the other hand, bitterness may have been acquired due to hybridization. We have evidence from a spontaneous hybrid that *S. edule* probably is able to hybridize with *S. compositum* (Newstrom, 1986). In fact, Guatemalan farmers complain whenever *S. compositum* grows near their chayote plantations that the crop plants often become bitter, so they eradicate the weed whenever they find it (J. Leon, pers. comm.).

We also have evidence that at least one of the populations is interfertile with cultivated chayote. In October, 1984, crosses were made between chayote cultivars and a plant from the Tetla population growing in the Chapingo gene bank at Huatusco, Veracruz. Of the six artificial hybrids resulting from 18 trials (Newstrom, 1986), all had bitter fruits (A. Cruz Leon, pers. comm.), suggesting that bitterness is controlled at one or a few loci by a single dominant allele(s). Backcrosses to verify that the bitterness is controlled by a single recessive gene have not yet been carried out. Crosses with the remaining populations have not been attempted as yet.

In three of these populations, there is morphological evidence for hybridization with a wild species. The population at Manzanillo, Veracruz, in particular, is most likely a hybrid between the cultivated chayote and *S. compositum* because it has fruits variable in shape, size and denseness of spines; but this population is similar to chayote cultivars. Its flowers, however, are more similar to *S. compositum*, with thin, straight anther branches.

The two populations, from Chiltepec and Valle National, are also potential hybrids between the cultivated crop and *S. compositum*. They have anther branches like *S. compositum*, but fruit morphology like *S. edule*. The fruit sugar ratios, and percent sugar (w/w) are more similar to *S. compositum* than to *S. edule*. One of the populations has notched petals, a character not found in either *S. edule* or *S. compositum*. J. Dieterle, an expert on MesoAmerican cucurbits (Dieterle, 1976), identified that same population as *S. compositum*, but she may not have seen the fruits. Further exploration of the variability in *S. compositum* is necessary to determine if smooth fruits with a distal cleft, so similar to chayote cultivars, also are typeical of *S. compositum*. Thus, these two populations may be morphs of *S. compositum* or the result of hybridization between the cultivated crop and *S. compositum.*

CONCLUSIONS AND SUMMARY

Cultivated chayote, which originated in Mexico and Guatemala, easily becomes naturalized where it is introduced, perhaps because it has a generalist pollination syndrome and is self-compatible. The seed lacks dormancy, regularly germinating with the shoot emerging from the fruit on the parent plant. To determine if this is teratological vivipary, or true vivipary, the relative frequency of retained fruits with protruding embryos in the different cultivars and in different environments should be studied. The coordination of the timing of the fruit abscission layer with emergence of the embryo from the fruit needs to be investigated. Vivipary in cultivated chayote is most likely the result of selection for rapid and even germination which is common in crop evolution. It is not known if vivipary occurs in wild populations of chayote or in the related *S. compositum.*

The status of recently discovered populations of chayote found outside cultivation is an important part of understanding the domestication of this species. Of the five populations studied, the evidence to date suggests that the two with primitive fruit types are possibly the wild progenitors of chayote. However, the inheritance of bitterness, thought to be due to the presence of cucurbitacins, and the amount of gene flow between the crop and these populations, need to be investigated before conclusions can be made. Three of the populations show morphological and chemical evidence of hybridization with *S. compositum.* This pattern of periodic gene exchange between crops and their wild or weedy relatives is a common feature of evolution in crops. However, it is by no means universal, as Kiang and Chiang's studies of soybeans (this volume) indicate. Since the five populations grow in disturbed habitats and appear to be capable of hybridization with the crop, they may form a weed-crop complex which has undergone cycles of hybridization and differentiation (Harlan, 1975; Hawkes, 1983).

LITERATURE CITED

Albone, K. S., P. Gaskin, J. MacMillan, and V. M. Sponsel. 1984. Identification and localization of gibberellins in maturing seeds of the cucurbit *Sechium edule,* and a comparison between this cucurbit and the legume *Phaseolus coccineus.* Planta 162: 560-565.

Anderson, E. 1953. Introgressive hybridization. Biol. Rev. 28: 280-307.

Anderson, E., and G. L. Stebbins. 1954. Hybridization as an evolutionary stimulus. Evolution 8: 378-388.

Aung, L. H., and G. J. Flick. 1976. Gibberellin induced seedless fruit of chayote Sechium edule Swartz. HortScience 11: 460-462.

Backer, C. A., and R. C. Bakhuizen van der Brink. 1963. Flora of Java. Vol. 1. N. V. P. Noordhoff-Groningen, The Netherlands.

Baker, H. G. 1970. Plants and Civilization. Wadsworth Publishing Co, Belmont, CA.

Borchers, E. A., and R. T. Taylor. 1988. Inheritance of fruit bitterness in a cross of Cucurbita mixta X C. pepo. HortScience 23: 603-604.

Brucher, H. 1977. Tropische Nutzpflanzen: Ursprung, Evolution und Domestikation. Springer-Verlag, New York.

Bueno, R., R. S. de Moura, and O. M. da Fonseca. 1970. Preliminary studies on the pharmacology of Sechium edule leaves' extract. Ann. Acad. Brasil. Cienc. 42 (Suppl.): 285-289.

Bukasov S. M. 1981. Las Plantas Cultivadas de Mexico, Guatemala, y Colombia. Proyecto C. A. T. I. E. - G. T. Z. de Recursos Geneticos. Turrialba, Costa Rica.

Casseres, E. 1980. Produccion de Hortilizas. Instit. Interam. Cienc. Agric. San Jose, Costa Rica.

Ceccarelli, N., and R. Lorenzi. 1983. Gibberellin biosynthesis in endosperm and cotyledons of Sechium edule seeds. Phytochemistry 22: 2203-2205.

Cook, O. F. 1901. The Chayote: a Tropical Vegetable. U.S.D.A. Div. of Bot., Bull. No 18.

de Cordenoy, E. J. 1895. Flore de l'Ile de la Réunion. Paris.

Coria, O., and E. M. Engleman. 1983. Anatomia de la Testa de Sechium edule Sw. Chapingo 8: 28-30..

da Costa, C. P., and C. M. Jones. 1971. Cucumber beetle resistance and mite susceptibility controlled by the bitter gene in Cucumis sativus L. Science 172: 1145-1146.

Crovetto, R. Martinez 1948. Anormalidades florales en Sechium edule. Lilloa 12: 49-60.

Cruz Leon, A., and D. Querol Lipcovich. 1985. Cataloga de Recursos Geneticos de Chayote (Sechium edule Sw.) en el Centro Regional Universitario Oriente de la Universidad Autonoma Chapingo. Universidad Autonoma Chapingo. Huatusco, Veracruz, Mexico.

Dieterle, J. V. A. 1976. Cucurbitaceae. In D. L. Nash [ed.]. Flora of Guatemala. Fieldiana 24 (XI, 4): 306-395.

Engels, J. M. 1983. Variation in Sechium edule Sw. in Central America. J. Amer. Hort. Sci. 108: 706-710.

Enslin, P. R. 1954. Bitter principles of the Cucurbitaceae. I. Observations on the chemistry of cucurbitacin A. J. Sci. Food Agric. 5: 410-416.

Ferguson, J. E., and R. L. Metcalf. 1985. Cucurbitacins. Plant-derived defense compounds for diabroticites (Coleoptera: Chrysomelidae). J. Chem Ecol. 11: 311-318.

Fidel, J., and E. Tristan. 1931. Bau und Bestäubung der Blüte von Sechium edule Sw. Biol. Gen. 7: 334-343.

Giusti, L., M. Resnik, T. del V. Ruiz, and A. Grau. 1978. Notas Acerca de la Biologia de Sechium edule (Jacq.) Swart. (Cucurbitaceae). Lilloa 35: 5-13.

Gnagnarini, M. R., and R. Lorenzi. 1985. Acido abscissico e sviluppo del seme in Sechium edule. Sw. Riv. Ortoflorofrutt. It. 69: 241-249.

Guha, J., and S. P. Sen. 1975. The cucurbitacins - a review. Plant Biochem. J. 2: 12-28.

Guppy, H. B. 1912. Studies in Seeds and Fruits. Williams and Norgate, London.

Harlan, J. R. 1975. Crops and Man. American Society of Agronomy and Crop Science. Madison, WI.

Harrington, J. F. 1972. Seed storage and longevity. In T. T. Kozlowski [ed.]. Seed Biology III.

Hawkes, J. G. 1983 The Diversity of Crop Plants. Harvard University Press, Cambridge, MA.

Heiser, C. B. 1988. Aspects of unconscious selection and the evolution of domesticated plants. Euphytica 37: 77-81.

Henckel, P. A. 1979. On the cognition of viviparity in the plant world. Zhurnal Obshchei Biologii 40: 60-66. (Russian with English summary).

Hoover, L. G. 1923. The Chayote: Its Culture and Uses. U.S. Department of Agriculture. Dept. Circ. 286.

Howe, W. L., J. R. Sanborn, and A. M. Rhodes. 1976. Western corn rootworm adult and spotted cucumber beetle associations with Cucurbita and cucurbitacins. Environ. Entomol. 5: 1043-1048.

Joshi, A. C. 1933. A suggested explanation of the prevalence of vivipary on the sea-shore. J. Ecol. 21: 209-212.

Juncosa, A. M. 1982. Developmental morphology of the embryo and seedling of Rhizophora mangle L. (Rhizophoraceae). Amer. J. Bot. 69: 1599-1611.

Juncosa, A. M. 1984. Embryogenesis and developmental morphology of the seedling in Bruguiera exaristata Ding Hou (Rhizophoraceae). Amer. J. Bot. 71: 180-191.

King, M. W. and E. H. Roberts. 1979. The Storage of Recalcitrant Seeds: Achievements and Possible Approaches. International Board for Plant Genetic Resources, Rome Italy. 96 pp.

Lee, Chi Won, and J. Janick. 1978. Inheritance of seedling bitterness in *Cucumis melo* L. HortScience 13: 193-194.

Leon, J. 1968. Fundamentos Botánicos de los Cultivos Tropicales. Inst. Interam. de Cience. Agric. de la O. E. A., San José, Costa Rica. 1st ed.

Littman, M. D., A. Stolar, and J. R. Blake. 1981. Chilling disorders in fruit of the choko (*Sechium edule*). Queensland J. Agric. and Animal Sci. 38: 65-69.

Lopes, J. F. 1979. Banco Ativo de Germoplasma de Chuchu. Congr. Bras. de Olericult. 19. Fpolis, SC. EMPASC.

Lordello, L. G. E. 1972. Um nematóide nocivo ao chuchu. Rev. Agr. (Piracicaba) 47: 30.

Lorenzi, R., and N. Ceccarelli. 1983. Endogenous gibberellins in endosperms and cotyledons of *Sechium edule* during seed growth and maturation. Phytochemistry 22: 2189-2191.

Lorenzi, R. and N. Ceccarelli. 1986. Isolation and characterization of conjugated gibberellins in maturing seeds of *Sechium edule*. Phytochemistry 25: 817-822.

MacGregor, S. E. 1976. Insect Pollination of Crops. Agriculture Handbook No 496. Agricultural Research Service, Wash., D.C.

Maffioli, A. 1981. Recursos Geneticos de Chayote, *Sechium edule* (Jacq.) Swartz. (Cucurbitaceae). C.A.T.I.E. Turriabla, Costa Rica.

Martínez, M. 1959a. Las Plantas Medicinales de Mexico. 4th ed. Ediciones Botas, Imp. Leon Sanchez, Mexico.

Martínez, M. 1959b. Plantas Utiles de la Flora Mexicana. Ediciones Botas, Imp. M. Leon Sanchez, Mexico.

Medina, J. M. 1974. Una planta totalmente comestible? Surco Latinoamericano (Mexico) 74: 17.

Merola, A. 1949. La Germinazione endocarpica del *Sechium edule* Sw. Delpinoa 2: 147-176.

Messiaen, C. M. 1979. Las Hortilizas: Tecnicas Agricoles y Producciones Tropicales. Blume, Mexico. (Trans. from French.)

Metcalf, R. L. 1986. Coevolutionary adaptations of rootworm beetles (Coleoptera: Chrysomelidae) to cucurbitacins. J. Chem. Ecol. 12: 1109-1124.

Metcalf, R. L., R. A. Metcalf, and A. M. Rhodes. 1980. Cucurbitacins as kairomones for diabroticite beetles. Proc. Nat. Acad. Sci. USA 77: 3769-3772.

508

Metcalf, R. L., A. M. Rhodes, R. A. Metcalf, J. Ferguson, E. R. Metcalf, and Po-Yung Lu. 1982. Cucurbitacin contents and diabroticite (Coleoptera: Chrysomelidae) feeding upon *Cucurbita* spp. Environ. Entomol. 11: 931-937.

Morton, J. F. 1981. The chayote, a perennial, climbing, subtropical vegetable. Proc. Fla. State Hort. Soc. 94: 240-245.

Murray, D. R. 1984. The seed and survival. In D. R. Murray [ed.]. Seed Physiology, Vol. 1, Development. Academic Press, Sydney, Australia.

Newstrom, L. E. 1986. Studies in the origin and evolution of chayote, *Sechium edule* (Jacq.) Sw. (Cucurbitaceae). PhD Thesis, Department of Botany, University of California, Berkeley.

Penzig, O. 1922. Pflanzen-Teratologie (Systematisch geordnet.) Verlag von Gebrüder Borntraeger, Berlin.

Pereira, A. Lima G., and A. G. Zagatto. 1973. Ocorrência de galhas em chuchu (*Sechium edule* Sw.) causadas por *Agrobacterium tumefaciens* (Erwin F. Smith and C. O. Towsend). Conn no Estado de São Paulo. Rev. dos Tech. de Inst. Biol. São Paulo, Brasil 39: 17.

Ponte, J. J. da. 1973. Parasitismo de *Meloidogyne incognita* emchuchu. Rev. Agric. 48: 109-110.

Purseglove, J. W. 1968. Topical Crops. Dicotyledons. Vol. II. Longmans, London.

Rabinowitz, D. 1978. Dispersal properties of mangrove propagules. Biotropica 10: 47-57.

Rao, V. G. 1964. Some new market and storage diseases of fruits and vegetables in Bombay - Maharashtra. Myocopath. et Mycol. Appl. 23: 304.

Reiche, K. F. 1921. Zur Kenntnis von *Sechium edule* Sw. Flora 114 : 232-248.

Rhem, S., and J. H. Wessels. 1957. Bitter principles of the Cucurbitaceae. VIII. Cucurbitacins in seedlings - occurrence, biochmeistry and genetical aspects. J. Sci. Food Agric. 8: 687-691.

Rhodes, A. M., R. L. Metcalf, and E. R. Metcalf. 1980. Diabroticite beetle responses to cucurbitacin kairomones in *Cucurbita* hybrids. J. Amer. Soc. Hort. Sci. 105: 838-842.

Ruberté, R., and F. W. Martin. 1975. Hojas Comestible del Tropico. Antillian College Press, Mayaquez, Puerto Rico.

Rymal, K. S., O. L. Chambliss, M. D. Bond, and D. A. Smith. 1984. Squash containing toxic cucurbitacin compounds occurring in California and Alabama. J. Food Prot. 47: 270-271.

Singh, D. 1965. Ovule and seed of *Sechium edule* Sw.: a reinvestigation. Curr. Sci. 34(24): 696-697.

Sornay, P. de 1921. Les cucurbitacées tropicales (*Sechium edule* Sw.) Chayotte - chouchou. Agronomie Coloniale 6: 81-85, 151-156, 198-226.

Sullia, S. B., and K. R. Khan. 1980. Fungal diseases of certain vegetables and fruits marketed in Bangalore City. Nat. Acad. Sci. Letters 3: 137.

Whitaker, T. W., and G. N. Davis. 1962. Cucurbits: Botany, Cultivation and Utilization. World Crop Books. Interscience Publishers Inc., New York.

Wille, E., and E. Orozco. 1983. Polinización del chayote *Sechium edule* (Jacq.)Swartz en Costa Rica. Rev. Biol. Trop. 31: 145-154.

BIOGEOGRAPHY AND CONSERVATION

As the human population grows and its demands become more unreasonable in the context of the finite resources of this planet, the need for conservation becomes ever more pressing. However, we cannot conserve or preserve everything. Choices will be made. These choices will be dictated by various criteria, and biogeographical criteria are among the most important: biological and geographical rarity of plants, animals and ecosystems will contribute to decision-making.

The Bocks' paper gives a biogeographical definition of the Great Plains, and then discusses the premier importance of drought, grazing, and fire to the ecology and evolution of the world's grasslands, including those of North America. Using species distributions of Gomphocerine grasshoppers in combination with similar data for grasses, the Bocks' extend the Great Plains' boundary farther south and west than usually is acknowledged. From their work with bison in South Dakota, they link grazing patterns of this North American native ungulate to grassland fires, and demonstrate the role of fire in selectively excluding woody species from the grasslands of eastern Montana. The Bocks' work was included in this section because the most abused and neglected natural resource within the boundaries of the continental U.S. may very well be its grasslands. Before they were described and documented, North America's grasslands were gone, or so changed that we can scarcely reconstruct their prehistoric condition. Of course, this pattern of grassland destruction is not unique to the New World. The root cause of recent famines in parts of Africa and Asia is human overpopulation, but the misuse and over-exploitation of once productive grasslands is certainly the penultimate cause for this misery and deprivation. It is a great irony that grassland ecosystems, our evolutionary homeland, have been so altered by our presence that we residents of the temperate zones risk losing these particular natural laboratories we ought to find most interesting and hold most dear.

The ecosystems of greatest global concern to conservationists are the tropical rainforests of the world. Their degradation is occurring at such a rapid rate that it must be seen to be fully appreciated. In the opinion of most students of tropical ecology, evidence now exists that there are life threatening consequences to humans and to other species everywhere at the present time because of extant policies. The enormous losses to biodiversity already are well documented. The concommitant atmospheric changes possess potential to permanently change life on this planet for the worse. Herbert Baker has been a leader in founding and contributing to ecological work going on at present in the world's tropics. The two concluding papers illustrate the passion with which we view the "plight of the tropics" today. Both Frankie and Janzen have invested their considerable talents and their personal resources into conservation, including practical demonstrations of some possible solutions.

Opler's paper is the first in this section to deal with the New World tropics. He presents a survey of the interactions between a small pierid butterfly and the many species of plants it pollinates in a variety of habitats. His data were collected over a span of several years in Guanacaste Province, Costa Rica. This longitudinal study leads to an appreciation of the ways in which a pollinator modifies its behavior to survive abrupt shifts in seasonal conditions. Because this species is so phenotypically flexible, it can pollinate a wide array of plant species. The complexity of one small animal's role in nature emphasizes the importance of the biogeographical setting. This work also illustrates that the study of natural history continues to play a central role in contemporary research in plant reproductive biology. It is unlikely that the role of *Eurema daira* could be described or appreciated without such efforts.

Frankie's work is based upon his more than two decades of field experience in Costa Rica. He shows how a dozen sympatric species of bees in the genus *Centris* can live together in a deciduous forest at the Barbudal Biological Reserve. These *Centris* species feed upon and pollinate a rather broad spectrum of flowering plants. They live on male selected territories, whose only known function is attracting females. Among these bee species there are several patterns of chemical behavior for identifying territorial boundaries. Coexistence is facilitated by species-specific patterns of colonial vs. solitary nesting and living in the soil vs. living in pre-formed tree nest sites. Unlike some other documentations of niche partitioning in closely allied organisms, here the animals appear not to be separated by food requirements. These species have similar food preferences and feeding behaviors which suggests that the bees are not food limited. They may be nest-site limited, and their territoriality, loose though it is, may be related to nest-site protection. Frankie has seen a general decline in *Centris* populations of Costa Rica in the last decade. The principal threats to them are anthropogenic: fires and agriculture. His contribution provides excellent documentation for the philosophy of conservation put forward in Janzen's paper.

Janzen brings this section and this volume to a fitting close with a discussion of his philosophy for conservation. His examples are tropical because that is where his work has centered from student days until now. His thoughts provide a punctuation point for all contributions in this book because we all are field biologists who have devoted our careers to studying plants and animals in nature. The prevalent management patterns for tropical forests today threaten not just those ecosystems, but all life as we know it. Janzen argues properly that parks and preserves too often use the one species approach to justify their establishment and continuance. Parks have tended to focus on one rare plant or one photogenic animal, rather than the preservation of functional units of nature. His studies of insect ecology have shown him that a more sensible philosophy would focus preservation efforts upon protecting and conserving "lowly" insect species and their associated plants. Janzen, Frankie and Opler demonstrate that this strategy requires a contiguous area large enough to contain habitats for all stages of the insects' life histories, as well as their migration routes. The animals depend upon their plant associates for food and shelter. All these organisms need to exist in large and varied populations in order to have the opportunity to survive.

The health of our biosphere depends upon the maintenance of temperate and tropical grasslands and forests. For those of us who are working in these places, our sense of urgency for understanding and protecting them is based upon our knowledge of evolutionary ecology. The information presented in this section should inspire good citizens everywhere to proceed with putting our cumulative knowledge of evolutionary ecology to purposeful use.

ECOLOGICAL AND BEHAVIORAL ASPECTS OF SEASONAL DIPHENISM IN *EUREMA DAIRA* (PIERIDAE, LEPIDOPTERA)

Paul A. Opler

INTRODUCTION

Eurema daira (L), a small pierid butterfly (Figs. 1-8), frequents open habitats, including pasture, savanna, and strand, from northern Argentina northward through South America, Central America, the Antilles, and the southeastern United States. It is found from the lowlands to moderate elevations (1500 m in Costa Rica), and is often the most abundant butterfly where it occurs. This adaptable insect exhibits strong geographic variation, is sexually dimorphic, and moreover, is seasonally diphenic in locations with prolonged dry seasons.

This study focuses on seasonal diphenism. Specifically, I examined *E. daira* seasonal forms for differences in adult behavior, longevity, and ecological characteristics. The study site was in the seasonally dry, tropical lowlands of Guanacaste Province, Costa Rica. Particular attention was given to those characteristics of a seasonally variable environment which may generate selective pressures leading to the evolution of seasonally diphenic populations.

MATERIALS AND METHODS
Study Site

The work was carried out at Hacienda La Pacifica, 50 meters elevation, 5km NW Cañas, Guanacaste Province, Costa Rica. The habitat consists of a mosaic of orchards, pasture, second growth, as well as deciduous and riparian forests. The focus of the local *Eurema daira* populations was in regularly grazed pastures where its larval host plant, *Aeschynomene americana* (Fabaceae), occurred.

On this ranch, pastures used by *E. daira* were surrounded by small patches of forest. Small shrubs, used by *E. daira* as wet season roosting sites, and various nectar sources were scattered throughout the pastures, while additional nectar sources were common at the pasture-forest margins. Forest edges, orchards, and water courses served as foci for the butterfly's dry season aggregations.

516

Figures 1-4. Wet season phenotypes.
1. Male, dorsal surface. **2.** male, vental surface. **3.** female, dorsal surface. **4.** female, ventral surface

Figures 5-8. Dry season phenotypes.
5. Male, dorsal surface. **6.** male, ventral surface. **7.** female, dorsal surface. **8.** female, ventral surface

Climate

At La Pacifica precipitation averages about 1600mm annually, but falls almost exclusively during a six month rainy season extending from May through October. During the dry season little precipitation occurs, while persistent trade winds and slightly higher temperatures lead to more extreme drought conditions than might be expected on the basis of precipitation alone. The climate has been described in greater detail by Daubenmire (1972), Frankie et al. (1974), and Turner (1975).

Techniques

Multiple mark-recapture experiments were employed to study survivorship, population fluctuations, and movements; most were carried out in the same pasture at La Pacifica. During four separate experiments, (wet season (2), dry season (2)) 4290 captures, including recaptures, were made.

Estimates of population size, survivorship and recruitment were calculated by the method of Jolly (1963, 1965), as modified by Scott (1973), while survival rates were calculated by the method of Fisher and Ford (1947). Life expectancy was calculated directly from the daily survival rate using the formula:

$$\text{Life expectancy (days)} = \frac{1}{\log \text{ survival rate}}$$

Maximum longevity was defined as the longest interval between initial capture date and the latest recapture date of that individual in a given experiment.

To study the effect on longevity of different chemical components of nectar, freshly emerged *Eurema daira* were placed in four large wire mesh cages with artificial flowers containing different "nectars." Ten male and ten female dry season morphs and two male wet season morphs were placed in each cage. The butterflies did not utilize the artificial flowers, although in another similar experiment *Anartia fatima* fed readily under these conditions (Opler and Baker, unpubl. manu.). Thus, the experiment inadvertently became one which estimated longevity in the absence of normal sustenance, and eliminated the possibility of emigration.

Local population movement was studied by two-day mark-recapture experiments once in each season. In these experiments at least 100 individuals were marked at three or four sites, 50 to 200 meters apart, on one day. The following day recapture efforts were made at each site.

Behavior was studied in different seasons by tallying the numbers of individuals of each sex engaged in different activities at one or two hour intervals throughout the day. Behavioral modes were nocturnal roosting, sunning, resting in shade, taking nectar, directed flight, puddling, patrolling (males), oviposition (females), courting (males), mating, and interacting (male-male or female-female).

Energetic investment in reproductive and somatic tissue was crudely approximated by bomb calorimetry. Freshly emerged individuals of each morph were collected on the same day during early January 1973. Each individual, after oven drying, was separated into abdomen versus remainder of body portions. Heat of combustion, expressed as calories/gram, was calculated for each run and a ratio (abdomen/head and thorax) was used to indicate whether differential investment of reproductive activity occurred between sexes or seasonal forms.

RESULTS
Phenotype

Eurema daira is sexually dimorphic (Figs. 1-8). Both sexes have black forewing apexes above and black markings on the outer hindwing margin above. Males have a broad black bar along the forewing margin above. In Costa Rican populations, the dorsal forewing ground color is yellow and the dorsal hindwing ground color is white, while for females the dorsal ground color is white on both wings.

Seasonal diphenism of color pattern also is evidenced (Figs. 1-8). The ventral hindwing of wet season individuals is pure white, while that of dry season individuals is mottled yellow tan with two small discal black dots. The extent of black markings on both sexes is reduced in dry season individuals and dry season females are more extensively yellowish above. Two morphological characters, forewing length - an indicator of relative size - and proboscis length, also varied strikingly between dry and wet season individuals, while within seasons forewing lengths differed between the sexes (Tables 1, 2). Females were significantly larger than males in both seasons. Dry season individuals were significantly larger than those of the wet season. Although no significant differences in proboscis length were found between the sexes (t wet = 0.53, $p < 0.5$; t dry = 1.03, $p < 0.2$), there was a highly significant seasonal difference for sexes combined (t. = 6.63, $p < 0.001$; t = 4.93 $p < 0.001$. After correcting for size differences, by dividing proboscis length by forewing length, male proboscises are relatively longer (by 6 - 8%) than those of females (Table 2). Even more significant are the relative

Table 1. Comparison of mean forewing length (cm) (± 1 S.D.) between *E. daira* seasonal forms and sexes. Significant differences indicated by Student's T statistic.

	Wet	Dry	
Male	1.57 ±0.07	1.74 ± 0.08	T = 4.46, P < 0.001
Female	1.66 ± 0.06	1.83 ± 0.09	T = 8.15, P < 0.001
	T = 2.84, p < 0.01	T = 5.25, p < 0.001	

Table 2. Comparison of proboscis length (cm), forewing length (cm), and their ratio between *E. daira* sexes and seasonal forms.

	Male	Female
Wet		
Proboscis (cm)	0.71 ± 0.04	0.77 ± 0.06
Forewing (cm)	1.57 ± 0.07	1.66 ± 0.06
Ratio - P:F	0.45	0.46
Dry		
Proboscis (cm)	0.98 ± 0.05	0.99 ± 0.12
Forewing (cm)	1.74 ± 0.08	1.83 ± 0.09
Ratio - P:F	0.56	0.54

differences between seasonal forms, with the relative proboscis length of each sex 20% longer among dry season individuals.

Seasonal Phenology

At the onset of the wet season in mid-May, the first individuals of the wet season emerge coincidentally with the foliation of the host plant. These individuals are fully reproductive and a succession of generations takes place through the wet season until early November. During that time no individuals of dry season phenotype are found and no emigratory movements take place. During November, host plant leaf growth ceases as the dry season begins. *E. daira* adults continue to emerge through mid-December. The first individuals of dry seasonal phenotype emerge in late November, and the proportion of such individuals continues to increase until mid-December when virtually all eclosing individuals represent the dry season morph. During December, population structures are unstable and several events take place more or less simultaneously. Many individuals of wet season phenotype continue their normal reproductive behavior in pastures, but by the end of the month all such activity has virtually ceased. Emerging dry season morphs, together with a number of male wet season morphs, form aggregations along shady pasture edges or openings in riparian forest. Finally, extensive upslope emigration by wet season morphs, especially females, takes place. By January, only aggregations of diapausing adults, predominantly of the dry season phenotype, remain. These aggregations persist through March and most of April, but there are extensive inter-aggregation movements through the dry season, probably in response to fluctuating nectar resource levels that the aggregations rely upon for maintenance. By mid-March, some dry season morphs begin reproductive activity, and by early May all dry season morphs have mated, laid eggs and died. By early May, only the immature stages of *E. daira* are extant. The first individuals of wet season form result from these, and the yearly cycle continues.

Daily Activity

Eurema daira partitions the day between different behavioral activities (Table 3). In the wet season, most individuals perch either in a resting posture or in a lateral basking position, prior to 08:00 hr.

By 08:00, both sexes begin to visit flowers for nectar. Such feeding continues until about 15:00 hr, reaching its peak between 11:00 and 13:00 hr. During this time, all other activities are

Table 3. Partitioning of behaviors by *E. daira* at 10:00 - 10:30 h. on 2 dates in 1972 (September 9 and November 20) - both wet season and one date in1973 (January 4 - dry season). Numbers indicate percent of individuals seen exhibiting each behavior. M = male, F = female, D = dry season phenotype, W = wet season phenotype.

| Behavior | Sept. 9 | | Nov. 20 | | Jan. 4 | | |
	M - W	F - W	M - W	F - W	M-D	M-W	F-D
Rest (Perch)	11.7	30.5	1.3	2.2	71.4	87.6	86.9
Fly (Search)	48.6	11.1	72.0	6.5	14.3	0.0	0.0
Nectaring	31.4	47.2	5.3	41.3	9.5	11.7	9.4
Interact	8.6	0.0	13.3	8.7	3.2	0.7	0.0
Court	2.9	0.0	8.0	0.0	1.6	0.0	0.0
Oviposit	-	11.1	-	41.3	-	-	3.6

interspersed with nectaring. Males fly actively from 8:00-14:00 hrs in search of receptive mates. Both sexes were wet season morphs in 15/17 pairs; in the other two, the females were the dry season form, the males were of the wet season form. Butterfly observations spanned 06:30-16:30 h. All mated pairs were observed between 12:08 and 14:30 hrs (N = 17), with almost half (8/17) seen between 14:00 and 14:30. Oviposition occurred only between 8:30 and 13:30, with a strong peak between 10:00 and 12:00. Only at the onset of the dry season (January 4, 1973) were any females observed to oviposit prior to 10:00. Freshly emerged males, and rarely a few females, may visit wet sandy or muddy areas adjacent to streams or roads, where individuals may be observed to take up moisture. This activity, referred to as "mud puddling" in the literature, is most frequent during mid-day, particularly when conditions are windless and humid.

During the dry season, small to large aggregations, with individuals, sometimes numbering in the thousands, are found in shade along the edges of large openings in forest or at pasture-forest interfaces, usually near streams or rivers. Individuals of these aggregations spend more than 80% of each day resting. The only other activities engaged in during this period are flower visitation or local movements (see below). It is likely that most individuals in an aggregation spend a short time, perhaps about 30 minutes, each day taking nectar.

Movements

Emigration is important in *E. daira*. At the onset of the dry season in Guanacaste, many individuals, particularly wet season morphs and among those especially females, emigrate up-slope.

These emigrations may take individuals 40-50 km and as much as 2,000 m higher in elevation, and produces reproductive populations or resting aggregations of *E. daira* in localities where the species is absent in the wet season. Short distance emigratory movements of dry season morphs continue through the dry season. These movements, which may be mass, unidirectional up-slope or down-slope flights or only random shifts of individual aggregations, seem to be in response to humidity changes in the case of the former and to changes in local nectar resource abundance in the case of the latter. During the wet season, consecutive day mark-recapture efforts failed to detect inter-pasture movement. During this season the species must be extremely sedentary.

Nectar Utilization

Nectar feeding is the most critical activity engaged in by adult *Eurema daira* throughout the year (Table 3). Wet season individuals alternate all activities with flower visits, and for dry season adults, flower visitation is often their primary overt daily activity (Table 3). At La Pacifica, wet season adults, with proboscises averaging slightly less than 0.8 cm, usually nectared at various *Sida* species (Malvaceae). In contrast, dry season adults, with their longer proboscises averaging close to 1.0 cm, frequently visited flowers whose corolla lengths were 1 cm or longer. *Eurema daira* visits the widest array of nectar sources of any butterfly in Guanacaste Province (Opler, unpublished). The adults will take nectar from many different flowers they happen upon in the course of their daily activities (Table 4). From 1970-74 I found that *E. daira* nectared at more than 100 plant species in Guanacaste, and I observed that as many as 24 plant species were visited during one day at a single site on January 4, 1973. The specific species used can differ between sites. In the area I studied there, *Melanthera nivea, Eupatorium odoratum* (both Asteracea), *Cnidscola urens* (Euphorbiaceae), and *Lantana camara* (Verbenaceae) were favored plants, but at another aggregation less than one km distant *Blechum brownei* (Acanthaceae), *Oxalis* sp. (Oxalidaceae) and *Tridax procumbens* (Asteraceae) were the primary nectar sources. The butterflies rarely visited any flowers more than 1.5 m above ground level.

During the wet season, pasture plants such as *Sida* spp. (Malvaceae), particularly *S. acuta*, were most frequently visited, but some other plants such as *Baltimora recta, Tridax procumbens* (both Asteraceae), and *Cordia* species (Boraginaceae) also were visited often. In some instances, females were observed to

Table 4. Use of floral resources by *E. daira* on 3 days. Numbers 1-7 are ranks of visitation frequency; numbers in parentheses are number of individuals recorded visiting flowers of each plant.

Species	Family	Jan. 4, 1973	Sept. 9, 1973	Nov. 20, 1972
Melanthera nivea	Asteraceae	1 (68)		4 (2)
Sida acuta (orange fl.)	Malvaceae	2 (33)	2 (54)	1 (35)
Eupatorium odoratum	Asteraceae	3 (29)		
Cnidiscola urens	Euphorbiaceae	4 (8)	7 (1)	
Melochia nodiflora	Sterculiaceae	5 (4)		
Aeschynomene americana	Fabaceae	6 (3)	6 (3)	
Lantana camara	Verbenaceae	6 (3)		
Eupatorium sinclairii	Asteraceae	6 (3)		
Sida acuta (white fl.)	Malvaceae	7 (1)	4 (22)	5 (1)
Sida rhombifolia	Malvaceae		5 (11)	3 (4)
Baltimora recta	Asteraceae	7 (1)	1 (81)	
Cordia pringlei	Boraginaceae		3 (24)	
Emelia sonchifolia	Asteraceae		6 (3)	
Calopogonium mucunoides	Fabaceae			2 (7)
Oxalis neeae	Oxalidaceae			4 (2)
Hyptis suaveolens	Labiatae			4 (2)

alternate oviposition with visits to flowers of *Aeschynomene americana*, their larval host.

The ability to survive the dry season appears to be keyed to ability to obtain floral nectar from a wide variety of plants. This interpretation is based on the facts that (1) flower visitation is the primary activity engaged in during much of the dry season, and (2) the proboscises of dry season morphs are disproportionately longer than the butterflys' larger size would indicate.

Survival

One of the most striking differences between the two seasonal forms is in survival rate (Table 5). Percentages of recapture over time for the wet and dry seasons are shown in Figures 9 and 10, respectively. Survival during the dry season was 5 to 7 times longer than that during the wet season. Moreover, the failure to detect emigration in the wet season and the extreme day to day variation in recapture rate during the dry season indicate that the wet season statistics are reasonably accurate longevity estimates, while those for the dry season are conservative estimates, as they reflect a combination of loss to mortality and loss to inter-aggregation movements.

The maximum longevity, as indicated by the greatest interval between dates of initial marking and last recapture, differed even more strikingly between seasons (Table 5). Among the 1,755 individuals captured and marked in the wet season the maximum longevities for both sexes were 9 days. In the dry season 1,976 individuals were marked, and maximum longevities of 74 to 78 days for male and female dry season forms, respectively, were obtained. One male wet season form, the only one recaptured during the dry season, had a longevity of 19 days, more than twice that expected for the same form during the wet season. Since all dry season individuals appear to have emerged by the first of January, it may be that sufficient dry season individuals of both sexes must live until early April, i.e. as long as 120 days, about 15 times longer than wet season individuals.

The results of the caging experiment were similar, but of a different magnitude. Upon being placed in the cages, all wet season morphs continually flew against the sides of the cages in an apparent attempt to escape, while the dry season morphs immediately perched on the sides. Within 24 hours, all 8 wet season male were dead, while the last dry season males and females survived for as long as 4 to 15 days, respectively. This is especially remarkable since none of the butterflies fed during the entire time.

Table 5. *Eurema daira* survival and longevity statistics calculated by Jolly stochastic formula for 4 mark-recapture experiments. G statistic was used to determine the probability that survival rates were obtained by chance.

	Maximum Longevity d[1] (sex)	Survival Rate	Life Expectancy (d)	G	P
Wet Season					
Experiment #1 (Nov. 23 - Dec. 4, 1972)	9 (F) 9 (M)	0.58	1.9	0.268	0.05
Experiment #2 (Sept. 9 - 21, 1973)	8 (F) 7 (M)	0.61	2.0	9.437	0.05
Dry Season					
Experiment #3 (Mar. 20-May 1, 1973	25 (F) 25 (M)	0.93	13.3[2]	53.512	0.005
Experiment #4 (Jan. 10 - May 1, 1973)	74 (F) 78 (M) 19 (F-wet)[3]	0.90	9.7	4.813	0.05

[1] Longest period from initial mark to last recapture
[2] All individuals were at least 79 days old on March 20
[3] Only individual of wet form recaptured during dry season

Table 6. Results of bomb calorimetry (cal/g) for freshly emerged *Eurema daira*, 10 of each sex and phenotype combination, collected during early January 1973.

	Wet Form ♂	Wet Form ♀	Dry Form ♂	Dry Form ♀
Head + thorax + wings	3367	7918	3802	3493
Abdomen	4040	6213	4972	4176
Total Body	7407	14,131	8774	7669
Abdomen/Thorax ratio	0.83	1.27	0.76	0.84

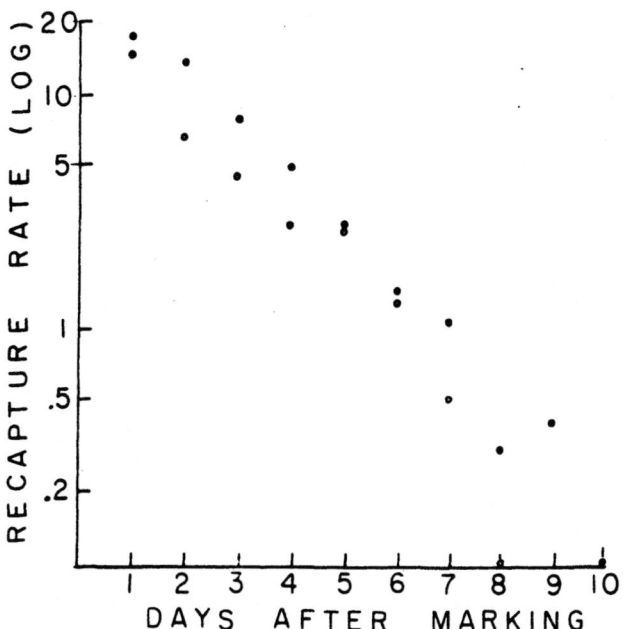

Figure 9. Change in recapture rate over time during two wet season experiments at Hacienda La Pacifica, Guanacaste Pr., Costa Rica, males - closed circles, females - open circles.

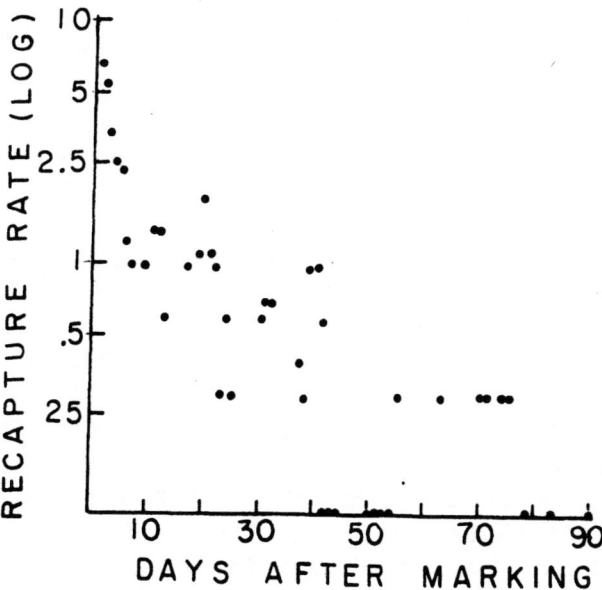

Figure 10. Change in recapture rate over time during two dry season experiments at Hacienda La Pacifica, Guanacaste, Pr., Costa Rica.

Energy Partitioning

Wet season females differed significantly from all other morphs in their energy allocations (Table 6). Note that shortly after this sample was collected, almost all wet season females emigrated to higher elevations or the Atlantic slope where continuously reproductive dry season colonies are established. The caloric content of these females probably was typical of fully reproductive *E. daira*. The caloric concentration for female abdomens was much larger for wet season as for dry season morphs which emerge with atrophied ovarioles (O.R. Taylor, pers. comm.). Wet season males, enigmatically, had the lowest caloric concentration, but still had proportionately more caloric concentration in their abdomens than did dry season males. The ratio of caloric content of abdomen over that of thorax plus head, a rough indicator of reproductive partitioning, was greater in the wet season morph of each sex. A major caloric component of abdomens, probably especially so for dry season individuals, is fat body. Future workers should carefully separate somatic and reproductive tissues when repeating the above experiment.

DISCUSSION
Phenotype

The existence of seasonal color morphs of tropical butterflies has been known for many decades (e.g. Butler, 1895). During this time, the seasonal forms of both temperate and tropical polyphenic butterflies, including several *Eurema* species, were often regarded as differing subspecies or species (Haskin 1933). Mather (1965) and Smith et al. (1982) demonstrated that many of the named subspecies and forms of *Eurema daira* are actually various expressions of the species' two seasonal phenotypes.

The wet season phenotype is adapted to a life of continual reproductive activity, while the dry season phenotype's relatively cryptic ventral hindwing pattern (Figs. 6,8) enhances background resemblance when the butterflies rest on dry foliage, and is presumably a factor in avoiding or damping predation.

The production of seasonal phenotypes in tropical pierid butterflies has been shown to be under photoperiodic control. Pease (1962) first demonstrated this for the forms of *Ascia monuste*, while Douglas and Grula (1978) have shown the photoperiodic control of *Nathalis isole's* seasonal forms. In both species, short photophase is responsible for production of the dry season form.

The unexpectedly greater length of the dry season phenotype's proboscis (Table 2) has not been previously reported in the seasonal morphs of any other insect. It may be important in increasing the survival and life expectancy of the dry season individual because it increases the array of floral resources available to them (Table 4).

Lantana camara, whose corolla length is 1.0 cm, is one clear example of a plant whose flowers are readily visited in the dry season, but not at all in the wet season, even though the plants are in flower during both seasons.

Seasonal Phenology

The seasonal phenology of most homodynamic (continuously brooded) lowland neotropical pierid butterflies is very similar to that of *E. daira*. Most species have a continuous sequence of generations in the wet season, followed by a dry season long-lived diapausing generation, with emigration a common activity in this period. It also appears that most if not all of the dry season forms possess a reproductive diapause that correlates with the changed behaviors observed. O.R. Taylor (personal communication) has demonstrated the occurrence of such reproductive arrest and, presumably, diapause, among dry season phenotypes in *Eurema daira* in Guanacaste Province, Costa Rice.

The occurrence of a parallel seasonal phenology in Australian *Eurema* has been shown by Jones et al. (1985), while Reinks (1985) found that the dry season morph of *Eurema laeta* had reproductive arrest.

Behavior

The seasonal differences in behavior are correlated with the appearance of the seasonal phenotypes, i.e. the wet season phenotype is continually reproductive during the wet season, and the dry season phenotype enters into aggregations and ceases courtship and reproduction for much of the dry season. An exception to this pattern is the appearance of a few wet season phenotypes among the diapausing dry season aggregations. Under severe dry season conditions, survival of these atypical wet season morphs must be greatly reduced (Table 5). The fact that wet season individuals are almost continuously active during the day (e.g. Table 3) is a major factor in explaining their low survival rate. They are continually exposed to predation and they continuously expend their energy resources. In contrast, dry season individuals minimize their exposure to predation and conserve their energy by resting quietly most of each day (Table 3).

Confirmation of the impact of such behavioral differences was provided serendiptiously by the caging experiment.

Individuals of the two morphs that had emerged in the field at the same time exhibited very different behaviors under caged conditions, with very different resultant longevities.

Movements

Both local and long-distance movements apparently play an important role in *Eurema daira's* overall survival and reproduction. During several dry seasons in Guanacaste I saw several periodic brief episodes of either local up-slope or down-slope movements of *Eurema daira* dry season morphs. These movements seemed to be in response to marked changes in humidity due to the passage of weather fronts. Although I obtained few quantitative data, both kinds of movements were observed each year of my residence in Guanacaste Province (1971-74). At the onset of each dry season, I observed a massive up-slope departure of wet season morphs. In a separate study, Haber (personal comment) has documented that *Eurema daira*, along with other species, crosses the Continental Divide and establishes temporary reproductive populations on the wet Atlantic slope during the Guanacaste Dry Season. I observed no *E. daira* populations on the Atlantic slope during the Guanacaste wet season. These observations demonstrate that under varying moisture conditions individuals with the same genotype may produce either seasonal morph (see also Haskin, 1933).

The observed local movements by the butterflies were probably attempts to adjust, through the sum of individual movements and density of their aggregations, to daily quantity of floral resources-the levels of which probably show sharp differences under different humidity regimes, even within the dry season.

Energy Partitioning

The differences in proportional energetic content in abdomens versus remainder of body demonstrates differential investment in reproductive tissues between the two seasonal forms. This difference, coupled with size differences, behavioral contrasts, and presumed reproductive rates, is an important highlight to the contrast in adaptiveness of the two forms to strikingly different environmental conditions.

Evolutionary and Genetic Implications

The production of two different seasonal morphs by many tropical butterflies, as exemplified by *Eurema daira*, seems to have been a frequent evolutionary response in homodynamic insects to the differential selective pressures posed by the two-season environments in many lowland tropical areas. The two seasonal morphs of *Eurema daira* have evolved two complexes of behaviors, morphological differences, and energy investments.

Nectar Utilization and Its Consequences

There are several different nectar visitation strategies among the butterfly fauna at Guanacaste. The variation includes partitioning by flower size, flower color, inflorescence structure, time of anthesis, plant height, and habitat. For some plants, butterflies appear to be the sole or primary flower visitors; in these instances, butterflies have probably played an important role in their evolution and ecology. In most cases, however, butterflies share visitation with other sorts of insects, usually Hymenoptera, and it is unlikely that butterflies have played an important role in their evolution and ecology.

Eurema daira is the most catholic of Guanacaste butterflies in the wide array of flowers it visits. The angiosperm taxa it visited were adapted for many kinds of pollinators, but several floral types appeared to be primarily butterfly-adapted. These included Asteraceae: *Baltimora recta, Lycoceris grande, Melanthera nivea* and *Tridax procumbens*; Boraginaceae: *Cordia pringlei*; and Verbanaceae: *Lantana camara*. For all of these *E. daira* is the most frequent visitor at times, and undoubtedly it or related pierids have played a role as selective factors in the plants' evolution. *Lycoceris grande* is dioecious and *Cordia pringlei* is heterostylous, and the cross-transfer of pollen by butterflies is necessary for genetic exchange and fruit formation by the plants.

SUMMARY AND CONCLUSIONS

The seasonal diphenism shown by *Eurema daira* is accompanied by changes in behavior as an adaptation to the strong seasonal changes in tropical dry forest and similar habitats. An adaptive complex involving color pattern, morphology, diapause, and behavior allows the species to be successful in a situation of differential selective factors - continuously available caterpillar host plant (*Aeschynomene*) and nectar sources during the wet season (May-October), versus no available caterpillar food and shifting nectar availability during the dry season (November-April). Emigration out of the area with establishment of temporary populations in wet highland and Atlantic localities is another part of *E. daira's* arsenal against the harsh 6-month dry season.

Altered behavior, diapause, and increased size and proboscis length permit dry season *E. daira* to survive the long dry season by resting and taking advantage of nectar resources that would be unavailable to their short-tongued wet season counterparts.

Eurema daira is typical of a number of tropical butterflies and other insects that use seasonal diphenism and diapause to survive in seasonally dry tropical environments. The existance of butterflies with "wet season phenotype" but "dry season diapause and behavior" suggests the presence of a complex interaction between environmental conditions and genetically-controlled features of color pattern, morphology, diapause, and behavior.

ACKNOWLEDGMENTS

Irene Baker carried out the bomb calorimetry runs. James A. Scott ran the Jolly Stochastic and Fisher-Ford analyses. Ronald E. Kirby, James A. Scott, and Arthur M. Shapiro critiqued the manuscript. Kate Gilbert typed the final draft.

LITERATURE CITED

Butler, A. G. 1895. Notes on seasonal dimorphism in certain African butterflies. Trans. Ent. Soc. London 1895: 519-522.

Daubenmire, R. 1972. Phenology and other characteristics of tropical semi-deciduous forest in north-western Costa Rica. J. Ecol. 60: 147-170.

Douglas, M. M., and J. W. Grula. 1978. Thermoregulatory adaptations allowing ecological range expansion by the pierid butterfly, Nathalis iole Boisduval. Evolution 32: 776-783.

Fisher, R. A., and E. B. Ford. 1947. The spread of a gene in natural conditions in a colony of the moth Panaxia dominula L. Heredity 1: 143-174.

Frankie, G. W., H. G. Baker, and P. A. Opler. 1974. Comparative phenological studies of trees in tropical wet and dry forests in the lowlands of Costa Rica. J. Ecol. 62: 881-919.

Haskin, J. R. 1933. The life histories of Eurema demoditas, Lycaena theonus and L. hanno. Ent. News 44: 153-156.

Jolly, G. M. 1963. Estimates of population parameters from multiple recapture data with both death and dilution - deterministic model. Biometrika 50: 113-128.

Jolly, G. M. 1965. Explicit estimates from capture-recapture data with both death and immigration-stochastic model. Biometrika 52: 225-247.

Jones, R. E., J. H. Rienks, and L. Wilson. 1985. Seasonally and environmentally induced polyphenism in Eurema laeta lineata (Lepidoptera: Pieridae). J. Aust. Ent. Soc. 24: 161-167.

Mather, B. 1965. Eurema daira daira in Mississippi. Lepid. News 10: 204-206.

Pease, R. W., Jr. 1962. Factors causing seasonal forms in *Ascia monuste* (Lepid.). Science 137: 987-988.

Rienks, J. H. 1985. Phenotypic response to photoperiod and temperature in a tropical pierid butterfly. Aust. J. Zool. 33: 837-848.

Scott, J. A. 1973. Convergence of population biology and adult behavior in two sympatric butterflies, *Neominois ridingsii* (Papilionoidea: Nymphalidae) and *Ambryscirtes simius* (Hesperioidea: Hesperiidae). J. Anim. Ecol. 42: 633-672.

Smith, D. S., D. Leston, and B. Lenczewski. 1982. Variation in *Eurema daira* (Lepidoptera: Pieridae) and the status of *palmira* in southern Florida. Bull. Allyn. Mus. 70: 1-8.

Turner, D. C. 1975. The Vampire Bat. Johns Hopkins Univ. Press, Baltimore.

ECOLOGICAL AND EVOLUTIONARY SORTING OF 12 SYMPATRIC SPECIES OF *CENTRIS* BEES IN COSTA RICAN DRY FOREST

Gordon W. Frankie, S. B. Vinson, and Howard Williams

INTRODUCTION

Several species of *Centris* bees (Family: Anthoporidae) are known to occur sympatrically in many areas of the Costa Rican dry forest. In some areas, up to a dozen species of these solitary bees can be found together. Their "togetherness" stems from the various pollen and nectar resources they share and may compete for . The bees are considered one of the most important pollinator groups for a wide variety of plant species in the dry forest. Their importance is directly related to their high relative abundance and efficiency as outcrossing agents (Bawa, 1974; Frankie et al., 1976; Heithaus, 1979abc; Frankie et al., 1983).

A research project was initiated in 1976 to investigate how sympatric *Centris* "sort out" reproductively in a single dry forest site. The work was conducted primarily at the Lomas Barbudal Biological Reserve and vicinity, Guanacaste Province, where 12 species were known to occur (Table 1). Three major research projects were pursued: male territorial behavior, male chemical ecology, and female nesting biology. A synopsis of the findings from several relevant studies is presented in this paper.

STUDY SITE

The study was conducted primarily at the Lomas Barbudal Reserve. The center of the 2,400 ha property is 15 km SW of the town of Bagaces in the Pacific lowland watershed; elevation 10-180 m. About 3 km south of Lomas Barbudal is the northernmost boundary of the national wildlife reserve of Palo Verde (Frankie et al., 1983). The site can be broadly classified as tropical deciduous forest. Despite varying disturbances from hunting and fire, especially between 1984-1986, the forest remains largely intact. Detailed descriptions of the site, its habitats, vegetation and climate are reported in Frankie et al. (1976, 1988a).

Table 1. Subgenera and Species of *Centris* at the Lomas Barbudal Biological Reserve. See Snelling (1984) for taxonomic treatment of *Centris*.

Centris		*Heterocentris*:	
C. (C.)	*adanae*	C. (H.)	*analis*
C.	*aethyctera*	C.	*bicornuta*
C.	*flavifrons*	*Trachina*:	
C.	*inermis (=segregata)*	C. (T.)	*fuscata*
Hemisiella:		C. (T.)	*heithausi*
C. (H.)	*nitida*	*Xanthemisia*:	
C.	*trigonoides*	C. (X.)	*lutea*
C.	*vittata*		

MALE TERRITORIAL BEHAVIOR

Results from studies of *Centris* male territoriality are summarized in Table 2. Males of 10 species were observed displaying territorial behavior. Each could be recognized according to plant association, size of territory, and pattern of chemical marking on plant substrate. Differences in bee size and color patterns also allowed easy identification of most species.

Interspecific overlap of territories was rarely observed despite similarities in plant preferences and height/size of territories among selected *Centris* species. For example, *C. flavifrons, C. intermis,* and *C. nitida* had similar preferences; however, each tended to prefer certain individual trees and certain sites within a tree. The bases for these highly specific preferences are unknown. *Centris flavifrons*, largest of the *Centris* studied, preferred hollows or openings in the tops of crowns of selected trees of *Byrsonima crassifolia*. The smaller *Centris nitida* was only rarely observed in *B. crassifolia* and never in association with *C. flavifrons*. *Centris inermis*, intermediate in size between the other two species, was never observed in the same *B. crassifolia* trees as *C. flavifrons* or *C. nitida*. When *C. inermis* displayed territorial behavior in *B. crasifolia* trees it did so at the lower crown levels. On one occasion, *C. nitida* was observed displacing *C. inermis* from a single tree of *Cassia grandis*. Displacement occurred through persistent occupation of the same territorial space.

Centris adanae has been studied the most to date (Frankie et al., 1980). It established territories in grass patches and/or in close proximity to small shrubs (Table 2). When first set up, territories were at their maximum size and marking was frequent

Table 2. Territory Characteristics of *Centris* Species.

Species	Plant Assoc. 1/	Terr. height above ground	Max. size 2/ aerial terr. (m²)	Scent marking behavior 3/	Common tree spp. 4/ used for terr.
C. (C.) adanae	Gr, Sh	.5 m	3 - 4	+++	N/A
aethyctera	Sm Tr	3 & up	8 - 9	-	Ai, Cg, Ce, Co, Ca, Dr, Pr, Tb spp.
flavifrons	Sh, Sm Tr	Sh: .5 Tr: 2-4	Sh: 6 Tr: 12	+++	By
inermis	Sm Tr	2 - 4	9	+++	By
C. (H.) nitida	Sm & Lg Tr	2 - 15	8 - 9	++	By, Cg
trigonoides	Gr	2 - 3	15	++	N/A
vittata	Lg Tr	5 - 7	9	+	Cg
C. (T.) fuscata	Sh, Sm Tr	Sh: 2 Tr: 3-5	Sh: 4-6 Tr: 9	rare	Cb, Co
heithausi	Sm & Lg Tr	3 - 7	18 - 20	+	Cg, My
C. (X.) lutea	Gr, Sh	.5 - 2	20 - 30	++	N/A

1/ Gr = grass; Sh = shrub; Tr = tree; Sm = small; Lg = large.

2/ When first established.

3/ +++ = common and frequent; ++ = periodic; + = infrequent; - = never observed.

4/ Ai = Andira inermis   Ca = Cordia alliodora
By = Byrsonima crassifolia   Dr = Dalbergia retusa
Ce = Caesalpinia eriostachys   My = Myrospermum frutescans
Cb = Cassia biflora (shrub)   Pr = Pterocarpus rohrii
Cg = Cassia grandis   Tb. spp. Tabebuia species
Co = Cochlospermum vitifolium

and intensive; marking was interspersed with occasional patrols. Through time, territory size was reduced and marking/patrolling declined. This activity pattern was also common in other species such as *C. flavifrons* and *C. inermis*.

Centris nitida, *C. lutea* and *C. trigonoides* displayed periodic, but predictable, scent marking behavior. Compared to the frequent marking behavior seen in *C. adanae*, *C.flavirons* and *C. inermis* (Table 2), periodic marking was characterized by a rare and rapid flurry of markings separated by periods of 30 or more minutes. During these periods, bees patrol occasionally. Through time marking bouts and patrols were greatly reduced.

Centris heithausi and *C. lutea* were distinct in having very large territories. In some situations, *C. heithausi* also moved back and forth between territories in two or more closely-spaced trees. In contrast, *Centris aethyctera* rarely moved between trees, maintaining temporary territories in each.

Evidence to date indicates that territories served to attract females for mating (Frankie et al., 1980). However, the males did not defend females or their food resources; rather the territories appeared to be symbolic (Alcock, 1980). A combination of male characteristics for each species was consistent with the male dominance polygyny system described by Emlen and Oring (1977).

CHEMICAL ECOLOGY

Specialized glands in males for marking territories were found in 11 of the 12 *Centris* species (Table 3). The glands were of two types: mandibular glands located in the head (Vinson et al., 1982, 1984) and tibial glands located in the hind leg (Williams et al., 1984) (Fig. 1). Most species possessed one type of well developed gland. Although all species possessed mandibular glands, they were poorly developed in two groups (subgenera *Hemisiella* and *Xanthemisia*) and did not appear to function in territorial marking. *Centris aethyctera* was unique in having poorly developed mandibular glands (Fig. 1A) and no evidence of a tibial gland. Further, this species was never observed marking vegetation, although it maintained territories in association with the greatest number of plant species (Table 2).

Mandibular Glands

These glands (Fig. 1B) were found mainly in the subgenus *Centris* and *Heterocentris*. In the former subgenus they produced mostly monoterpene compounds. When emitted and applied to plant substrate, these compounds were sweet and aromatic, and to humans they smelled like citronella. About half the species with

SOURCE OF MALE Centris BEHAVIORAL
CHEMICALS

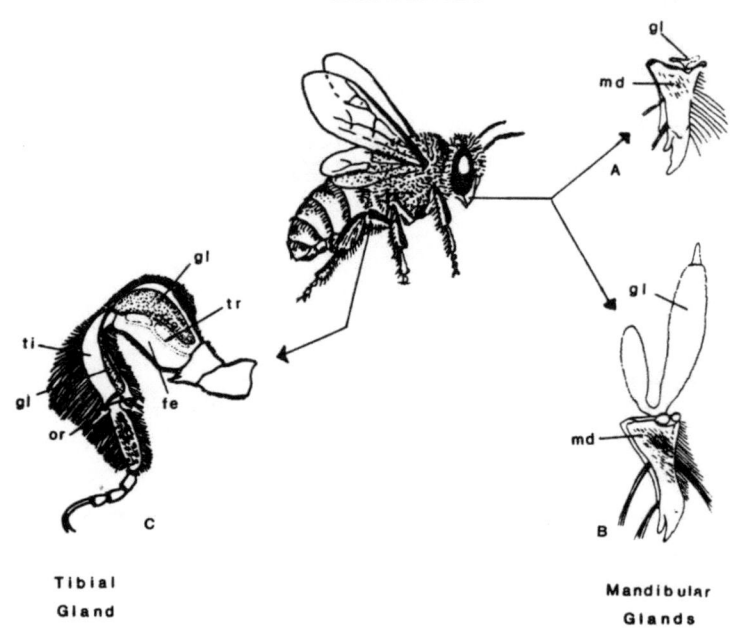

Tibial
Gland

Mandibular
Glands

Figure. 1. Generalized *Centris* bee showing location of glands
(gl). A = undeveloped mandibular gland, B = developed
mandibular gland, and C = tibial gland .

Table 3. The glands and general classes of pheromes identified for *Centris* species at Lomas Barbudal

Species	Gland Type[1]	Chemicals[2]
C. (C.) adanae	MG	TERPENE-OH>TERPENE-ESTER >>FA ESTER
aethyctera	MG	-
flavifrons	MG	TERPENE-OH>>TERPENE-AL
inermis	MG	TERPENE-AL>TERPENE-OH
C. (H.) nitida	TG	KEYTONE
trigonoides	TG	KEYTONE>>FA ESTER
vittata	?	?
C. (H.) analis	MG	?
bicornuta	MG	?
C. (T.) fuscata	MG	-
heithausi	TG	HYD>>FA ESTER
C. (X.) lutea	TG	HYD

[1] MG = mandibular gland; TG = tibial gland.

[2] Pheromone chemistry for many of the species is published in Vinson et al. (1982, 1984) and Williams et al. (1984). Chemistry for *C. flavifrons* and *C. lutea* is in prep. FA = fatty acid; HYD = hydrocarbons; AL = Aldehyde; OH = Alcohol.

well-developed mandibular glands marked consistently and regularly (Table 2). An example of the pattern of chemical marking of *C. adanae* in grass is provided (Fig. 2).

Tibial Glands
These glands (Fig. 1C) were characteristic of the subgenera *Hemisiela* and *Xanthemisia* and were present, although poorly developed, in *C. heithausi* and absent in *C. fuscata*, both members of the subgenus *Trachina*. Tibial glands released a mixture of compounds that consisted of esters of straight chain alcohols, methyl ketones, and some alkaines and alkenes. When released, they produced a musky odor. The best examples of tibial-gland

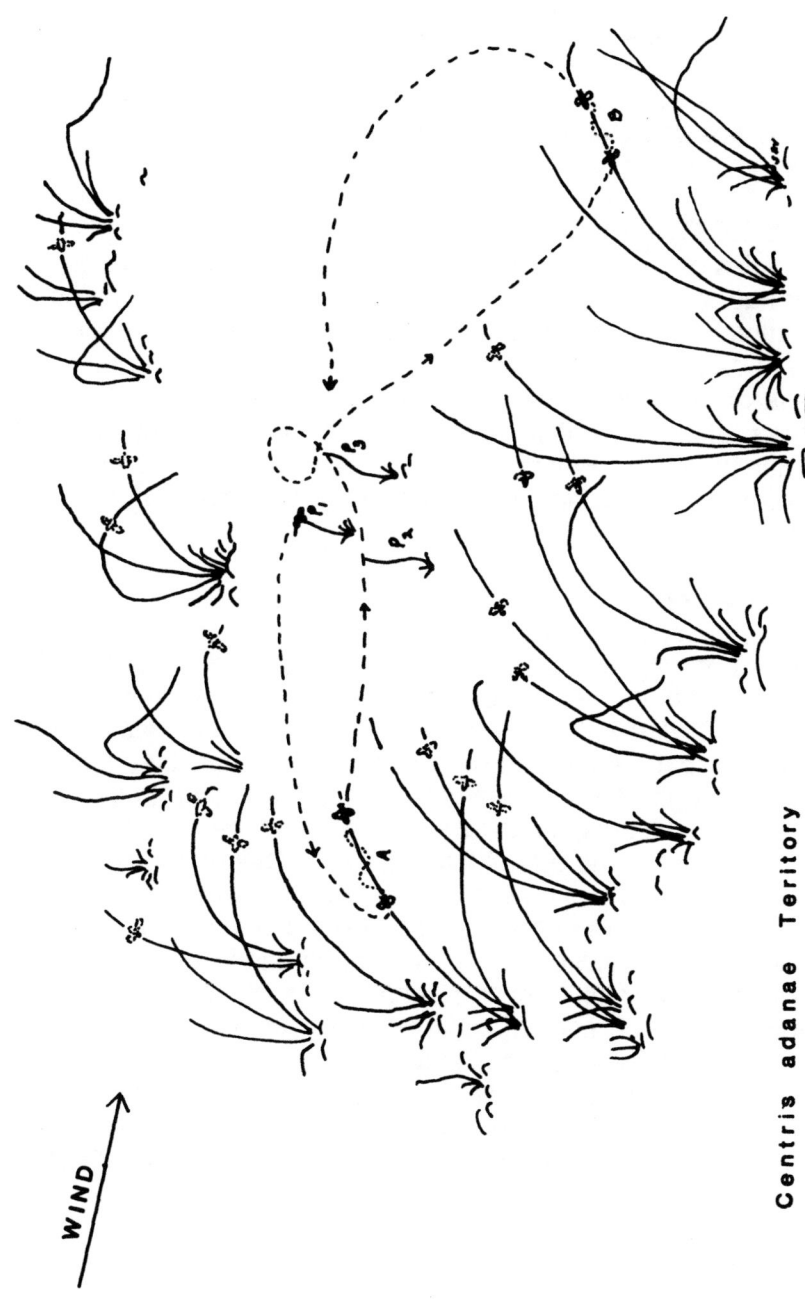

WIND

Centris adanae Territory

Figure 2. A male territory is depicted with perches (P1-P3) being used during rest periods in the center of the territory. Males mark plant stem (A and B) and hover or perch. Small dotted bees indicate various locations where marking occurred.

marking were observed in *C. nitida* and *C. lutea*, both of which had a predictable marking schedule (Table 2).

There were similarities in the chemical composition of each gland's secretions, but most species also had some distinct compounds and qualitative differences that were species specific. Separating species taxonomically, using a chemical ecological approach, was possible based on gland location and chemical contents (Frankie et al., in prep.).

FEMALE NESTING BIOLOGY

Studies of nesting biology demonstrated that each *Centris* species had specific requirements. The most obvious difference was observed at the subgeneric level where members of subgenera *Centris* and *Trachina* were exclusively ground nesters, and members of the other three subgenera were tree-hole nesters, choosing preformed cavities, most of which were made by wood-boring beetle larvae. Within each of these two general categories there were many specific differences that clearly separated species.

Ground Nesters

Differences in solitary or colonial behavior, habitat preferences, soil type, and nest architecture served to separate the six species in the two subgenera (Table 4). For example, *C. adanae* and *C. flavifrons* build single-cell nests. Both species were found widely scattered in sandy loam soils; however, they differed in the length of the tunnel leading to the cell (entrance to cell cap). The tunnel leading to *C. adanae* was short, generally less than 6 cm, whereas the tunnel leading to *C. flavifrons* was in excess of 50 cm and possessed many turns. *Centris aethyctera*, another widely scattered species, was also found in sandy loam soils (Vinson and Frankie, 1977) and also in sandy to rocky sandy soils (Vinson and Frankie, unpub. data). It usually excavated a short tunnel in which several cells were constructed on top of another, but in rocky soil, females sometimes excavated a chamber where several cells were constructed side by side. The multiple-cell nests of *C. inermis* (=*segregata*) often were aggregated in alluvium of riparian habitat (Coville et al., 1983).

Tree-Hole Nesters

Tree-hole nesters were conveniently divided into two groups: those using sand as nest construction material (3 species) and those using wood chips (3 species). Preferred habitats, hole diameters, and differences in cell size and cell cap

Table 4. Characteristic of ground-nesting *Centris*.

Species	Observed nesting habitats[1]	Observed soil type[2]	Nest depth (cm)	No. cells	Outside cell dimension (mm)	
					width	length
C. (C.) *adanae*	DDF, Sav	SL, SS	3 - 10	1	-	-
aethyctera[3]	Sav, FE	SL, SS, SRS	8 - 14	3 - 6	10	15.0 - 18.0
flavifrons	Sav	SL & Preexisting soil cavities	50 - 108	1	14.0 - 15.5	20.2 - 26.2
inermis[3]	DDF, Sav Rip (river banks)	SL	7 - 16	1 - 6	11.9 - 12.5	18.9 - 20.2
C. (T.) *fuscata*	DDF	LC	-	3	-	-
heithausi	DDF	SRS	5 - 8	2 - 3	11.4 - 13.3	19.0 - 20.5

1 DDF = dry deciduous forest; Sav = savanna; Rip = riparian; FE = forest edges.

2 SL = sandy loam; SS = sandy soil; LC = light clay; SRS = mix of small rock & sand.

3 Colonial nesters under some circumstances.

construction easily separated species within the two groups (Table 5). For example, *Centris nitida* preferred 9.5 mm holes in oak forest (Frankie et al., 1988a and unpub.). *Centris trigonoides* always preferred smaller sized holes, whereas *C. vittata* was found in larger holes. Further, *C. vittata* sealed its outer nest plug with a partially-formed oil deposit. Nests of *C. analis* and *C. bicornuta* were quite similar; however, thickness of outer plug wall and differences in color of oil deposit on the plugs easily separated these species. Little is known about nesting in *C. lutea* except that it used wood chips to construct large sized cells. This clearly distinguished it from *C. analis* and *C. bicornuta*.

PATTERNS OF SEASONALITY

During the study of female nesting biology, information on seasonal nesting periods also was gathered. Previous survey work by Heithaus (1979a) and Frankie et al. (1983) suggested that *Centris* generally nest during most of the dry season (January-May), with a few species such as *C. nitida, C. vittata* and *C. inermis* experiencing limited nesting during the first part of the wet season (June-July). Ongoing monitoring of nesting activity through trap nests (Frankie et al., 1988a) and rearing of tree-hole *Centris* indicates that considerable seasonality exists for at least some species (Frankie and Vinson, unpub. data).

Centris analis, C. bicornuta and *C. vittata* seem to nest primarily from late January through March, with the specific periods for each species appearing to depend on local abundance and flowering phenology of certain key food plants (see below). The primary nesting period of *C. nitida* occurs over a long period from January through part of April. All of the tree-hole *Centris* pass through diapause in the late larval stage from late July to November.

Seasonal studies on ground-nesting *Centris* have been difficult for lack of good monitoring techniques(s). Three exceptions, however, have provided some seasonal information. The "*inermis* form" of *C. inermis* is known to nest primarily from late December through part of February. The "*segregata* form" of this species nested mainly from mid-February to April, with a few individuals nesting into early July. Two ground-nesting *Centris* species outside the Lomas Barbudal area have been observed nesting during specific periods of the dry season. Both species, *C. flavofasciata* (Vinson et al., 1987) and *C. aethiocesta* (unpub.), are known to nest in localized Pacific Coast beach strand habitats primarily from December through February.

Table 5. Nest characteristics and habitat preferences of the six tree-hole nesting *Centris* species. Table modified from Frankie et al. (1988).

Centris species	Accepted nest home diam. (mm)	Cell wall material[1]	Outer Plug(s)		oily deposit[3]	Habitat preference[4]
			no. (1 or 2)	thick. (mm)[2]		
nitida	8, 9.5, 11	sand	1 or 2	- 1	none	OF
trigonoides	8	sand	1	1	none	?
vittata	11	sand	1	1	clear & cloudy mix; medium visc.	Rip
analis	6.5, 8	FWC; walls thin	2	- .5	golden high visc.	?
bicornuta	6.5, 8, 9.5	CWC; walls thicker than *analis*	2	1 - 12	white, cloudy; high visc.	Rip, MS, OF
lutea	?	CWC	1	?	none	Sav?

[1] FWC = fine wood chips; CWC = coarse wood chips.

[2] Thickness measured at center of plug wall.

[3] Condition of deposit on outer plug wall shortly after deposition; visc. = viscosity.

[4] Rip = riparian; MS = mesic; OF = oak forest; Sav = savanna.

FLORAL RESOURCE UTILIZATION

Centris bees used a wide variety of nectar sources in the Costa Rican dry forest (Frankie et al., 1983). A detailed account of acceptable and preferred pollens by *Centris* is lacking at this time; however, it appears that a wide variety of pollen types are taken, with an emphasis on common buzz flowers such as *Cochlospermum vitifolium* and several *Cassia* species (trees and shrubs).

Another floral resource, oil from selected species, has proved to be a key resource for *Centris* bees. The oil is found in epithelial glands (elaiophores *sensu* Vogel, 1969) in a number of plants (Buchmann, 1987). In the Costa Rican dry forest, oil was found primarily in species of Malpighiaceae, especially the tree *Byrsonima crassifolia* and to a lesser extent in viney species of *Stigmatophyllon*. Because *B. cassifolia* is habitat restricted and has the potential to produce much oil, it was considered a locally rich and key resource for *Centris*. Bees are dependent on the oil to construct cells and/or provision brood. Areas having poor representation of *B. crassifolia* had noticeably low population levels of *Centris* bees.

CONSERVATION CONSIDERATIONS

Insects are important components of tropical dry forest communities. As Janzen notes in his chapter, these animals need more recognition as objects of conservation efforts. During the course of our bee studies which began in 1972, we have become increasingly aware of the general decline in populations of several *Centris* and other bee species in the vicinity of Lomas Barbudal. This decline, which has been gradual, is most noticeable at preferred floral resources (pollen, nectar, and oil), and at territorial and nesting sites. Comparison of pre-1972 population levels of selected bee species with 1986-88 levels is currently in progress. The focus is on eight *Centris*, several *Epicharis*, two *Xylocopa*, and *Gaesischia exul* (all in the family Anthoporidae).

Several human activities are responsible for the observed decline in bee populations. The two most important activities are fire and the conversion of adjacent forest land into agricultural production, especially for crops. Details on these activities and the affected bee species are presented in Frankie et al. (1988b).

Despite the relatively intact habitats of Lomas Barbudal Reserve, the site is largely an island surrounded by agriculture. The Reserve, and the forest that once surrounded it, supported large populations of bees that probably moved opportunistically through the environment, foraging intensively on different host plants as they came into flower. The 2,400 ha of conserved dry forest may at first seem adequate for maintaining good bee populations. However, now that some of the factors which limit bee populations have been identified (*i.e.*, fire, agricultural production, and limited floral and nesting habitats), it is becoming clear why bee numbers have been substantially reduced (Frankie et al., 1988a,b).

In the case of the Lomas Barbudal Reserve, there are steps that can be taken to reverse the decline in numbers of bees (and for other organisms as well). Fire often can be greatly suppressed with appropriate preparedness and action, and this was demonstrated in 1987-88 (GWF, Fire Report to World Wildlife Fund 1988 Year Book). Securing poorly represented habitats that contain critical floral resources (*e.g. Byrsonima crassifolia*) and critical nesting sites is a realistic goal because some intact poorly-represented habitats still exist around and in the near vicinity of the Reserve. Recently, the Xerces Society (Oregon) raised funds to purchase 85 ha of critical *B. crassifolia* savanna habitat, which is also prime nesting habitat for many ground-nesting *Centris* and other solitary bees. Other poorly-represented habitats, such as riparian environments, are available; however they await future fund-raising efforts. Moist, riparian environments support a wide variety of bees (Frankie et al., 1988a) and other wildlife as well. Finally, the reforestation of agricultural land surrounding the Reserve is a possibility. However, at this time there is no progress to report.

SUMMARY AND CONCLUSIONS

A research project initiated in 1976 to investigate the reproductive ecology of 12 sympatric *Centris* bee species (fam. Anthophoridae) sort out reproductively in a Costa Rican dry forest reserve, Lomas Barbudal. Three major research subprojects were pursued: male territoriality, female chemical ecology and females' nesting biology. We were especially interested in learning how habitat partitioning can allow these species to coexist. The studies demonstrated that males of each species had characteristic territorial behavior and specific chemicals for scent marking their territories. Females of each species constructed distinct nests. There was no evidence to suggest interspecific competition for territories, chemical scent marking and nesting sites.

The studies also yielded information about floral resource utilization and limiting factors such as fire, agricultural crop production and critical habitats. The results are now being used for decision-making in the conservation, management and protection of *Centris* bees at the Lomas Barbudal Reserve.

ACKNOWLEDGMENTS
We thank the National Science Foundation and the National Geographic Society for major support for this research. The California and Texas Agricultural Experiment Stations also provided support. We thank John Barthell, Rollin Coville and Jutta Frankie for the many hours they spent making detailed field observations on *Centris* bee behavior. The University of California Research Expeditions Program (at Berkeley) was instrumental in securing participants who also assisted with field observations.

LITERATURE CITED
Alcock, J. 1980. Natural selection and the mating systems of solitary bees. Amer. Sci. 68: 146-153.

Bawa, K.W. 1974. Breeding systems of tree species of a lowland tropical community. Evolution 28: 85-92.

Buchmann, S. L. 1987. The ecology of oil flowers and their bees. Ann. Rev. Ecol. Syst. 18: 343-369.

Coville, R. E., G. W. Frankie and S. B. Vinson. 1983. Nests of *Centris segregata* (Hymenoptera: Anthophoridae) with a review of the nesting habits of the genus. J. Kans. Entomol. Soc. 56: 109-122.

Emlen, S. T., and L. W. Oring. 1977. Ecology, sexual selection, and the evolution of mating systems. Science 197: 215-223.

Frankie, G. W., P. A. Opler and K. S. Bawa. 1976. Foraging behavior of solitary bees: Implications for outcrossing of a neotropical forest tree species. J. Ecol. 64: 1049-1057.

Frankie, G. W., S. B. Vinson, and R. E. Coville. 1980. Territorial behavior of *Centris adani* and its reproductive function in the costa Rican dry forest (Hymenoptera: Anthophoridae). J. Kansas. Ent. Soc., 53: 837-857.

Frankie, G. W., W. A. Haber, P. A. Opler, and K. S. Bawa. 1983. Characteristics and organization of the large bee pollination system in the Costa Rican dry forest. In Handbook of Pollination Ecology. E. C. Jones and R. J. Little [eds.]. Van Nostrand - Reinhold, New York.

Frankie, G. W., S. B. Vinson, L. E. Newstrom, and J. F. Barthell. 1988a. Nest site and habitat preferences of *Centris* bees in the Costa Rican dry forest. Biotropica 20: 301-310.

Frankie, G. W., S. B. Vinson, L. E. Newstrom, J. F. Barthell, W. A. Haber and J. K. Frankie. 1988b. Plant phenology, pollination ecology, pollinator behavior and conservation of pollinators in Neotropical dry forest. In K. S. Bawa and M. Handley [eds.] Reproductive Ecology of Tropical Forest Plants. Cambridge Univ. Press, Cambridge, U. K.

Heithaus, E.R. 1979a. Community structure of neotropical flower visiting bees and wasps: diversity and phenology. Ecology 60: 190-202.

Heithaus, E. R. 1979b. Flower-feeding specialization in wild bee and wasp communities in seasonal neotropical habitats. Oecologia 42: 179-194.

Heithaus, E. R. 1979c. Flower visitation records and resource overlap of bees and wasps in northwest Costa Rica. Brenesia 16: 9-52.

Snelling, R. R. 1984. Studies on the taxonomy and distribution of American Centridine bees (Hymnenoptera: Anthophoridae). Contrib. No. 347 Nat. Hist. Mus. Los Angeles Co., CA

Vinson, S.B., and G. W. Frankie. 1977. Nests of *Centris aethyctera* (Hymnenoptera: Apoidea: Anthophoridae) in the dry forest of Costa Rica. J. Kans. Entomol. Soc. 50: 301-311.

Vinson, S. B., H. J. Williams, G. W. Frankie, J. W. Wheeler, M. S. Blum and R. E. Coville. 1982. Mandibular glands of male *Centris adani*, their function in scent marking and territorial behavior (Hymnenoptera: Anthophoridae). J. Chem. Ecol.. 8: 319-327.

Vinson, S. B., H. J. Williams, G. W. Frankie, and R. E. Coville. 1984. Comparative morphology and chemical contents of the mandibular glands of males of several *Centris* species (Hymnenoptera: Anthophoridae) in Costa Rica. Comp. Biochem. Physiol. 77A: 685-688.

Vinson, S. B., H. J. Williams, and R. E. Coville. 1987. Nesting habits of *Centris flavofasciata* Freise (Hymnenoptera: Apoidea: Anthophoridae) in Costa Rica. J. Kans. Entomol. Soc. 60: 249-263.

Vogel, S. 1969. Flowers offering oil instead of nectar. XI Bot. Congr. Abstr. Seattle, WA

Williams. H. J., S. B. Vinson, G. W. Frankie, R. E. Coville, and G. W. Ivie. 1984. Description, chemical contents and function of the tibial gland in *Centris nitida* and *Centris trigonoides subtarsata* (Hymenoptera: Anthophoridae) in the Costa Rican dry forest. J. Kans. Entomol. Soc. 57: 50-54.

ECOLOGY AND EVOLUTION IN THE GREAT PLAINS

Jane H. Bock and Carl E. Bock

INTRODUCTION

Grasslands are the ecological arena where human genetic and cultural evolution occurred. This has proven to be a mixed blessing at best for these ecosystems. We indisputably are the most domineering, most extraordinary grassland species ever to have evolved. Yet, by our own numerical and technological success, we have tragically despoiled most grassland ecosystems. Our ignorance of the origins, tolerances, and replacement patterns of grassland species is greater than for some other, much rarer ecosystems. Almost before Europeans had time to describe them, America's grasslands were gone, or so changed that today we can scarcely reconstruct their primitive condition. The continuing, major goal of our research at the University of Colorado has been to reach a better understanding of the origins, history, current status, and future prospects for the plants and animals of the western Great Plains. Herbert Baker taught us that the reward for ecological research is to be able to be able to reach defendable, evolutionary conclusions. This reward makes the game worthwhile.

Our field work has been carried out for the most part in the grasslands west of the 100th meridian, but east and south of the Great Basin, with southeastern Montana as a northern boundary and southeastern Arizona as the southern one. Our comments will be centered within this area. In this chapter we report on three grassland studies that focus on biogeography, plants and fire, and herbivory and fire, respectively. These studies differ in the problems they address and, therefore, in approach; but we intend that each contributes to an understanding of the ecology and evolution of western grasslands.

BIOGEOGRAPHY

Introduction

Any consideration of the evolutionary ecology of the Great Plains must include a description of their geographic extent and biogeographical affinities. This is a challenging problem for two reasons. First, the plains have been so heavily disturbed historically that it is difficult to reconstruct their primitive condition. Second, this great grassland apparently did not develop its maximum treeless extent until recent (post-Pleistocene) times, and there are few endemic species to indicate its limits (Dort and Jones, 1970; Axelrod, 1985; Wells and Stewart, 1987). Even the dominant grasses frequently are distributed far beyond traditionally described plains boundaries.

Herbivorous animals are an important component of grassland ecosystems, and they probably were a significant evolutionary force shaping the character and distribution of the Great Plains (e.g., Stebbins, 1981). Large mammals such as *Bison bison* have received particular attention in this regard, because of their abundance and because of paleontological and historical data available about their evolution and distribution (McDonald, 1981; Mack and Thompson, 1982). However, the large mammal fauna of the Great Plains is species-poor, and has been drastically altered within the past 150 years.

In addition to mammals, grasshoppers (Orthoptera: Acrididae) are dominant above-ground herbivores in grassland ecosystems (Otte, 1981). Members of the subfamily Gomphocerinae seem especially restricted to grass foods and grassland habitats (e.g. Otte and Joern, 1977; Joern, 1979). As such, their distributions should be reflective of grassland history and geography. Seventy-one gomphocerine species are known from North America north of Mexico, and excellent range maps have been published for each (Otte, 1981).

We have attempted to provide new insights into American grassland biogeography by analyzing the known distributions of the Gomphocerinae, with particular emphasis on the Great Plains. We applied multivariate analytical techniques to this problem by 1) dividing the continent into numerous small blocks, 2) characterizing each block by the gomphocerines known to occur in it, 3) computing faunal similarities between all pairs of blocks, and then 4) subjecting the blocks to cluster analysis based upon the matrix of similarity values. Our principal goal was to define objectively the geographic extent of the Great Plains, based upon distributional patterns shown by this diverse group of grassland specialists. Secondarily, we used the same data set to describe faunal similarities between the plains and other parts of North America, as well as to examine major biogeographic patterns within the plains themselves.

Methods

This analysis was restricted to the U.S. and Canada, due to the relative scarcity of collection records from Mexico and Central America. The study area was divided into 253 blocks, 132 of which included records for at least 5 gomphocerines (Fig. 1). The

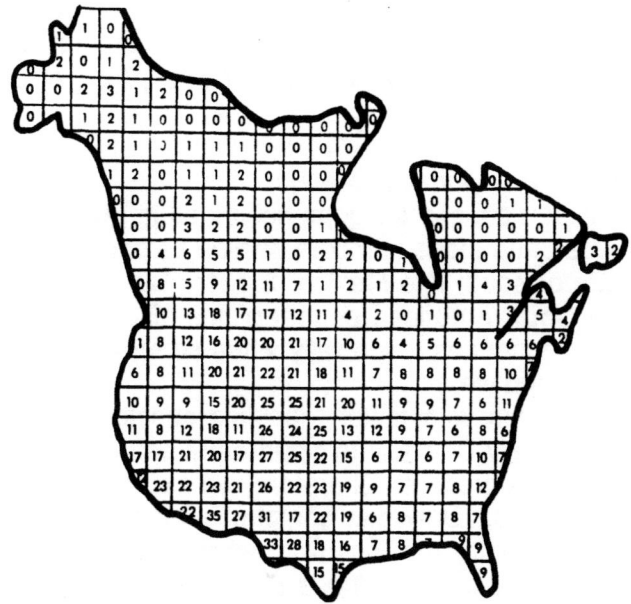

Fig. 1. Number of species of gomphocerine grasshoppers recorded for each of 253 blocks in North America north of Mexico (data from Otte, 1981). Blocks were subjected to cluster analyses based upon similarities of their gomphocerine faunas.

presence or absence of each of the 71 species was recorded for each block by laying a transparent grid over the maps printed in Otte (1981); gomphocerine nomenclature also is taken from this source. Data were analyzed using the program NT-SYS (Rohlf et al., 1972). A matrix of similarity coefficients was computed comparing each pair of blocks, using the Coefficient of Jaccard (Sneath and Sokal, 1973). This measure is calculated as:

$$S_j = \frac{a}{a + u}$$

where a = the number of species found in both blocks being compared, and u = the number of species found in either but not both blocks. Blocks then were clustered using the unweighted pair-group method with arithmetic averaging (UPGMA). The result was a dendrogram showing faunal similarities among sets of blocks. A variety of similarity indices have been proposed for numerical geographic analyses (Peters, 1971). Most, however, give similar results, and the S_j coefficient has proven insightful in previous studies similar to the present one (Bock et al., 1978; Bock et al., 1981).

Major Faunal Subdivisions

The species richness for gomphocerines is greatest in parts of the Southwest, and these numbers decline as one moves north toward Canada or south into Mexico (Fig. 1; also, see Otte, 1981, Fig. 17). The North American central plains also support a large number of gomphocerines, especially when compared to adjacent areas to the east and west.

Northern blocks with fewer than five species (Fig. 1) clustered only weakly due to the apparently patchy distributions of those few species present. Therefore, we excluded these blocks from further analyses. Only one gomphocerine (*Aeropedellus arcticus*) is endemic to the far north, while several others (e.g. *Chloealtis abdominalis, Chorthippus curtipennis* and *Stethophyma lineata*) are distributed variously southward into southern Canada and the U.S..

A dendrogram showing relationships between the remaining 132 blocks is itself far too large for publication. Instead, we determined from the dendrogram those groups of blocks that clustered at the 50% similarity level (Fig. 2), and then considered that part of the dendrogram showing relationships among these 15 Primary Regions (Fig. 3). For the most part the Primary Regions included groups of adjacent blocks, and these in turn clustered into contiguous larger areas. Gomphocerine biogeographic regions and clusters of regions corresponded strikingly with certain patterns of climate and vegetation, as variously described by Daubenmire (1978, Fig. 55), Küchler (1964), Shelford (1963) and others. This suggests that gomphocerine distributions are accurate indicators of important phytogeographic patterns.

Primary Region B (Fig. 2) includes Küchler's (1964) oak-hickory-pine and southern mixed forest types of the southeastern Coastal Plain. The gomphocerine fauna of this area was distinctive, but most similar to that of the savannahs of eastern Texas (Region A, Figs. 2 and 3). Region D corresponds to the mixed deciduous/coniferous forests of the Midwest and Appalachia, while Region C includes the eastern portion of the tallgrass prairie and especially its intergradation with deciduous forests. Together, regions A-D comprised one of the two major branches of the block dendrogram (Fig. 3). The boundary separating regions

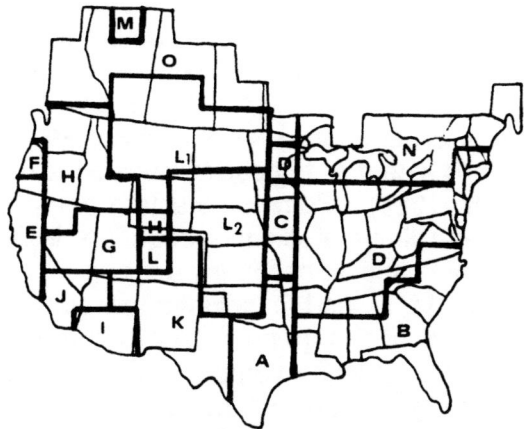

Fig. 2. Locations of 15 clusters of blocks (Primary Regions) that combined in the block dendrogram at ≥ 50% similarity, based upon their gomphocerine grasshopper faunas. At the 55% similarity level, Region L was divided into northern and southern halves, as shown.

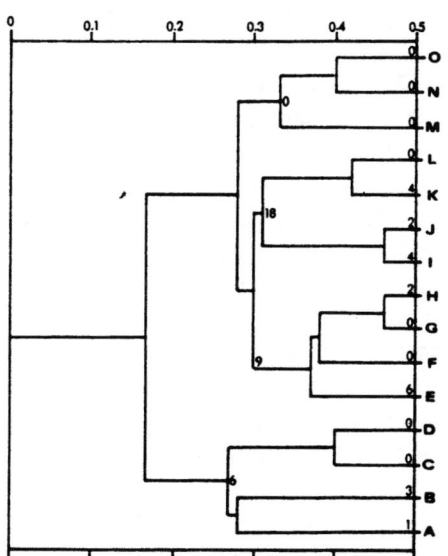

Figure 3. Dendrogram showing hierarchical relationships among Primary Regions A through O (Fig.2), based upon similarities of their gomphocerine grasshopper faunas. Regions are shown on the vertical axis, similarity levels on the horizontal axis. Cophenetic correlation = 0.86. Numbers on the diagram indicate how many out of the 71 possible species are endemic to the area included in a particular stem of the dendrogram. Species listed as endemic to regions A, I, and K extend southward into Mexico in similar habitats (see maps in Otte, 1981).

A-D from the rest of the study area corresponds strikingly with: 1) a 32 inch (81.3 cm) isopleth of mean annual total precipitation (Baldwin, 1973); 2) the eastern and southern limits of certain xeromorphic grasses such as blue grama (*Bouteloua gracilis*; Stubbendieck et al., 1986); and 3) the northern and western extent of temperate deciduous forests and savannahs (Küchler, 1964). Common western gomphocerines reaching their eastern limits along this boundary include *Amphitornus coloradus* and *Ageneotettix deorum*.

Primary regions M, N, and O included few gomphocerines. The southern boundary of this group of blocks corresponds closely with the southern limits of white spruce (*Picea glauca*; Daubenmire, 1978, Fig. 9), and to the distributional limits of boreal forest generally (Shelford, 1963). Region O also encompasses the aspen (*Populus tremuloides*) parklands of south-central Canada.

Another major stem of the regional diagram included blocks comprising the Great Basin and Pacific Coast (Primary Regions E through H, Fig. 2). The two intermontane regions (G and H) meet near a boundary between Küchler's sagebrush steppe unit to the north (including the Palouse Prairie) and his Great Basin sagebrush or saltbush-greasewood units to the south. Regions E and F included a variety of endemic Pacific coastal gomphocerines, but these areas also showed affinities with Great Basin shrubsteppe (Fig. 3). Most gomphocerines that occurred generally through regions E - H were distributed beyond this part of the continent (e.g., *Chorthippus curtipennis, Psoloessa deliculata, Aulocara elliotti*). In fact, only one species (*Stenobothrus shastanus*) was restricted to but coextensive with these four regions. As a unit, then, regions E - H were distinguished primarily by the absence of species found throughout other parts of the West, such as *Phlibostroma quadrimaculatum*.

Two primary Regions (L and K) comprised the central plains of North America, the former also including U.S. portions of the Chihuahuan Desert. These results show the strong affinities of the entire plains to Chihuahuan grasslands. Gomphocerine faunas of regions K and L were similar to those of the Sonoran and Mojave deserts (Regions I and J, Fig. 2). Common species distributed through regions I, J, K, and L included *Boopedon nubilum* and *Eritettix simplex*.

Subdivisions within the Great Plains

Two physical environmental gradients characterize the plains, one of increasing precipitation from west to east, the other of increasing temperature from north to south (Baldwin, 1973). In the vegetation, the moisture gradient is reflected in the tallgrass prairie in the east. Cool climate C_3 grasses in genera such as *Stipa* and especially the wheatgrasses (*Agropyron* spp.) dominate northern regions, while warm season C_4 genera such as grama (*Bouteloua*), buffalograss (*Buchloë*), and bluestem (*Andropogon*) are more abundant to the south. Gomphocerine distributions reflect both of these patterns, but the north-south gradient seems more important. The boundary between Primary Regions K and L (Figs. 2 and 3) corresponds approximately with the northern and eastern limits of Küchler's (1964) grama-buffalo grass vegetational unit. A subdivision of Primary Region L (as shown in Fig. 2) separates the central from the northern plains in a line across southern Montana and northern South Dakota.

Endemism

The average gomphocerine occupied 26 of the 132 blocks, while the 15 Primary Regions (Fig. 2) averaged only 8.9 blocks. That is, while each Primary Region or group of regions was characterized by a unique set of species, most species ranged in individually distinctive ways beyond these regional boundaries. Nevertheless, 22 of 71 species were endemic to one Primary Region, and three parts of the study area showed much higher levels of endemism than the others. These appear to represent refugia or centers of evolution during the Pleistocene or earlier. Grasslands of interior and coastal California have long been isolated from other North American steppe environments, and they include a variety of endemic grasses such as *Stipa pulchra* and *S. cernua* (Wester, 1981). Five gomphocerines endemic to this area (Region E, Fig. 2) are among the most narrowly distributed North American species: *Chloealtis dianae, C. gracilis, Esselania vanduzeei, Eupignodes megacephala,* and *E. sierranus*

Primary Regions J, I, and K (in the south) correspond approximately to the three southwestern deserts: Mojave, Sonoran, and Chihuahuan, respectively. During colder glacial periods these deserts were geographically less extensive and may have been important refugia (Daubenmire, 1978). A number of gomphocerines today remain endemic to these deserts and desert grasslands. *Xeracris snowi* and *X. minimus* are largely restricted to the Mojave. *Ageneotettix brevipennis* is a

Chihuahuan species, while *A. salutator* is largely Sonoran. *Procorypha snowi* is restricted to grasslands at higher elevations within the Sonoran region. *Ligurotettix planum* is a Chihuahuan grasshopper. Several other Gomphocerinae (*e.g. Cibolacris parviceps* and *Bootettix argentatus*) are distributed across all three desert grassland regions. A third area of some endemism that also may reflect a Pleistocene refuge is the southeastern U.S. Coastal Plain (Region B, Fig. 2). Gomphocerinae restricted to this area include *Eritettix obscurus, Mermiria intertexta,* and *Achurum carinatum.*

The flora of the Great Plains is considered to be recently derived from the herbaceous components of adjacent wooded ecosystems, with little endemism even among the dominant grasses (Wells, 1970; Axelrod, 1985). Ross (1970) likewise noted that only 3 of 108 grasshopper species (Acrididae) occurring on the plains were restricted to them. However, at least for the Gomphocerinae, the degree of endemism depends upon how one defines the boundaries of the Great Plains (Figs. 2 and 3). If we let the gomphocerines themselves define those boundaries, then the following conclusions emerge. First, the eastern margins of tallgrass prairie (Regions A and C, Fig. 2) as well as the prairie peninsula of Illinois and Indiana are more appropriately considered as part of eastern U.S. woodlands. While the remaining central and northern plains (Region L) have no endemic gomphocerines, the southern plains (Region K) have four endemic species, and an additional six are restricted to other southwestern desert grasslands. Furthermore, fully 18 (25%) species are restricted to some parts of the central plains and Southwestern deserts as a unit (Fig. 3), though some extend into related habitats in Mexico (Otte, 1981) This is by far the highest level of endemism of any of the four major faunal regions revealed by the cluster analysis (Fig. 3). Notable among species widely distributed but "endemic" to the Great Plains, if they are defined as including Southwestern grasslands, are *Phlibostroma quadrimaculatum, Boopedon nubilum,* and *Mermiria texana.*

If gomphocerines were excluded from the central and northern plains during the Pleistocene, it seems likely that most modern plains' species took refuge or perhaps evolved in grasslands of the Southwest and northern Mexico. The Southwestern, especially Chihuahuan, affinity of the plains' gomphocerine fauna seems clear (Figs. 2 and 3). However, this need not imply that gomphocerines were absent from the plains during glacial maxima. They may have lived in scattered grassland clearings perhaps maintained by fires (e.g. Wendorf, 1970;

Axelrod, 1985; Wells and Stewart, 1987). Alexander and Hilliard (1969) studied grasshopper distributions across an elevational gradient in the Colorado Rocky Mountains. Grasshoppers were locally abundant in montane meadows, in habitats seemingly much like those that characterized the plains during the Pleistocene. Alexander and Hilliard found such typical plains' gomphocerines as *Ageneotettix deorum*, *Amphitornus coloradus*, and *Eritettix simplex* to be abundant in montane meadows. These species also range well into Southwestern desert grasslands, and there is no reason to suppose they might not have exhibited similar ecological amplitude during the Pleistocene.

Summary

Grasshoppers in the acridid subfamily Gomphocerinae are grass and grassland specialists, although most are rather wide-ranging through appropriate habitats. Narrowly distributed forms are concentrated in California, the Southwest, or the southeastern Coastal Plain. Distributional analysis reveals four major gomphocerine faunal regions in North America: 1) the northern boreal forest and associated grasslands, 2) Great Basin shrub-steppe, Palouse Prairie, and the Pacific Slope, 3) eastern and southeastern U.S. forests and grasslands, including the Coastal Plain and eastern margins of the tallgrass prairie, and 4) the central plains and Southwest. The fourth area is relatively species-rich; 25% of North American species are restricted to parts or all of it, including related habitats in Mexico. The gomphocerine fauna of the central U.S. and Canada is most similar to that of the Southwest. From an ecological and especially an evolutionary perspective, results of this study suggest that the traditional boundaries of the Great Plains should be expanded to include Southwestern arid and semiarid grasslands. Chihuahuan Desert grasslands in particular may have been important evolutionary centers and possible Pleistocene refugia for species now characterizing the Great Plains as a whole. Our analysis of grasshopper distributions confirms other work based on vegetation, bison, and dung beetles, that grasslands of the Great Basin have a distinctly different ecology and evolutionary history (Mack and Thompson, 1982).

FIRE AND VEGETATION
Introduction

Many students of grasslands have listed three major environmental forces common to all such ecosystems. These are fire, herbivory, and drought (for example, see Risser et al., 1981; McNaughton et al., 1983). Two of these factors, fire and herbivory, readily lend themselves to small and large scale experimental manipulations. Studies of the effects of fire and herbivory on grasslands are reviewed in numerous recent works (for example, Risser et al., 1981; Stebbins, 1981; Mack and Thompson, 1982; Wright and Bailey, 1982; Crawley, 1983; Axelrod, 1985; Joss et al., 1986; Westoby, 1986).

The role of fire in maintaining western grasslands has been debated. Some authors have suggested that fire is the primary factor preventing woodland invasion into grassland. There is considerable agreement with this hypothesis for the eastern boundary of the Great Plains. Other workers have pointed out that west of the hundredth meridian, including the area of our field studies, a "grassland climate" exists where evapo-transpiration exceeds precipitation for much of the growing season (U.S. Geologic Survey, 1970). However, within this area, invasions of grasslands by shrubs and small trees are common. Fire suppression by humans and overgrazing by domestic herds often are cited as probable causes for these invasions. Workers have found that some of the invasive shrub species are killed outright by fire. In other species, the shrubs can lose all above-ground biomass, but quickly will regrow following the fire (e.g. *Baccharis pteronioides* D.C. as described in Kenney et al., 1986).

From 1984 through 1986 we studied post-fire succession on grassland near Crow Agency, Bighorn County, at Custer Battlefield National Monument in south-central Montana. The Battlefield's historic value is well appreciated, because on June 25, 1876, Colonel George Armstrong Custer and his troops were annihilated by assembled members of the Cheyenne and Sioux nations. The site has since been afforded an unusually high level of protection from human-caused disturbance. The surrounding area is used as rangeland for cattle or is plowed and planted to winter wheat. Grazing livestock have been excluded from the Battlefield since 1891. The Battlefield had remained unburned from the time of the Battle until August, 1983, when 90% of the Monument was burned in a wildfire. This fire provided a unique opportunity for studying post-fire responses in a northern high plains ecosystem. The National Park Service was interested in learning details of post-fire succession in this relatively undisturbed grassland, in order to better plan for its management. In spring, 1984, we commenced a three year study of post-fire responses in vegetation.

Küchler (1964) placed the lowlands of southeastern Montana in a transitional area between relatively pure grasslands to the northeast, and sage-dominated shrub steppe of inter-mountain basins to the southwest. We have chosen to call it northern mixed grass prairie, after the International Biological Program designation (Risser et al., 1981). We report here on fire responses of the five most common shrubs on the Battlefield: *Artemisia tridentata* Nutt. (big sagebrush), *A. cana* Pursh (known locally as sweet sage), *Prunus virginiana* L. (choke cherry), *Rosa arkansana* Porter (prairie rose), and *Symphoricarpos occidentalis* Hook. (western snowberry). *Artemisia tridentata* Nutt., big sagebrush, was the most common shrub on the Battlefield prior to the 1983 fire (Figs. 4, 5, and 6).

Methods

A limitation to this study was the absence of vegetation data prior to the wildfire. Our data collection commenced in May, 1984, following the August 1983 fire. We selected our controls from the 10% of the Battlefield which did not burn in 1983, and used our data from this area as a substitute for data about the condition of the pre-burned vegetation. In May, 1984 we established five 20 x 20-m plots on the unburned land, and set up 20 similar plots on the burned area. Custer Battlefield has been mapped and staked into a grid at 100-m intervals, with each grid intersection marked by a tagged steel bar. We used these markers to establish the southeastern corners of each of our 20 x 20-m plots. All shrubs were counted each year on all plots from 1984 through 1986. No trees were found on the Battlefield (Table 1). Shrub density data were analyzed in several ways, depending upon how the individual species appeared to be responding to the burn. *Artemisia tridentata* densities were compared between treatments within year using the Mann-Whitney U statistic. We used the Chi-square contingency statistic to test for independence of numbers of *A. cana* between years and treatments. The other three shrubs were locally common only in swales. Unfortunately, no sites similar to these remained unburned. Therefore, data for these species come only from the burned area. For these we applied Chi-square goodness-of-fit tests to burned plot data, testing null expectations of equal numbers for each post-fire year.

Results and Discussion

Artemisia cana appeared to be stimulated to reestablish on the burn by seed and sprouting during the second and third post-fire growing seasons (Table 1). The other three common shrubs, *Prunus*, *Rosa*, and *Symphoricarpos* plants survived the fire, although their above-ground densities were reduced dramatically

Table 1. Average number of shrubs per $400m^2$ plot on burned (n = 20) vs. unburned (n = 5) portions of the Custer National Battlefield.

Species	Treatment	1984	1985	1986
Artemisia	burned	0.2	0.2	0
tridentata [a]	unburned	144.2	167.3	172.0
Artemisia cana [b]	burned	0.2	147.7	143.5
	unburned	7.2	79.4	48.4
Prunus virginiana [c]	burned	1.7	15.9	26.2
	unburned	0	0	0
Rhus trilobata	burned	0.2	0.6	0.7
	unburned	0.4	0.4	0.4
Rosa arkansana [c]	burned	0.5	83.9	23.9
	unburned	0	0	0
Sarcobatus	burned	0.3	2.1	1.5
vermiculatus	unburned	0.4	0.4	0.4
Symphoricarpos	burned	1.0	470.9	262.8
occidentalis [c]	unburned	0	0	0

[a] $P < 0.001$, Mann-Whitney U tests comparing abundances per plot between treatments within years.

[b] $P < 0.001$, Chi-square contingency test for independence of treatments and years, based on total shrubs per treatment per year.

[c] $P < 0.001$, Chi-square goodness-of-fit tests of total shrubs per year on burned plots.

Figure 4. Custer Battlefield, Montana, as it appeared three years after the Battle in 1879. Note scattered big sagebrush (*Artemisia tridentata*) among the horse bones and other Battle relicts. (Photo printed by Rob Cross, negative from U.S. National Park Service.)

564

Figure 5. Custer Battlefield, Montana, as it appeared on the 10th anniversary of the Battle, June 25, 1886. Note the dense sagebrush cover. (Photo printed by Rob Cross, negative from U.S. National Park Service).

565

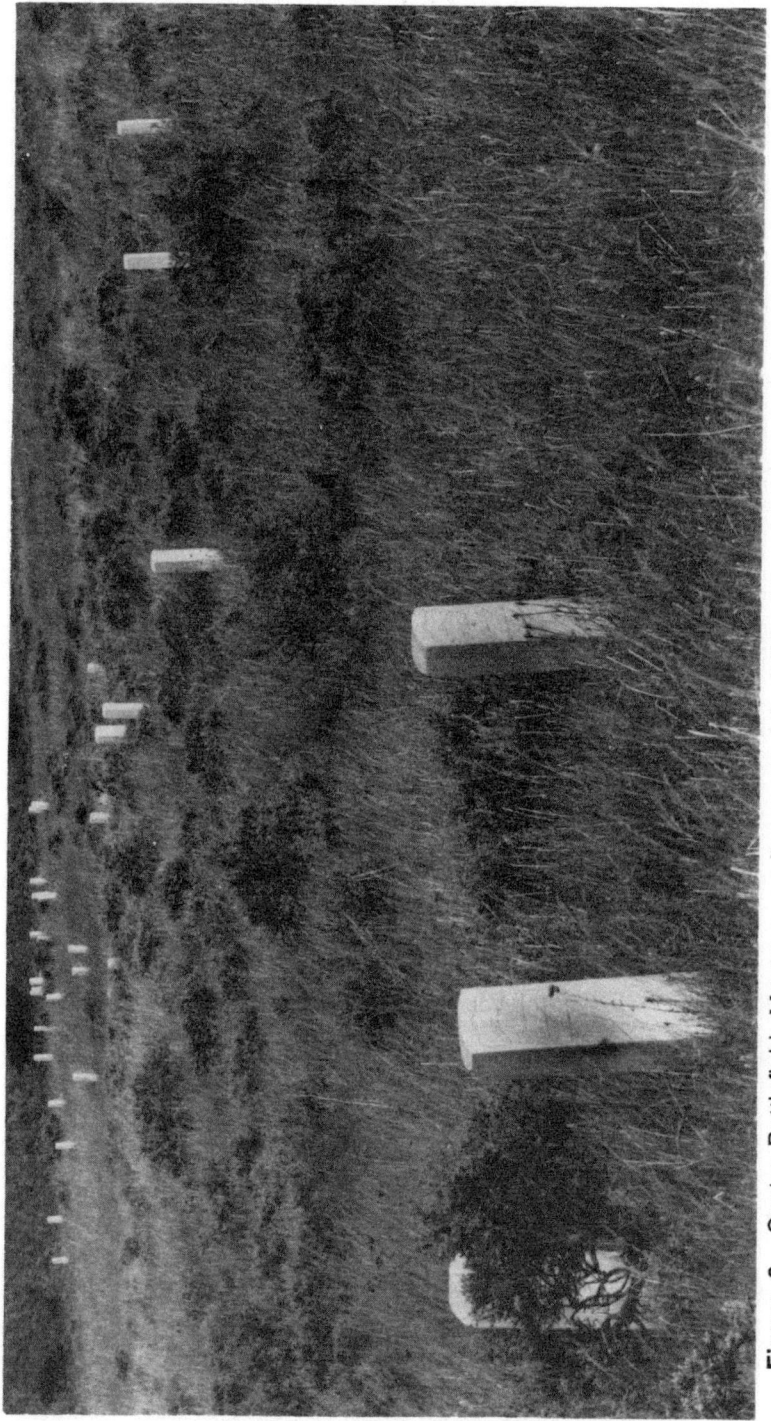

Figure 6. Custer Battlefield, Montana, as it appeared in 1965. The sagebrush cover persists. (Photo printed by Rob Cross, negative from U.S. National Park Service.)

Figure 7. Custer Battlefield, Montana, as it appeared in 1985, two years after the fire. Photo duplicates view seen in Fig. 6. (Photo by Carl Bock)

during the first post-fire year. By 1985 they had resprouted in very large numbers (Table 1).

Big sagebrush had been the dominant shrub on the unburned portions of Battlefield prior to the 1983 fire (Figs. 4, 5, and 6), but no individuals were found alive in the censuses of shrubs on the twenty burned plots or elsewhere on the burned portion of the Battlefield (Table 1). Big sagebrush showed no sign of regrowth throughout three post-fire growing seasons (Fig. 7). These data and an examination of old photos of the Battlefield convince us that sage mortality was the single most dramatic result of the 1983 fire.

Big sagebrush is well known to show extreme sensitivity to fire (Pickford, 1932; Blaisdell, 1953; Beardall and Sylvester, 1976; Wright and Bailey, 1982; Wright, 1986). Based upon a rich photographic record (1878 to the present), we believe Custer Battlefield had been continuously covered with big sagebrush from 1876 until 1983. The reestablishment of big sagebrush on burned areas in northern grasslands appears to be a very slow process. In Idaho, recovery took approximately 30 years (Harniss and Murray, 1973).

Whether big sagebrush will colonize the Battlefield is of considerable interest to the Monument's staff because they wish to maintain the site as it appeared at the time of the Battle. Our evidence strongly suggests the area was covered with big sagebrush on 25 June, 1876. Early photographs (e.g. Figs. 4 and 5) plus anecdotes from Battle survivors and U.S. troops who visited the site soon after the Battle helped us document big sagebrush's presence. During the Battle, combatants used *Artemisia tridentata* for cover.

Because big sagebrush is killed outright by fire, it must colonize by seeds following a burn. Plenteous seed reservoirs from living *A. tridentata* plants exist on the 10% of the Battlefield which did not burn, as well as on adjacent rangelands. Ordinarily, the invasion of big sagebrush is facilitated by grazing of large herbivores (Crawley, 1983, p.300). Such grazing activities open up the grass cover and reduce competition for the colonizing seedlings. The 1983 fire offered us an opportunity to study patterns of *Artemisia tridentata* invasion in the absence of such large herbivores. To date (1987) there are no new plants, and it appears that re-establishment will be a very slow process.

HERBIVORY AND FIRE
Introduction

Herbivorous animals are an important component of grassland ecosystems, and they probably were a significant evolutionary force affecting vegetation of the Great Plains in particular (e.g., Stebbins, 1981). Large mammals such as *Bison bison* have received particular attention in this regard because of their abundance, and because of paleobotanical and historical data available about their evolution and distribution (e.g., McDonald, 1981; Mack and Thompson, 1982).

Between 1979 and 1981 we carried out field studies in Wind Cave National Park, located on the southern edge of the Black Hills in southwestern South Dakota. The general purpose of this study was to evaluate the effects of prescribed fires on plants and animals of the grassland-ponderosa pine (*Pinus ponderosa*) ecotone. Emphasis was on forest understory vegetation, pine seedlings, rodents, breeding birds, and bison. All except the last have been reported on elsewhere (Bock, C. and J. Bock, 1983; Bock, J. and C Bock, 1984).

In the absence of fire, ponderosa pines readily invade adjacent grasslands in the Black Hills. For most of this century, fires were suppressed there. We have a photographic record of the effects of this policy. Under the command of Col. George Armstrong Custer, the U.S. Army in 1874 carried out an Expedition to the Black Hills. The Expedition was well documented by its offical photographer, W.H. Illingsworth, and by accompanying news reporters. Also, Custer himself regularly sent press dispatches about the Expedition to eastern newspapers (Krause and Olson, 1974). Nearly 100 years later, Progulske and Sowell (1974) retraced the Expedition and duplicated many of the photos taken by Illingsworth. This pictorial record illustrates dramatically the invasion of pine into the grassland resulting from a century of fire suppression in the region.

Bison once ranged widely across North American plains and prairies (McDonald, 1981 and references therein), and doubtless experienced fire as an ecological and evolutionary force. Fire and bison grazing in turn must have played a major role in shaping the ecology and distribution of the Great Plains' vegetation. Our goal for this work was to learn about the effects of fire on the plants and animals inhabiting the grassland-ponderosa pine boundary. Here, we report on the responses to burned vegetation by the largest indigenous herbivore, the bison.

Methods

We collected data on vegetation and bison prior to a prescribed fire in 1979, both on the site to be burned and on adjacent comparable control sites. The burn was carried out by National Park Service personnel on 17 October, 1979. Data collections were repeated through two post-fire growing seasons

(1980 and 1981). Four 20 x 33 m study plots were established on the "burned site" and four comparable sites were placed just off the burn. The prescription called for a 63 hectare burn on wooded and grassland areas. The management goals for this burn were to thin young ponderosa pines from the forested area and to kill young ponderosa pines which were invading the grassland area. These goals were achieved (Bock and Bock, 1984).

Vegetation was sampled by means of belt transects. All plants were identified to species and their heights recorded. Bison frequently grazed and rested in the vicinity of the burn, allowing us to compare their utilization of burned vs. unburned habitats. In 1980 and 1981 we recorded whether a particular grass or sedge plant had been clipped (grazed). We are confident that most clipped plants had in fact been grazed by bison, because: 1) plants with only small amounts of foliage removal, usually along the leaf margins and as might be caused by insects, were excluded; and 2) bison were observed frequently grazing the hillsides, while other herbivorous vertebrates were uncommon.

The fire tower in Wind Cave National Park offered an excellent overview of the burned site and its surrounding unburned grassland. Fortuitously, the fire tower was occupied during the summer months by fire watchers who recorded numbers of bison and their locations at half-hour intervals during daylight hours. Counts were made in summer: on 21 days in 1979, 78 days in 1980, and 45 days in 1981. While efforts between years, and between days within years, were highly variable, at all times the observations were equally distributed between burned and unburned sites. Clearly, the individual sightings were not independent of one another, since the bison usually travelled in groups and sometimes remained on burn vs. control for several observation periods. Until the onset of the mating season (August) the mature bulls ordinarily moved about the Park in small groups of two to six animals, or occasionally alone. The year-round cow/calf herds were so large, up to 200 animals, that they frequently occupied burned and unburned sites simultaneously. This meant that the bison group could not be assigned to one treatment or the other as a unit. We simply tallied the total sightings of individual animals on each area in each summer.

Vegetation and grazing data were analyzed with Chi-square contingency or goodness-of-fit tests as appropriate, with critical values set at $P<0.01$. We applied the Chi-square contingency statistic to the bison sightings in order to test for independence of treatments and years.

Results and Discussion

The controlled burn was ignited at mid-day (17 October, 1979). On the previous day, soil moisture was about 4% by weight, while litter moisture was about 28%. Temperature at ignition was 14.4 C, relative humidity was 45%, and winds were about 16 km/hr. The fire was relatively cool, but it burned slowly and completely, covering the area in about 4 hours. Virtually all above-ground grassland vegetation and litter were combusted. The fire boss declared the fire was "out" 48 hours after ignition.

Big and little bluestem were the dominant graminoids (Table 2), followed by sedges and grama grasses (mostly *Bouteloua hirsuta*). Total graminoid canopy cover ranged from a low of 61% on the burn in 1980 (368 of 600 points) to a high of 84% on the control in 1981. Cover increased on both sites over the three years.

Little bluestem canopy was reduced by the fire (X^2 contingency = 13.99, $P < 0.001$), particularly in the first post-fire year (Table 2). Needle and thread grass (*Stipa comata*) was stimulated by the fire ($X^2 = 12.81$, $P < 0.005$), apparently through two post-fire growing seasons. Sedges, other grasses, and total graminoid canopy cover were not significantly affected by the fire.

Bison grazed total graminoids differently on the two treatments in the two years (X^2 contingency = 10.71, $P < 0.005$), clearly concentrating on the burn in the first post-fire year, but not the second (Table 2). In that first post-fire year (1980), bison grazed much more big bluestem and little bluestem on the burn vs. the control than would be expected based upon relative availability in the two areas (X^2 goodness-of-fit = 52.49 and 48.46, respectively, $P < 0.001$). This also was true for sedges, but to a lesser degree ($X^2 = 10.34$, $P < 0.005$). Grazing did not differ on burn vs. control in 1981 for any category.

We tested for selectivity in grazing within treatment and year by comparing numbers of each taxon grazed against expected levels of clipping if it occurred in proportion to abundance. Grazing was highly selective on the burn in 1980 ($X^2 = 56.64$, $P < 0.001$), with the bluestems being clipped much more often than expected by chance. Data in the other year and/or treatment were limited due to lower overall grazing levels (Table 2), but they suggest a general preference for big bluestem.

We suspect that palatability and/or nutritional value of all graminoids, but particularly the bluestems, must have been

Table 2. Canopy cover and grazing intensity for common graminoids on burned vs. unburned grassland, Wind Cave National Park, South Dakota. Data are the numbers of transect points occupied, out of 600 points sampled per treatment per year. Data were collected in mid-June; burning occurred in October, 1979. In parentheses for 1980 and 1981 are the numbers of each taxon that showed signs of having been grazed by large herbivores in that season.

Canopy cover and grazing intensity

Taxon	Treatment	1979	1980	1981
Big bluestem	burned	119	142 (81)	176(11)
(*Andropogon gerardii*)	unburned	122	170 (18)	152(11)
Little bluestem	burned	108	65 (35)	84 (1)
(*Schizachyrium scoparium*)	unburned	05	127 (6)	144 (1)
Sedges	burned	57	68 (13)	78 (1)
(*Carex* spp.)	unburned	60	82 (2)	87 (1)
Grama grasses	burned	69	79 (3)	115 (3)
(*Bouteloua* spp.)	unburned	45	45 (1)	70 (0)
Needle & thread grass	burned	0	5 (0)	18 (1)
(*Stipa comata*)	unburned	10	9 (0)	10 (0)
Other graminoids	burned	33	9 (0)	9 (0)
	unburned	49	25 (0)	38 (0)
TOTAL GRAMINOIDS	burned	386	368 (132)	480 (17)
	unburned	391	458 (27)	501 (13)

Table 3. Total numbers (and percentages) of bison counted on burned vs. unburned grassland in Wind Cave National Park, South Dakota. Data are the cumulative numbers counted at 0.5 hour intervals on 21 days in 1979, 78 days in 1980, and 45 days in 1981, from a fire tower occupied during the summer months. The fire occurred in October, 1979. Each area was about 40ha.

Total numbers (and percentages) counted per year

	Year	Burned area	Unburned area
Bulls[a]	1979	290 (39.9)	436 (60.1)
	1980	1843 (50.0)	1845 (50.0)
	1981	797 (32.8)	1630 (67.2)
Cow/calf[b]	1979	2757 (34.4)	5262 (65.6)
	1980	15627 (68.3)	7266 (31.7)
	1981	1057 (23.0)	3540 (77.0)

[a] Chi-square = 177.6 ($P \ll 0.001$), contingency test for independence of treatments and years.

[b] Chi-square = 4,900.6 ($P \ll 0.001$), as above.

higher on the burn for one post-fire year. This hypothesis could be confirmed by anatomical and biochemical analyses.

Small groups of bull bison travelled independently of larger cow/calf herds for most of the summer, so we analyzed data separately for these two categories (Table 3). Both sexes, but particularly cows and their calves and yearlings, showed an overwhelming preference for the burn in 1980 (the first post-fire year) compared to the pre-burn or second post-burn years.

From our studies we conclude that bison responded dramatically and positively, but only temporarily, to fire. On a larger scale, we can imagine that bison herds once moved over the Great Plains in search of recently-burned grassland. The smoke of huge pre-settlement prairie fires would have been visible over great distances then, as they are today. We observed that bison were attracted by fire. The animals walked over the smoldering grass, just after the fire had passed. They frequently licked the warm ashes, perhaps for the minerals present. In our South

Dakota work, we have shown that burned areas are the preferred feeding sites for bison during the first post-fire growing season. If prairie fire attracted bison historically, it also must have kept the herds moving over the unfenced, open grasslands because this attractiveness was transitory. A spatial and temporal mosaic of burns on the Great Plains may have prevented bison from over-exploiting any one portion of them, a pattern strikingly different from that shown by the fenced herds of domestic herbivores in this same habitat today.

SUMMARY AND CONCLUSION

Conservation, restoration, enlightened management, and scientific studies of the few intact remnants of the Great Plains are vital. We have been privileged to work at a few of these places, as have some of our co-authors in this book (see chapters by Cruden and Lyon, Keeler, and Uno). Much remains to be learned. Although the Great Plains are of recent origin, they nevertheless evolved largely in the absence of humans, and so have been particularly vulnerable to their intrusions. The first people in North America may have hunted important herbivores to extinction (Martin and Klein, 1984; Martin, 1987). Certainly the plains were unprepared for the advent of the cow and the plow. These alien forces disrupted a natural fire regime, allowed woody plants to invade and sometimes completely replace grassland, and encouraged the spread of exotic vegetation (e.g. Humphrey, 1958; Baker, 1974; Mack and Thompson, 1982; Bock et al, 1986; and Joss et al., 1986).

Results of our own work are consistent with those of many other studies, both ecological and paleoecological, and they lead us to the following two major conclusions.

First, the Great Plains as an ecological and evolutionary unit properly should include much of the Southwest. This conclusion is based upon our grasshopper work, as well as phytogeograpical studies by others (e.g., Axelrod, 1985). Thus redefined, the plains have a higher level of endemism than frequently is acknowledged. Grasslands of the Intermountain West and Pacific Slope have had a different history, and are ecologically distinct (Mack and Thompson, 1982)..

Second, Great Plains grasslands are inherently vulnerable to invasion by shrubs and trees. Plains' boundaries always have been unstable, whether under the modern influence of human activities, or as determined prehistorically by interactions of climate, fire, and herbivores. Cycles of drought and cooling, operative since the Pleistocene, have caused the plains to vary from scattered forest openings to vast treeless expanses. These expanses were labeled (accurately, in our opinion) as the "Great American Desert" by Major Stephen Long on a hand-drawn map of the region during his 1819-20 expedition. This designation, now considered obsolete, persisted into the early part of the 20th century. We feel it is appropriate to consider all of the western plains as a desert, because the region is characterized by frequent but unpredictable drought. Great, irreparable damage was done to the plains by humans with plows and misconceptions about the fragility of the land they were "settling."

Like fire, climate is a major environmental force keeping the plains free of woody vegetation. Increases in moisture and herbivore activity can reduce the frequency and intensity of fire. There have now been a large number of studies that carefully document the impacts of fire on plains' grasslands. Less is known about the long-term effects of climate change in these ecosystems, or about the ecological and evolutionary relationships between the grasses and their herbivores. These are and should be the subjects of much present and future research.

ACKNOWLEDGMENTS

We gratefully acknowledge the contributions of our students to this work. Especially, we thank R. Amerman, A. Fernald, E. Gawith, J. Gersh, C. Jolls, W. Kenney, J. Old Coyote, K. Jepson-Innes, J. Ortega, M. O'Shea-Stone, and D. Stone. We thank those who supported this research through funding over the years, including The Charles A. Lindbergh Foundation, the Earthwatch Foundation and Center for Field Research, the U.S. Forest Service, the National Audubon Society, the National Geographic Society, the U.S. National Park Service, the National Science Foundation, Joe and Helen Taylor, and the University of Colorado.

LITERATURE CITED

Alexander, G. and J. R. Hilliard, Jr. 1969. Altitudinal distribution of Orthoptera in the Rocky Mountains of northern Colorado. Ecol. Monogr. 39: 385-431.

Axelrod, D. I. 1985. Rise of the grassland biome, central North America. Bot. Rev. 51: 163-201.

Baker, H. G. 1974. The evolution of weeds. Ann. Rev. Ecol. Syst. 5: 1-24.

Baldwin, J. L. 1973. Climates of the United States. U.S. Department of Commerce, Washington, D.C.

Beardall, L. E., and V. E. Sylvester. 1976. Spring burning for removal of sagebrush competition in Nevada. Proc. Tall Timbers Fire Ecol Conf. 14: 529-547.

Blaisdell, J. P. 1953. Ecological effects of planned burning of sagebrush grass range on the upper Snake River Plains. USDA Tech. Bull.1075. Washington, D.C.

Bock, C. E., and J.H. Bock. 1983. Responses of birds and deer mice to burning in ponderosa pine. J. Wildlife Manage. 47: 836-840.

Bock, C. E., J. H. Bock, K. L. Jepson, and J. C. Ortega. 1986. Ecological effects of planting African lovegrasses in Arizona. Nat. Geogr. Res. 2: 456-463.

Bock, C. E., J. B. Mitton, and L. W. Lepthien. 1978. Winter biogeography of North American Fringillidae (Aves): a numerical analysis. Syst. Zool. 27: 411-420.

Bock, J. H., and C. E. Bock. 1984. Effect of fires on woody vegetation in the pine-grassland ecotone of the southern Black Hills. Amer. Midl. Nat. 112: 35-42.

Bock, J. H., C. E. Bock, and R. J. Fritz. 1981. Biogeography of Illinois reptiles and amphibians: a numerical analysis. Amer. Midl. Nat. 106: 258-270.

Crawley, M. J. 1983. Herbivory. University of California Press, Berkeley.

Daubenmire, R. 1978. Plant Geography. Academic Press, New York.

Dort, W., Jr., and J. K. Jones, Jr. [eds.] 1970. Pleistocene and Recent Environments of the Central Great Plains. University of Kansas Press, Lawrence.

Harniss, R. O., and R. B. Murray. 1973. Thirty years of vegetal change following burning of sagebrush-grass range. J. Range Manage. 26: 322-325.

Humphrey, R. R. 1958. The desert grassland: a history of vegetational change and an analysis of causes. Bot. Rev. 24: 193-252.

Joern, A. 1979. Feeding patterns in grasshoppers (Orthoptera: Acrididae): factors influencing diet specialization. Oecologia 38: 325-327.

Joss, P. J., P. W. Lynch, and O. B. Williams [eds.]. 1986. Rangelands: a Resource under Siege. Cambridge University Press, Cambridge.

Kenney, W. H., J. H. Bock, and C. E. Bock. 1986. Patterns of post-grazing responses of Baccharis pteronioides in the semi-desert grassland of southeastern Arizona. Amer. Midl. Nat. 116: 429-431.

Krause, H. and G. D. Olsen. 1974. Prelude to Glory. A Newspaper Accounting of Custer's 1874 Expedition to the Black Hills. Brevet Press, Sioux Falls, South Dakota.

Küchler, A. W. 1964. Potential Natural Vegetation of the Conterminous United States. American Geographical Society, Special Publication No. 36. New York, New York.

McDonald, J. N. 1981. North American Bison: their Classification and Evolution. University of California Press, Berkeley.

McNaughton, S. J., L. L. Wallace, and M. B. Coughenour. 1983. Plant adaptation in an ecosystem context: effects of defoliation, nitrogen and water on growth of an African C4 sedge. Ecol. 64: 307-318.

Mack, R. N., and J. N. Thompson. 1982. Evolution in steppe with few large, hooved mammals. Amer. Nat. 119: 757-773.

Martin, P. S. 1987. Clovisia the beautiful! Nat. Hist. 96: 10-13

Martin, P. S. and R. G. Klein. 1984. Quaternary Extinctions: a Prehistory Revolution. University of Arizona Press, Tucson.

Otte, D. 1981. The North American Grasshoppers. Volume I. Acrididae: Gomphocerinae and Acridinae. Harvard University Press, Cambridge, Massachusetts.

Otte, D., and A. Joern. 1977. On feeding patterns in desert grasshoppers and the evolution of specialized diets. Proc. Acad. Nat. Sci., Phila. 128: 89-126.

Peters, J. A. 1971. A new approach in the analysis of biogeographic data. Smithsonian Contrib. Zool. 107: 1-28.

Pickford, G. D. 1932. The influence of continued heavy grazing and of promiscuous burning on spring-fall ranges in Utah. Ecology 13: 159-171.

Progulske, D. R., and R. H. Sowell. 1974. Yellow Ore, Yellow Hair, Yellow Pine. Bull. 616, Agri. Exper. Sta., South Dakota State University, Brookings.

Risser, P., E. Birney, H. Blocker, S. May, W. Parton, and J. Wiens. 1981. The True Prairie Ecosystem. Hutchinson Ross Publ. Co., Stroudsburg, Pennsylvania.

Rohlf, F. J., J. Kishpaugh, and D. Kirk. 1972. NT-SYS: Numerical Taxonomy System of Multivariate Statistical Programs. State University of New York, Stoney Brook.

Ross, H. H. 1970. The ecological history of the Great Plains: evidence from grassland insects. In Dort, W., Jr., and J. K. Jones, Jr. [eds.] Pleistocene and Recent Environments of the Central Great Plains. University of Kansas Press, Lawrence.

Shelford, V. E. 1963. The Ecology of North America. University of Illinois Press, Urbana.

Sneath, P. H. A., and R. R. Sokal. 1973. Numerical Taxonomy. W.H. Freeman and Company, San Francisco, California.

Stebbins, G. L. 1981. Coevolution of grasses and herbivores. Ann.of the Missouri Botan. Gard. 68: 75-86.

Stubbendieck, J., S. L. Hatch, and K. J. Hirsh. 1986. North American Range Plants. University of Nebraska Press, Lincoln.

U.S. Geologic Survey. 1970. The National Atlas of the United States of America. U.S. Government Printing Office, Washington, D.C.

Wells, P. V. 1970. Vegetational history of the Great Plains: a post=glacial record of coniferous woodland in southeastern Wyoming. In Dort, W., Jr., and J.K. Jones, Jr. [eds.] Pleistocene and Recent Environments of the Central Great Plains. University of Kansas Press, Lawrence.

Wells, P. V., and J. D. Stewart. 1987. Cordilleran-boreal taiga and fauna on the central Great Plains of North America, 14,000-18,000 years ago. Amer. Midl. Nat. 118: 94-106.

Wendorf, F. 1970. The Lubbock subpluvial. In Dort, W., Jr., and J.K. Jones, Jr. [eds.] Pleistocene and Recent Environments of the Central Great Plains. University of Kansas Press, Lawrence.

Wester, L. L. 1981. Composition of the native grasslands of the San Joaquin Valley. Madroño 28: 231-241.

Westoby, M. 1986. Mechanisms influencing grazing success for livestock and wild herbivores. Amer. Nat. 128: 940-941.

Wright, H.A. 1986. Effects of fire on arid and semi-arid ecosystems - North American continent. In Joss, P.J., P.W. Lynch, and O.B. Williams [eds.]. Rangelands: A Resource under Siege. Cambridge University Press, Cambridge.

Wright, H. A. and A. W. Bailey. 1982. Fire Ecology. John Wiley and Sons, New York.

AFFIRMATIVE ACTION FOR INSECTS IN TROPICAL NATIONAL PARKS

Daniel H. Janzen

The conservation of a tropical wildland into perpetuity demands that it offer a product to its neighbors and other users. This product must be offered with a minimal loss of biodiversity in the conserved wildland. All uses involve some interference and impact on individual organisms and ecosystems under examination, study, manipulation etc; this is the tax that conserved wildlands must pay for their existence, just as the rugs in a library wear out and some borrowed books never return. In general: a) the larger the population of the focal organism, b) the faster its replacement rate, c) the wilder its swings in density in the natural circumstance, and d) the less bothered it is by the presence of a large mammal, the more likely is the human population to be able to make use of it without interfering with its conservation, either in isolation or in its interactions with other organisms. Insects (and other invertebrates) rank much higher on this scale of potential use than do vertebrates. That is to say, the conservation problems generated by human-insect interactions in a user-friendly national park are potentially much less than are those generated by interactions between humans and other vertebrates.

Tropical national parks have had their historical roots primarily in their forests, big woolly mammals, and birds. Insects have long received short shrift; it's time that they get their dues. Tropical conservationists have always had insects in the backs of their minds; insects are those little creatures that get conserved along with the Brazil nuts and the tigers. But explicit conservation efforts for insects are almost non-existent except when some gaudy butterfly gets on an endangered species list. Yes, I know that insects often get honorary mention in the plans, but when it gets down to the nuts and bolts of tropical conservation, who ever gave a half a million dollars to save the Alexander beetle?

What is an entomological emphasis in a national park? It is affirmative action for the conservation and non-destructive use of invertebrates, along with the other organisms and habitats. This action takes the form, in a large part, of fine-tuning to the biological needs of insects and to the peculiarities of getting to know and use insects. These needs and peculiarities have generally escaped the traditional tropical national park. Taking insects into account implies a broadening and even modification of the way the park is viewed.

Size of the Park (and Related Considerations)

Humans have better instincts and data for the amount of area needed to support vertebrate populations than they do for invertebrate populations. Since almost all individual insects are thought of as needing much less resource than even the smallest vertebrate, it is implicit in contemporary park design that an area big enough to maintain the vertebrates will automatically pick up the invertebrates. Equally, it is implicit that an insect population contains so many individuals that it is always much further from extinction than is a vertebrate population. While there is much truth to these two generalizations, there are many important exceptions:

1. Entire populations of invertebrates often move seasonally from one area to another - wasps, bees, dragonflies butterflies, moths, beetles, flies, etc.. This makes them appear analogous to birds in needing either a very large area or at least two areas, but insect migration distances, routes and targets are often not the same as are those of the birds from the same source area, and therefore are not automatically covered by considerations of bird, bat or megafaunal migration. Worse, one group of insect species may migrate 2 km north of a park to a seasonal refugium, another group may migrate 10 km downward to leave an inimical season behind, and a third group may move 100 km east - and even do it to escape a different season. That is to say, among the tens of thousands of species of insects in a tropical national park, there are many more kinds of movement patterns and seasons than among the vertebrates. Park planning that takes these movements into account is a severe challenge, and park size may have to be large not so much to allow the persistence of a large population as to include the several habitats that large population uses in the course of a year. It is like planning for migratory waterfowl with one single chain of parks from Canada to Mexico, but simultaneously doing it for other life forms migrating in other directions.

2. Easily 50% of the species of insects in a tropical habitat can be herbivores. Owing to the extreme dependency of many herbivorous insects on particular species of plants, what may well matter is not whether the area is large but whether it contains appropriate individuals of particular species of plants. Very large areas of undisturbed habitat may be nothing more than

wastelands to a monophagous moth species - even if the adult moth can be captured throughout the habitat.

3. The more generalist insects are often thought of as being able to, like deer and coyotes, range over a variety of habitats for their resources. However, many insects have complex life cycles, demanding one very special habitat for one portion of the cycle (e.g. a particular nesting site for a bee or wasp) even if another portion of the life cycle can range far and wide in a very generalist manner.

4. Tropical insects often fluctuate wildly in density between and among years. They are noticed when common ('ah, see, they are so common that we need not worry about them being in danger of extinction'), but simply forgotten about in a bad year ('ah, insects always fluctuate in abundance'). However, the question is whether the park is large or complex enough to sustain the population in its "bad years." A three year run of dry years may easily eliminate the springs that are needed to keep an aquatic insect population alive in the park during the dry season. A smaller park designed such that it includes some ever-running springs may be much better than a much larger park with many springs that are, however, susceptible to regional drought. For vertebrates, a spring, a river or a stock watering tank may be nothing more than a drink of water, while to an aquatic insect they are three vastly different habitats.

5. Even a very small area may well be large enough to maintain a healthy population of a given species of insect, provided that its host plants and other needs are present. Even peninsulas in park margins and very linear parks may be of value in this context, despite that they have lost many of their vertebrate populations. A monophagous insect may persist in a habitat despite very severe reduction in the species richness of other organisms, if its host plant is hanging on. These considerations are especially critical in restoration of ecosystems. hHabitat fragments may contain many of their original invertebrate populations; these population fragments that can later expand into more 'normal' populations if the habitat is given back.

6. It is commonplace to note that a small park carved out of a large habitat gradually "decays" through the loss of species and changes in relationships between species. However, such an insular fragment of habitat will eventually come into some kind of equilibrium, just as naturally small habitats have their equilibrium faunas. If the focus is just on the vertebrates and the large plants, such a shrinking jewel may seem far too damaged to be worth saving. However, from the invertebrate standpoint, it still may be very species-rich and quite worth bothering with. A

damaged 15th century cathedral is better than none at all; from the viewpoint of the vertebrate conservationist, the cathedral may be a pile of rubble, while from the entomological viewpoint, it may still be standing and even serving a community function. Equally, a park that is too small to maintain significant vertebrate populations may well be worth saving just for its invertebrate and small plant populations; this concept has long been in place in extra-tropical habitats, but is conspicuously lacking as a major rational for tropical national parks and preserves.

Biocultural Restoration

In today's tropics, humanity is very rapidly extinguishing its remnant rich knowledge of the natural world. A national park offers a major opportunity to re-enrich the cultural heritage of a tropical farming region. Such a view places a major emphasis on demonstration of intellectually and emotionally interesting aspects of the park. This means intellectual and physical contact with organisms. Plants sit quietly feeding and breathing during most of the year. Big woolly mammals are often hard to find and get close to (and once you have seen one lion and zebra, you have seen them all). Birds are often no more than feathered sounds that flee from any group of people greater than one. That leaves us with the lowly caterpillar, the hunting beetle and the soaring butterflies. And the moths at the lights at night, the bees on the flowers, and the columns of army ants. And the dung beetles with their dungy cakes. The list of coming attractions is almost endless.

Insects are not just decorations on the great green natural cake. Sex aggression, competition, mutualism, hunting, fleeing, allurement, deception, parasitism and all those other very human traits can be very easily examined with insects. A pair of copulating beetles is a very convenient device to begin a discussion with park visitors of why there is sex in the first place, and those beetles often stay put for all to see. There are a multitude of very human traits, traits that badly need understanding by their human practioners, that can be most instructively examined by seeing them first in other organisms. And insects certainly can be cooperative about performing right in front of the visitor's nose. On the other hand, because they are insects, the visitor may need more guidance than with mammals or other vertebrates so as to be able to begin to interpret what is being seen. Is a hovering bee waiting for females, searching for a nest site, waiting for you to remove your foot from its nest entrance, or deciding to attack? It may take a gentle prod to get past 8 or 30 years of fear of caterpillars.

What does a tropical national park offer its neighbors? It offers, among other things, a chance to regain an understanding of the world that comes from a much more diverse teacher than are a school and set of neighbors/parents. And when the visitor walks into the green library of a tropical forest, it is the insects that are the books most easily opened. Why are leaves shaped like they are? Tough question. Why are butterfly wings shaped like they are? Much easier. Why do the big cats hunt where they hunt? Just try following one some time. Why do army ants hunt where they hunt? Much easier.

The diversity of the easily observable insect world is vastly greater than is that of other groups. All that is required is that we take some affirmative action towards its display, and devote the same kinds of time to learning its natural history that has long been bestowed on vertebrates and plants. A few things come immediately to mind:

a. Big crawlly caterpillars are almost begging to be passed around among small school children. A large net over a host treelet, with a gravid female saturniid moth tossed in on occasion, can guarantee a nearly year-round supply of large caterpillars for public discussion, photographs and destruction of aversion to insects.

b. The visitor to a national park often is frustrated by seeing hardly more than a brilliant blue or yellow dot sailing far overhead, or flitting off through the shady understory. This is the place to recognize the museum-display component of a national park. A visitor center with a large collection, for example, of the butterflies likely to be seen on nature trails is highly appropriate. But not just a display - a micro-discourse on their biology as well is appropriate. Mimicry, flash colors, crypsis, etc., are all well and good to be seen in the field, but there is nothing like an accompanying static (and colorful) display to orient the visitor.

c. We need affirmative action to think small. Many visitors bumble along looking for the monkey or parrot in distant tree crowns, but are not offered the suggestion that if you look through your binoculars backwards, they make excellent magnifying glasses. Searching the foliage for trailside insects may result in moving only 100 m in a morning, but what was the hurry anyway?

d. Field guides to the birds and mammals have long been in hand, but somehow field guides to the conspicuous insects have been very slow to appear in Neotropics. While taxonomic incompleteness may be a legitimate excuse for the lack of field guides to many groups of tropical insects, for others such as butterflies and moths, there is no scientific excuse for their absence.

A New Role for Taxonomists

Taxonomic knowledge of tropical plants and vertebrates is now to the point where the taxonomist can say, for many parks, well, if tropical parks want their field guides, all they have to do is take their photographs and tie them up with taxonomic literature. That is a job for writers, for teachers. Things have not progressed to this stage with insects, and are not likely to in the near future. There are not more than 5 new species of birds to be described from Costa Rica; there are at least 50,000 undescribed species of insects in Costa Rica.

The world of insect taxonomy has evolved from general collecting and species descriptions to today's strong emphasis on generic and familial revisions coupled with development of theory that is of use to such revisions. While it is evident that generic and high-level revisions are necessary and will eventually appear for all groups, they are expensive in time and resources. We do not have a century to revise the 610 species of a genus of parasitic wasp ranging from Canada to Paraguay before offering a reliable taxonomic literature for the 52 species of this genus that occur in a single Panamanian national park. The work-management, education, basic biology - done with these 52 species will be a solid brick in the park edifice, the edifice that will have to withstand the continuous social pressure to turn the park into the production of more directly edible materials. That is to say, an insect taxonomist's focused contribution to the field guide library of the park now will be a significant component of whether those 52 species exist as anything other than museum specimens when the revision is finished a century later.

This affirmative action for an entomological national park requires far more than a simple personal decision by a taxonomist to take a half a year out of the revisionary work flow to produce a field guide for a select park. It requires that the social and administrative structure that supports that taxonomist accept such an action as a legitimate and rewardable activity. The ledger should be clear. If it takes ten years to do a serious revision of the 1100 members of a Neotropical moth subfamily, an additional year spent along the way to produce a field guide to the 30-70 species of moths of that subfamily in each of five geographically distinct entomological national parks is clearly a good conservation investment. There is even a conservation spinoff. The insect fauna of a given neotropical national park has much in common with faunas of other neotropical national parks. That is to say, a field guide for one national park's 28 species of dung beetles, for example, will apply to perhaps 20 species of dung beetles even tens of degrees of latitude away. Furthermore, experience with regional field guides to birds suggests that the presence of an incomplete field guide (because it was done for a different area) is a strong incentive for the production of a complete guide written specifically for the park in question.

It should be clear that we have now entered, once again, into an area where the entomologist cannot do the task alone. Maximally efficient use of the taxonomist's expertise is attained by a writer or other park-oriented person teaming up with the taxonomist and providing the perspective, the budget, the illustrations and the leg-work to generate park-specific field guides and reference collections.

Insects Are Glue

As people set up to preserve the vertebrates and the plants, there is a strong tendency to simply forget the trophic and social roles played by the insects, or to feel that insects occur in such large numbers that we need not worry about them. After all, there always seem to be enough *Trigona* bees to eat your lunch, so what's the problem, anyway?

When a biologist notes that there are many species of migrant moths that fly in and out of a tropical national park, if there is any reaction at all, it is to ask if their loss would mean a serious reduction in available pollinators. The answer is certainly yes, but that is a very incomplete answer. The migrant moths are often very abundant during their 1-2 generations in the park; many of their larvae are a major part of the food for vertebrate and invertebrate predators. This is not the place to treat this in detail, but many of these insect species are not interchangeable; because of their species-level traits, or because of the portion of the habitat in which they live, a kilo of sphingid moth caterpillars does not equal a kilo of noctuid moth caterpillars. The same applies to bees. Yes, bees visit flowers, but certainly not all flowers. Social bees, largely pollen and nectar robbers, are no substitute for solitary bees. The loss of a few hectares of nesting habitat for a group of solitary bees may result in no pollinations for many square kilometers of plants.

Planning in relation to aquatic systems in tropical national parks always ignores the insects. Insectivorous game fish are introduced without a thought. Rivers are often chosen as park boundaries, thereby almost guaranteeing that no aquatic fauna will be preserved as the agrochemical application by the park's neighbors moves into full swing. Parks established to protect rivers and drainage basins often forget that even the smallest feeder streams must be included. While sprayplane over flights of national parks are greeted with cries of outrage, little or no thought is given to surface runoff into the feeder streams that later deliver to the park.

Economic Value

The mass of insects in a national park are, quite simply, the biotic gene and species bank that supports those few species that will eventually turn out to be critical to the sophisticated agriculture that will be coming down the road. The insects support the invertebrate carnivores - parasitoids, viruses, bacteria, predators - that will be major tools in the game of biotic pest controls. The wildland insect population may also even be the agroecosystem designer's only chance to locate genomes not yet resistant to pesticides, carnivores and climate modification.

Today the tropics often seems nothing more than a 1940's throwback, complete with DDT, pesticide blooms on the produce, and sprayplane pilots with their bare arms glistening with spray. But it is fast changing - economics and better health education are demanding it. And with genetic engineering of better crop plants bustling down the road, the parks will be evermore small islands of the natural world dotted over a very artificial landscape. This already is a landscape of intense agriculture and intensive investment. As the yield goes up and the investment goes up, so also rises the impact of an insect outbreak. It's all well and good to get the tropical act together to produce multi-million dollar rice crops in Costa Rica, but it's a multi-million dollar loss when a new strain of rice borer goes to work on it.

The parks are going to be the libraries to which that high-tech, fine-tuned agricultural system turns when it is searching for the next wave of genetic engineering materials. And the insects will be among the most used books. A moth fauna of 3000 species is 3000 little packages of multi-specific genomes capable of detoxifying about as many kinds of plant defenses; that's what caterpillars are.

Administration

Affirmative action for the insect conservation in a tropical national park does introduce some administrative peculiarities. Perhaps the most outstanding is generated by the fact that insect taxonomy is far, far behind that of vertebrates and plants. This means that we are still in the days where simple collecting is mandatory. Worse, many insects are small enough that they must be collected in large numbers, and studied as such, before they can be taxonomically well enough understood to explain and identify for the park user.

A national park can often be made legitimately off limits to the collector of vertebrates. It may be able to develop with minimal plant collecting. However, the administrative attitude towards the world of insect taxonomy must be "come and use us." Get series of our specimens into every major collection, so that we get considered in each and every regional and higher order taxonomic revision. And then bring some of specimens back to us or to national collections as identified reference material, and help to produce the first field guides. A park administration can largely ignore the services and administrative complications of close ties with a major vertebrate museum or herbarium; most of the needed taxonomic work can be done with the use of already extant field guides and some plant collecting by park personnel. However, a national park is helpless to deal with the tens of thousands of insect and other invertebrate species that it contains. A mutualism between the administration and the community of taxonomists, with all its implications, is mandatory.

Manipulative research, like taxonomic collections, is much more possible in the context of conservation biodiversity if done with insects than with vertebrates (due to our primitive understanding of the conservation needs of invertebrates, it is also much more needed). To remove the 11 resident tapirs from a valley bottom is likely to have dramatic and long-reaching impact on the very biodiversity that the park conserves; to remove (or double) all the caterpillars on ten individuals of a single individual host plant may both illuminate the relationship between the caterpillar and the host, and do no more than mimic something that occurs annually throughout the habitat. A hundred individuals of common butterfly may be removed for a starvation experiment, with no measurable or real effect on the habitat overall, whereas the removal of 100 squirrels for the same experiment may well alter the site for decades.

A consideration of the entomological nature of a national park also increases the justification for resource expenditure on other seemingly disparate activities. Rain gauges, floras, and flowering phenology may all be justified solely in their facilitation of insect studies, as well as be scientific activities in their own right. A mouse live-trapping study gives the opportunity for the study of ectoparasite population dynamics. A road or trail through unexceptional vertebrate habitat may offer exceptional opportunities in insect-watching. An anthropogenic fire, of no threat to highly mobile or subterranean vertebrates (and even producing food for them in the form of sucker shoots), may be worth extinguishing because it consumes enormous numbers of dormant and active insects, as well as much of the food for the litter community (which may in turn have been a major food item for the armadillos on the site).

In Closing

This essay was written in the midst of the social and conservation turmoil of Costa Rica's strenuous efforts to bring its economic and conservation development into harmony. When Herbert Baker and I arrived in Costa Rica in early 1965, the abundant forest was a biological inspiration, a meat market and lumber mill to all of us. A little more than two decades later it is almost all gone except for the 20% with explicit reserve status. The challenge today is not how to set aside more parks, but how to make the parks so user-friendly that they are still here, with their biodiversity as intact as biologically possible, a century from now. Insects play a dormant but potentially great role in the answer to this challenge.

Twenty-five years ago, Herbert and Irene Baker aggressively sought out the entomology department at the University of California at Berkeley, as a source of different thinking about plants. We all profited from that. We all played in Herbert and Irene's garden, and we still do. Their botany was user-friendly. There is a lesson there.

INDEX